AF388881

Collagen from Marine Biological Source and Medical Applications

Collagen from Marine Biological Source and Medical Applications

Editors

Azizur Rahman
Tiago H. Silva

MDPI • Basel • Beijing • Wuhan • Barcelona • Belgrade • Manchester • Tokyo • Cluj • Tianjin

Editors
Azizur Rahman
University of Toronto
Canada

Tiago H. Silva
University of Minho
Portugal

Editorial Office
MDPI
St. Alban-Anlage 66
4052 Basel, Switzerland

This is a reprint of articles from the Special Issue published online in the open access journal *Marine Drugs* (ISSN 1660-3397) (available at: https://www.mdpi.com/journal/marinedrugs/special_issues/Collagen_Source_Medical_Applications).

For citation purposes, cite each article independently as indicated on the article page online and as indicated below:

LastName, A.A.; LastName, B.B.; LastName, C.C. Article Title. *Journal Name* **Year**, *Volume Number*, Page Range.

ISBN 978-3-0365-3663-7 (Hbk)
ISBN 978-3-0365-3664-4 (PDF)

Contents

About the Editors

Azizur Rahman, Ph.D., is currently serving as one of the Research Heads/Directors of the Center for Climate Change Research (CCCR) and President of Environmental Solutions under the entrepreneurship program of ICUBE, University of Toronto. He received his M.S. and Ph.D. degrees from the University of Ryukyus, Japan, and conducted his postdoctoral studies at the Department of Chemistry, University of Ryukyus, funded by the Japan Society for the Promotion of Sciences (JSPS). He also served as a Humboldt Fellow (funded by Alexander Von Humboldt Foundation) at the University of Munich, Germany, for two years. After the completion of his Humboldt fellowship, he joined at the University of Toronto in 2013. He pursues a broad range of research interests, including freshwater/marine biology, drug discovery, environment and climate change, biomineralization, inorganic chemistry, analytical chemistry, biotechnology, and protein biochemistry. His interdisciplinary research has been recognized by over 20 major national and international awards, including several best paper awards and a gold medal. He is a recipient of the Presidential Honorary Award for Scientists, the highest honor given by the University of Ryukyus in Japan for outstanding novel research findings. Dr. Rahman also received the Top Peer-Reviewers Award (2017–2019) recognized by Web of Science Group (powered by Publons). Dr. Rahman is passionate about science and its promotion in communities through innovation and entrepreneurship and by applying his diverse background. He is also a serial entrepreneur who has founded a number of companies and non-profit organizations strapped to social innovation and the science community.

Tiago H. Silva has a PhD in chemistry (Faculty of Sciences, University of Porto, Portugal, 2006) and is currently a Principal Researcher in the 3B's Research Group, I3Bs Research Institute on Biomaterials, Biodegradables and Biomimetics, at the University of Minho (Portugal), a member of the ICVS/3B's Associate Laboratory. He has been working on the valorization of marine resources and by-products and the cross-talk between blue and red biotechnologies via the development of marine inspired biomaterials based on biopolymers such as marine collagens, chitosan, and fucoidan as well as on ceramics such as calcium phosphates and biosilica for regenerative medicine strategies and other advanced therapies, namely for cancer and diabetes. He also studies the use of marine invertebrates, particularly marine sponges, as inspiration for the development of biomedical applications.

 marine drugs

Editorial

Collagens from Marine Organisms towards Biomedical Applications

Azizur Rahman [1,2,*] **and Tiago H. Silva** [3,4,*]

1   Centre for Climate Change Research, University of Toronto, ONRamp, Toronto, ON M5G 1L5, Canada
2   A.R. Environmental Solutions, ICUBE-University of Toronto, Mississauga, ON L5L 1C6, Canada
3   3B's Research Group, I3Bs–Research Institute on Biomaterials, Biodegradables and Biomimetics, University of Minho, Headquarters of the European Institute of Excellence on Tissue Engineering and Regenerative Medicine, AvePark-Parque de Ciência e Tecnologia, Zona Industrial da Gandra, 4805-017 Barco, Guimarães, Portugal
4   ICVS/3B´s-PT Government Associate Laboratory, 4806-909 Guimarães, Braga, Portugal
*   Correspondence: mazizur.rahman@utoronto.ca (A.R.); tiago.silva@i3bs.uminho.pt (T.H.S.)

Citation: Rahman, A.; Silva, T.H. Collagens from Marine Organisms towards Biomedical Applications. *Mar. Drugs* **2022**, *20*, 170. https://doi.org/10.3390/md20030170

Received: 3 February 2022
Accepted: 14 February 2022
Published: 25 February 2022

Publisher's Note: MDPI stays neutral with regard to jurisdictional claims in published maps and institutional affiliations.

Collagen is the main fibrous structural protein in the extracellular matrix and connective tissue of animals. It is a primary building block of bones, tendons, skin, hair, nails, cartilage, and all joints in the body. It is also considered a "glue" that holds the body together. In this regard, it receives great attention in healthcare and wellbeing, both as a functional ingredient in different formulations and as a component of several products, such as medical devices and pharmaceutical systems. The production of collagen begins to slow down, and cell structures start losing their strength as we get older, and supplementing with collagen is being explored as a vital way to help our body revive and stay youthful. Indeed, a wide range of products comprising collagen (or collagen hydrolysates, in most cases) can be found, such as lotions, creams, face masks and even nutraceuticals. On the other hand, collagen-based biomedical materials have also been developed, with important and clinically effective materials gaining wide acceptance.

However, collagen extraction from land animal sources is complex, time-consuming and expensive. Moreover, there are some concerns over adverse inflammatory and immunologic response and the prevalence of various diseases among land animals which causes health complications, resulting in ethical and regulatory constraints, pushing industry—and scientists—to look for alternatives. Hence, marine sources have started to be researched and have been found to be quite convenient and safe for obtaining collagen. The main advantages being claimed over the land animal sources are (1) a high content of collagen; (2) environmentally friendly; (3) the presence of biological contaminants and toxins being almost negligible; (4) low inflammatory response; (5) greater absorption due to low molecular weight; (6) less significant religious and ethical constraints; (7) minor regulatory and quality control problems.

Marine resources for the production of collagen include both invertebrates and vertebrates, such as sponges, coralline red algae, sea urchins, octopi, squids, jellyfish, cuttlefish, starfish, sea anemones, and prawns; in addition to the different species of fish, the latter has the advantage that much biomass being used for collagen extraction is a by-product of fish processing for food, such as fish skins, scales, etc., thus offering both economic and environmental benefits. Moreover, several applications for marine collagens have been studied and proposed, promising a great contribution to marine biotechnology products and medical applications in the short term. Aware of the significant scientific relevance of marine collagens and the pivotal role they can assume in human health, we edited this Special Issue comprising a series of original studies and reviews on the biological sources of these proteins, including production methodologies and on their promising applications in medical and related fields.

1

In particular, the collagens that can be found in the different marine invertebrates and ways to extract them have been overviewed in a review paper [1], followed by another review focusing particularly on the marine sponges [2], an ancestral group of animals widely studied for their biological role played in marine ecosystems, and additionally for being an untapped source of bioactive compounds. Herein, a special focus can be found on three sponge species: *Axinella cannabina*, *Suberites carnosus* and *Chondrosia reniformis*. The former two were the subject of a study by Tziveleka et al. [3], where insoluble collagen, intercellular collagen, and spongin-like collagen were extracted (Figure 1) and characterized envisaging biotechnological application.

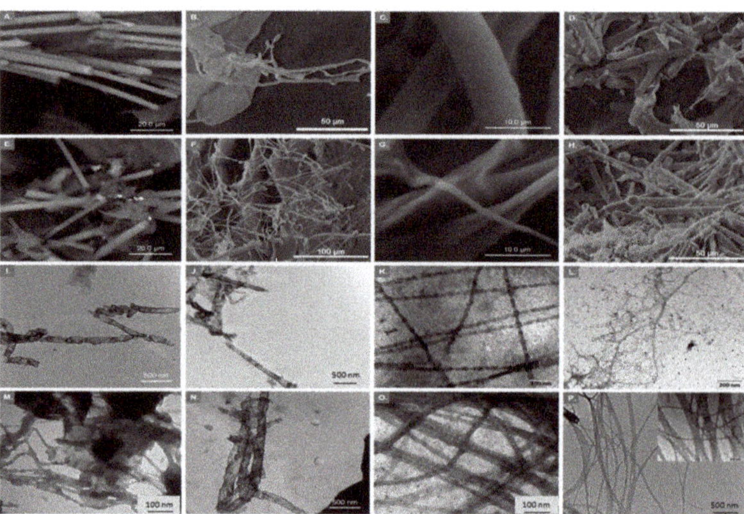

Figure 1. SEM micrographs of insoluble collagen (InSC; **A,E**), intercellular collagen (ICC; **B,C,F,G**), and spongin-like collagen (SlC; **D,H**) from *Axinella cannabina* (row 1) and *Suberites carnosus* (row 2), respectively. TEM micrographs of insoluble collagen before (InSC; **I,M**) and after (SF-InSC; **J,N**) spicule removal, intercellular collagen (ICC; **K,O**) and spongin-like collagen (SlC; **L,P**) from *Axinella cannabina* (row 3) and *Suberites carnosus* (row 4), respectively.

The latter sponge species, *Chondrosia reniformis*, is well known for its richness in collagen, and different methodologies have been proposed for its extraction, enabling the production of hydrogel-like materials with interesting rheological behavior and sticky features [4]. Moreover, collagen membranes derived from sponges were assessed for their water-binding capacity, antioxidant activity, and cytocompatibility envisaging skincare application [5]. Considering the biomedical relevance promised by this sponge collagen, efforts have been made on the aquaculture of *Chondrosia reniformis* [6], aiming to establish a sustainable route providing the biomass needed for collagen production. Still, in invertebrates, other phyla have been also researched, namely Cnidaria, in which jellyfish have been particularly explored considering the biomass availability resulting from blooms. In this regard, Bernhardt et al. [7] studied the use of *Rhopilema esculentum* jellyfish collagen for the development of biphasic scaffolds (interestingly, in combination with another marine collagen, from salmon skin) for osteochondral tissue engineering. Other cnidarians also inspiring scientists are corals, with Benayahu et al. [8] studying the unique collagen fibers identified in *Sarcophyton* soft corals and further combining them with alginate to produce biocomposite hydrogels for biomedical application.

Despite the advancements achieved with marine invertebrate collagens and the intriguing features exhibited by some of these organisms, such as the dynamic collagenous tissues observed in echinoderms, fish (vertebrates) are, by far, the most explored group of marine animals for collagen production. Different species have been studied, researching several parts of the animal, namely skin, scales, bones, swim bladders and fins. In this Special Issue, studies can be found addressing collagen from giant croaker (*Nibea japonica*) [9], codfish (*Gadus morhua*) [10], Nile tilapia (*Oreochromis niloticus*) [11,12], pufferfish (*Takifugu flavidus*) [13], small-spotted catshark (*Scyliorhinus canicula*) [14], common smooth-hound (*Mustelus mustelus*) [15] and blue shark (*Prionace glauca*) [16]. These species are, in general, already known for the production of collagen, with the herein-reported studies exploring advances in the extraction methodologies and the evaluation of applicability of these collagens, mostly regarding human health. In this perspective, the optimization of extraction conditions, such as acid and enzyme concentrations, biomass-to-solution ratio, temperature or time of reaction, has been pursued [9], including by the application of specific statistical tools, such as response surface methodology (RSM) [14]. In addition, new purification techniques have been also proposed, such as electrodialysis that promises economic and environmental advantages over traditional dialysis [13]. Moreover, thorough characterization of collagen was implemented by different authors to understand the main properties of the isolated materials, namely the molecular weight, by assessing the electrophoretic profile, the denaturation temperature, extract purity, and effect on cell behavior, in which a positive effect was observed over the cell adhesion or proliferation of fibroblasts [10], endothelial cells [11], osteoblasts [11,16] and mouse bone marrow–mesenchymal stem cells [16].

The field of biomaterials and biomedical applications is expanding at an unprecedented pace, trying to face the challenges of a growing population and longer life expectancy, and collagen-based devices are playing a relevant role. In this perspective, this Special Issue includes studies using marine collagen for the development of different types of products, such as wound dressings [12], bioactive films [15], membranes [5] and scaffolds for tissue regeneration [7,8]. With the same rationale, the use of collagen on bone substitutes for facial bone reconstruction has been systematically reviewed by Cicciù et al. [17] and, although not clarifying the advantages of biomaterials vs. autologous bone, the use of marine collagens seems to provide dimensional stability and growth factor carrier versatility to the constructs.

In addition to the use of the integral protein, the use of collagen hydrolysates and peptides is also of great interest for healthcare and wellbeing, as illustrated by the study reported by Ito et al. [18], in which a pilot study in humans has shown the benefits of a supplement comprising fish collagen peptide in the improvement of skin conditions, including elasticity and hydration.

Overall, we hope that this Special Issue elucidates the importance of marine collagens in biomedicine, appearing as an alternative to mammal collagens. It demonstrates the wide biodiversity that can be explored as a raw material for the production of collagen. Additionally, it provides representative examples of biomedical products and devices being proposed for wound healing, tissue regeneration and globally improving human health. This is a growing field, not only among scientists, but also in industry and public policies, and although delivering valuable information, this Special Issue reveals only a small part of the puzzle, hopefully triggering the curiosity of the readers to research further and contribute to the new knowledge being built.

Funding: This research received no external funding.

Conflicts of Interest: The authors declare no conflict of interest.

References

1. Rahman, M.A. Collagen of Extracellular Matrix from Marine Invertebrates and Its Medical Applications. *Mar. Drugs* **2019**, *17*, 118. [CrossRef] [PubMed]
2. Ehrlich, H.; Wysokowski, M.; Żółtowska-Aksamitowska, S.; Petrenko, I.; Jesionowski, T. Collagens of Poriferan Origin. *Mar. Drugs* **2018**, *16*, 79. [CrossRef] [PubMed]
3. Tziveleka, L.-A.; Ioannou, E.; Tsiourvas, D.; Berillis, P.; Foufa, E.; Roussis, V. Collagen from the Marine Sponges Axinella cannabina and Suberites carnosus: Isolation and Morphological, Biochemical, and Biophysical Characterization. *Mar. Drugs* **2017**, *15*, 152. [CrossRef] [PubMed]

4. Fassini, D.; Duarte, A.R.C.; Reis, R.L.; Silva, T.H. Bioinspiring Chondrosia reniformis (Nardo, 1847) Collagen-Based Hydrogel: A New Extraction Method to Obtain a Sticky and Self-Healing Collagenous Material. *Mar. Drugs* **2017**, *15*, 380. [CrossRef] [PubMed]

5. Pozzolini, M.; Scarfi, S.; Gallus, L.; Castellano, M.; Vicini, S.; Cortese, K.; Gagliani, M.C.; Bertolino, M.; Costa, G.; Giovine, M. Production, Characterization and Biocompatibility Evaluation of Collagen Membranes Derived from Marine Sponge Chondrosia reniformis Nardo, 1847. *Mar. Drugs* **2018**, *16*, 111. [CrossRef] [PubMed]

6. Gökalp, M.; Wijgerde, T.; Sarà, A.; De Goeij, J.M.; Osinga, R. Development of an Integrated Mariculture for the Collagen-Rich Sponge Chondrosia reniformis. *Mar. Drugs* **2019**, *17*, 29. [CrossRef] [PubMed]

7. Bernhardt, A.; Paul, B.; Gelinsky, M. Biphasic Scaffolds from Marine Collagens for Regeneration of Osteochondral Defects. *Mar. Drugs* **2018**, *16*, 91. [CrossRef] [PubMed]

8. Benayahu, D.; Sharabi, M.; Pomeraniec, L.; Awad, L.; Haj-Ali, R.; Benayahu, Y. Unique Collagen Fibers for Biomedical Applications. *Mar. Drugs* **2018**, *16*, 102. [CrossRef] [PubMed]

9. Yu, F.; Zong, C.; Jin, S.; Zheng, J.; Chen, N.; Huang, J.; Chen, Y.; Huang, F.; Yang, Z.; Tang, Y. Optimization of Extraction Conditions and Characterization of Pepsin-Solubilised Collagen from Skin of Giant Croaker (*Nibea japonica*). *Mar. Drugs* **2018**, *16*, 29. [CrossRef] [PubMed]

10. Carvalho, A.M.D.M.P.E.F.; Marques, A.P.; Silva, T.H.; Reis, R.L. Evaluation of the Potential of Collagen from Codfish Skin as a Biomaterial for Biomedical Applications. *Mar. Drugs* **2018**, *16*, 495. [CrossRef] [PubMed]

11. Song, W.-K.; Liu, D.; Sun, L.-L.; Li, B.-F.; Hou, H. Physicochemical and Biocompatibility Properties of Type I Collagen from the Skin of Nile Tilapia (*Oreochromis Niloticus*) for Biomedical Applications. *Mar. Drugs* **2019**, *17*, 137. [CrossRef] [PubMed]

12. Chen, J.; Gao, K.; Liu, S.; Wang, S.; Elango, J.; Bao, B.; Dong, J.; Liu, N.; Wu, W. Fish Collagen Surgical Compress Repairing Characteristics on Wound Healing Process In Vivo. *Mar. Drugs* **2019**, *17*, 33. [CrossRef] [PubMed]

13. Chen, J.; Li, M.; Yi, R.; Bai, K.; Wang, G.; Tan, R.; Sun, S.; Xu, N. Electrodialysis Extraction of Pufferfish Skin (*Takifugu flavidus*): A Promising Source of Collagen. *Mar. Drugs* **2019**, *17*, 25. [CrossRef] [PubMed]

14. Blanco, M.; Vázquez, J.A.; Pérez-Martín, R.I.; Sotelo, C.G. Collagen Extraction Optimization from the Skin of the Small-Spotted Catshark (*S. canicula*) by Response Surface Methodology. *Mar. Drugs* **2019**, *17*, 40. [CrossRef] [PubMed]

15. Ben Slimane, E.; Sadok, S. Collagen from Cartilaginous Fish By-Products for a Potential Application in Bioactive Film Composite. *Mar. Drugs* **2018**, *16*, 211. [CrossRef] [PubMed]

16. Elango, J.; Lee, J.W.; Wang, S.; Henrotin, Y.; De Val, J.E.M.S.; Regenstein, J.M.; Lim, S.Y.; Bao, B.; Wu, W. Evaluation of Differentiated Bone Cells Proliferation by Blue Shark Skin Collagen via Biochemical for Bone Tissue Engineering. *Mar. Drugs* **2018**, *16*, 350. [CrossRef] [PubMed]

17. Cicciù, M.; Cervino, G.; Herford, A.S.; Famà, F.; Bramanti, E.; Fiorillo, L.; Lauritano, F.; Sambataro, S.; Troiano, G.; Laino, L. Facial Bone Reconstruction Using both Marine or Non-Marine Bone Substitutes: Evaluation of Current Outcomes in a Systematic Literature Review. *Mar. Drugs* **2018**, *16*, 27. [CrossRef] [PubMed]

18. Ito, N.; Seki, S.; Ueda, F. Effects of Composite Supplement Containing Collagen Peptide and Ornithine on Skin Conditions and Plasma IGF-1 Levels—A Randomized, Double-Blind, Placebo-Controlled Trial. *Mar. Drugs* **2018**, *16*, 482. [CrossRef] [PubMed]

 *marine drugs*

Review

Collagen of Extracellular Matrix from Marine Invertebrates and Its Medical Applications

M. Azizur Rahman [1,2]

[1]   Department of Chemical & Physical Sciences, University of Toronto, Mississauga, ON L5L 1C6, Canada;
      aziz@climatechangeresearch.ca or mazizur.rahman@utoronto.ca; Tel.: +1-647-892-4221

[2]   Center for Climate Change Research, Toronto, ON M4P 1J4, Canada

Received: 22 December 2018; Accepted: 5 February 2019; Published: 14 February 2019

Abstract: The extraction and purification of collagen are of great interest due to its biological function and medicinal applications. Although marine invertebrates are abundant in the animal kingdom, our knowledge of their extracellular matrix (ECM), which mainly contains collagen, is lacking. The functions of collagen isolated from marine invertebrates remain an untouched source of the proteinaceous component in the development of groundbreaking pharmaceuticals. This review will give an overview of currently used collagens and their future applications, as well as the methodological issues of collagens from marine invertebrates for potential drug discovery.

Keywords: collagen; chitin; corals; extracellular matrix; marine invertebrates; marine proteins

1. Introduction

Collagen is one of the most abundant proteins in the extracellular matrix of animal bodies. This protein is the main fibrous, structural protein and supports the formation of all joints in the body. Supplementing collagen is an important way to keep our body healthy. Nowadays, collagen-based biomedical materials are used for the treatment of many human diseases (e.g., bone tissue regeneration). The challenge currently facing scientists is to find a suitable source of collagen, and the extraction and purification of collagen, which would be appropriate for applying to medical applications.

There is a huge source of collagen from marine organisms, and recent research has demonstrated that the marine source is the most convenient and safest way to obtain it, with invertebrates and crustose coralline algae [1–9] being the most abundant and potential sources (see Figure 1A for examples). A marine source also has lots of advantages over land animals such as being environmentally friendly, having a high quantity of collagen, having biological toxins that are almost negligible, having better absorption due to low molecular weight, having a minimal inflammatory response, having less religious and ethical constraints, being metabolically compatible, and having few regulatory and quality control problems.

In this review, I included crustose coralline algae (CCA) because they have similar characteristics of proteinaceous components and mineralization processes like calcifying marine invertebrates. CCA are rock-hard calcareous with two key functional roles in coral reef ecosystems: (1) reef calcification and cementation and (2) inducing the larval settlement of many benthic organisms. CCA contain calcium carbonate with hard skeletons and minerals (e.g., calcite) similar to coral skeletons. In addition, CCA have a high content of organic matrix skeletal proteins, including chitin and collagen [5,9]. CCA are abundant and are found in marine waters all over the world. I therefore introduce these abundant marine sources with the invertebrates presented in this review, which might have a high potential for the extraction of collagens, and moreover use for medical applications.

Invertebrates make up almost 95% of the animal kingdom, but our knowledge of their extracellular matrices, in particular, the polymer collagen is very weak. The information on the biology of collagen within the extracellular matrix is scanty. A large number of marine invertebrates produce polysaccharides and extracellular matrices [10–14] within their connective tissues, and their molecular structures and functions are similar to humans [15,16]. Moreover, polysaccharides extracted from marine calcifiers that contain extracellular matrices have an enormous assortment of structures (Figure 1B), and they can be considered an extraordinary source of biochemical variety. We therefore discuss the studies of collagens of invertebrates (including related marine calcifiers) and their plausible medical application.

Treatment of bone defects such as replacing tissue or regeneration requires biomaterials with similar mechanical integrity to natural bone, which can adapt and contribute to the tissue growth processes. From an applicable biomaterials point of view, the mineralized extracellular matrix of collagen in marine invertebrate structures has a vast richness for tissue engineering [17,18]. The skeletons in marine invertebrates are classic bio-resources that have tailored architectures to give structural support, and their functions are feasible for human tissue regeneration and repair. Marine calcifiers, for example, coralline, sea urchin, and coral, have interconnected porous structures that are enriched with bioactive elements and medical materials that could be used for tissue engineering and drug design applications [5,11,12,19–23]. The main purpose of this review is to provide an overview of currently used collagens from marine invertebrates and related calcifying organisms, and their medicinal potential, as well as the technical issues in purifying collagen from them.

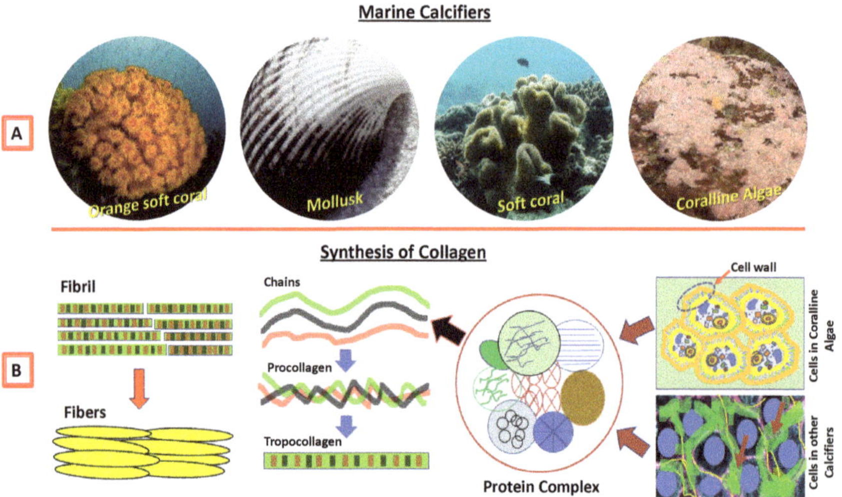

Figure 1. Marine calcifiers and their collagens. (**A**) Examples of marine calcifiers/invertebrates. (**B**) A model image on the biological synthesis of collagens from the marine invertebrates and crustose coralline algae.

2. Current State of Collagen Research and its Medical Application

There has been significant progress in the research of marine natural products in the purpose of medical application nowadays. Marine invertebrates are the main source of this purpose; however, finding collagen for the treatment of bone-related diseases is not well established yet. Many research groups have been studying collagen in some marine calcifier tissues with a focus on structure and functional relationships [1–10]. The biology of the extracellular matrix (Figure 2), particularly of collagen in invertebrates is essential to understanding the continuing research in the field of marine natural products. One of the key components of the structure of the collagen is a glycoprotein (see Figure 2, left panel), and many organisms in the CCA and invertebrate such as corals (especially, soft corals), coralline algae, and jewelry corals [5,24–26] have already demonstrated this key molecule (see Figure 3 for some examples). Helman et al. [24] reported collagen production in the ECM of both soft (*Xenia elongata*) and hard (*Montipora digitata*) corals. They clearly demonstrated the presence of glycoprotein in the ECM of corals, which means the presence of collagen must exist if the glycoprotein is present in the species. This is an indicator for species that contain collagen molecules.

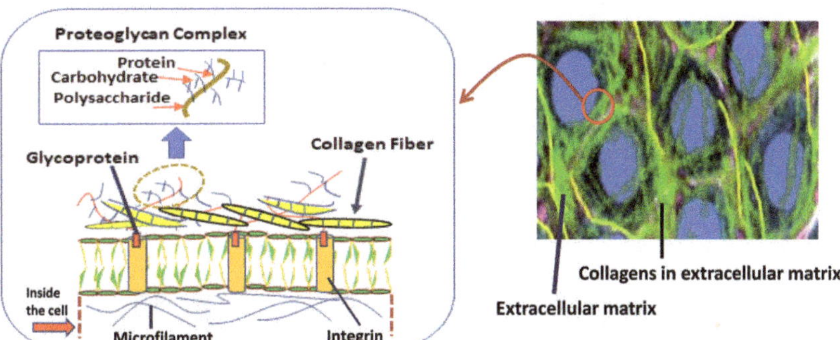

Figure 2. Collagen of extracellular matrix and its biology in invertebrates. The right panel shows a model of cells. The left panel shows the structural components of the extracellular matrix, which are involved in the formation of collagen in marine invertebrates.

To date, collagen has been identified in corals, sponges, sea urchin, salmon, jellyfish, mollusk, and coralline red algae [2–5,7–10,20,27–33], among others. Most of these organisms have also been applied for use in tissue engineering [34]. Collagen from marine invertebrates and related calcifiers have been discussed in numerous review papers [5,6,31,35–42] where the authors highlighted details regarding the structure and application of the collagen of this abundant marine source. It is a great possibility to use the huge source of marine invertebrates for extracting and purifying collagen, not only for the medical application and bone-related disease but also for use in cosmetics and anti-aging [43–48].

Recently, our group explored collagen in coralline red algae [5,9]. The research is now continuing, and a high number of collagens have now been extracted from this organism (papers in preparation). Some portions of this organism contain both chitin and collagen (Figures 4 and 5). Because of the huge number of these organisms available in shallow water of the sea, it would be an easy way to collect this marine group for extracting collagen. However, purification of collagen from these organisms has been a problem, and this issue has already been mostly solved (see Section 3 for details). This is a new group of marine organisms, which could get special attention for the extraction of collagen molecules in the near future.

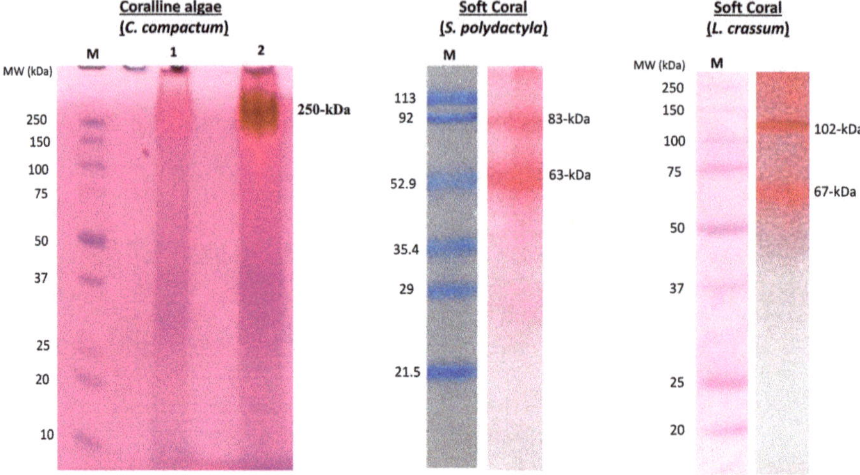

Figure 3. Collagen associated glycoproteins in marine calcifiers. Coralline red algae: Sodium dodecyl sulfate-polyacrylamide gel electrophoresis (SDS-PAGE) with a periodic acid-schiff (PAS) staining to detect glycoprotein in the extracellular matrix of *Clathromorphum compactum*. M, protein ladder. Lane 1 and 2, high molecular weight (250 kDa) of a glycoprotein. Soft coral (*Sinularia polydactyla*): SDS-PAGE with a PAS staining to detect glycoprotein in the extracellular matrix of *S. polydactyla*. M, protein ladder. Two glycoproteins (83 and 63 kDa) were identified in this species. Soft Coral (*Lobophytum crassum*): SDS-PAGE with a PAS staining. The PAS staining to detect glycoprotein in the extracellular matrix of *L. crassum*. M, protein ladder. Two glycoproteins (102 and 67 kDa) were identified in this soft coral species. The Precision Plus SDS-PAGE protein ladder (Bio-Rad) was used for the electrophoresis analysis of all above-mentioned glycoproteins. The glycoproteins presented here were reproduced from Rahman [5] for the coralline red algae and Rahman et al. [25] for the two soft corals (*S. polydactyla*, *L. crassum*).

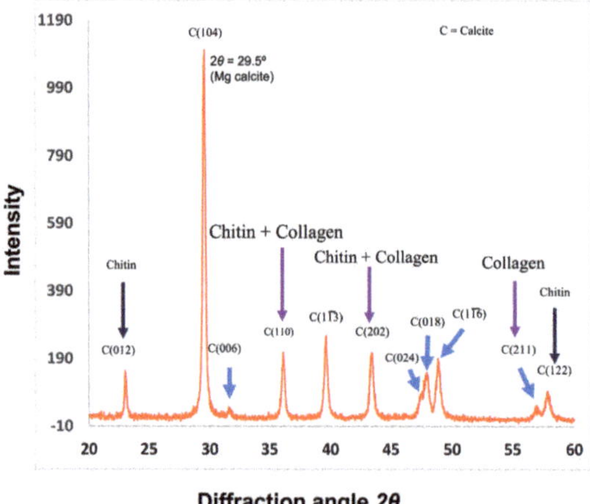

Diffraction angle 2θ

Figure 4. X-ray diffraction (XRD) analysis of *C. compactum*. The 2θ scan identifies the mineral form of CaCO$_3$ crystal planes, which were nucleated by chitin and collagen matrices. Purple arrows show the collagen bands. Reproduced with permission from Rahman and Halfar [9].

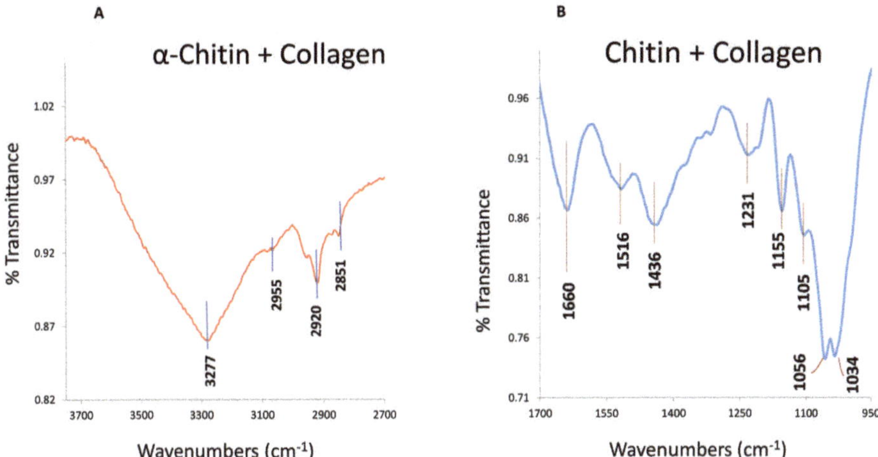

Figure 5. Infrared (IR) of collagens in *C. compactum*. Attenuated total reflection (ATR)–Fourier–transform infrared spectroscopy (FTIR) spectra reveal the collagen bands in both soluble (**A**) and insoluble (**B**) organic matrix fractions. [Reproduced from Rahman and Halfar (9)].

It is assumed that half of all marine-derived biomaterials are sourced from marine sponges, which might be the highest number of organisms in the invertebrates currently being used for the extraction of collagen. In sponges, collagen fibers have an interesting structural feature [45,49], and the molecules isolated from this group have a wide range of activities that can be used for promising biomedical applications [4,6], especially collagenous marine sponge skeletons, which are extremely strong, highly absorbent, elastic, and resistant to bacterial attack. A recent review by Ehrlich et al. [6] described details about collagen and collagen-like structural proteins from sponges. They also highlighted the prospects and trends of collagen extracted from sponges in biomedical applications, materials science, and technology. From the same research group [1], a hydroxylated fibrillar collagen containing an unusual motif of "Gly–3Hyp–4Hyp" was isolated from the glass sponge (Hexactinellida). The authors hypothesized that this motif in fibrillar collagen subject is a silica precipitation and a template for biosilicification. Recently, Tziveleka et al. [4] isolated and characterized the collagens from the marine demosponges *Suberites carnosus* (Suberitidae) and *Axinella cannabina* (Axenillidae) and found three different collagen-insoluble collagen (InSC), spongin-like collagen (SlC), and intracellular collagen (ICC) for biomedical applications. Collagen was isolated from many other marine sponges, for instance, *Chondrosia reniformis* [47], *Microciona prolifera*, *Spongia graminea*, *Haliclona oculata* [42], *Cacospongia scalaris*, *Hippospongia communis* [49], *Chondrosia reniformis* [50,51], *Geodia cydonium* [52], and several *Ircinia species* [53].

Corals are an abundant source of biologically and structurally active compounds. Coral skeletons have interconnected pores and are composed of $CaCO_3$, with appropriate porosity and pore sizes, making them a suitable material for bone implant application [17]. Regarding its interesting structural formation, coral has been in use commercially since the 1990s and is available as interpore and bio-coral [21]. There are several studies that have been found for such kinds of application, e.g., a three-dimensional coral skeleton structure endorsed the hard tissue growth and was totally replaced by new bone [22]. Similarly, a coral skeleton was used in human grafting [23]. Because of the structural compositions of coral, it absorbed $CaCO_3$ very quickly in growing new bone tissue, allowing for a formation of a scaffold. These reports indicate that the corals might have collagenous molecules, which can be applied as a treatment for bone-related disease.

However, the research for collagen on corals, especially for soft corals, has remarkably improved. Over the last several years [2,3,44,54], very interesting findings on collagen molecules have been demonstrated from soft corals. Also, a number of collagen-associated glycoproteins have been detected in soft corals [14,24,25,55,56]. The researchers found unique collagen fibers from the soft coral *Sarcophyton ehrenbergi* [2,3]. These fibers expose a 3D structure and hyper-elastic behavior, which are analogous to natural human tissues. The peculiarity of these fibers is too long (9 ± 0.37 μm). The research also demonstrated the collagen I and II types. The structural characterization of these collagen fibers reveals a highly suitable biomaterial for medical applications. Benayahu et al. [54] invented an interesting patent from the same soft coral species *S. ehrenbergi*. The inventors claimed that "(1) the collagen fibers from the soft coral have high adjustable extensibility compared with mammalian collagen fibers and (2) the stiffness of the collagen fibers isolated from this species is at the top range of the reported stiffness range of mammal collagen fibers". Another study [24] demonstrated the structural differentiation of collagen production in the ECM of soft corals; however, the authors did not investigate the types of collagen. Besides the above-mentioned marine organisms, sea urchin, marine fish, and mollusk organisms have been used for extracting different types of collagen, and the evidence showed that these collagen molecules have a strong role in the treatment of bone-related disease [20,27,29–31,33].

As mentioned above, research on the medical application of marine invertebrate collagen is currently progressing well. However, most collagen research findings from marine calcifiers/invertebrates are used in the application of bone-related disease (34), but the research in this field is still suffering from various complications. The medical application of marine collagens has been highlighted in recent reviews [6,31,35,57,58]. A review report by Cicciù et al. [57] suggested the facial bone reconstruction defect by applying marine collagen. During this review, the authors conducted a search using the MEDLINE and EMBASE databases (2007 to 2017), and their search results suggested that marine collagen can support the stability of the bone graft and could be an excellent carrier for growth factors. There are some recent reports of marine sources (coral, sponge, sea urchin, and fish) focused on the medical application (including bone tissue engineering and related diseases) of collagen available in the literature [3,6,17,31,59–61]. Moreover, collagen derived from mollusks, echinoderms, and sponges was reported [62–71], with some other important medical applications.

3. Purification Technique of Collagens from Marine Invertebrates

The molecules in invertebrates are complex, and therefore the purification of any specific molecule from this group is tricky. An individual species is required to apply different techniques, as the characterization of their components is multifaceted. For instance, soft corals have sclerites and soft tissue (unlike the stony corals) comprising complex organic matrices [14]. For these complexities, it was difficult to purify molecules; however, our group successfully purified the molecules [11,14], including the functional extracellular matrix proteins (e.g., ECMP-67), enzymes, calcium-binding proteins, and glycoproteins (see Figure 3 for examples). Glycoprotein in the extracellular matrix protein is a key component of collagen (Figure 2) that plays the main role in the biological process of collagen in invertebrates. Applying similar techniques, we recently investigated coralline algae, which have a high concentration of both chitin and collagen biopolymers and are functional in both soluble and insoluble organic matrix fractions (Figures 4 and 5) [5,9].

Coralline algal concentrations of the soluble organic matrix (0.9%) and insoluble organic matrix (4.5%) fractions are significantly higher than those of other marine invertebrates such as soft corals, with a soluble organic matrix and insoluble organic matrix of 0.03% and 0.05%, respectively [56,72]. The evidence of purified collagen in the coralline skeletons was also shown by X-ray diffraction (XRD) analysis (Figure 4). Jiang et al. [73] identified mineral crystals in collagen fibrils in a different marine invertebrate. The findings by Jiang et al. support our XRD results, and this technique has been revealed as a promising tool in analyzing collagens in the mineralization process. The results obtained by XRD demonstrated that XRD will become an important tool to study biological materials like collagen from the ECM of invertebrates. Such a high concentration of collagen present in the organic matrices of marine calcifiers presents the opportunity for future drug development in bone-related disease, and, moreover, both chitin and collagen present in the same species can take a significant role in drug design of other related diseases, because these two polymers are commonly used in drug design [74–83].

At present, the methods for the isolation and purification of collagens from the octocorals have been significantly improved. A patented protocol on the collagen purification from the soft coral [54] is now on the market. Since this method is patented, it is not open to the public. However, there are several publications by the same research group that currently exist in the literature, in which they established the methods in purifying collagens (including collagen types I and II) from the soft coral [2,3]. The development of these new technologies, along with the technologies established by our group as mentioned above, will be extremely beneficial for purifying functional collagens from these marine organisms.

Despite the importance of collagenous marine sponge skeletons being documented, the techniques for the purification of collagens from this group are not well-established yet because of their insolubility and mineralization, which might cause difficulties in its separation and characterization [84,85]. However, researchers are trying to resolve these issues, and numerous investigations have so far been reported in this group [47,50,53,86,87]. Recently, Pozzolini et al. [47] established several new methods to purify collagenous fibrillar suspensions from the *Chondrosia reniformis* demosponge. The authors demonstrated that the obtained fibrillar collagens are extensively useful for tissue engineering and regenerative medicine, as well as in antioxidant activity.

There are some techniques that have been established in purifying collagens from the invertebrates; however, a proteomic approach might be a useful tool to learn more about the collagen and its functions in detail. Proteomics have already been established as an important tool for the detection, characterization, and analysis of pharmaceutically useful proteins from marine organisms, and this approach provides the most precise evaluation of protein identities, abundance, composition, and protein expression profiling [5,26,88–90]. Therefore, in regard to collagen, the proteomics approach could be a promising toolkit in the near future. The overview regarding marine collagen of invertebrates stated above allows us to understand some newly developed techniques and suitable methods for extracting and purifying collagen, as well as for applying proteomics approach for medical applications.

4. Future Applications of Invertebrate Collagens in Medical Field

The marine ecosystem provides suitable and numerous diversified resources for human health in comparison to the terrestrial ecosystem. In the last few decades, marine resources, especially invertebrates, have been recognized to be a promising source for many drugs (e.g., Cytarabine, Vidarabine, and Halichondrin B) [91]. According to the discussion above (Sections 1–3), marine invertebrates and related calcifying organisms such as soft and hard corals, sponge, mollusk, sea urchin, and coralline algae could be a major source of medicines over the next decades. However, extraction and purification of collagen for the purpose of medical application of these resources is still under investigation developing. Despite some impressive work having been performed on collagenous sponges and corals [1–4,6,7,15–17,47,54,91–93], an intensive study is necessary with these two groups and other invertebrates to use these huge apposite resources in future years. The potential of marine invertebrates for collagen could be realized by developing new technologies; indeed, there are many methods such as proteomics, computer-aided design, bioinformatics, and combinatorial synthesis that are now being applied.

The biological diversity of marine invertebrates and complex protein and peptide components direct us toward discovery of many new drugs for various therapeutic areas, including bone-related disease (e.g., osteoporosis) [94]. Besides cancer, microbial infections, and inflammation, drug discovery for bone-related disease is the biggest challenge of the current century, and collagen extraction from marine invertebrates shows new promise in fighting against this and other related diseases.

5. Concluding Remarks

In this review, the current state of research on collagen extracted from the ECM of invertebrates and its applications in the medical field have been discussed, and some light has been shed on future perspectives of this important marine material. The methodological issues of collagen purification from invertebrates, which the researchers are currently struggling with, have also been highlighted. The discussion concerning the purification techniques in this review could be of tremendous help in the extraction of purified collagen from invertebrates. The extracellular matrix, which is one of the key components in invertebrates and is responsible for producing collagen in this marine group, has been elaborated with informative imaging. In addition, the glycosylation activity with the formation of glycoproteins (size of the protein, which varies from species to species) in invertebrates, whose biological processes are involved in producing collagen, has been discussed for the first time in this review. The obtained results demonstrate the potential for marine invertebrates to generate new drugs, especially for bone tissue regeneration.

Conflicts of Interest: The authors declare no conflict of interest.

References

1. Ehrlich, H.; Deutzmann, R.; Brunner, E.; Cappellini, E.; Koon, H.; Solazzo, C.; Yang, Y.; Ashford, D.; Thomas-Oates, J.; Lubeck, M. Mineralization of the metre-long biosilica structures of glass sponges is templated on hydroxylated collagen. *Nat. Chem.* **2010**, *2*, 1084–1088. [CrossRef] [PubMed]

2. Orgel, J.P.; Sella, I.; Madhurapantula, R.S.; Antipova, O.; Mandelberg, Y.; Kashman, Y.; Benayahu, D.; Benayahu, Y. Molecular and ultrastructural studies of a fibrillar collagen from octocoral (Cnidaria). *J. Exp. Biol.* **2017**, *220*, 3327–3335. [CrossRef] [PubMed]

3. Benayahu, D.; Sharabi, M.; Pomeraniec, L.; Awad, L.; Haj-Ali, R.; Benayahu, Y. Unique Collagen Fibers for Biomedical Applications. *Mar. Drugs* **2018**, *16*, 102. [CrossRef] [PubMed]

4. Tziveleka, L.-A.; Ioannou, E.; Tsiourvas, D.; Berillis, P.; Foufa, E.; Roussis, V. Collagen from the Marine Sponges *Axinella cannabina* and *Suberites carnosus*: Isolation and Morphological, Biochemical, and Biophysical Characterization. *Mar. Drugs* **2017**, *15*, 152. [CrossRef] [PubMed]

5. Rahman, M.A. An Overview of the Medical Applications of Marine Skeletal Matrix Proteins. *Mar. Drugs* **2016**, *14*, 167. [CrossRef] [PubMed]

6. Ehrlich, H.; Wysokowski, M.; Żółtowska-Aksamitowska, S.; Petrenko, I.; Jesionowski, T. Collagens of Poriferan Origin. *Mar. Drugs* **2018**, *16*, 79. [CrossRef] [PubMed]

7. Ehrlich, H. Chitin and collagen as universal and alternative templates in biomineralization. *Int. Geol. Rev.* **2010**, *52*, 661–699. [CrossRef]

8. Ehrlich, H.; Heinemann, S.; Heinemann, C.; Simon, P.; Bazhenov, V.V.; Shapkin, N.P.; Born, R.; Tabachnick, K.R.; Hanke, T.; Worch, H. Nanostructural organization of naturally occurring composites—Part I: Silica-collagen-based biocomposites. *J. Nanomater.* **2008**, 623838. [CrossRef]

9. Rahman, M.A.; Halfar, J. First evidence of chitin in calcified coralline algae: New insights into the calcification process of Clathromorphum compactum. *Sci. Rep.* **2014**, *4*, 6162. [CrossRef]

10. Rodriguez-Pascual, F.; Slatter, D.A. Collagen cross-linking: Insights on the evolution of metazoan extracellular matrix. *Sci. Rep.* **2016**, *6*, 37374. [CrossRef]

11. Rahman, M.A.; Fujimura, H.; Shinjo, R.; Oomori, T. Extracellular matrix protein in calcified endoskeleton: A potential additive for crystal growth and design. *J. Cryst. Growth* **2011**, *324*, 177–183. [CrossRef]

12. Laurienzo, P. Marine polysaccharides in pharmaceutical applications: An overview. *Mar. Drugs* **2010**, *8*, 2435–2465. [CrossRef] [PubMed]

13. Rahman, M.A.; Oomori, T. In vitro regulation of CaCO3 crystal growth by the highly acidic proteins of calcitic sclerites in soft coral, *Sinularia polydactyla*. *Connect. Tissue Res.* **2009**, *50*, 285–293. [CrossRef] [PubMed]

14. Rahman, M.A.; Oomori, T.; Worheide, G. Calcite formation in soft coral sclerites is determined by a single reactive extracellular protein. *J. Biol. Chem.* **2011**, *286*, 31638–31649. [CrossRef] [PubMed]

15. Green, D.W.; Padula, M.P.; Santos, J.; Chou, J.; Milthorpe, B.; Ben-Nissan, B. A therapeutic potential for marine skeletal proteins in bone regeneration. *Mar. Drugs* **2013**, *11*, 1203–1220. [CrossRef] [PubMed]

16. Cooper, E.L.; Hirabayashi, K.; Strychar, K.B.; Sammarco, P.W. Corals and their potential applications to integrative medicine. *Evid. Based Complement. Altern. Med.* **2014**, *2014*, 184959. [CrossRef] [PubMed]
17. Macha, I.J.; Ben-Nissan, B. Marine Skeletons: Towards Hard Tissue Repair and Regeneration. *Mar. Drugs* **2018**, *16*, 225. [CrossRef]
18. Webster, T.J.; Ahn, E.S. Nanostructured Biomaterials for Tissue Engineering Bone. In *Tissue Engineering II: Basics of Tissue Engineering and Tissue Application*; Lee, K., Kaplan, D., Eds.; Springer: Berlin/Heidelberg, Germany, 2007; pp. 275–308.
19. Senni, K.; Pereira, J.; Gueniche, F.; Delbarre-Ladrat, C.; Sinquin, C.; Ratiskol, J.; Godeau, G.; Fischer, A.M.; Helley, D.; Colliec-Jouault, S. Marine polysaccharides: A source of bioactive molecules for cell therapy and tissue engineering. *Mar. Drugs* **2011**, *9*, 1664–1681. [CrossRef]
20. Omura, Y.; Urano, N.; Kimura, S. Occurrence of fibrillar collagen with structure of $(\alpha_1)_2\alpha_2$ in the test of sea urchin *Asthenosoma ijimai*. *Comp. Biochem. Phys. B* **1996**, *115*, 63–68. [CrossRef]
21. Doherty, M.J.; Schlag, G.; Schwarz, N.; Mollan, R.A.; Nolan, P.C.; Wilson, D.J. Biocompatibility of xenogeneic bone, commercially available coral, a bioceramic and tissue sealant for human osteoblasts. *Biomaterials* **1994**, *15*, 601–608. [CrossRef]
22. Guillemin, G.; Patat, J.L.; Fournie, J.; Chetail, M. The use of coral as a bone graft substitute. *J. Biomed. Mater. Res.* **2004**, *21*, 557–567. [CrossRef] [PubMed]
23. Marchac, D.; Sandor, G. Use of coral granules in the craniofacial skeleton. *J. Craniofac. Surg.* **1994**, *5*, 213–217. [CrossRef] [PubMed]
24. Helman, Y. Extracellular matrix production and calcium carbonate precipitation by coral cells in vitro. *Proc. Natl. Acad. Sci. USA* **2007**, *105*, 54–58. [CrossRef] [PubMed]
25. Rahman, M.A.; Isa, Y.; Uehara, T. Studies on two closely related species of octocorallians: Biochemical and molecular characteristics of the organic matrices of endoskeletal sclerites. *Mar. Biotechnol.* **2006**, *8*, 415–424. [CrossRef] [PubMed]
26. Rahman, M.A.; Karl, K.; Nonaka, M.; Fujimura, H.; Shinjo, R.; Oomori, T.; Worheide, G. Characterization of the proteinaceous skeletal organic matrix from the precious coral *Corallium konojoi*. *Proteomics* **2014**, *14*, 2600–2606. [CrossRef]
27. Cluzel, C.; Lethias, C.; Garrone, R.; Exposito, J.Y. Distinct maturations of *n*-propeptide domains in fibrillar procollagen molecules involved in the formation of heterotypic fibrils in adult sea urchin collagenous tissues. *J. Biol. Chem.* **2004**, *279*, 9811–9817. [CrossRef]
28. Bernhardt, A.; Paul, B.; Gelinsky, M. Biphasic Scaffolds from Marine Collagens for Regeneration of Osteochondral Defects. *Mar. Drugs* **2018**, *16*, 91. [CrossRef]
29. Bairati, A. The Collagens of the Mollusca. In *Biology of Invertebrate and Lower Vertebrate Collagens. NATO ASI Series (Series A: Life Sciences)*; Bairati, A., Garrone, R., Eds.; Springer: Boston, MA, USA, 1985.
30. Maria, F.; Maria, F.; Andronescu, E.; Voicu, G.; Ficai, D.; Albu, M.G.; Ficai, A. Mollusc shell/collagen composite as potential biomaterial for bone substitutes. *Rom. J. Mat. (RRM)* **2010**, *40*, 359–364.
31. Goh, K.L.; Holmes, D.F. Collagenous Extracellular Matrix Biomaterials for Tissue Engineering: Lessons from the Common Sea Urchin Tissue. *Int. J. Mol. Sci.* **2017**, *18*, 901. [CrossRef]
32. Benedetto, C.D.; Barbaglio, A.; Martinello, T.; Alongi, V.; Fassini, D.; Cullorà, E.; Patruno, M.; Bonasoro, F.; Barbosa, M.A.; Carnevali, M.D.C.; et al. Production, Characterization and Biocompatibility of Marine Collagen Matrices from an Alternative and Sustainable Source: The Sea Urchin *Paracentrotus lividus*. *Mar. Drugs* **2014**, *12*, 4912–4933. [CrossRef]
33. Hoyer, B.; Bernhardt, A.; Lode, A.; Heinemann, S.; Sewing, J.; Klinger, M.; Notbohm, H.; Gelinsky, M. Jellyfish collagen scaffolds for cartilage tissue engineering. *Acta Biomater.* **2014**, *10*, 883–892. [CrossRef]
34. Clarke, S.; Walsh, P.; Maggs, C.; Buchanan, F. Designs from the deep: Marine organisms for bone tissue engineering. *Biotechnol. Adv.* **2011**, *29*, 610–617. [CrossRef] [PubMed]
35. Silva, T.H.; Moreira-Silva, J.; Marques, A.L.P.; Domingues, A.; Bayon, Y.; Reis, R.L. Marine origin collagens and its potential applications. *Mar. Drugs* **2014**, *12*, 5881–5901. [CrossRef] [PubMed]
36. Adams, E. Invertebrate collagens. *Science* **1978**, *202*, 591–598. [CrossRef] [PubMed]
37. Engel, J. Versatile collagens in invertebrates. *Science* **1997**, *277*, 1785–1786. [CrossRef] [PubMed]
38. Bailey, A.F. The nature of collagen. In *Comprehensive Biochemistry, Extracellular and Supporting Structures*; Florkin, M., Stotz, E.H., Eds.; Elsevier: Amsterdam, The Netherlands, 1968.

39. Exposito, J.Y.; Cluzel, C.; Garrone, R.; Lethias, C. Evolution of collagens. *Anat. Rec.* **2002**, *268*, 302–316. [CrossRef] [PubMed]

40. Garrone, R. Evolution of metazoan collagens. *Prog. Mol. Subcell. Biol.* **1999**, *21*, 119–139.

41. Tanzer, M.L. The biological diversity of collagenous proteins. *Trends Biochem. Sci.* **1978**, *3*, 15–17. [CrossRef]

42. Gross, J.; Sokal, Z.; Rougvie, M. Structural and chemical studies on the connective tissue of marine sponges. *J. Histochem. Cytochem.* **1956**, *4*, 227–246. [CrossRef]

43. Hyun, C.H.; Joo, H.H.; Hee, K.J.; Yoo, S.J.; Su, J.Y.; Dae, H.L.; Hye, M.P. Preparing collagen useful in cosmetic composition, involves hydrolyzing fish by-products using enzyme. Korea Patent KR2013094989-A, 27 July 2013.

44. Thao, N.P.; Luyen, B.T.T.; Lee, S.H.; Jang, H.D.; Kiem, P.V.; Minh, C.V.; Kim, Y.H. Antiosteoporotic and antioxidant activities of diterpenoids from the Vietnamese soft corals *Sinularia maxima* and *Lobophytum crassum*. *Med. Chem. Res.* **2015**, *24*, 3551–3560. [CrossRef]

45. Aizenberg, J.; Weaver, J.C.; Thanawala, M.S.; Sundar, V.C.; Morse, D.E.; Fratzl, P. Skeleton of *Euplectella* sp.: Structural Hierarchy from the Nanoscale to the Macroscale. *Science* **2005**, *309*, 275–278. [CrossRef] [PubMed]

46. Nicklas, M.; Schatton, W.; Heinemann, S.; Hanke, T.; Kreuter, J. Preparation and characterization of marine sponge collagen nanoparticles and employment for the transdermal delivery of 17beta-estradiol-hemihydrate. *Drug Dev. Ind. Pharm.* **2009**, *35*, 1035–1042. [CrossRef] [PubMed]

47. Pozzolini, M.; Scarfi, S.; Gallus, L.; Castellano, M.; Vicini, S.; Cortese, K.; Gagliani, M.C.; Bertolino, M.; Costa, G.; Giovine, M. Production, Characterization and Biocompatibility Evaluation of Collagen Membranes Derived from Marine Sponge *Chondrosia reniformis* Nardo, 1847. *Mar. Drugs* **2018**, *16*, 111. [CrossRef]

48. Latire, T.; Legendre, F.; Bigot, N.; Carduner, L.; Kellouche, S.; Bouyoucef, M.; Carreiras, F.; Marin, F.; Lebel, J.-M.; Galéra, P.; et al. Shell Extracts from the Marine Bivalve *Pecten maximus* Regulate the Synthesis of Extracellular Matrix in Primary Cultured Human Skin Fibroblasts. *PLoS ONE* **2014**, *9*, e99931. [CrossRef] [PubMed]

49. Junqua, S.; Robert, L.; Garrone, R. Biochemical and morphological studies on the collagens of horny sponges. *Ircinia* filaments compared to spongines. *Connect. Tissue Res.* **1974**, *2*, 193–203. [CrossRef] [PubMed]

50. Garrone, R.; Huc, A.; Junqua, S. Fine structure and physicochemical studies on the collagen of the marine sponge *Chondrosia reniformis* Nardo. *J. Ultrastruct. Res.* **1975**, *52*, 261–275. [CrossRef]

51. Swatschek, D.; Schatton, W.; Kellermann, J.; Müller, W.E.G.; Kreuter, J. Marine sponge collagen: Isolation, characterization and effects on the skin parameters surface-pH, moisture and sebum. *Eur. J. Pharm. Biopharm.* **2002**, *53*, 107–113. [CrossRef]

52. Diehl-Seifert, B.; Kurelec, B.; Zahn, R.K.; Dorn, A.; Jeričevic, B.; Uhlenbruck, G.; Müller, W.E.G. Attachment of sponge cells to collagen substrata: Effect of a collagen assembly factor. *J. Cell Sci.* **1985**, *79*, 271–285.

53. Pallela, R.; Bojja, S.; Janapala, V.R. Biochemical and biophysical characterization of collagens of marine sponge, *Ircinia fusca* (Porifera: Demospongiae: Irciniidae). *Int. J. Biol. Macromol.* **2011**, *49*, 85–92. [CrossRef]

54. Benayahu, Y.; Benayahu, D.; Kashman, Y.; Rudi, A.; Lanir, Y.; Sela, I.; Raz, E. Coral Derived Collagen and Methods of Farming Same. U.S. Patent US20110038914A1, 17 February 2011.

55. Rahman, M.A.; Shinjo, R.; Oomori, T.; Worheide, G. Analysis of the proteinaceous components of the organic matrix of calcitic sclerites from the soft coral *Sinularia* sp. *PLoS ONE* **2013**, *8*, e58781. [CrossRef]

56. Rahman, M.A.; Isa, Y. Characterization of proteins from the matrix of spicules from the alcyonarian, *Lobophytum crassum*. *J. Exp. Mar. Biol. Ecol.* **2005**, *321*, 71–82. [CrossRef]

57. Cicciù, M.; Cervino, G.; Herford, A.S.; Famà, F.; Bramanti, E.; Fiorillo, L.; Lauritano, F.; Sambataro, S.; Troiano, G.; Laino, L. Facial Bone Reconstruction Using both Marine or Non-Marine Bone Substitutes: Evaluation of Current Outcomes in a Systematic Literature Review. *Mar. Drugs* **2018**, *16*, 27. [CrossRef] [PubMed]

58. Felician, F.F.; Xia, C.; Qi, W.; Xu, H. Collagen from Marine Biological Sources and Medical Applications. *Chem Biodivers.* **2018**, *15*, e1700557. [CrossRef] [PubMed]

59. Chen, J.; Gao, K.; Liu, S.; Wang, S.; Elango, J.; Bao, B.; Dong, J.; Liu, N.; Wu, W. Fish Collagen Surgical Compress Repairing Characteristics on Wound Healing Process in Vivo. *Mar. Drugs* **2019**, *17*, 33. [CrossRef] [PubMed]

60. Carvalho, A.M.; Marques, A.P.; Silva, T.H.; Reis, R.L. Evaluation of the Potential of Collagen from Codfish Skin as a Biomaterial for Biomedical Applications. *Mar. Drugs* **2018**, *16*, 495. [CrossRef] [PubMed]

61. Elango, J.; Lee, J.W.; Wang, S.; Henrotin, Y.; De Val, J.E.M.S.; Regenstein, J.M.; Lim, S.Y.; Bao, B.; Wu, W. Evaluation of Differentiated Bone Cells Proliferation by Blue Shark Skin Collagen via Biochemical for Bone Tissue Engineering. *Mar. Drugs* **2018**, *16*, 350. [CrossRef] [PubMed]

62. Pozzolini, M.; Millo, E.; Oliveri, C.; Mirata, S.; Salis, A.; Damonte, G.; Arkel, M.; Scarfi, S. Elicited ROS Scavenging Activity, Photoprotective, and Wound-Healing Properties of Collagen-Derived Peptides from the Marine Sponge *Chondrosia reniformis*. *Mar. Drugs* **2018**, *16*, 465. [CrossRef]

63. Jridi, M.; Bardaa, S.; Moalla, D.; Rebaii, T.; Souissi, N.; Sahnoun, Z.; Nasri, M. Microstructure, rheological and wound healing properties of collagen-based gel from cuttlefish skin. *Int. J. Biol. Macromol.* **2015**, *77*, 369–374. [CrossRef]

64. Uriarte-Montoya, M.H.; Arias-Moscoso, J.L.; Plascencia-Jatomea, M.; Santacruz-Ortega, H.; Rouzaud-Sández, O.; Cardenas-Lopez, J.L.; Marquez-Rios, E.; Ezquerra-Brauer, J.M. Jumbo squid (*Dosidicus gigas*) mantle collagen: Extraction, characterization, and potential application in the preparation of chitosan-collagen biofilms. *Bioresour. Technol.* **2010**, *101*, 4212–4219. [CrossRef]

65. Ferrario, C.; Leggio, L.; Leone, R.; Di Benedetto, C.; Guidetti, L.; Coccè, V.; Ascagni, M.; Bonasoro, F.; La Porta, C.A.; Carnevali, M.D.C.; et al. Marine-derived collagen biomaterials from echinoderm connective tissues. *Mar Environ. Res.* **2017**, *128*, 46–57. [CrossRef]

66. Green, D.; Howard, D.; Yang, X.; Kelly, M.; Oreffo, R.O. Natural marine sponge fiber skeleton: A biomimetic scaffold for human osteoprogenitor cell attachment, growth, and differentiation. *Tissue Eng.* **2003**, *9*, 1159–1166. [CrossRef] [PubMed]

67. Kim, M.M.; Mendis, E.; Rajapakse, N.; Lee, S.H.; Kim, S.K. Effect of spongin derived from *Hymeniacidon sinapium* on bone mineralization. *J. Biomed. Mater. Res. B Appl. Biomater.* **2009**, *90*, 540–546. [CrossRef] [PubMed]

68. Lin, Z.; Solomon, K.L.; Zhang, X.; Pavlos, N.J.; Abel, T.; Willers, C.; Dai, K.; Xu, J.; Zheng, Q.; Zheng, M. *In vitro* evaluation of natural marine sponge collagen as a scaffold for bone tissue engineering. *Int. J. Biol. Sci.* **2011**, *7*, 968–977. [CrossRef] [PubMed]

69. Pallela, R.; Venkatesan, J.; Janapala, V.R.; Kim, S.K. Biophysicochemical evaluation of chitosan-hydroxyapatite-marine sponge collagen composite for bone tissue engineering. *Biomed. Mater. Res. A.* **2012**, *100*, 486–495. [CrossRef] [PubMed]

70. Nandi, S.K.; Kundu, B.; Mahato, A.; Thakur, N.L.; Joardar, S.N.; Mandal, B.B. *In vitro* and *in vivo* evaluation of the marine sponge skeleton as a bone mimicking biomaterial. *Integr. Biol. (Camb).* **2015**, *7*, 250–262. [CrossRef] [PubMed]

71. Granito, R.N.; Custódio, M.R.; Rennó, A.C. Natural marine sponges for bone tissue engineering: The state of art and future perspectives. *J. Biomed. Mater. Res. B Appl. Biomater.* **2017**, *105*, 1717–1727. [CrossRef]

72. Rahman, M.A.; Oomori, T. Structure, crystallization and mineral composition of sclerites in the alcyonarian coral. *J. Cryst. Growth* **2008**, *310*, 3528–3534. [CrossRef]

73. Jiang, H.; Ramunno-Johnson, D.; Song, C.; Amirbekian, B.; Kohmura, Y.; Nishino, Y.; Takahashi, Y.; Ishikawa, T.; Miao, J. Nanoscale Imaging of Mineral Crystals inside Biological Composite Materials Using X-Ray Diffraction Microscopy. *Phys. Rev. Lett.* **2008**, *100*, 038103. [CrossRef]

74. Zhang, J.; Xia, W.; Liu, P.; Cheng, Q.; Tahirou, T.; Gu, W.; Li, B. Chitosan modification and pharmaceutical/biomedical applications. *Mar. Drugs* **2010**, *8*, 1962–1987. [CrossRef]

75. Da Sacco, L.; Masotti, A. Chitin and chitosan as multipurpose natural polymers for groundwater arsenic removal and AS$_2$O$_3$ delivery in tumor therapy. *Mar. Drugs* **2010**, *8*, 1518–1525. [CrossRef]

76. Khoushab, F.; Yamabhai, M. Chitin research revisited. *Mar. Drugs* **2010**, *8*, 1988–2012. [CrossRef]

77. Aam, B.B.; Heggset, E.B.; Norberg, A.L.; Sorlie, M.; Varum, K.M.; Eijsink, V.G. Production of chitooligosaccharides and their potential applications in medicine. *Mar. Drugs* **2010**, *8*, 1482–1517. [CrossRef] [PubMed]

78. Chaudhary, P.M.; Tupe, S.G.; Deshpande, M.V. Chitin synthase inhibitors as antifungal agents. *Mini Rev. Med. Chem.* **2013**, *13*, 222–236. [PubMed]

79. Ruiz-Herrera, J.; San-Blas, G. Chitin synthesis as target for antifungal drugs. *Curr. Drug Targets Infect. Disord.* **2003**, *3*, 77–91. [CrossRef] [PubMed]

80. Reese, T.A.; Liang, H.E.; Tager, A.M.; Luster, A.D.; van Rooijen, N.; Voehringer, D.; Locksley, R.M. Chitin induces accumulation in tissue of innate immune cells associated with allergy. *Nature* **2007**, *447*, 92–96. [CrossRef] [PubMed]

81. Addad, S.; Exposito, J.Y.; Faye, C.; Ricard-Blum, S.; Lethias, C. Isolation, characterization and biological evaluation of jellyfish collagen for use in biomedical applications. *Mar. Drugs* **2011**, *9*, 967–983. [CrossRef] [PubMed]

82. Natali, I.; Tempesti, P.; Carretti, E.; Potenza, M.; Sansoni, S.; Baglioni, P.; Dei, L. Aragonite crystals grown on bones by reaction of CO_2 with nanostructured $Ca(OH)_2$ in the presence of collagen. Implications in archaeology and paleontology. *Langmuir* **2014**, *30*, 660–668. [CrossRef] [PubMed]

83. Yang, T.L. Chitin-based materials in tissue engineering: Applications in soft tissue and epithelial organ. *Int. J. Mol. Sci.* **2011**, *12*, 1936–1963. [CrossRef]

84. Imhoff, J.M.; Garrone, R. Solubilization and characterization of *Chondrosia reniformis* sponge collagen. *Connect. Tissue Res.* **1983**, *11*, 193–197. [CrossRef]

85. Ehrlich, H.; Hanke, T.; Simon, P.; Goebel, C.; Heinemann, S.; Born, R.; Worch, H. Demineralisation von natürlichen Silikat-basierten Biomaterialien: Neue Strategie zur Isolation organischer Gerüststrukturen. *Biomaterialien* **2005**, *6*, 297–302. [CrossRef]

86. Heinemann, S.; Ehrlich, H.; Douglas, T.; Heinemann, C.; Worch, H.; Schatton, W.; Hanke, T. Ultrastructural studies on the collagen of the marine sponge *Chondrosia reniformis* Nardo. *Biomacromolecules* **2007**, *8*, 3452–3457. [CrossRef] [PubMed]

87. Gökalp, M.; Wijgerde, T.; Sarà, A.; De Goeij, J.M.; Osinga, R. Development of an Integrated Mariculture for the Collagen-Rich Sponge *Chondrosia reniformis*. *Mar. Drugs* **2019**, *17*, 29. [CrossRef] [PubMed]

88. Molloy, M.P.; Witzmann, F.A. Proteomics: Technologies and applications. *Brief Funct. Genomics Proteomics* **2002**, *1*, 23–39. [CrossRef]

89. Rahman, M.A. The medicinal potential of promising marine organisms: A review. *Blue Biotechnol. J.* **2012**, *1*, 318–333.

90. Dauphin, Y. Comparative studies of skeletal soluble matrices from some Scleractinian corals and molluscs. *Int. J. Biol. Macromol.* **2001**, *28*, 293–304. [CrossRef]

91. Mayer, A.M.; Glaser, K.B.; Cuevas, C.; Jacobs, R.S.; Kem, W.; Little, R.D.; McIntosh, J.M.; Newman, D.J.; Potts, B.C.; Shuster, D.E. The odyssey of marine pharmaceuticals: A current pipeline perspective. *Trends Pharmacol. Sci.* **2010**, *31*, 255–265. [CrossRef]

92. Peng, B.-R.; Lu, M.-C.; El-Shazly, M.; Wu, S.-L.; Lai, K.-H.; Su, J.-H. Aquaculture Soft Coral *Lobophytum crassum* as a Producer of Anti-Proliferative Cembranoids. *Mar. Drugs* **2018**, *16*, 15. [CrossRef]

93. Ye, F.; Zhu, Z.-D.; Gu, Y.-C.; Li, J.; Zhu, W.-L.; Guo, Y.-W. Further New Diterpenoids as PTP1B Inhibitors from the Xisha Soft Coral *Sinularia polydactyla*. *Mar. Drugs* **2018**, *16*, 103. [CrossRef]

94. Chaugule, S.R.; Indap, M.M.; Chiplunkar, S.V. Marine Natural Products: New Avenue in Treatment of Osteoporosis. *Front. Mar. Sci.* **2017**, *4*, 384. [CrossRef]

Review

Collagens of Poriferan Origin

Hermann Ehrlich [1,*], **Marcin Wysokowski** [2], **Sonia Żółtowska-Aksamitowska** [2], **Iaroslav Petrenko** [1] **and Teofil Jesionowski** [2]

[1]   Institute of Experimental Physics, TU Bergakademie Freiberg, Leipziger str. 23, 09599 Freiberg, Germany; iaroslavpetrenko@gmail.com
[2]   Institute of Chemical Technology and Engineering, Faculty of Chemical Technology, Poznan University of Technology, Berdychowo 4, 61131 Poznan, Poland; marcin.wysokowski@put.poznan.pl (M.W.); sonia.zoltowska-aksamitowska@doctorate.put.poznan.pl (S.Ż.-A.); teofil.jesionowski@put.poznan.pl (T.J.)
*   Correspondence: hermann.ehrlich@physik.tu-freiberg.de; Tel.: +49-3731-39-2867

Received: 30 December 2017; Accepted: 28 February 2018; Published: 3 March 2018

Abstract: The biosynthesis, structural diversity, and functionality of collagens of sponge origin are still paradigms and causes of scientific controversy. This review has the ambitious goal of providing thorough and comprehensive coverage of poriferan collagens as a multifaceted topic with intriguing hypotheses and numerous challenging open questions. The structural diversity, chemistry, and biochemistry of collagens in sponges are analyzed and discussed here. Special attention is paid to spongins, collagen IV-related proteins, fibrillar collagens from demosponges, and collagens from glass sponge skeletal structures. The review also focuses on prospects and trends in applications of sponge collagens for technology, materials science and biomedicine.

Keywords: collagen; spongin; collagen-related proteins; sponges; scaffolds; biomaterials

---

1. Introduction

Collagens constitute a superfamily of long-lived extracellular matrix structural proteins of fundamental evolutionary significance, found in both invertebrate and vertebrate taxa. They are among the most studied proteins due to their important functions in mammals, including humans. In addition to their structural function in cartilage and skin formation [1,2], as well as in the biomineralization of hard tissues [3] including bone [4] and dentine [5], collagens are involved in the regulation of diverse cellular functions and processes. During the last 60 years, research into collagens has evolved from the discovery of the structure of collagen [6,7], through studies on its chemistry and biochemistry [8–10], to present-day applications in cell therapy [11], biomedicine [12–14], cosmetics [15], and the food industry [16]. A rod-like triple-helical domain is the typical structural element in all collagens. However, they differ in their size, dislocations of the globular domains and imperfections within the triple helix, self-assembly behavior, and functional roles. The classification of collagens is based on structural and functional features of vertebrate collagens. For example, 28 collagen types have so far been identified and characterized at the molecular level in mammals (see for review [1,17]). Collagens are also divided into subfamilies based on their supramolecular assemblies: fibrils, beaded filaments, anchoring fibrils, and networks [11]. Usually, the amino acid sequences in collagens are responsible for the corresponding functional properties: energy storage capacity, stiffness, or elasticity [18]. Even the type of amino acid motif within the tropocollagen molecule of a collagen can significantly affect its mechanical properties. Consequently, it can be hypothesized that the diversity of collagen polyforms determines their future functions, even within the same organism.

Marine vertebrate collagens have attracted scientific attention, mostly as products of fisheries [19]. In particular, fish-sourced collagens from skins and scales [20–22] have been studied and used as alternative collagen sources to avoid the potential risks associated with mammalian collagen due to bovine spongiform encephalopathy and the swine influenza crisis [23].

In contrast to marine vertebrate collagens, similar structural proteins found in marine invertebrates represent one of the most ancient protein families within Metazoa. Marine invertebrate collagens arose earlier than their vertebrate analogs, and possess diverse unique structural features, including very special structure–function interrelations. Collagens from poriferans, coelenterates, annelids, mollusks, echinoderms, and crustaceans have been discussed in detail in several review papers (e.g., [24–32]) and books (e.g., [33,34]). The limiting factors that have hindered progress in this field of research are the difficulty of purifying marine invertebrate collagens and their relative species-dependent complexity. However, there are more than enough examples in practically every order of marine invertebrates to inspire experts in materials science and biomedicine, especially because the similarities in structure and biosynthesis between vertebrate and invertebrate collagens appear to be more impressive than the differences [24].

Sponges (Porifera) are the most simple and ancient multicellular organisms on our planet, and mostly live attached to a suitable substratum (rock, sandy sediments) on the seabed. Poriferans diverged from other Metazoans earlier in evolutionary history than any other known animal phylum, extant or extinct [35], with the first fossilized sponge remnants found in 1.8 billion-year-old sediments [36–41]. The phylum Porifera is divided into four classes: Hexactinellida, Demospongiae, and Homoscleromorpha, with silica-based skeletons; and Calcarea, with a skeletal network made of calcium carbonates [42]. According to Exposito et al. [27], before the divergence of the sponge and eumetazoan lineages took place, the genes which were responsible for the synthesis of some kind of ancestral fibrillar collagen arose at the dawn of the Metazoa. The duplication events leading to the formation of the A, B, and C clades of the fibrillar collagens occurred before the eumetazoan radiation. Interestingly, the similarity in the modular structure of sponges and humans is preserved only in the B clade of fibrillar collagens. This phenomenon correlates well with the hypothesis of the primordial function of type V/XI fibrillar collagens in initiating the formation of collagen fibrils [27].

Different systems of terminology relating to poriferan collagens are found in the literature, as sponges also display considerable polymorphism with respect to their collagenous structures. The insolubility of most poriferan collagens has been the main obstacle to carrying out any detailed biochemical analysis. Studies on the morphology and nanotopography of the collagenous fibrils have shown that they are dispersed throughout the intracellular matrix within the skeletons of sponges. Cuticular structures have been found in some sponges, but their molecular composition has not been determined [43].

It was accepted very early that collagen fibers in sponges can possess quite different morphological features [44]. Gross et al. isolated two distinct forms of collagen from *Spongia graminea*, which they called spongin A and spongin B [29]. The first corresponds to fine intercellular collagen fibrils, visible only by electron microscopy. The second, spongin B, forms macroscopically-visible rigid fibers which are characteristic of keratosan demosponges [43]. This was probably the moment when the terminological divergence arose with regard to the term *spongin*, which was initially proposed by Städeler [45] to denote the skeletal fibrous matter of bath sponges, and was also used for spongins A and B defined by Gross et al. [29]. Up to the present, the authors of numerous publications—especially those on applications of spongin-based scaffolds in tissue engineering [31,46–51]—have used the term collagen for spongin, or even defined spongin as "collagenic skeleton" [52]. Very recently, Tziveleka et al. [53] studied collagen from the marine demosponges *Axinella cannabina* and *Suberites carnosus*, and proposed three different terms: insoluble collagen (InSC), intercellular collagen (ICC), and spongin-like collagen (SlC). It is worth noting that the isolation of each form of collagen from demosponges is based on the selectivity of the method used. Data on collagen extraction methods from diverse mineralized sponges (Hexactinellida, Demospongiae) and sponges which lack mineralized skeletons (the subclass Keratosa)—including yields of the extracted collagens—may be found in the relevant papers.

In this review, we focus on the structural diversity of collagens and collagen-like proteins in selected sponges, with particular focus on their origin, structural features, and applications in biomedicine and technology, including materials science and biomimetics. The review has the ambitious goal of providing thorough and comprehensive coverage of poriferan collagens (Figure 1) as a multifaceted topic with controversial hypotheses and numerous open questions. We begin with a brief description of spongins and their practical applications. Next, we examine the collagen IV-related proteins in diverse representatives of Porifera. Special attention is paid to *Chondrosia* sp. collagens and their applications in marine biotechnology, biomedicine, and cosmetics. Finally, we discuss the current state of work related to the unique hydroxylated collagen discovered in anchoring siliceous spicules of psychrophilic deep-sea glass sponges. We are optimistic that both the attempts to establish implications for poriferan collagens and the numerous open questions raised in this review will inspire the scientific community to carry out research into collagens and collagen-related proteins from sponges, as ancient and intriguing structural biopolymers.

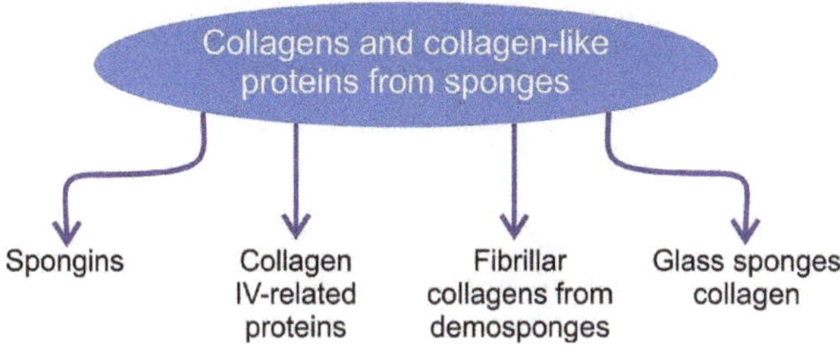

Figure 1. Schematic overview of the collagens and collagen-like structural proteins of poriferan origin described in this review.

2. Spongins as Enigmatic Structural Proteins in Sponges

It is recognized that so-called spongioblasts—derived from the epithelium of sponges—are responsible for the formation of spongin. Minchin claims that the fibers of skeletal spongin are formed extracellularly; however, the cuticular spongin fibrils are of intracellular origin [19]. In contrast to such structural proteins as collagen, fibroin (silk), elastin, resilin, and keratin, the chemistry and molecular biology—including the sequences—of spongins so far remain unknown. It seems that spongin is the last enigmatic proteinaceous biopolymer, although it is of very ancient origin and has undergone more than 300 years of investigations. Spongin in the form of cell-free skeletons of diverse bath sponges (Figures 2 and 3) has been used for more than 3000 years [54,55] for painting, bathing, and cleaning, as padding for battle armor, for medical purposes, and as a vessel for drinking water [56]. A brief overview of the practical applications of spongin from bath sponges in biomedicine and technology in recent times is given in the next section.

A suggestion of a similarity between silk and bath sponge skeletal fibers was reported for the first time by Geoffroy in 1705 [57], and was based on his chemical experiments. After that, attention was paid to practical applications of sponges in pharmacology due to the presence of iodine in their skeletons. For example, in 1819, Andrew Fyfe—a professor of chemistry in Aberdeen—identified large quantities of iodine in the marine sponge *Spongia usta*, the "Coventry Remedy", which was used even in ancient China [58]. In 1841, bath sponges were described as those in which the essential base of the skeleton consists of keratose fibrous matter. At that time, the structural and chemical similarity between horny fibers of sponges and silk was again suggested by Croockewit [59]. It would appear that the horny matter of sponge is closely analogous to silk and related proteins, differing from them only in that it contains additional halogens. According to Croockewit, the chemical formula of horny matter must be as follows: $20(C_{39}H_{62}N_{12}O_{17}) + I_2S_3P_{10}$ [59]. Schlossberger [60], however, reported the very slight solubility of the fibrous matter in ammoniacal solution of copper hydroxide. Additionally, treatment with diluted sulfuric acid leads to the identification of leucine and glycocoll, in contrast to the isolation of tyrosine and serine from sericin under similar conditions. Städeler in 1859 obtained similar results [45] and introduced for the first time the scientific term *spongin* for this horny matter. Then, in 1864, von Kölliker [61] carried out the first histological studies on sponges, including investigations of the structural features of fibrous spongin. Diverse iodine-containing sponges and the matter termed as "*Jodospongin*" were discussed by Hundeshagen in 1895 [62]. The organic origin of iodine in bath sponge skeleton was suggested by Harnack [63]. He estimated the concentration of iodine in spongin at 1.1–1.2%, and demonstrated that superheated steam destroys the organic portion of spongin fibers completely, liberating iodine.

Figure 2. The mineral- and cell-free skeleton of commercial *Hippospongia communis* bath sponge is an example of a 3D spongin scaffold.

In 1898, Harnack isolated the "*Jodspongin*" and characterized it as an albuminoid-like product, containing over 8.5% iodine and 9.4% nitrogen [63]. In 1926, Clancey carried out a critical analysis of the literature to-date relating to the identification of spongin by other authors. In contrast to other physiologists, he suggested that the origin of the skeletal spongin in Euceratosa was not the same as that of the spongin which surrounds the spicules in the Pseudoceratosa [64]. At that time, the common bath sponge *Hippospongia equina* and the "Turkey cup sponge" *Euspongia officinalis* were the sponges most studied with respect to spongin. The results published in various papers [65,66] showed remarkable differences, due to insufficiently effective analytical methods and the use of commercial sponges that had been variously prepared and bleached. Consequently, different results on the chemical nature of spongins from particular species were obtained.

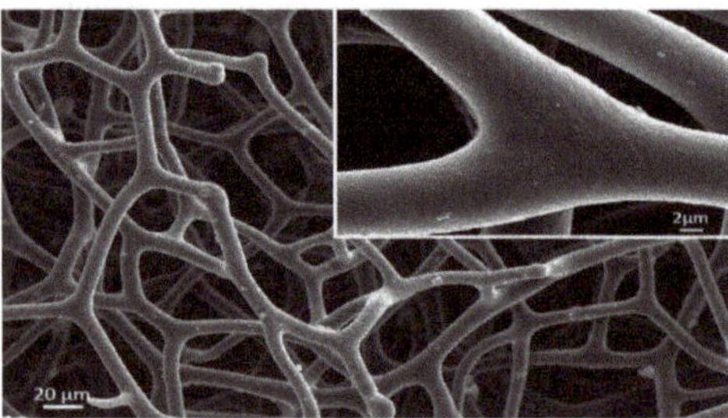

Figure 3. Scaning electron microscopy (SEM) image of anastomosed spongin fibers from the demosponge *H. communis*, which are organized as sets of unconnected structures with dendritic architecture.

For example, Clancey [64] isolated up to 7% of iodogorgonic acid besides the other amino acids in acidic hydrolysates of spongin. Clancey [64] did not identify hydroxyproline in spongin fibers of *Hippospongia equina* which had been treated with acid and alkali. It should be noted here that in natural collagen, a 3(S)-hydroxy-L-proline (3-Hyp) residue occurs together with a 4-Hyp residue, which is known to markedly increase the conformational stability of the collagen triple helix [67]. Hydroxyproline is found almost exclusively in collagen [8]. Thus, Clancey found a remarkably high amount of glutamic acid (18.4%), as well as 14% glycine, 5.7% proline, 2.8% tyrosine, 11% tryptophan or histidine, and a trace of cystine. Block and Bolling [68] presented the following results on the chemistry of spongin (Table 1).

Table 1. Amino acid composition of spongin.

Constituent	Content (%)
Nitrogen	13.0–14.8
Sulfur	0.7
Iodine	0.84–1.46
Histidine	0–0.2
Lysine	3–3.6
Arginine	4.3–5.9
Cystine	2.8
Tyrosine	0–0.8
Tryptophan	0
Phenylalanine	3.3
Glycine	13.9–14.4
Diiodotyrosine	4.7
Molecular ratio of lysine to arginine	4:6

The content of glycine (about 14%) in this spongin is significantly lower than in collagen (between 25% and 33%) [8]. Thus, until the identification of two different spongins by Gross et al. in 1956, spongin was recognized, for the most part, as a halogenated scleroprotein (see Table 2) [69–71] or neurokeratin-like protein [68] due to the presence of cystine.

Table 2. The chemistry of spongin according to Ackerman and Burkhard [69].

C	H	N	I	Br	S	Cl	Ashes
47.00	6.28	16.06	1.41	2.93	0.87	0	1.16

Consequently, it is very curious that the two morphologically-distinct forms of spongin fibers—termed spongin A and spongin B—were classified by Gross et al. [29] as members of the collagen family. This was probably because such an analysis was supported by electron microscopy and X-ray diffraction, and by their general amino acid pattern, including corresponding glycine and hydroxyproline content. Ratios of glycine to hydroxyproline were 1.6 and 1.8 for spongins A and B, respectively. The results obtained with small-angle X-ray diffraction and electron microscopy showed the diameter of the spongin A unbranched fibril to be on the order of 20 nm, with an axial period of about 650 Å. The large and branched fibers of spongin B were 10–50 μm in width and composed primarily of bundles of thin unbranched filaments less than 10 nm wide [29]. Both fiber types and the amorphous matrix contain hexosamine, hexose, pentose, and uronic acid. Glucosamine, galactosamine, glucose, galactose, mannose, fucose, arabinose, and uronic acid were identified chromatographically in both spongin A and in the amorphous substance. It was shown that spongin B contains a small amount of amino sugar plus glucose and galactose. In contrast to mammalian collagen, neither spongin can be dissolved at all by collagenase (*Clostridium hystolyticum*) or pepsin, nor were they dissolved to any appreciable extent in alkali solutions or dilute acid [29]. In a paper by Katzmann et al. [72], it was reported that spongin B accounts for over 70% of the dry weight of the bath sponge *H. gossypina*, and contains approximately 7% by weight of glucosylgalactosylhydroxylysine but a negligible amount of other sugars.

Recently, Langasco et al. [52] isolated glycosaminoglycans (GAGs) from sponginous skeletons of selected bath sponges. Total GAG content—expressed as μg hexuronate/mg dry weight—shows some variability among the tested species, being 0.171 ± 0.021, 0.367 ± 0.028, and 0.460 ± 0.081 for *H. communis*, *Spongia officinalis*, and *S. lamella*, respectively. The data obtained suggest that these sponge GAGs are structurally divergent from vertebrate GAGs [52].

Thus, it seems that spongin chemistry is made very complex by the presence of diverse halogens (I, Br) which have never been reported in natural collagens or keratins. This may explain the very high resistance of this proteinaceous biopolymer to enzymatic treatment. Its unique resistance to various enzymes—including amylases, lysozymes, trypsin, pronase, collagenases, and other proteases—is well reported [44,72]. On the other hand, in the natural environment diverse bacteria are able to destroy spongin enzymatically and lead to extremely high levels of damage to the structure of the spongin-based skeletal fibers (see for details [73]). The isolation and purification of such special *"sponginases"* remain a challenge for future research, and will provide a key way to obtain peptides that will be useful for detailed proteomic analysis and the sequencing of spongin.

Understanding of the nature and origin of spongins—especially in keratosan demosponges (the orders Verongiida, Dictioceratida, and Dendroceratida)—changed dramatically after the discovery of chitin as a second structural component of the skeletal fibers of demosponges in the order Verongiida by the Ehrlich Group in 2007 [74–76]. It was shown that anastomosing and macroporous skeletons of diverse verongiids are made of some kind of spongin–chitin biocomposites. The content of chitin in such composites ranges between 10% and 60% depending on the sponge species [76]. The isolation and characterization of chitin in these composites was possible due to the well-known resistance of chitin to dissolution in alkaline solutions [77–79], in contrast to spongins, which are quickly dissoluble in alkali [72,80]. Consequently, all publications prior to 2007 on spongins found in Verongiida sponges must be re-examined. The only existing and up-to-date classification of spongins is that proposed by Garrone in 1978 [43]. He states that the following types of spongins can be defined and discussed (see Figure 4). The first spongin is to be found in the form of spiculated fibers. These structures are associated with the endogenous mineralized skeleton of the sponge. It is also responsible for the formation of wide fibers which include only a very thin mineral element in the core. This kind of spongin is also resistant to mild acid or alkaline hydrolysis, as well as to pepsin and diverse bacterial collagenases. However, this spongin can be partially destroyed by cuprammonium hydroxide treatment at room temperature.

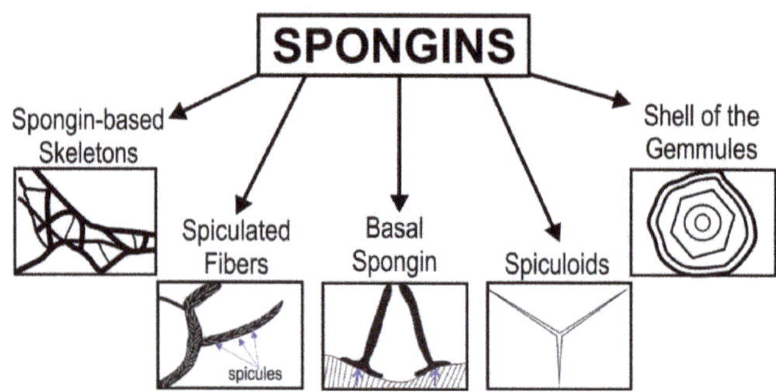

Figure 4. Diversity of spongins according to [43].

Second are the spongin fibers which form the skeleton of the keratosan demosponges: the abundance and compactness of the spongin and the almost complete lack of its own inclusions—which are replaced with foreign particles—testify to the originality of the spongin in this group. A typical example of such spongin can be found in the genus *Ircinia*, characterized by spongin fibers cored with foreign debris (sand microparticles) [81]. Recently, Castritsi-Catharios et al. [82–84] described the chemical elements and the physical properties of such skeletal spongin from diverse commercial sponges before and after chemical treatment.

The importance of the so-called basal spongin is evident for all sponges as sessile animals. In sponges with no organized internal skeleton, the organism is attached to the substratum by a more or less continuous layer of external spongin. This spongin is secreted by the basopinacocytes. The basal spongin is continuous with the internal spongin only in poriferans with an organized skeleton, formed either of spongin fibers or spiculated fibers. Due to the function of the basal spongin in such demosponges as *Chondrosia reniformis* (a species lacking spicules and internal spongin), the animal is attached strongly to its substratum. The basal spongin is discontinuous in erect sponges, where it forms the starting points of the internal organized skeleton. However, in the endemic fresh water demosponge *Lubomirskia baicalensis*, the holdfast which is responsible for attaching the sponge body to the hard substratum contains both basal spongin and chitin [77].

The extremely flexible and elastic organic structures which are morphologically similar to mineralized spicules are known as spiculoids [85]. They have been described in representatives of the genera *Darwinella* and *Igernella* (order Dendroceratida), where they are either free or partly joined to the fibers of the skeleton. They are compressible and can be easily torn apart. Finally, spongin may be responsible for the protection of gemmule shells. Gemmules are formed within the tissues of most freshwater and some marine sponges, and represent morphologically diverse asexual reproductive spherical bodies a few tenths of a millimeter to more than 1 mm in diameter, composed of a dense mass of identical cells and surrounded by an organic coat called the shell. The shell of gemmules is fortified with siliceous spicules and gemmoscleres, which are embedded into a matrix composed of both chitin and a collagenous protein. This collagen has been referred to as spongin [43].

Trends in the Applications of Spongins

The history of studies on the chemistry, molecular biology, biochemistry, and bioinspired materials science of spongins remains relevant today, partly due to the poorly understood basis of ecological disaster in the case of sponge diseases, but mostly due to recent progress in the direct applications of sponge skeletons as 3D spongin scaffolds in tissue engineering and biomimetics. Additionally, the marine ranching of bath sponges worldwide is a crucial factor in the adoption of spongins as renewable naturally prestructured proteinaceous scaffolds.

The spongin-based skeletons of bath sponges appear to possess a number of unique and useful properties, which had been exploited long before such scientific fields as tissue engineering and bioengineering were proposed. As reviewed by Szatkowski et al. [86], from the 18th century commercial bath sponges were valued in medicine due to their softness, high compressive strength, ability to retain shape, and high sorption rates. For these reasons, they were used as compression bandages for pressing open sinuses, in overcoming strictures of body passages (including the rectum), for dilation of the cervix uteri [86–90], and in the form of sponge tents applied in the uterus to expand the cavity and enable examination. More intriguingly, fragments of sponge skeleton were used as small prostheses in early "plastic surgery" [91]. Revolutionary results were obtained by Hamilton in 1881. In a paper entitled *"On sponge-grafting"* [92], he reported the following case. A woman underwent surgery for removal of a mammary tumor, during which a large area of skin was removed. The skin was replaced with a thin slice of an aseptic sponge skeleton, which ten days after the surgery was observed to be vascular, and three months later was covered with epithelial tissue (Figure 5).

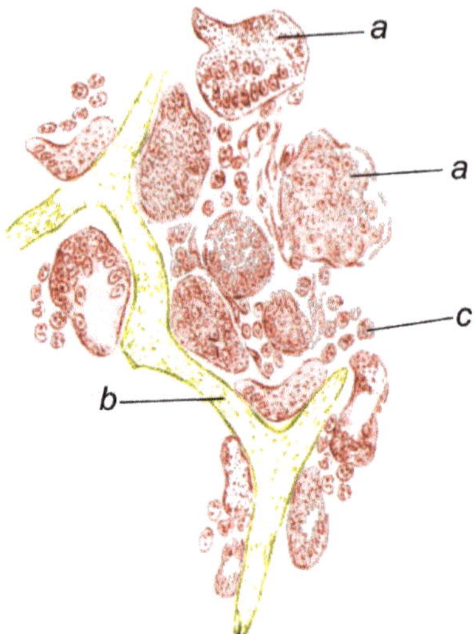

Figure 5. Sketch of a fragment of spongin framework (**b**) surrounded by a great number of living cells (**a**,**c**) in a sponge-grafting application (adapted from [92]).

Today, spongin-based scaffolds are actively used in diverse applications related to tissue engineering. Positive results have been reported with human osteoprogenitor cells on the skeleton of *S. officinalis* [46], with osteoblast-like MG-63 cells growing on spongin from *Hymeniacidon sinapium* [93] and with mouse primarily osteoblasts on spongin from *Callyspongiidae* marine demosponges [49]. Recently, Nandi et al. [51] have proposed that the skeleton of the marine sponge *Biemna* sp.—alone and in combination with growth factors—is a promising biomaterial for bone repair and bone augmentation.

Besides applications in the biomedical field, spongin-based scaffolds have been successfully used as adsorbents of diverse dyes [94,95] and as supports for enzyme immobilization [96]. It was recently shown that spongins are thermostable up to 260 °C [86,97,98]. This property opens the door for applications of spongin-based scaffolds with 3D architecture in such novel scientific disciplines as Extreme Biomimetics [99], with the aim of developing novel advanced composite materials.

3. Collagen IV and Related Proteins in Sponges

It is now well established that collagens are key to the structural integrity and biomechanical properties of various tissues of Metazoans. One of them, the basement membrane-forming collagen IV, is extremely ancient. Collagen IV networks have a polygonal architecture that endows basement membranes (BMs) with a tensile strength sufficient to protect tissues from mechanical stress, in addition to serving as important regulators of the dynamic events associated with cell adhesion, signaling, and survival [100]. According to the modern view [101], only the presence of the collagen IV gene was precisely correlated with the emergence of BMs in animals. Thus, the triple helical collagen IV was required for the development of BMs.

BMs underlie the epithelia in Metazoa from sponges to humans [102]. Interestingly, until 1996, basement membrane structures and type IV collagen were known to be present in all multicellular animal species except sponges. In Porifera, BMs are associated with the basal surfaces of polarized epithelial cells [103]. After the first report on the identification of type IV collagenous sequences in the homoscleromorph sponge *Pseudocorticium jarrei* by cDNA and genomic DNA [103], this collagen has been found in diverse poriferans. For example, in corresponding transcriptome data from a calcareous sponge (*Sycon coactum*) and another homoscleromorph sponge (*Corticium candelabrum*), two new type IV collagen genes were found in each [104]. Homologs of important components of basement membrane genes, including type IV collagen, have been found in the Demospongiae *Spongilla lacustris*, *Ircinia fasciculata*, and *Chondrilla nucula* [105]. The discovery of type IV collagen in Calcarea and Demospongiae is very important, because nowhere in this group has a BM-like structure been noted. The presence of type IV collagen in glass sponges (Hexactinellida) remains to be detected. Polyclonal antibodies have detected type I (but not type IV) collagen in the anchoring spicules of the *Hyalonema sieboldii* glass sponge [3].

The relationship between type IV collagen and the so-called spongin short chain collagen (SSCC) [106] is still under investigation [101]. SSCC has been considered as ancestral to type IV collagen [107]. Like type IV collagen, SSCC also has NC1 domains which produce the globular heads particular to type IV collagen and which are required for assembly of the unique scaffold of the BM (see for review [104]). It is suggested that collagen IV and its spongin variant are primordial components of the extracellular microenvironment, where collagen IV especially was a key player in the evolution of epithelial tissues in Metazoa, including sponges, due to the transition to multicellularity [101].

Interestingly, collagen IV from the demosponge *Chondrosia reniformis* has recently been patented as a source of special membranes for biomedical applications [108]. The collagen was isolated with an extraction solution of 100 mM Tris-HCl, 10mM EDTA, 8 M urea, and 100 mM 2-mercaptoethanol, rendering the protein in the form of a precipitate. This was used for the development of stable and non-cytotoxic type IV collagen membranes, which can be applied in tissue engineering and regenerative medicine approaches for epithelial repair, regeneration, or replacement. The technology includes the re-epithelialization of any single and stratified epithelium, with emphasis on the skin.

4. Fibrillar Collagens in the Mesohyl of Demosponges

The mesohyl includes a noncellular colloidal mesoglea with embedded collagen fibers, spicules, and various cells, being as such a type of mesenchyme. It is currently debated whether the mesohyl and pinacoderm layers in sponges are true tissues [109]. Collagens serve several functions in sponges [27,106,110]. The formation of mesohyl certainly involves the activity of fine fibrils made of fibrillar collagen. The collagen fibrils both mediate cell–matrix interactions via membrane receptors and provide the structure of the extracellular matrix (ECM), a situation observed in vertebrates. The increase in the structural diversity of fibrillar collagen chains, their different forms of maturation, and interactions with other ECM components appeared during the process of evolution [111]. The diversity of sponges which contain high amounts of fibrillar collagen within their mesohyl has been described previously (see for review [110–112]).

Fibrillar bundles, formed by the association of several hundred collagen fibrils, have been observed in diverse species of *Tethya*, *Chondrosia*, *Chondrilla*, *Jaspis*, and *Suberites* (see for details [112,113]). The densely packed bundles of collagen fibrils are secreted exclusively by the highly polarized lophocyte cells [43,111]. These are actively moving cells, pulling behind them a bundle of regularly arranged collagen fibrils.

Another kind of collagen-producing cell has been discovered in the mesohyl of the demosponge *Suberites domuncula* [114,115], in which the expression of collagen genes is controlled by silicate and myotrophin [116]. SEM observations have revealed the complex collagen network surrounding the spicules within the mesohyl of adult specimens (Figure 6).

Collagen fibers have also been identified in the mesohyl of the demosponge *Haliclona rosea* [116]. Collagen has also been reported in the mesohyl of such Calcarea sponges as *Leucosolenia* sp. and *Leucandra* sp. [117].

Collagen fibril content is also high in the external asexual buds that occur in *Tethya lyncurium* [118]. Similarly, the buds of *T. sychellensis* contain a dense collagen matrix [119]. Buds consist of cellular masses that sprout out from the surface of adults and are able to develop into new functional individuals [119].

Recently, special attention has been focused on fibrillar collagens in the mesohyl of *C. reniformis*. This species is the only sponge which has been experimentally proven to contain a dynamic collagenous mesohyl capable of stiffening upon being manipulated [120]. It was shown that the different physiological states recorded in laboratory experiments are expressions of the mechanical adaptability of the collagenous mesohyl of *C. reniformis*, and suggest that stiffness variability in this sponge is under cellular control [121].

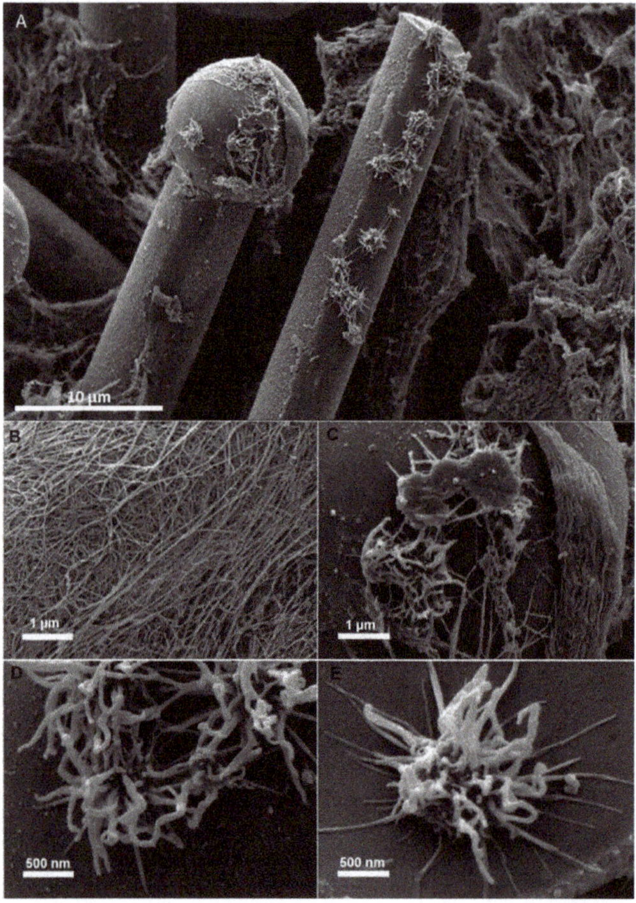

Figure 6. SEM view through the collagenous mesohyl of the demosponge *S. domuncula*. Layers of collagen fibrils (**A,B**) are a result of the activity of the unique collagen-producing cells which are seen to line up along the surface of the spicules (**C–E**). The line of cells (**A**) can move from left to right along the spicule, depositing a rough, nanofibrillar collagenous layer in their wake (**C**) (see also [114]).

5. *Chondrosia* Collagens

Collagens from the demosponge *Chondrosia reniformis* (Nardo 1847) have received attention from researchers since 1970 [113,117–119] due to their diversity (type IV collagen, fibrillar and nonfibrillar collagens) [120] and interesting structural [121], physicochemical [122], and ecophysiological properties [123–127]. For example, slices of fibrillar collagen incubated with collagenase are not modified even after 48 h of incubation, and do not show any changes in the aspect, consistency, or fine structure of the fibrils. No kind of enzymatic damage was observed by electron microscope on the isolated collagen fibrils after collagenase treatment.

The mechanical properties of this collagen have been partially described by Garrone et al. [117]. The cortex of *Chondrosia* sponges is less resistant than calf skin, but has mechanical properties of the same order as those of bovine nasal cartilage (Young's modulus 150–250 kg/cm^2 and 100–250 kg/cm^2 respectively). Probably due to special mechanical features, the body of *Chondrosia* can slowly become flat and slide to avoid compression or stretch itself into a slender thread under continuous stress. Such *creeping* behavior of a fibrous and living material provides a remarkable example for the study of mechanical stresses as morphogenetic factors [117]. Although the nanomorphology of *C. reniformis* collagen fibrils has now been well investigated (Figure 7) [121], there is still a lack of knowledge about the relationship between the ultrastructural features of this collagen and its mechanical and physicochemical properties.

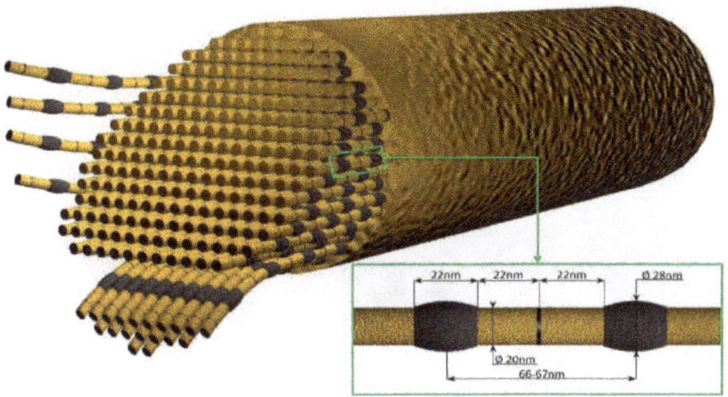

Figure 7. Schematic diagram of *C. reniformis* collagen fiber with numerous nanofibrils with characteristic nanotopography. Along the fibril, one characteristically thick segment (protrusion) about 28 nm in diameter is followed by two equal thinner and closer conjoined segments (interband) about 20 nm in diameter. The average distance between the protrusions is about 67–69 nm. The distance between two consecutive peaks of the interband regions or between a protrusion and an adjacent interband region is about 21–23 nm. The average step height between the protrusions and the interband regions was calculated to be about 4 nm (see for review [121]).

The well-known biocompatibility of *C. reniformis* fibrillar collagen has stimulated many studies on its possible applications in cosmetics and pharmacology (see for review [31]), including in transdermal drug delivery [128].

The production and selection of a triple transformed *Pichia pastoris* yeast strain expressing a stable P4H tetramer derived from *C. reniformis* sponge and a hydroxylated nonfibrillar procollagen polypeptide from the same organism have recently been reported by Giovine et al. [118]. The obtained recombinant sponge P4H has the ability to hydroxylate its natural substrate in both X and Y positions in the Xaa-Yaa-Gly collagenous triplets. It is suggested that the *Pichia* system could be used for the large-scale production of hydroxylated sponge- or marine-derived collagen polypeptides, which have high pharmacological potential [118].

The possibility of the application of *Chondrosia* fibrillar collagen as an organic template for in vitro silicification has been confirmed in several studies [121,129,130]. There are no doubts that the mechanical properties of biomimetically-inspired hybrid composites can be significantly improved with the presence of this special collagen.

6. Glass Sponge Collagen

Collagen is known as a universal template in biomineralization, including both calcification and silicification. It is proposed that this biopolymer functions as a fundamental template in biomineralization, inasmuch as it is very ancient from an evolutionary point of view and is common to many species and biological systems with a global distribution [131]. The identification of diverse collagens in demosponges as described above suggests that they may also be found within skeletal structures in the sister group, the glass sponges. Hexactinellida Schmidt (Porifera), with more than 700 species, consists exclusively of marine glass sponges. These are psychrophilic organisms which can produce huge biosilica-based skeletons and anchoring spicules at temperatures between $-2\,^{\circ}\mathrm{C}$ and $4\,^{\circ}\mathrm{C}$ [132].

The challenging task of isolating and identifying collagen in the skeletal structures of diverse glass sponges was completed successfully only in 2010 [3], following numerous attempts at gentle demineralization [133,134]. Studies in this area have been motivated by the great flexibility of the glassy spicules, which allows researchers to tie a spicule into a bundle (Figure 8). It has been suggested that this peculiar feature of spicules in the hexactinellids must be due to the presence of a structural carcass of organic nature both on the surface (Figure 9) and within the spicules [133].

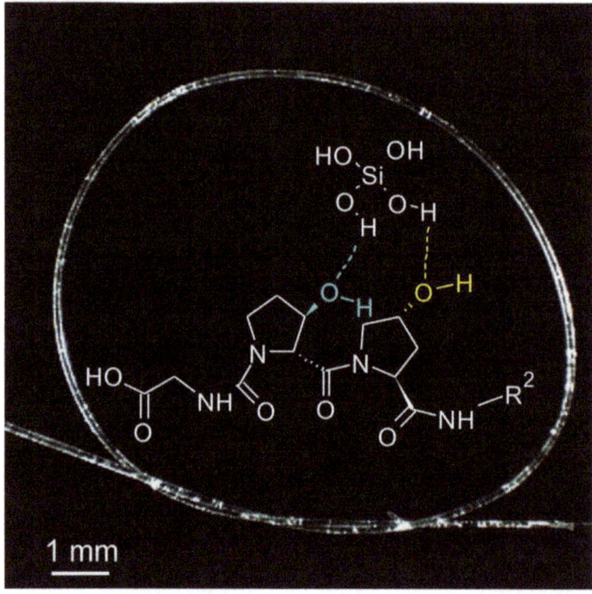

Figure 8. Photograph demonstrating the unique flexibility of the *H. sieboldi* anchoring spicule, and schematic view of the role of special hydroxylated collagen in silica condensation in this natural basilica structure (for details see [3]).

The organic phase has been identified as a highly hydroxylated fibrillar collagen which contains an unusual [Gly–3Hyp–4Hyp] motif predisposed for silica precipitation, and provides a novel template for biosilicification in nature [3]. This collagen presents a layer of hydroxyl groups that can undergo condensation reactions with silicic acid molecules with consequent loss of water. As a result, the initial layer of condensed silicic acid will be held fixed to the collagenous template in a geometric arrangement that will favor further polymerization of silicic acid. It therefore appears that collagen was a novel template for biosilicification that emerged at an early stage during metazoan evolution, and that the occurrence of additional trans-3-Hyp plays a key role in stabilizing silicic acid molecules and initiating the precipitation of silica.

Collagen has also been reported as the main organic component of the spicules of the glass sponge *Monorhaphis* sp. [135] (Figure 10). Results of the amino acid analysis of protein extracts isolated from demineralized spicules of this sponge showed an amino acid content typical for collagens of the same origin. Comparison with the Microsatellite Database (MSDB) protein database led to the identification of alpha 1 collagen in two high-MW bands. In contrast to its analog in *H. sieboldi*, collagen isolated and identified from *Monorhaphis* sp. was matched only to the type I collagen pre-pro-alpha (I) chain (COL1A1) from dog (AAD34619) (MW 139,74) [135].

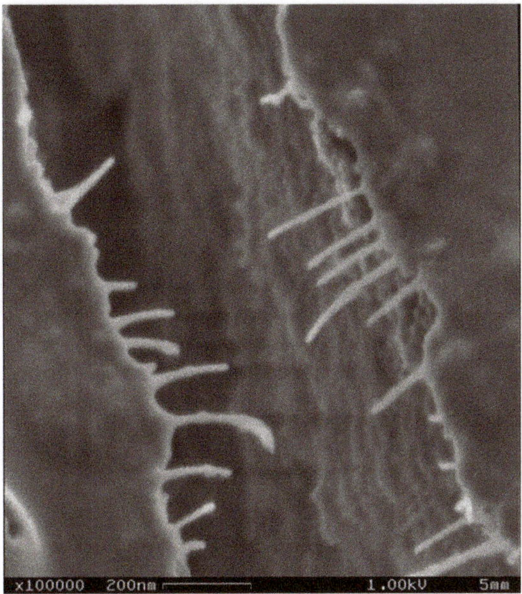

Figure 9. SEM image of the nanofibrillar collagenous layer on the surface of an *H. sieboldi* glass sponge anchoring spicule.

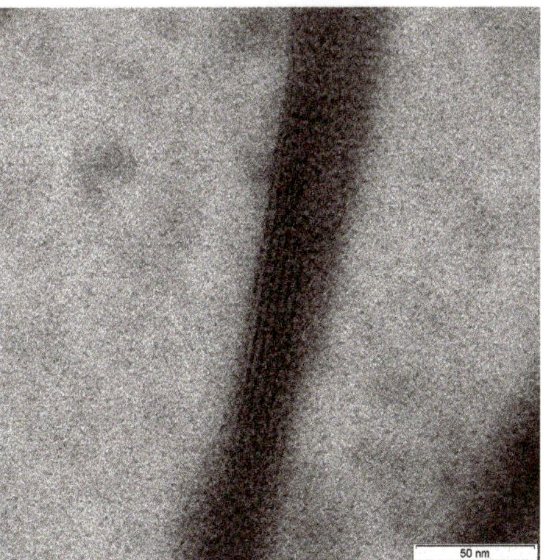

Figure 10. High-resolution transmission electron microscope image of a fragment of *M. chuni* collagen nanofibril isolated from the glassy spicule (for details see [135]). The nanomorphology of such fibrils is similar to that from *H. sieboldi* glass sponge collagen [3], but different from the striated collagen fibrils from the demosponge *C. reniformis* [121].

The existence of naturally occurring collagen–silica-based composites in the form of spicules of glass sponges stimulated material scientists to develop analogous hybrid materials. Due to the limited available amounts of glass sponge collagen for the development of silica-based composite materials, fibrillar collagen from the demosponge *C. reniformis* has been successfully used as an alternative by the Ehrlich research group [121,129]. More recently, a new concept in biosilica material synthesis which does not require phosphate supplements and is based on the fusion of stabilized polysilicic acid into a fluidic precursor phase upon infiltration into polyamine-enriched collagen has been proposed by the Tay research group [136–138]. It has recently been shown that silicified collagen scaffolds produced by infiltrating collagen matrices with intrafibrillar amorphous silica exhibit angiogenic and osteogenic potential and can be used in tissue engineering [139]. In work by Aime et al. [140], collagen triple helices have been confined on the surface of sulfonate-modified silica particles in a controlled manner. This gives rise to hybrid building blocks with well-defined surface potentials and dimensions. Additionally, oligomeric collagen-fibril matrices with tunable microstructural properties have been used to template and direct the formation of biocompatible mesoporous sol–gel silica to develop nanostructured hybrid organic–inorganic composites [141]. It was experimentally confirmed that silica mineralization kinetics are critical for the precision-tuning of properties of the hybrid materials, including porous microstructure, mechanical strength, depth of silica penetration, and mass transport properties. It has also been shown that microstructural properties of the collagen-fibril template are preserved in the silica surface of hybrid materials [142]. Such novel silica-collagen hybrid materials may be useful, for example, in the regeneration of bone tissue or in cellular microencapsulation [141].

7. Conclusions

We have presented here only a brief discussion of selected examples, which nonetheless provide novel data concerning poriferan collagens. In spite of the progress made in this research field, numerous open questions remain. For example, additional investigations are necessary to obtain understanding of the nature and origin of halogenated spongins. It is still not clear how many collagen and/or keratin domains they contain. Additionally, the unique resistance of these biopolymers against diverse chemicals and enzymes remains poorly investigated. The possible role of collagens in the spiculogenesis of demosponges and formation of axial filaments must also be researched. The discovery of crystalline proteins within amorphous biosilica-based structures in sponges is ground-breaking in the understanding of biomineralization. What can be discovered about the crystallinity of collagen within biosilica-related structures in sponges? The existence of collagen-based crystals within siliceous biominerals could revolutionize our understanding of the origin and evolution of collagens, from the point of view of biomineralization in sponges as the first multicellular organisms on Earth. Further, the relationship of collagen and chitin in the skeletal structures of diverse sponge classes and orders is entirely unknown. Consequently, we believe that the use of modern X-ray imaging techniques based on the "diffraction before destruction" principle is the best way forward to gain understanding of the principles of the unique organization of collagen within both fossil and recent collagen-based biomineralized constructs.

Novel approaches must be proposed which will bring together modern bioanalytical and molecular biology methods for better understanding of the fundamental principles of collagen fibrillogenesis and the mechanisms of its cross-linking in sponges, as well as details of the structural organization of poriferan collagens at the molecular and atomic levels. The best way to address this challenging task on these levels is by coherent synergetic collaboration using explicit reasoning and well-tested explanatory principles of multidisciplinary knowledge, experience, and new technologies. Finally, we suggest that studying the processes of marine farming of the collagen-producing demosponges has implications for a variety of practical large-scale applications, ranging from the design of highly effective extraction techniques to the development of novel collagen-containing composites for biomedicine and technology.

Acknowledgments: This work was partially supported by DFG Project HE 394/3-2 and PUT Research Grant no. 03/32/DSPB/0806. M.W. is grateful for financial support from the Foundation for Polish Science: START 097.2017, and S.Ż.-A. for support from the DAAD and Erasmus Plus programs.

Author Contributions: Hermann Ehrlich and Marcin Wysokowski researched the literature and wrote the manuscript; Iaroslav Petrenko prepared diagrams and images and edited the manuscript; Soni Żółtowska-Aksamitowska and Teofil Jesionowski discussed ideas and edited the manuscript.

Conflicts of Interest: The authors declare no conflict of interest.

References

1. Luo, Y.; Sinkeviciute, D.; He, Y.; Karsdal, M.; Henrotin, Y.; Mobasheri, A.; Önnerfjord, P.; Bay-Jensen, A. The minor collagens in articular cartilage. *Protein Cell* **2017**, *8*, 560–572. [CrossRef] [PubMed]
2. Tzaphlidou, M.; Berillis, P. Structural alterations caused by lithium in skin and liver collagen using an image processing method. *J. Trace Microprobe Tech.* **2002**, *20*, 493–504. [CrossRef]
3. Ehrlich, H.; Deutzmann, R.; Brunner, E.; Cappellini, E.; Koon, H.; Solazzo, C.; Yang, Y.; Ashford, D.; Thomas-Oates, J.; Lubeck, M.; et al. Mineralization of the metre-long biosilica structures of glass sponges is templated on hydroxylated collagen. *Nat. Chem.* **2010**, *2*, 1084–1088. [CrossRef] [PubMed]
4. Shahar, R.; Weiner, S. Open questions on the 3D structures of collagen containing vertebrate mineralized tissues: A perspective. *J. Struct. Biol.* **2018**, *201*, 187–198. [CrossRef] [PubMed]
5. Cai, J.; Palamara, J.E.A.; Burrow, M.F. Effects of collagen crosslinkers on dentine: A literature review. *Calcif. Tissue Int.* **2018**, *102*, 265–279. [CrossRef] [PubMed]
6. Ramachandran, G.N.; Kartha, G. Structure of collagen. *Nature* **1955**, *176*, 593–595. [CrossRef] [PubMed]
7. Rich, A.; Crick, F.H.C. The structure of collagen. *Nature* **1955**, *175*, 863–864. [CrossRef]
8. Shoulders, M.D.; Raines, R.T. Collagen structure and stability. *Annu. Rev. Biochem.* **2009**, *78*, 929–958. [CrossRef] [PubMed]
9. Squire, J.M.; Parry, D.A. Fibrous protein structures: Hierarchy, history and heroes. *Subcell. Biochem.* **2017**, *82*, 929–958.
10. Bella, J.; Hulmes, D.J. Fibrillar collagens. *Subcell. Biochem.* **2017**, *82*, 457–490. [PubMed]
11. Ricard-Blum, S. The collagen family. *Cold Spring Harb. Perspect. Biol.* **2011**, *3*, a004978. [CrossRef] [PubMed]
12. Porfírio, E.; Fanaro, G.B. Collagen supplementation as a complementary therapy for the prevention and treatment of osteoporosis and osteoarthritis: A systematic review. *Rev. Bras. Geriatr. Gerontol.* **2016**, *19*, 153–164. [CrossRef]
13. Pawelec, K.M.; Best, S.M.; Cameron, R.E. Collagen: A network for regenerative medicine. *J. Mater. Chem. B* **2016**, *4*, 6484–6496. [CrossRef] [PubMed]
14. Conway, J.R.W.; Vennin, C.; Cazet, A.S.; Herrmann, D.; Murphy, K.J.; Warren, S.C.; Wullkopf, L.; Boulghourjian, A.; Zaratzian, A.; Da Silva, A.M.; et al. Three-dimensional organotypic matrices from alternative collagen sources as pre-clinical models for cell biology. *Sci. Rep.* **2017**, *7*, 16887. [CrossRef] [PubMed]
15. Avila Rodríguez, M.I.; Rodríguez Barroso, L.G.; Sánchez, M.L. Collagen: A review on its sources and potential cosmetic applications. *J. Cosmet. Dermatol.* **2018**, *17*, 20–26. [CrossRef] [PubMed]
16. Hashim, P.; Mohd Ridzwan, M.S.; Bakar, J.; Hashim, M. Collagen in food and beverage industries. *Int. Food Res. J.* **2015**, *22*, 1–8.
17. Raspanti, M.; Requzzoni, M.; Protasoni, M.; Basso, P. Not only tendons: The other architecture of collagen fibrils. *Int. J. Biol. Macromol.* **2018**, *107*, 1668–1674. [CrossRef] [PubMed]
18. Uzel, S.G.; Buehler, M.J. Nanomechanical sequencing of collagen: Tropocollagen features heterogeneous elastic properties at the nanoscale. *Integr. Biol.* **2009**, *1*, 452–459. [CrossRef] [PubMed]
19. Ehrlich, H. *Biological Materials of Marine Origin. Vertebrates*; Springer Science + Business Media: Dordecht, The Netherlands, 2015.
20. Alves, A.L.P.; Marques, A.L.; Martins, E.; Silva, T.H.; Reis, R.L. Cosmetic potential of marine fish skin collagen. *Cosmetics* **2017**, *4*, 39. [CrossRef]
21. Venkatesan, J.; Anil, S.; Kim, S.K.; Shim, M.S. Marine fish proteins and peptides for cosmeceuticals: A review. *Mar. Drugs* **2017**, *15*, 143. [CrossRef] [PubMed]

22. Berillis, P. Marine Collagen: Extraction and applications. In *Research Trends in Biochemistry, Molecular Biology and Microbiology*; Saxena, M., Ed.; SM Group open access eBooks: Dover, DE, USA, 2015.

23. Zhang, J.; Sun, Y.; Zhao, Y.; Wei, B.; Xu, C.; He, L.; Oliveira, C.L.P.; Wang, H. Centrifugation-induced fibrous orientation in fish-sourced collagen matrices. *Soft Matter* **2017**, *13*, 9220–9228. [CrossRef] [PubMed]

24. Adams, E. Invertebrate collagens. *Science* **1978**, *202*, 591–598. [CrossRef] [PubMed]

25. Bailey, A.F. The nature of collagen. In *Comprehensive Biochemistry, Extracellular and Supporting Structures*; Florkin, M., Stotz, E.H., Eds.; Elsevier: Amsterdam, The Netherlands, 1968.

26. Engel, J. Versatile collagens in invertebrates. *Science* **1997**, *277*, 1785–1786. [CrossRef] [PubMed]

27. Exposito, J.Y.; Cluzel, C.; Garrone, R.; Lethias, C. Evolution of collagens. *Anat. Rec.* **2002**, *268*, 302–316. [CrossRef] [PubMed]

28. Garrone, R. Evolution of metazoan collagens. *Prog. Mol. Subcell. Biol.* **1999**, *21*, 119–139.

29. Gross, J.; Sokal, Z.; Rougvie, M. Structural and chemical studies on the connective tissue of marine sponges. *J. Histochem. Cytochem.* **1956**, *4*, 227–246. [CrossRef] [PubMed]

30. Tanzer, M.L. The biological diversity of collagenous proteins. *Trends Biochem. Sci.* **1978**, *3*, 15–17. [CrossRef]

31. Silva, T.H.; Moreira-Silva, J.; Marques, A.L.P.; Domingues, A.; Bayon, Y.; Reis, R.L. Marine origin collagens and its potential applications. *Mar. Drugs* **2014**, *12*, 5881–5901. [CrossRef] [PubMed]

32. Goh, K.L.; Holmes, D.F. Collagenous extracellular matrix biomaterials for tissue engineering: Lessons from the common sea urchin tissue. *Int. J. Mol. Sci.* **2017**, *18*, 901. [CrossRef] [PubMed]

33. Betancourt-lozano, M.; Gonzalez-farias, F.; Garcıa-gasca, A. Variation of antimicrobial activity of the sponge *Aplysina fistularis* (Pallas, 1766) and its relation to associated fauna. *J. Exp. Mar. Biol. Ecol.* **1998**, *223*, 1–18. [CrossRef]

34. Pallela, R.; Ehrlich, H. *Marine Sponges: Chemicobiological and Biomedical Applications*; Springer India: New Delhi, India, 2016.

35. Reitner, J.; Mehl, D. Monophyly of the Porifera. *Verh. Naturwiss. Ver. Hambg.* **1996**, *36*, 5–32.

36. Nichols, S.; Wörheide, G. Sponges: New views of old animals. *Integr. Comp. Biol.* **2005**, *45*, 333–334. [CrossRef] [PubMed]

37. Moldowan, J.M. C_{30} steranes, novel markers for marine petroleums and sedimentary rocks. *Geochim. Cosmochim. Acta* **1984**, *48*, 2767–2768. [CrossRef]

38. Moldowan, J.M.; Seifert, W.K.; Gallegos, E.J. Relationship between petroleum composition and depositional environment of petroleum source rocks. *Am. Assoc. Pet. Geol. Bull.* **1985**, *69*, 1255–1268.

39. Moldowan, J.M.; Fago, F.J.; Lee, C.Y.; Jacobson, S.R.; Watt, D.S.; Slougui, N.-E.; Jeganathan, A.; Young, D.C. Sedimentary 24+propylcholestanes, molecular fossils diagnostic of marine algae. *Science* **1990**, *247*, 309–312. [CrossRef] [PubMed]

40. Wainright, P.O.; Hinkle, G.; Sogin, M.L.; Stickel, S.K. Monophyletic origins of the Metazoa: An evolutionary link with fungi. *Science* **1993**, *360*, 340–342. [CrossRef]

41. Wood, R. Reef-building sponges. *Am. Sci.* **1990**, *78*, 224–235.

42. Van Soest, R.W.M.; Boury-Esnault, N.; Vacelet, J.; Dohrmann, M.; Erpenbeck, D.; de Voogd, N.J.; Santodomingo, N.; Vanhoorne, B.; Kelly, M.; Hooper, J.N.A. Global diversity of sponges (Porifera). *PLoS ONE* **2012**, *7*, e35105. [CrossRef] [PubMed]

43. Garrone, R. *Phylogenesis of Connective Tissue: Morphological Aspects and Biosynthesis of Sponge Intercellular Matrix*; John Wiley & Sons: Hoboken, NJ, USA, 1978.

44. Junqua, S.; Robert, L.; Garrone, R. Biochemical and morphological studies on the collagens of horny sponges. *Ircinia* filaments compared to spongines. *Connect. Tissue Res.* **1974**, *2*, 193–203. [CrossRef] [PubMed]

45. Städeler, G. Untersuchungen über das Fibroin, Spongin und Chitin, nebst Bemerkungen über den tierischen Schleim. *Eur. J. Organ. Chem.* **1859**, *111*, 12–28.

46. Green, D.; Howard, D.; Yang, X.; Kelly, M.; Oreffo, R.O.C. Natural marine sponge fiber skeleton: A biomimetic scaffold for human osteoprogenitor cell attachment, growth, and differentiation. *Tissue Eng.* **2003**, *9*, 1159–1166. [CrossRef] [PubMed]

47. Green, D.W.; Lee, J.-M.; Jung, H.-S. Marine structural biomaterials in medical biomimicry. *Tissue Eng. Part B Rev.* **2015**, *21*, 438–450. [CrossRef] [PubMed]

48. Green, D.W.; Padula, M.P.; Santos, J.; Chou, J.; Milthorpe, B.; Ben-Nissan, B. A therapeutic potential for marine skeletal proteins in bone regeneration. *Mar. Drugs* **2013**, *11*, 1203–1220. [CrossRef] [PubMed]

49. Lin, Z.; Solomon, K.L.; Zhang, X.; Pavlos, N.J.; Abel, T.; Willers, C. In vitro evaluation of natural marine sponge collagen as a scaffold for bone tissue engineering. *Int. J. Biol. Sci.* **2011**, *7*, 968–977. [CrossRef] [PubMed]

50. Pallela, R.; Venkatesan, J.; Bhatnagar, I.; Shim, Y.; Kim, S. Application of marine collagen–based scaffolds in bone tissue engineering. In *Marine Biomaterials: Isolation, Characterization and Applications*; Kim, S.-K., Ed.; CRC-Taylor & Francis: Boca Raton, FL, USA, 2011; pp. 519–528.

51. Nandi, S.K.; Kundu, B.; Mahato, A.; Thakur, N.L.; Joardar, S.N.; Mandal, B.B. In vitro and in vivo evaluation of the marine sponge skeleton as a bone mimicking biomaterial. *Integr. Biol.* **2015**, *7*, 250–262. [CrossRef] [PubMed]

52. Langasco, R.; Cadeddu, B.; Formato, M.; Lepedda, A.J.; Cossu, M.; Giunchedi, P.; Pronzato, R.; Rassu, G.; Manconi, R.; Gavini, R. Natural collagenic skeleton of marine sponges in pharmaceutics: Innovative biomaterial for topical drug delivery. *Mater. Sci. Eng. C* **2017**, *70*, 710–720. [CrossRef] [PubMed]

53. Tziveleka, L.A.; Ioannou, E.; Tsiourvas, D.; Berillis, P.; Foufa, E.; Roussis, V. Collagen from the marine sponges *Axinella cannabina* and *Suberites carnosus*: Isolation and morphological, biochemical, and biophysical characterization. *Mar. Drugs* **2017**, *15*, 152. [CrossRef] [PubMed]

54. De Laubenfels, M.; Storr, J. The taxonomy of American commercial sponges. *Bull. Mar. Sci.* **1958**, *8*, 99–117.

55. Schulze, F.E. Untersuchungen über den Bau und die Entwicklung der Spongien. *Z. wiss. Zool.* **1877**, *28*, 1–48.

56. Cresswell, E. *Sponges: Their Nature, History, Modes of Fishing, Varieties, Cultivation, etc.*; Sir Isaac Pitman & Sons Ltd.: London, UK, 1922.

57. Geoffroy, C.J. Analyse chimique de l'eponge de la moyenne espece. *Hist. Acad. R. Sci.* **1705**, 660–661.

58. Fyfe, A. Account of some experiments, made with the view of ascertaining the different substances from which iodine can be produced. *Edinb. Philos.* **1819**, *1*, 254–258.

59. Croockewit, J.H. Ueber die Zusammensetzung des Badeschwammes. *Eur. J. Organ. Chem.* **1843**, *48*, 43–56. [CrossRef]

60. Schlossberger, J. Ueber Fibroin und die Substanz des Badeschwamms. *Arch. Pharm.* **1859**, *147*, 62–65. [CrossRef]

61. Von Kölliker, A. *Der feinere Bau der Protozoen*; Wilhelm Engelman: Leipzig, Germany, 1864.

62. Hundeshagen, F. Über jodhaltige Spongien and Jodospongin. *Angew. Chem. Int. Ed.* **1895**, *8*, 473–476. [CrossRef]

63. Harnack, E. Ueber das iodospongin, die jodhaltige eiweissartige Subsanz aus dem Badeschwamm. *Z. Physiol. Chem.* **1898**, *25*, 412–424. [CrossRef]

64. Clancey, V.J. The constitution of sponges. The common bath sponge, *Hippospongia equina*. *Biochem. J.* **1926**, *20*, 1186–1189. [CrossRef] [PubMed]

65. Abderhalden, E.; Strauss, E. Die Spaltprodukte des Spongins mit Sauren. *Hoppe-Seyler's Z. Physiol. Chem.* **1906**, *48*, 49–53. [CrossRef]

66. Strauss, E. *Studien ueber die Albuminoide mit Besonderer Berucksichtigung des Spongins und der Keratine*; Carl Witer's Universitatsbuchhandlung: Munchen, Germany, 1904.

67. Jenkins, C.L. Insights on the conformational stability of collagen. *Nat. Prod. Rep.* **2002**, *19*, 49–59. [PubMed]

68. Block, R.J.; Bolling, D. The amino acid composition of keratins. *J. Biol. Chem.* **1939**, *127*, 685–693.

69. Ackermann, D.; Burchard, C. Zur Kenntnis der Spongine. *Hoppe-Seyler's Z. Physiol. Chem.* **1941**, *271*, 183–189. [CrossRef]

70. Ackermann, D.; Müller, I. Über das Vorkommen von Dibromotyrosin neben Dijodtyrosin im Spongin. *Hoppe-Seyler's Z. Physiol. Chem.* **1941**, *269*, 146–157. [CrossRef]

71. Low, E.M. Halogenated amino acids of the bath sponge. *J. Mar. Res.* **1951**, *10*, 239–245.

72. Katzman, R.L.; Halford, M.H.; Reinhold, V.N.; Jeanloz, R.W. Invertebrate connective tissue. IX. Isolation and structure determination of glucosylgalactosylhydroxylysine from sponge and sea anemone collagen. *Biochemistry* **1972**, *11*, 1161–1167. [CrossRef] [PubMed]

73. Gaino, E.; Pronzato, R. Ultrastructural evidence of bacterial damage to *Spongia officinalis* fibres (Porifera, Demospongiae). *Dis. Aquat. Organ.* **1989**, *6*, 67–74. [CrossRef]

74. Ehrlich, H.; Maldonado, M.; Spindler, K.; Eckert, C.; Hanke, T.; Born, R.; Simon, P.; Heinemann, S.; Worch, H. First evidence of chitin as a component of the skeletal fibers of marine sponges. Part I. Verongida (Demospongia: Porifera). *J. Exp. Zool. Part B* **2007**, *356*, 347–356. [CrossRef] [PubMed]

75. Ehrlich, H.; Ilan, M.; Maldonado, M.; Muricy, G.; Bavestrello, G.; Kljajic, Z.; Carballo, J.L.; Schiaparelli, S.; Ereskovsky, A.; Schupp, P.; et al. Three-dimensional chitin-based scaffolds from Verongida sponges (Demospongiae: Porifera). Part I. Isolation and identification of chitin. *Int. J. Bol. Macromol.* **2010**, *47*, 132–140. [CrossRef] [PubMed]

76. Wysokowski, M.; Bazhenov, V.V.; Tsurkan, M.V.; Galli, R.; Stelling, A.L.; Stöcker, H.; Kaiser, S.; Niederschlag, E.; Gärtner, G.; Behm, T.; et al. Isolation and identification of chitin in three-dimensional skeleton of *Aplysina fistularis* marine sponge. *Int. J. Biol. Macromol.* **2013**, *62*, 94–100. [CrossRef] [PubMed]

77. Ehrlich, H.; Kaluzhnaya, O.V.; Tsurkan, M.V.; Ereskovsky, A.; Tabachnick, K.R.; Ilan, M.; Stelling, A.; Galli, R.; Petrova, O.V.; Nekipelov, S.V.; et al. First report on chitinous holdfast in sponges (Porifera). *Proc. R. Soc. Lond. B* **2013**, *280*, 20130339. [CrossRef] [PubMed]

78. Ehrlich, H.; Kaluzhnaya, O.V.; Brunner, E.; Tsurkan, M.V.; Ereskovsky, A.; Ilan, M.; Tabachnick, K.R.; Bazhenov, V.V.; Paasch, S.; Kammer, M.; et al. Identification and first insights into the structure and biosynthesis of chitin from the freshwater sponge *Spongilla lacustris*. *J. Struct. Biol.* **2013**, *183*, 474–483. [CrossRef] [PubMed]

79. Wysokowski, M.; Petrenko, I.; Stelling, A.; Stawski, D.; Jesionowski, T.; Ehrlich, H. Poriferan chitin as a versatile template for extreme biomimetics. *Polymers* **2015**, *7*, 235–265. [CrossRef]

80. Shavandi, A.; Silva, T.H.; Bekhit, A.A.; Bekhit, A.E. Keratin: Dissolution, extraction and biomedical application. *Biomater. Sci.* **2017**, *5*, 1699–1735. [CrossRef] [PubMed]

81. Cerrano, C.; Calcinai, B.; Gioia, C.; Camillo, D.; Valisano, L.; Bavestrello, G. How and why do sponges incorporate foreign material? Strategies in porifera. In *Porifera Research: Biodiversity, Innovation and Sustainability*; Custódio, M.R., Lôbo-Hajdu, G., Hajdu, E., Muricy, G., Eds.; Série Livros 28; Museu Nacional: Rio de Janeiro, Brazil, 2007; pp. 239–246.

82. Castritsi-Catharios, J.; Zaoutsos, S.P.; Berillis, P.; Zouganelis, G.D.; Ekonomou, G.; Kefalas, E.; Pantelis, J. Kalymnos, the island which made history in sponge fishery. Data on physical parameters, elemental composition and DNA barcode preliminary results of the most common bath sponge species in Aegean Sea. *Reg. Stud. Mar. Sci.* **2017**, *13*, 71–79. [CrossRef]

83. Castritsi-Catharios, J.; Zaoutsos, S.P.; Ekonomou, G.; Berillis, P. Physical parameters and chemical composition of four sponge species. Preliminary results. In Proceedings of the HydroMedit 2016, Messolonghi, Greece, 10–12 November 2016; pp. 156–159.

84. Castritsi-Catharios, J.; Magli, M.; Vacelet, J. Evaluation of the quality of two commercial sponges by tensile strength measurement. *J. Mar. Biol. Assoc. UK* **2007**, *87*, 1765–1771. [CrossRef]

85. Dandy, A. On the occurrence of gelatinous spicules, and their mode of origin, in a new genus of siliceous sponges. *Proc. R. Soc. Lond. B* **1916**, *89*, 315–322. [CrossRef]

86. Szatkowski, T.; Jesionowski, T. Hydrothermal synthesis of spongin-based materials. In *Extreme Biomimetics*; Ehrlich, H., Ed.; Springer International Publishing: Cham, Switzerland, 2017; pp. 251–274.

87. Von Raimann, E. Über den Nutzen des Bade- oder Waschschwammes bey heftigen Blutun-gen. *Med. Jahrbücher* **1939**, *27*, 306–307.

88. White, C. An account of the successful use of the sponge, in the stoppage of an Haemorrhage, occasioned by amputation below the knee; and of the remarkable effects of that application in preventing the absorption of matter. In *Cases in Surgery with Remarks: Part 1*; Johnston: London, England, 1770; pp. 151–158.

89. Zschiesche, P. *Die Anwendung des Pressschwammes in der Gynaekologie und Seine Gefahren*; Universitaet Greifswald: Greifswald, Germany, 1873.

90. Haussmann, D. Kann die Erweiterung des Verengten Muttermundes durch Pressschwamm die Empfangniss erleichtern? Gesellschaft für Geburtshilfe und Gynäkologie Berlin: Berlin, Germany, 1878; pp. 311–327.

91. Petrus, C. *Naauwkeurige Afbelding en Beschryving van eene Geheel en al Verloorene, Maar door Kunst Herstelde Neus en Verhemelte*; J.C.SEPP Boekverkooper: Amsterdam, The Netherlands, 1771.

92. Hamilton, D.J. On sponge-grafting. *Edinb. Med. J.* **1881**, *27*, 385–413.

93. Kim, M.M.; Mendis, E.; Rajapakse, N.; Lee, S.H.; Kim, S.K. Effect of spongin derived from *Hymeniacidon sinapium* on bone mineralization. *J. Biomed. Mater. Res. B* **2009**, *90*, 540–546. [CrossRef] [PubMed]

94. Norman, M.; Bartczak, P.; Zdarta, J.; Ehrlich, H. Anthocyanin dye conjugated with *Hippospongia communis* marine demosponge skeleton and its antiradical activity. *Dyes Pigments* **2016**, *134*, 541–552. [CrossRef]

95. Norman, M.; Bartczak, P.; Zdarta, J.; Tomala, W.; Żurańska, B.; Dobrowolska, A.; Piasecki, A.; Czaczyk, K.; Ehrlich, H.; Jesionowski, T. Sodium copper chlorophyllin immobilization onto *Hippospongia communis* marine demosponge skeleton and its antibacterial activity. *Int. J. Mol. Sci.* **2016**, *17*, 1564. [CrossRef] [PubMed]

96. Zdarta, J.; Norman, M.; Smułek, W.; Moszyński, D.; Kaczorek, E.; Stelling, A.L.; Ehrlich, H.; Jesionowski, T. Spongin-based scaffolds from *Hippospongia communis* demosponge as an effective support for lipase immobilization. *Catalysts* **2017**, *7*, 147. [CrossRef]

97. Szatkowski, T.; Siwińska-Stefańska, K.; Wysokowski, M.; Stelling, A.; Joseph, Y.; Ehrlich, H.; Jesionowski, T. Immobilization of titanium(IV) oxide onto 3D spongin scaffolds of marine sponge origin according to extreme biomimetics principles for removal of C.I. Basic Blue 9. *Biomimetics* **2017**, *2*, 4. [CrossRef]

98. Szatkowski, T.; Wysokowski, M.; Lota, G.; Pęziak, D.; Bazhenov, V.V.; Nowaczyk, G.; Walter, J.; Molodtsov, S.L.; Stöcker, H.; Himcinschi, C.; et al. Novel nanostructured hematite–spongin composite developed using an extreme biomimetic approach. *RSC Adv.* **2015**, *5*, 79031–79040. [CrossRef]

99. Ehrlich, H. *Extreme Biomimetics*; Springer International Publishing: Cham, Switzerland, 2017.

100. Chioran, A.; Duncan, S.; Catalano, A.; Brown, T.J.; Ringuette, M.J. Collagen IV trafficking: The inside-out and beyond story. *Dev. Biol.* **2017**, *431*, 124–133. [CrossRef] [PubMed]

101. Fidler, A.L.; Darris, C.E.; Chetyrkin, S.V.; Pedchenko, V.K.; Boudko, S.P.; Brown, K.L.; Gray Jerome, W.; Hudson, J.K.; Rokas, A.; Hudson, B.G. Collagen IV and basement membrane at the evolutionary dawn of metazoan tissues. *eLife* **2017**, *6*, e24176. [CrossRef] [PubMed]

102. Rodriguez-Pascual, F.; Slatter, D.A. Collagen cross-linking: Insights on the evolution of metazoan extracellular matrix. *Sci. Rep.* **2016**, *6*, 37374. [CrossRef] [PubMed]

103. Boute, N.; Exposito, J.Y.; Boury-Esnault, N.; Vacelet, J.; Nor, N.; Miyazaki, K.; Yoshizato, K.; Garrone, R. Type IV collagen in sponges, the missing link in basement membrane ubiquity. *Biol. Cell* **1996**, *88*, 37–44. [CrossRef]

104. Leys, S.P.; Riesgo, A. Epithelia, an evolutionary novelty of metazoans. *J. Exp. Zool. Part B* **2011**, *314*, 438–447. [CrossRef] [PubMed]

105. Riesgo, A.; Taboada, S.; Sánchez-Vila, L.; Solà, J.; Bertran, A.; Avila, C. Some like it fat: Comparative ultrastructure of the embryo in two demosponges of the genus Mycale (order Poecilosclerida) from Antarctica and the Caribbean. *PLoS ONE* **2015**, *10*, e0118805. [CrossRef] [PubMed]

106. Exposito, J.Y.; Le Guellec, D.; Lu, Q.; Garrone, R. Short chain collagens in sponges are encoded by a family of closely related genes. *J. Biol. Chem.* **1991**, *266*, 21923–21928. [PubMed]

107. Aouacheria, A.; Geourjon, C.; Aghajari, N.; Navratil, V.; Deléage, G.; Lethias, C.; Exposito, J.Y. Insights into early extracellular matrix evolution: Spongin short chain collagen-related proteins are homologous to basement membrane type IV collagens and form a novel family widely distributed in invertebrates. *Mol. Biol. Evol.* **2006**, *23*, 2288–2302. [CrossRef] [PubMed]

108. Rocha Moreira Da Silva, J.C.; Soares Diogo Carlos, G.; De Sousa E Silva Barros Prata, M.; Quinteiros Lopes Henriques Da Silva, T.J.; Pinto Marques, A.M.; Gonçalves Dos Reis, R.L.; Teixeira Cerqueira, M.; Vieira Pereira Ferreira, M.S. Marine-Sponge Type IV Collagen Membranes Its Production and Biomedical Applications Thereof. Patent WO2,015,186,118, 10 December 2015.

109. Cavalier-Smith, T. Origin of animal multicellularity: Precursors, causes, consequences—The choanoflagellate/sponge transition, neurogenesis and the Cambrian explosion. *Philos. Trans. R. Soc. Lond. B* **2017**, *372*, 20150476. [CrossRef] [PubMed]

110. Simpson, T.L. Collagen Fibrils, Spongin, Matrix Substances. In *The Cell Biology of Sponges*; Springer: New York, NY, USA, 1984.

111. Garrone, R.; Pottu, J. Collagen biosynthesis in sponges—Elaboration of spongin by spongocytes. *J. Submicrosc. Cytol.* **1973**, *5*, 199–218.

112. Bairati, A.; Garrone, R. *Biology of Invertebrate and Lower Vertebrate Collagens*; Plenum: New York, NY, USA, 1985.

113. Imhoff, J.M.; Garrone, R. Solubilization and characterization of *Chondrosia reniformis* sponge collagen. *Connect. Tissue Res.* **1983**, *11*, 193–197. [CrossRef] [PubMed]

114. Eckert, C.; Schröder, H.C.; Brandt, D.; Perovic-Ottstadt, S.; Muller, W.E.G. Histochemical and electron microscopic analysis of spiculogenesis in the demosponge *Suberites domuncula*. *Zootaxa* **2006**, *54*, 1031–1040. [CrossRef] [PubMed]

115. Ehrlich, H. *Biological Materials of Marine Origin. Invertebrates*; Springer Science + Business Media B.V.: Dordrecht, The Netherlands, 2010.

116. Krasko, A.; Lorenz, B.; Batel, R.; Schröder, H.C.; Müller, I.M.; Müller, W.E.G. Expression of silicatein and collagen genes in the marine sponge *Suberites domuncula* is controlled by silicate and myotrophin. *Eur. J. Biochem.* **2000**, *267*, 4878–4887. [CrossRef] [PubMed]

117. Garrone, R.; Huc, A.; Junqua, S. Fine structure and physicochemical studies on the collagen of the marine sponge *Chondrosia reniformis* Nardo. *J. Ultrastruct. Res.* **1975**, *52*, 261–275. [CrossRef]

118. Pozzolini, M.; Scarfi, S.; Mussino, F.; Ferrando, S.; Gallus, L.; Giovine, M. Molecular cloning, characterization, and expression analysis of a prolyl 4-hydroxylase from the marine sponge *Chondrosia reniformis*. *Mar. Biotechnol.* **2015**, *17*, 393–407. [CrossRef] [PubMed]

119. Ledger, P.W. Types of collagen fibres in the calcareous sponges *Sycon* and *Leucandra*. *Tissue Cell* **1974**, *6*, 385–389. [CrossRef]

120. Pozzolini, M.; Bruzzone, F.; Berilli, V.; Mussino, F.; Cerrano, C.; Benatti, U.; Giovine, M. Molecular characterization of a nonfibrillar collagen from the marine sponge *Chondrosia reniformis* Nardo 1847 and positive effects of soluble silicates on its expression. *Mar. Biotechnol.* **2012**, *14*, 281–293. [CrossRef] [PubMed]

121. Heinemann, S.; Ehrlich, H.; Douglas, T.; Heinemann, C.; Worch, H.; Schatton, W.; Hanke, T. Ultrastructural studies on the collagen of the marine sponge *Chondrosia reniformis* Nardo. *Biomacromolecules* **2007**, *8*, 3452–3457. [CrossRef] [PubMed]

122. Palmer, I.; Clarke, S.A.; Nelson, J.; Schatton, W.; Dunne, N.J.; Buchanan, F. Identification of a suitable sterilisation method for collagen derived from a marine Demosponge. *Int. J. Nano Biomater.* **2012**, *4*, 148–163. [CrossRef]

123. Bavestrello, G.; Cerrano, C.; Cattaneo-Vietti, R.; Sara, M.; Calabria, F.; Cortesogno, L. Selective incorporation of foreign material in *Chondrosia reniformis* (Porifera, Demospongiae). *Ital. J. Zool.* **1996**, *63*, 215–220. [CrossRef]

124. Bavestrello, G.; Benatti, U.; Calcinai, B.; Cattaneo-Vietti, R.; Cerrano, C.; Favre, A.; Giovine, M.; Lanza, S.; Pronzato, R.; Sara, M. Body polarity and mineral selectivity in the demosponge *Chondrosia reniformis*. *Biol. Bull.* **1998**, *195*, 120–125. [CrossRef] [PubMed]

125. Fassini, D. *Coordination Phenomena in the Marine Demosponge Chondrosia reniformis*; Università Degli Studi Di Milano: Milan, Italy, 2013.

126. Fassini, D.; Parma, L.; Lembo, F.; Candia Carnevali, M.; Wilkie, I.; Bonasoro, F. The reaction of the sponge *Chondrosia reniformis* to mechanical stimulation is mediated by the outer epithelium and the release of stiffening factor(s). *Zoology* **2014**, *117*, 282–291. [CrossRef] [PubMed]

127. Pozzolini, M.; Ferrando, S.; Gallus, L.; Gambardella, C.; Ghignone, S.; Giovine, M. Aquaporin in *Chondrosia reniformis* Nardo, 1847 and its possible role in the interaction between cells and engulfed siliceous particles. *Biol. Bull.* **2016**, *230*, 220–232. [CrossRef] [PubMed]

128. Nicklas, M.; Schatton, W.; Heinemann, S.; Hanke, T.; Kreuter, J. Preparation and characterization of marine sponge collagen nanoparticles and employment for the transdermal delivery of 17β-estradiol-hemihydrate. *Drug Dev. Ind. Pharm.* **2009**, *35*, 1035–1042. [CrossRef] [PubMed]

129. Heinemann, S.; Heinemann, C.; Ehrlich, H.; Meyer, M.; Baltzer, H.; Worch, H.; Hanke, T. A novel biomimetic hybrid material made of silicified collagen: Perspectives for bone replacement. *Adv. Eng. Mater.* **2007**, *9*, 1061–1068. [CrossRef]

130. Heinemann, S.; Ehrlich, H.; Knieb, C.; Hanke, T. Biomimetically inspired hybrid materials based on silicified collagen. *Int. J. Mater. Res.* **2007**, *98*, 603–608. [CrossRef]

131. Ehrlich, H. Chitin and collagen as universal and alternative templates in biomineralization. *Int. Geol. Rev.* **2010**, *52*, 661–699. [CrossRef]

132. Tabachnick, K.; Janussen, D.; Menschenina, L. Cold biosilicification in Metazoan: Psychrophilic glass sponges. In *Extreme Biomimetics*; Ehrlich, H., Ed.; Springer International Publishing: Cham, Switzerland, 2017; pp. 53–80.

133. Ehrlich, H.; Ereskovskii, V.; Drozdov, L.; Krylova, D.D.; Hanke, T.; Meissner, H.; Heinemann, S.; Worch, H. A modern approach to demineralization of spicules in glass sponges (Porifera: Hexactinellida) for the purpose of extraction and examination of the protein matrix. *Russ. J. Mar. Biol.* **2006**, *32*, 186–193. [CrossRef]

134. Ehrlich, H.; Worch, H. Sponges as natural composites. In *Porifera Research Biodiversity, Innovation and Sustainability*; Custodio, M.R., Lobo-Hajdu, G., Hajdu, E., Muricy, G., Eds.; Série Livros 28; Rio de Janerio Museu National: Rio de Janeiro, Brazil, 2007; pp. 303–312.

135. Ehrlich, H.; Heinemann, S.; Heinemann, C.; Simon, P.; Bazhenov, V.V.; Shapkin, N.P.; Born, R.; Tabachnick, K.R.; Hanke, T.; Worch, H. Nanostructural organization of naturally occurring composites—Part I: Silica-collagen-based biocomposites. *J. Nanomater.* **2008**, *2008*, 623838. [CrossRef]

136. Niu, L.; Jiao, K.; Qi, Y.; Yiu, C.K.Y.; Ryou, H.; Arola, D.D.; Chen, J.; Breschi, L.; Pashley, D.H.; Tay, F.R. Infiltration of silica inside fibrillar collagen. *Angew. Chem. Int. Ed.* **2011**, *50*, 11688–11691. [CrossRef] [PubMed]

137. Luo, X.J.; Yang, H.Y.; Niu, L.N.; Mao, J.; Huang, C.; Pashley, D.H.; Tay, F.R. Translation of a solution-based biomineralization concept into a carrier-based delivery system via the use of expanded-pore mesoporous silica. *Acta Biomater.* **2016**, *31*, 378–387. [CrossRef] [PubMed]

138. Zhang, W.; Luo, X.; Niu, L.; Yang, H.; Yiu, C.K.; Wang, T.; Zhou, L.; Mao, J.; Huang, C.; Pashley, D.H.; Tay, F.R. Biomimetic intrafibrillar mineralization of type I collagen with intermediate precursors-loaded mesoporous carriers. *Sci. Rep.* **2015**, *5*, 11199. [CrossRef] [PubMed]

139. Sun, J.-L.; Jiao, K.; Niu, L.N.; Jiao, Y.; Song, Q.; Shen, L.J.; Tay, F.R.; Chen, J.H. Intrafibrillar silicified collagen scaffold modulates monocyte to promote cell homing, angiogenesis and bone regeneration. *Biomaterials* **2017**, *113*, 203–216. [CrossRef] [PubMed]

140. Aimé, C.; Mosser, G.; Pembouong, G.; Bouteiller, L.; Coradin, T. Controlling the nano–bio interface to build collagen–silica self-assembled networks. *Nanoscale* **2012**, *4*, 7127–7134. [CrossRef] [PubMed]

141. Kahn, J.L.; Eren, N.M.; Campanella, O.; Voytik-Harbin, S.L.; Rickus, J.L. Collagen-fibril matrix properties modulate the kinetics of silica polycondensation to template and direct biomineralization. *J. Mater. Res.* **2016**, *31*, 311–320. [CrossRef]

142. Rickus, J.L.; Harbin, S.L.; Kahn, J.L. Cell-Collagen-Silica Composites and Methods of Making and Using the Same. Patent WO2,016,172,365, 27 October 2016.

Article

Collagen from the Marine Sponges
Axinella cannabina and *Suberites carnosus*: Isolation and Morphological, Biochemical, and Biophysical Characterization

Leto-Aikaterini Tziveleka [1], Efstathia Ioannou [1], Dimitris Tsiourvas [2], Panagiotis Berillis [3], Evangelia Foufa [1] and Vassilios Roussis [1,*]

[1] Department of Pharmacognosy and Chemistry of Natural Products, Faculty of Pharmacy, National and Kapodistrian University of Athens, Panepistimiopolis Zografou, Athens 15771, Greece; ltziveleka@pharm.uoa.gr (L.-A.T.); eioannou@pharm.uoa.gr (E.I.); foufa_e@hotmail.com (E.F.)

[2] Institute of Nanosciences and Nanotechnology, NCSR "Demokritos", Aghia Paraskevi 15310, Attiki, Greece; d.tsiourvas@inn.demokritos.gr

[3] Department of Ichthyology and Aquatic Environment, School of Agricultural Sciences, University of Thessaly, Fytoko Str., Nea Ionia 38445, Magnesia, Greece; pveril@apae.uth.gr

* Correspondence: roussis@pharm.uoa.gr; Tel.: +30-210-727-4592

Academic Editors: Azizur Rahman and Tiago H. Silva

Received: 22 April 2017; Accepted: 25 May 2017; Published: 29 May 2017

Abstract: In search of alternative and safer sources of collagen for biomedical applications, the marine demosponges *Axinella cannabina* and *Suberites carnosus*, collected from the Aegean and the Ionian Seas, respectively, were comparatively studied for their insoluble collagen, intercellular collagen, and spongin-like collagen content. The isolated collagenous materials were morphologically, physicochemically, and biophysically characterized. Using scanning electron microscopy and transmission electron microscopy the fibrous morphology of the isolated collagens was confirmed, whereas the amino acid analysis, in conjunction with infrared spectroscopy studies, verified the characteristic for the collagen amino acid profile and its secondary structure. Furthermore, the isoelectric point and thermal behavior were determined by titration and differential scanning calorimetry, in combination with circular dichroism spectroscopic studies, respectively.

Keywords: *Axinella cannabina*; *Suberites carnosus*; sponges; marine collagen

---

1. Introduction

Collagen is an ubiquitous high molecular weight fibrous protein occurring in both invertebrate and vertebrate organisms, existing in more than 20 different types depending on its role in distinct tissues [1,2]. Its polypeptide chains are organized in a unique structure, in which three α-helices are intertwined forming a characteristic right-handed triple helix. These peptides are rich in glycine, proline, and hydroxyproline amino acids, all being crucial for the formation of the helical configuration [3].

Due to its high biocompatibility and biodegradability, collagen finds a plethora of applications, primarily in the sectors of cosmetics, pharmaceuticals, and medical care products [4,5]. Additionally, gelatin, the denatured form of collagen obtained by its partial hydrolysis, is used as an additive in the food processing industry and in nutraceuticals [6]. Its intrinsic low immunogenicity renders this natural biopolymer an ideal material for bone grafting, tissue regeneration, and construction of artificial skin [7,8]. Collagen destined for industrial use originates mainly from bovine and porcine sources, via an acid hydrolysis-based procedure [9]. Incidences of allergic reactions and connective tissue disorders, such as arthritis and lupus, as well as bovine spongiform encephalopathy and transmissible spongiform encephalopathy [10], have led to the reconsideration of cattle as the main source for collagen production. Furthermore, porcine collagen is prohibited for the Muslim and Jewish communities due to religious restrictions. Taking into account these two limitations, an alternative and safer source is currently actively sought.

Nowadays, collagen of marine origin as an alternative to mammalian sources is gaining ground, especially since the employment of recombinant technology is excluded due to its high cost [11,12]. Since collagen is the major constituent of the extracellular matrices of all metazoans, sponges are considered as one of the most promising sources [13–15]. Sponges, belonging in the phylum Porifera, composed of a mass of cells forming a porous skeleton made of organic (collagen fibers and/or spongin, especially in the case of the class Demospongiae) and inorganic (spicules) components, are the most primitive among multicellular animals (Metazoa) [16,17]. Marine sponges have been proven an inexhaustible source of secondary metabolites exhibiting diverse pharmacological properties [18–21]. In addition to these, macromolecules have gained interest since such biopolymers possess a wide range of bioactivities that can find applications in the biomedical sector. Collagen has been isolated from different marine sponges, e.g., *Spongia graminea*, *Microciona prolifera*, *Haliclona oculata* [22], *Hippospongia communis*, *Cacospongia scalaris* [23], *Geodia cydonium* [24] *Chondrosia reniformis* [25,26], and various *Ircinia* species [27], and in certain cases has shown high potency in tissue regeneration [28]. Although the importance of marine collagen has been recognized, only a few thorough investigations on marine sponges have so far been reported [25,27,29], probably due to its characteristic insolubility and mineralization, which cause difficulties in its isolation and characterization [30,31].

In the present study we report, for the first time, the isolation and characterization of collagens from the marine demosponges *Axinella cannabina* (Axenillidae) and *Suberites carnosus* (Suberitidae). By employing two different experimental approaches, the insoluble collagen (InSC), intercellular collagen (ICC), and spongin-like collagen (SlC) were obtained. The morphology of these collagens was analyzed by scanning electron microscopy (SEM), and their fibril formation and characteristic band periodicity was studied by transmission electron microscopy (TEM). Their secondary structure was evaluated based on their FT-IR spectra, while the amino acid composition of the ICCs was also determined. The thermal behavior of the ICCs was investigated by differential scanning calorimetry (DSC) and circular dichroism (CD) analyses.

2. Results

2.1. Isolation of Sponge Collagen

Two different procedures were used for the isolation of collagens from the demosponges *A. cannabina* and *S. carnosus*. The first method was initially introduced for the isolation of insoluble collagen (InSC) from *G. cydonium* [24] and *C. reniformis* [26] by employing an alkaline, both denaturing and reducing, homogenization buffer affording collagen in high yield. The second one utilizes a trypsin-containing extraction buffer, known to destroy the interfibrillar matrix and, therefore, releasing the collagen fibrils (ICC) [22,23]. After exhaustive water extraction, the remaining debris generally comprises the spongin/spongin-like collagen. In our case, since the specific sponges are deprived of spongin, the isolated samples are considered to contain spongin-like collagen (SlC).

The InSCs obtained by the application of the first method corresponded to 12.6% and 5.0% of the sponges' dry weight for *A. cannabina* and *S. carnosus*, respectively (Table 1). Application of the second method resulted in the isolation of ICC and SlC, leveling to 3.0% and 42.8% dry weight for *A. cannabina* and 1.9% and 21.8% dry weight for *S. carnosus*, respectively (Table 1). The percentages found for the ICC yield are in accordance with previously-reported results for *Hippospongia gossipina* [32].

The siliceous or calcareous sponges are characterized by a large number of inorganic spicules, which, in *Haliclona* and *Microciona*, are bound together by spongin [22]. In order to remove the expected siliceous spicules in the InSC, the samples were treated with an HF solution for 20 min at room temperature to obtain spicule-free insoluble collagen (SF-InSC). The spicules accounted for 32% and 49% (*w*/*w*) of the sponges' InSCs from *A. cannabina* and *S. carnosus*, respectively.

Table 1. Collagen composition (*w*/*w* %) [1] of the sponges *A. cannabina* and *S. carnosus*.

Isolated Collagen	*A. cannabina*	*S. carnosus*
Insoluble collagen (InSC)	12.6	5.0
Intercellular collagen (ICC)	3.0	1.9
Spongin-like collagen (SlC)	42.8	21.8

[1] Data are presented as the percent of sponge dry weight.

2.2. Examination of Surface Morphology

The collagenous nature of the isolated materials was investigated by SEM and TEM. Overall, the microscopically-observed structures (Figure 1) were similar to those already reported for collagen isolated from other sponges [25–27,29]. Figure 1A,E show the microstructure of the InSCs from *A. cannabina* and *S. carnosus*, respectively, as observed by SEM analysis. Smoothly wrinkled and folded sheets were evident. Additionally, the SEM pictures revealed that both sponges possess significant amount of spicules embedded in the very thin and soft sheet-like collagenous structure [22]. After removal of the spicules, the SF-InSCs appeared more as an amorphous matrix, while TEM depicted (Figure 1J,N) the collagenous material as appearing transparent, resembling those obtained before treatment with HF (Figure 1I,M). The complete removal of spicules was also confirmed by SEM (data not shown), where no silicate spicules were observed.

In the case of the SlCs, analogous structures were visible. In particular, siliceous spicules, known to support the sponges and provide defense against predation, were also detected (Figure 1D,H). On the other hand, ICCs presented the typical striation and sheet-like appearance of collagen fibers (Figure 1B,C,F,G), which conclusively proved the collagenous nature of the materials. Specifically, the ICCs from both sponges were observed as threads of various diameters along with the collagen sheets which is a combination of several collagen fibrils and fibers that are bundled together to form a fibril network and a dense pleated sheet-like structure. Sheets were smoothly wrinkled and folded, and appeared as very thin and soft (Figure 1C,G). Pleating of the sheets was visible at a magnification of 5000×.

The collagenous nature of the ICCs of *A. cannabina* and *S. carnosus* was further proved by TEM studies (Figure 1K,O). The obtained micrographs revealed the existence of filaments composed of striated collagen fibrils with repeated band periodicity, a characteristic feature of collagens, as observed earlier for sponge collagen fibrils [16]. Collagen fibrils were organized into bundles, while fibrils became aligned laterally in an ordered way, or curled into bundles consisting of up to 20 fibrils [24]. The individual fibrils displayed a visible, regular transverse banding pattern of about 300 Å periodicity (313 and 288 Å for *A. cannabina* and *S. carnosus*, respectively). These banding patterns are in accordance with the one reported for collagen fibrils isolated from *C. reniformis* [25,26]. The bundles revealed remarkable uniformity in the diameter of their constitutive fibrils (Table 2) with an average of 187 and 199 Å for *A. cannabina* and *S. carnosus*, respectively, in accordance with previously-reported data for other sponges [25,29].

The recorded TEM micrographs for the InSCs (Figure 1I,M) and the SlCs (Figure 1L,P) samples did not present a characteristic pattern. However, in the case of the SlC isolated from *S. carnosus* an area with clearly-striated collagen was detected (Figure 1P insert), most likely due to the nature of the preparation, composed of a mixture of ICC and SlC, also previously reported by Gross and coworkers [22]. The lack of a clear banding pattern might be attributed to the isolation, under the described conditions, of dominating collagenous structures presenting common characteristics with basement membrane (type IV) collagen. Transparent sheets of collagenous material were also previously observed for irciniid collagens, attributed to the non-fibrillar basement-type resembling collagens [27].

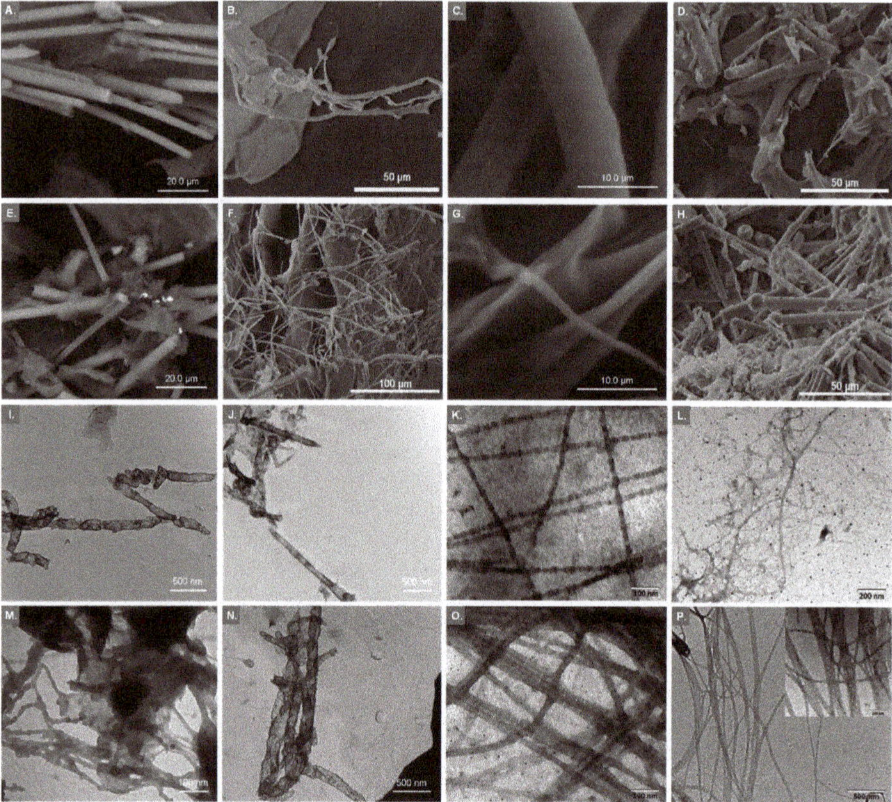

Figure 1. SEM micrographs of insoluble collagen (InSC; **A,E**), intercellular collagen (ICC; **B,C,F,G**), and spongin-like collagen (SlC; **D,H**) from *A. cannabina* (row 1) and *S. carnosus* (row 2), respectively. TEM micrographs of insoluble collagen before (InSC; **I,M**) and after (SF-InSC; **J,N**) spicule removal, intercellular collagen (ICC; **K,O**) and spongin-like collagen (SlC; **L,P**) from *A. cannabina* (row 3) and *S. carnosus* (row 4), respectively.

Table 2. Morphological characteristics (presented as means ± S.E.) of insoluble collagen (InSC), intracellular collagen (ICC) and spongin-like collagen (SlC) isolated from *A. cannabina* and *S. carnosus*.

Isolated Collagen	Period (nm) $n = 10$	Fibril Width (nm) $n = 10$
ICC from *A. cannabina*	31.29 ± 1.14 [1]	18.74 ± 1.27 [1]
ICC from *S. carnosus*	28.81 ± 1.73 [2]	19.91 ± 1.65 [1]
SlC from *S. carnosus*	26.60 ± 0.95 [3]	24.10 ± 1.54 [3]
InSC from *S. carnosus*		17.62 ± 2.91 [1,4]

[1,2,3,4] Data denoted by the same superscript are not significantly different ($p > 0.05$).

2.3. Infrared Spectroscopic Analysis

In the IR spectra of the isolated collagenous materials, all characteristic absorption bands of amides I, II, and III, as well as amides A and B (Table 3), indicative of the secondary structure of the different materials [33], were observed. The amide A absorption band, associated with the hydrogen-bonded N-H stretching vibration [34], was observed at lower frequencies (3279–3294 cm^{-1}), as opposed to the free N-H stretching vibration that appears in the range of 3400–3440 cm^{-1}. This peak is shifted at lower frequencies than the ones reported for the collagen of the marine sponge *C. reniformis* [25] and the calf skin type I collagen, indicating that the N-H group is involved in extensive hydrogen bonding, which stabilizes the helical structure of collagen [34–36]. On the other hand, the amide B band, related to the asymmetrical stretch of CH_2 and NH_3^+ [36,37] remained relatively constant (~2924 cm^{-1}), pointing to the absence of major differences in the lysine content in all of the examined samples [37].

Table 3. IR spectra peak position and assignments for insoluble collagen before (InSC) and after (SF-InSC) spicules removal, intracellular collagen (ICC), and spongin-like collagen (SlC) isolated from *A. cannabina* and *S. carnosus*. For comparison reasons, the respective peaks for bovine collagen (BOC) [34] are also included.

Region	Peak Wavenumber (cm^{-1})								
	A. cannabina				*S. carnosus*				BOC
	InSC	SF-InSC	ICC	SlC	InSC	SF-InSC	ICC	SlC	
Amide A	3288	3279	3294	3286	3282	3282	3292	3287	3295
Amide B	2924	2924	2922	2926	2924	2922	2922	2923	2933
Amide I	1622	1627	1654	1639	1622	1628	1652	1647	1635
Amide II	1543	1529	1547	1539	1535	1527	1543	1543	1545
Amide III	1232	1226	1238	1222	1234	1230	1238	1232	1235
C-O stretch	1055	1059	1078		1074	1066	1078		
	1035	1037	1028	1028	1028	1031	1035	1001	

The amide I band, mainly associated with the C=O stretching vibration coupled with the N-H bending vibration along the polypeptide backbone or with hydrogen bonding coupled with COO^-, C-N stretching, and CCN deformation, is the most intense band in proteins and, therefore, the most sensitive and useful marker for the analysis of the secondary structure of proteins with IR spectroscopy [38]. Normally resonating in the range of 1600–1700 cm^{-1} [39,40], bands around 1630 cm^{-1} indicate imide residues, and bands around 1660 and 1675 cm^{-1} are assigned to intermolecular crosslinks and b-turns, respectively [38]. In our samples, the amide I peaks are shifted to lower frequencies, indicative of higher hydrogen bonding potential [37], less intermolecular cross-linking, and decreased molecular order [39]. The lowest frequencies were observed in both InSCs, before and after treatment with HF, whereas the frequencies increased in the cases of the SlC and ICC samples, concomitantly to the molecular order increase. Additionally, the amide II band, associated with the N-H bending vibration coupled with the C-N stretching vibration, was also shifted to lower frequencies (1527–1547 cm^{-1}), indicative of the involvement of the N-H group in hydrogen bonding [35]. Finally, the amide III band, attributed to the C-N stretching vibration in combination with the N-H deformation, is considered as the collagen fingerprint because it is accredited to the characteristic collagen repeating tripeptide (Gly-X-Y) [38]. Furthermore, in the IR spectra of the isolated collagenous materials, additional bands at about 1030 cm^{-1} appeared, mostly attributed to C-O vibrations due to the presence of carbohydrates [25,41]. In the case of the SF-InSCs, in the recorded IR spectra (Figure 2) a less intense peak at ~1030 cm^{-1} appeared, possibly corresponding also to the Si-O-Si asymmetric bond stretching vibration, known to absorb in the range of 1030–1100 cm^{-1}.

The absorption intensity ratio between the amide III band (1238 cm^{-1}) and the band at approximately 1450 cm^{-1} was 0.88 and 0.89 for the ICCs of *A. cannabina* and *S. carnosus*, respectively, indicating that the triple helix has been adequately preserved during the isolation procedure [34,36,37,40]. Generally, a ratio of approximately 1 indicates that the triple helical structure of collagen is intact [42]. In the case of the InSCs, this ratio is low for both sponges (~0.7), indicating that the triple-helical structure might be slightly affected during the extraction procedure. It was shown earlier that this ratio might be lower when the collagen triple helix is affected by cleavage of telopeptides through pepsin digestion [36]. Moreover, upon treatment for the removal of spicules, the absorption intensity ratio between amide III band and the band approximately at 1450 cm^{-1} increased to 0.96 and 0.94 for *A. cannabina* and *S. carnosus*, respectively, demonstrating the removal of other impurities.

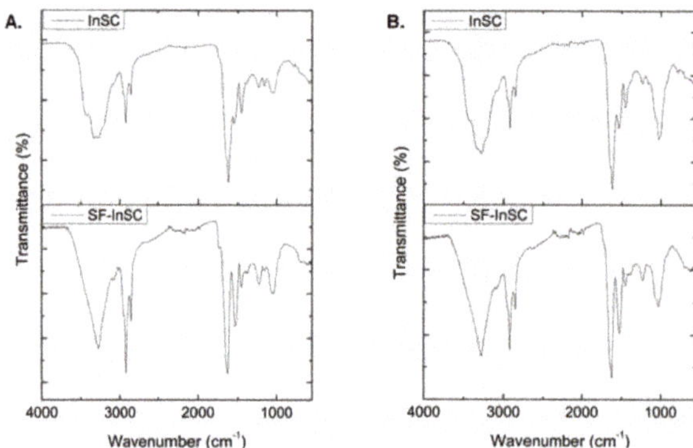

Figure 2. IR spectra of insoluble collagen before (InSC; upper) and after (SF-InSC; lower) spicule removal isolated from *A. cannabina* (**A**) and *S. carnosus* (**B**).

2.4. Isoelectric Point Determination

The SF-InSCs were subjected to titration for the determination of the acid-base properties and the isoelectric point. The titration curves are shown in Figure 3. The pH of freshly-prepared InSC dispersions were 3.48 and 3.54 for *A. cannabina* and *S. carnosus*, respectively. After the HF treatment, the pH of the SF-InSC dispersions were slightly altered (3.66 and 3.34 for *A. cannabina* and *S. carnosus*, respectively). These values are lower than those reported for *C. reniformis* [26], probably due to the higher content in acidic amino acids (aspartic acid or glutamic acid), as also supported by the high contents of Asx and Glx found in both sponges from the amino acid content analysis (Table 4). The isoelectric point was calculated to be approximately 6.7 and 6.3 for *A. cannabina* and *S. carnosus*, respectively. These results are in agreement with previously-reported data determining the isoelectric point of insoluble collagen at pH values around 7 [26].

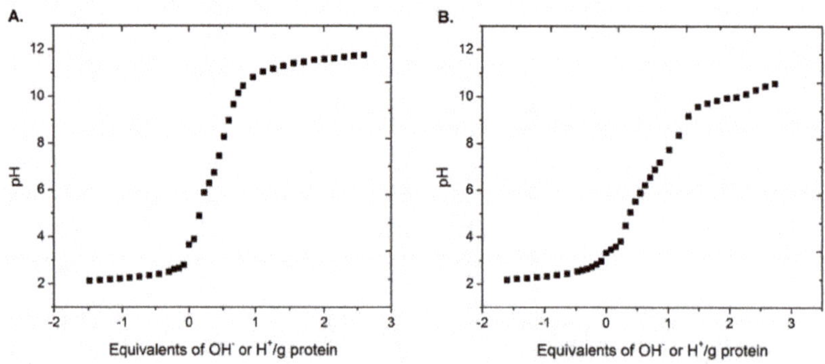

Figure 3. Titration curves of insoluble collagen after spicules removal (SF-InSC) isolated from *A. cannabina* (**A**) and *S. carnosus* (**B**).

2.5. Amino Acid Profile

The amino acid composition of collagen is one of the key factors affecting its properties. Therefore, the amino acid profile of the ICCs from both sponges was determined (Table 4). Their composition was analogous to that described for the sponges *G. cydonium*, *C. reniformis*, and *I. variabilis* [23–26]. Glycine (Gly) was found to be the major amino acid in both ICCs with 257 and 295 residues/1000 residues for *A. cannabina* and *S. carnosus*, respectively. This result is in accordance with the Gly-X-Y amino acid model in which Gly occurs in every third position. Relatively high contents of aspartic acid (Asx; 100 and 94 residues/1000 residues), glutamic acid (Glx; 82 and 84 residues/1000 residues), alanine (Ala; 72 and 89 residues/1000 residues), and proline (Pro; 58 and 56 residues/1000 residues) were observed for *A. cannabina* and *S. carnosus*, respectively. Both ICCs presented the characteristic high threonine (Thr) and serine (Ser) content (approximately 6% each) and low lysine (Lys) and hydroxylysine (Hyl) content, previously reported for *Ircinia* [23]. Low Hyl content (5–6 residues/1000 residues) has also been reported for acid- and pepsin-soluble collagens isolated from shark skin [39]. Moreover, the sum of Thr and Ser of both sponges' ICCs is similar to that of collagens reported for lower vertebrates and invertebrates. Additionally, no differences in the Lys content of the two different ICCs were observed, as already indicated from the same absorption bands in the IR spectra at 2922 cm^{-1} attributed to amide B (Table 3) [37].

Table 4. Amino acid composition (residue/1000) of intercellular collagen (ICC) isolated from *A. cannabina* and *S. carnosus*.

Amino Acid	*A. cannabina*	*S. carnosus*	Amino Acid	*A. cannabina*	*S. carnosus*
Hyp	38	47	Met	21	11
Asx [1]	100	94	Ile	37	24
Thr	56	56	Leu	62	48
Ser	63	57	Tyr	13	12
Glx [2]	82	81	Phe	33	22
Pro	58	56	Hyl	6	6
Gly	257	295	Lys	15	15
Ala	72	89	His	6	4
Val	44	43	Arg	37	43
Total imino acids	96	103			

[1] Asx: Asp + Asn. [2] Glx: Gln + Glu.

Compared to *S. carnosus*, ICC from *A. cannabina* contained higher amounts of methionine (Met), phenylalanine (Phe), leucine (Leu), and isoleucine (Ile), but lower amounts of Gly, Ala, and hydroxyproline (Hyp). The percentage of the remaining amino acids is in relatively good agreement to the above-mentioned studies, especially the amounts of Asx, Glx, Pro, His, and Ala. The number of sulfur-containing Met residues was significantly higher in the ICCs of both sponges (Table 4), as compared to collagen from porcine dermis (6 residues/1000 residues) [43].

Nevertheless, the overall percentages of Hyp were lower than those reported for other sponges [22]. Imino acids are involved in hydrogen bonding, therefore affecting the stability of the collagen triple helix and its thermal behavior [37,39,44]. The imino acid content value is usually lower in marine collagens in comparison to mammalian collagens, resulting in a lower thermal denaturation temperature [33].

The ICCs from both sponges contained approximately 12 tyrosine (Tyr) residues per collagen molecule, indicating that their nonhelical telopeptides, where all of the Tyr residues are located, were intact [45]. The reduced values for Gly, Hyp, and Hyl can also be attributed to the existence of glycoproteinaceous impurities, known to be strongly associated with collagen [46].

2.6. Thermal Behaviour

It is well established [47] that upon increasing temperature, thermal denaturation of collagen is taking place, during which hydrogen bonds break and helices unfold, leading to the formation of collagen coils. This process is accompanied with appreciable heat absorption and can, therefore, be monitored with DSC. Indeed, the DSC curves of the hydrated collagen samples (Figure 4) clearly indicate two major endothermic peaks. Previous DSC studies also revealed collagen's bimodal transition and concluded that the higher temperature peak was due to the helix-coil transition of collagen (denaturation of collagen), while the lower temperature peak originated from the breaking of the hydrogen bonds between collagen molecules or the defibrillation of the solubilized collagen fibrils [48,49]. This is attributed to the fact that the inter-triple helix hydrogen bonds responsible for the fibrillation are easier to break than the intra-triple hydrogen bonds that are responsible for helix formation [48].

In the present study, the thermal behavior of the ICCs isolated from *A. cannabina* and *S. carnosus* were monitored after removal of the entangled glycoconjugates [46]. The yield of the described procedure was 38% and 46% (w/w) for *A. cannabina* and *S. carnosus*, respectively. The low endothermic transition had its peak maximum transition temperature (T_{max}) at 25.4 °C (ΔH value 1.27 J g^{-1}) and 32.9 °C (ΔH value 5.74 J g^{-1}) for the ICCs from *A. cannabina* and *S. carnosus*, respectively (Figure 4). The high temperature endothermic peak had a T_{max} of 44.6 °C (ΔH value 0.37 J g^{-1}) and 51.6 °C (ΔH value 17.65 J g^{-1}) for the *A. cannabina* and *S. carnosus* ICCs, respectively (Figure 4). As is clearly evident from the examination of both reversing and non-reversing components of the thermograms, the total heat flow for the thermal denaturation of collagen involves a significant non-reversing component, while the reversing component is negligible. This is in line with previous studies that showed that collagen denaturation endotherms in fibers and in basement membranes are governed by an irreversible rate process [50,51] and not by equilibrium thermodynamics, as previously hypothesized. Given the irreversibility of the process within the time frame of temperature modulation (60 s), the transitions are registered as essentially a non-reversing event in temperature-modulated differential scanning calorimetry (TMDSC), although, in general, unfolding of proteins is a complex phenomenon that encompasses both reversible and irreversible steps [52].

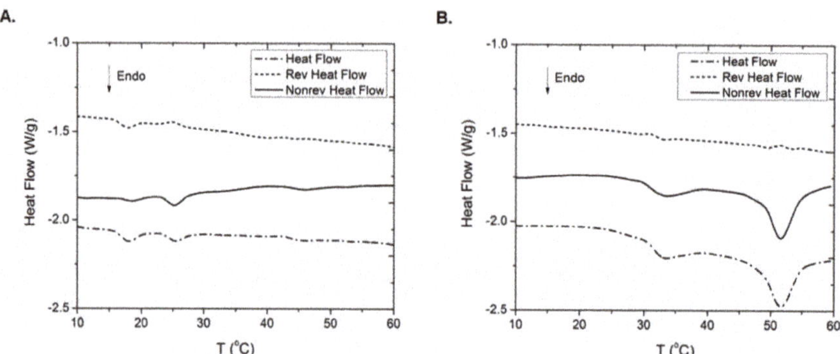

Figure 4. Temperature modulated DSC data of intercellular collagen (ICC) isolated from *A. cannabina* (**A**) and *S. carnosus* (**B**). The total (- - -), non-reversing (—) and reversing heat (---) flows are presented (curves are shifted vertically for clarity).

A rather low T_{max} value, as that observed for the ICC from *A. cannabina*, was reported earlier for collagen isolated from edible jellyfish (26.0 °C) [53]. On the other hand, T_{max} values around 31 to 33 °C, as that measured for the ICC from *S. carnosus*, have been observed for an array of collagens isolated from tropical fish [34,54]. Moreover, the ICC from *S. carnosus* exhibited a higher ΔH value

(5.74 J g^{-1}) than that of *A. cannabina* (1.27 J g^{-1}). It is widely accepted that T_{max} is directly correlated with imino acid content, body temperature of the specimen, and environmental temperature [55,56], whereas the enthalpy change (ΔH) can be influenced by molecular stability, directly correlated with the amino acid sequence in collagen.

In our case, the ICCs from both sponges contain low amount of iminoacids (Table 4) in comparison to that of terrestrial organisms (approximately 200 residues/1000 residues), with the ICC from *A. cannabina* displaying the lowest content (96 residues/1000 residues). The observed difference between the T_{max} of the ICC samples from the two sponges could, therefore, be attributed to the imino acid content difference, and especially to the Hyp content difference. This phenomenon might also be related to the superior stability of the ICC from *S. carnosus*, due to the high content of the Gly-X-Y sequence, as confirmed by the elevated percentage of Gly (17.9% vs. 15.0% *w/w* for *S. carnosus* and *A. cannabina*, respectively). This, in agreement with previous reports [33,57], might be an additional justification for the high value of T_{max} despite the low amount of imino acids. Additionally, as previously reported [58], the high Asp (pK $\approx$ 3.9) and Glu (pK $\approx$ 4.3) content can contribute to ion pair formation with the basic residues at neutral pH, resulting in increased stability, which might partially compensate for the decreased stability deriving from the low Hyp content (Table 4) [59]. Another possible reason might be the intensely-localized sulfur bonding interactions associated with the higher Met content [43].

Finally, an additional low temperature endothermic peak (T_{max} = 17.9 °C, ΔH = 1.65 J g^{-1}) was observed in the case of the ICC from *A. cannabina*. In contrast to the previous transitions discussed, the examination of both the reversing and non-reversing components of this specific transition suggests that this process is, to a great extent, reversible. Taking into consideration that during the denaturation of small proteins (for instance lysozyme) [60] the reversible unfolding has the largest contribution, whereas the irreversible process still remains well detectable, we tentatively ascribe this low temperature transition to the denaturation of small molecular weight collagen species that are present in this sample.

The CD spectra of the ICCs from the two sponges in the region of 190 to 250 nm are depicted in Figure 5A,B. Both samples showed a rotatory maximum at about 221 nm, a minimum at 193–196 nm, and a consistent crossover point (zero rotation) at about 212 nm. These spectral characteristics are typical of a collagen triple-helix structure [61–63]. The corresponding mean residue ellipticities, $[\theta]_{221}$, as a function of temperature, are shown in Figure 5C,D. The $[\theta]_{221}$ values decreased with temperature due to decomposition of the collagen triple helical structure, and indicated denaturation temperatures of 24.3 °C and 28.2 °C for the ICCs from *A. cannabina* and *S. carnosus*, respectively, in good agreement with the obtained results from the conducted DSC studies.

It has been earlier shown that thermal denaturation temperature of collagens from different sources correlates directly with the imino acid (Pro and Hyp) content [43,64]. Actually, higher imino acid content facilitates intra- and intermolecular crosslinking resulting in a more stable triple helical structure of the collagen molecule [44]. A good linear correlation was observed earlier when measured denaturation temperatures were plotted against the corresponding numbers of Hyp residues, this effect being less pronounced with respect to the Pro content [43]. The amino acid composition analysis of the investigated sponges (Table 4) confirms the above observations, since *A. cannabina* presents a lower Hyp, but equal Pro, content as compared to *S. carnosus* resulting, therefore, in a concurrently-reduced T_d. Interestingly, cold-water fish collagens have low T_d since their imino acid contents are very low [65], in contrast to the T_d of skin collagen of terrestrial mammals which are 37 °C and 40.8 °C, respectively [43], both possessing high imino acid content.

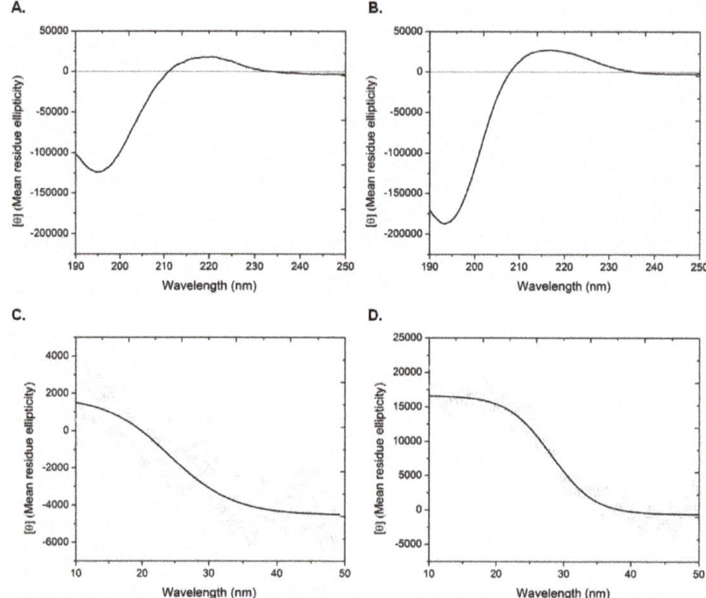

Figure 5. CD spectra in the region of 190–250 nm (recorded at 20 °C) and temperature effect on the CD spectra at 221 nm of intercellular collagen (ICC) isolated from *A. cannabina* ((**A**) and (**C**), respectively) and *S. carnosus* ((**B**) and (**D**), respectively).

3. Discussion

The presence of collagen in freshwater, as well as marine sponges, was unequivocally established more than 50 years ago by the work of Bronsted and Carlsen [66] and Gross and his coworkers [22]. Since then, many investigations regarding the fine structure and physicochemical properties of marine collagen have been performed. However, to the best of our knowledge, such extensive studies on the collagenous profile of sponge material have been conducted only on *C. reniformis* and *Ircinia* species [23,25–27,29]. In this context, the main purpose of the current study was the morphological characterization of various isolated collagenous materials (InSC, ICC, and SlC) from *A. cannabina* and *S. carnosus*, while further biochemical and biophysical characterization was undertaken only for the ICCs, given their relatively higher solubility and purity.

It has also been proven that in Demospongiae, collagen, constituting exclusively the intercellular organic framework, amounts to approximately 10% of the total organic matter [27,67]. In the present study, collagen content was experimentally calculated to amount for the 12.6% and 5.0% dry weight of *A. cannabina* and *S. carnosus*, respectively. The co-isolation of collagen with spicules is justified by the spicule formation procedure, generally accomplished by specialized cells that supply mineral ions or organic macromolecular particles, primarily consisting of proteins, carbohydrates, lipids, and seldom by nucleic acids [68].

Furthermore, the aforementioned characteristic insolubility has prevented the determination of the thermal behavior of sponge collagens. To our knowledge, only a few efforts have been made to determine their thermal behavior [25,69]. Our results corroborate to the existing knowledge that the thermal stability of marine collagens, which exhibit lower denaturation temperatures due to their lower content of imino acids, is generally lower than that of mammalian collagens. The low denaturation temperature of sponge collagen may also reflect the ambient temperature in which marine organisms live [70]. Moreover, the thermal stability of collagen is also directly correlated with the environmental and body temperatures of organisms [71].

Overall, the low denaturation temperature of sponge collagen observed in the present study enables gelatin extraction at lower temperature compared to mammalian gelatin, therefore providing an economic benefit for using marine sponges as a raw material of gelatin for the food industry [72].

All of our results point out that sponges contain collagen that retains its helical structure throughout the isolation procedure and all of its measured characteristics confirm the less crosslinked form, as verified by the IR, amino acid analysis, DSC, and CD data.

4. Materials and Methods

4.1. Animal Material

Specimens of *A. cannabina* were collected by SCUBA diving in Kea Island, Aegean Sea, Greece, at a depth of 15–20 m, whereas specimens of *S. carnosus* were collected by dredging at Kyllini Bay, Ionian Sea, Greece, at a depth of 50–70 m, and kept frozen until analyzed. Voucher specimens have been deposited at the animal collection of the Department of Pharmacognosy and Chemistry of Natural Products, National and Kapodistrian University of Athens (ATPH/MP0300 and ATPH/MP0106, respectively).

4.2. Chemicals

Tris(hydroxymethyl)aminomethane was from Mallinckrodt (Dublin, Ireland) and EDTA from Serva (Heidelberg, Germany). Urea and sodium carbonate were from Merck (Kenilworth, NJ, USA), while trypsin from bovine pancreas Type I, (~12,443 benzoyl L-arginine ethyl ester BAEE units/mg protein) was from Sigma (Darmstadt, Germany).

4.3. Isolation of Insoluble Collagen, Intercellular Collagen, and Spongin-Like Collagen

Sponge specimens were chopped and foreign inorganic and organic material was removed before washing with tap water. Before further processing the sponge tissues were immersed in EtOH for 24 h. InSC was isolated using an alkaline denaturing homogenization buffer (0.1 M Tris-HCl, pH 9.5, 0.01 M EDTA, 8 M urea, 0.1 M 2-mercaptoethanol), as previously described [24,26], whereas ICC and SlC were isolated using a trypsin-containing extraction buffer (0.1% trypsin in 0.1 M bicarbonate buffer, pH 8.0) [22,23]. All collagen samples were collected after centrifugation (at 20,000× g and 50,000× g, respectively) and lyophilization.

4.4. Removal of Spicules from Insoluble Collagen

In order to remove the siliceous spicules, InSC was treated with an HF solution (10% v/v) for 20 min at room temperature. The exact conditions were standardized using siliceous spicules isolated as previously described [73]. The material was rinsed with distilled-deionized water until pH ~ 6 was reached and the HF-treated spicule-free collagen (SF-InSC) was isolated by centrifugation at 12,000× g for 20 min and subsequently lyophilized.

4.5. Electron Microscopy

For SEM analysis lyophilized collagen samples were placed on stubs by using a double face adhesive tape, covered with a thin layer of gold using a Bal-tec SCD 004 sputter coater and examined under either a Cambridge Stereoscan 240 scanning electron microscope or a Philips Quanta Inspect (FEI Company) scanning electron microscope with a tungsten filament (25 kV).

For TEM analysis a small amount of sample dispersed in distilled-deionized water was placed on a Formvar-coated grid and stained with 2% aqueous solution of PTA (phosphotungstic acid hydrate; pH adjusted to 3.3 by using a solution of 1 N NaOH), which revealed the banding pattern of the fibrils, but not the whole periodicity. Each grid was examined under a Philips CM10 transmission electron microscope equipped with an Olympus Veleta digital camera.

4.6. Infrared Spectroscopy

IR spectra of lyophilized collagen samples were measured on a Bruker Tensor 27 FT-IR spectrometer using the attenuated total reflection (ATR) method, at room temperature, in the range of 500–4000 cm^{-1}.

4.7. Amino Acid Analysis

The amino acid profile analysis was performed at TAMU Protein Chemistry Lab (College Station, TX, USA). Finely-ground ICC samples (30–40 mg) were used for liquid HCl (6 N) hydrolysis. Hydrolyzed proteins were derivatized pre-column with o-phthalaldehyde and 9-fluoromethyl-chloroformate prior to separation and quantitation by reverse phase HPLC. The component amino acids were then separated by HPLC (Agilent 1260), detected by UV (Agilent G1365D) or fluorometry (Agilent G1321B), and quantitated. All system control and data analysis was performed by Agilent Chemstation software. Values are the means of two independent experiments that did not differ by more than 2.9%.

4.8. Titration

Samples (70 mg) of freeze-dried material (SF-InSC) were dispersed in 7 mL distilled-deionized water by ultrasonication (GeneralSonic GS3) [26]. One sample was titrated with 0.1 N NaOH and the other with 0.1 N HCl. After each titrant addition, the suspension was stirred for 10 min at room temperature and subsequently the pH was recorded (Jenway 3310 pHmeter). A blank sample without collagen was titrated under the same conditions. The resulting pHs versus the amount of NaOH and HCl were plotted within the pH range from 2 to 12.

4.9. Differential Scanning Calorimetry

ICCs were initially dispersed in 0.01 M EDTA, pH 8.0 and the resulting suspension remained under stirring overnight at 8 °C. The collagens were collected after centrifugation at 13,000 rpm for 15 min. Subsequently, a 1% SDS (w/v) solution was used for the removal of entangled glycoconjugates [46]. Finally, the collagens were collected after centrifugation, washed exhaustively with distilled-deionized water, and lyophilized.

TMDSC measurements were performed by employing a MDSC 2920 calorimeter (TA Instruments, New Castle, DE, USA) under nitrogen flow (20 mL/min), using a heating rate of 2 °C/min, a temperature modulation amplitude of 0.318 °C every 60 s, and an empty pan as a reference. In such experiments the linear heating rate is superimposed by a sinusoidal temperature variation and it is, thus, possible to separate the total signal (corresponding to that of a conventional DSC) into two different components, corresponding to the reversible and the irreversible heat flows. The TMDSC profiles were obtained only on heating. Heat and temperature calibrations were performed by using indium as a standard. The enthalpic content (ΔH) of each transition was calculated from the area under each peak, while the transition temperature was taken at the center of each transition. For each experiment ~2 mg of lyophilized collagen, weighted with an accuracy of ± 0.01 mg, was hydrated with distilled-deionized water at a collagen/water ratio of 1:20 (w/w) and placed in sealed aluminum pans. The samples were then kept at 4 °C for 48 h before analysis.

4.10. Circular Dichroism Spectroscopy

The molecular conformation and denaturation temperature (T_d) of ICCs, dissolved in distilled-deionized water to a concentration of 0.1 mg/mL, were assessed by CD spectroscopy using a Jasco J-715 circular dichroism spectropolarimeter equipped with a Peltier-type temperature control system (Jasco PTC-348Wi). CD spectra were recorded at 20 °C using a 0.1-cm path length quartz cell at 190–250 nm with a step size of 0.5 nm and a band width of 1.0 nm. Experiments were run in triplicate, and 10 scans for each spectrum were signal-averaged.

To determine the T_d, the rotatory angle at a fixed wavelength of 221 nm, $[\theta]_{221}$, was recorded with heating from 15 to 50 °C at a rate of 1 °C/min. The collagen concentration was adjusted to 0.1 mg/mL and the temperature was controlled. The T_d was determined as the midpoint temperature between native-folded and completely unfolded forms. The mean molecular ellipticity (θ) was calculated using the equation $[\theta] = 10^{-3} \theta$ M/LC (expressed in deg cm^2 dmol^{-1}), where θ is the measured ellipticity in degrees, L is the path length in mm, C is the concentration in mg/mL, and M is the average residue molecular weight of collagen equal to 91.2 [74].

5. Conclusions

In the present study, the collagenous content of the demosponges *A. cannabina* and *S. carnosus* was exhaustively examined. The insoluble, intercellular, and spongin-like collagens were isolated from *A. cannabina* and *S. carnosus*, representing 12.6%, 3.0%, and 42.8% dry weight for the former and 5.0%, 1.9%, and 21.8% dry weight for the latter sponge. SEM and TEM observations confirmed the characteristic fibrous structures, while IR spectroscopic analysis verified the characteristic absorption bands for proteins of the collagen class. Moreover, the acid–base properties of the insoluble collagen were investigated by titration, placing the isoelectric point approximately at pH 7. Marine sponge collagen, as compared to that derived from terrestrial animals and other marine collagen sources, has been reported to differentiate in its characteristics, such as amino acid composition, which consecutively affects collagen's thermal behavior, isoelectric pH, solubility, and many other properties. In our case, the measured low imino acid content for the intercellular collagen, already reported being low in marine sources and even lower, specifically, in sponges, results in thermal stability comparable to that determined for collagen isolated from edible jellyfish and tropical fish. Indeed, the denaturation temperatures of the intercellular collagen isolated from *A. cannabina* and *S. carnosus* were determined by DSC studies at 25.4 °C and 32.9 °C, respectively, the first one being relatively lower than that reported for other marine organisms, while the second one being comparable to values observed for an array of collagens isolated from tropical fish. CD spectra indicated the existence of helical structures and the fact that the denaturation temperatures were dependent on the amount of imino acids. Marine collagen is considered as an equivalent, although safer, biomaterial than the one from terrestrial sources dominating the market nowadays. Our results suggest that the sponges *A. cannabina* and *S. carnosus* can be considered as an alternative source of collagen.

Acknowledgments: The authors wish to thank Aggeliki Panagiotopoulou for recording the CD spectra at the Circular Dichroism Laboratory at the Institute of Biosciences and Applications of NCSR "Demokritos". This research did not receive any specific grant from funding agencies in the public, commercial, or not-for-profit sectors.

Author Contributions: Vassilios Roussis, Efstathia Ioannou, Leto-Aikaterini Tziveleka, and Dimitris Tsiourvas conceived and designed the experiments; Leto-Aikaterini Tziveleka, Dimitris Tsiourvas, Panagiotis Berillis, and Evangelia Foufa performed the experiments; Leto-Aikaterini Tziveleka, Vassilios Roussis, Efstathia Ioannou, and Dimitris Tsiourvas analyzed the data; Vassilios Roussis and Efstathia Ioannou contributed reagents/materials/analysis tools; and Leto-Aikaterini Tziveleka, Efstathia Ioannou, and Vassilios Roussis wrote the paper.

Conflicts of Interest: The authors declare no conflict of interest.

References

1. Gelse, K.; Pöschl, E.; Aigner, T. Collagens-structure, function, and biosynthesis. *Adv. Drug Deliv. Rev.* **2003**, *55*, 1531–1546. [CrossRef] [PubMed]
2. Myllyharju, J.; Kivirikko, K.I. Collagens, modifying enzymes and their mutations in humans, flies and worms. *Trends Genet.* **2004**, *20*, 33–43. [CrossRef] [PubMed]

3. Silva, T.H.; Moreira-Silva, J.; Marques, A.L.P.; Domingues, A.; Bayon, Y.; Reis, R.L. Marine origin collagens and its potential applications. *Mar. Drugs* **2014**, *12*, 5881–5901. [CrossRef] [PubMed]

4. Meena, C.; Mengi, S.A.; Deshpande, S.G. Biomedical and industrial applications of collagen. *J. Chem. Sci.* **1999**, *111*, 319–329. [CrossRef]

5. Berillis, P. Marine collagen: Extraction and applications. In *Research Trends in Biochemistry, Molecular Biology and Microbiology*; Madhukar, S., Ed.; SM Group: Dover, DE, USA, 2015; pp. 1–13.

6. Venugopal, V. *Marine Products for Healthcare: Functional and Bioactive Nutraceutical Compounds from the Ocean*; CRC Press: Boca Raton, FI, USA, 2009.

7. Patino, M.G.; Neiders, M.E.; Andreana, S.; Noble, B.; Cohen, R.E. Collagen: An overview. *Implant Dent.* **2002**, *11*, 280–285. [CrossRef] [PubMed]

8. Friess, W. Collagen-biomaterial for drug delivery. *Eur. J. Pharm. Biopharm.* **1998**, *45*, 113–136. [CrossRef]

9. Francis, G.; Thomas, J. Isolation and chemical characterization of collagen in bovine pulmonary tissues. *Biochem. J.* **1975**, *145*, 287–297. [CrossRef] [PubMed]

10. Lynn, A.K.; Yannas, I.V.; Bonfield, W. Antigenicity and immunogenicity of collagen. *J. Biomed. Mater. Res. B Appl. Biomater.* **2004**, *71*, 343–354. [CrossRef] [PubMed]

11. Arvanitoyannis, I.S.; Kassaveti, A. Fish industry waste: Treatments, environmental impacts, current and potential uses. *Int. J. Food Sci. Technol.* **2008**, *43*, 726–745. [CrossRef]

12. Leary, D.; Vierros, M.; Hamon, G.; Arico, S.; Monagle, C. Marine genetic resources: A review of scientific and commercial interest. *Mar. Policy* **2009**, *33*, 183–194. [CrossRef]

13. Garrone, R. The evolution of connective tissue. Phylogenetic distribution and modifications during development. *Prog. Clin. Biol. Res.* **1981**, *54*, 141–149. [PubMed]

14. Boot-Handford, R.P.; Tuckwell, D.S. Fibrillar collagen: The key to vertebrate evolution? A tale of molecular incest. *BioEssays* **2003**, *25*, 142–151. [CrossRef] [PubMed]

15. Simon, P.; Lichte, H.; Formanek, P.; Lehmann, M.; Huhle, R.; Carrillo-Cabrera, W.; Harscher, A.; Ehrlich, H. Electron holography of biological samples. *Micron* **2008**, *39*, 229–256. [CrossRef] [PubMed]

16. Garrone, R. The Collagen of the Porifera. In *Biology of Invertebrate and Lower Vertebrate Collagens*; Bairati, A., Garrone, R., Eds.; NATO ASI Series; Plenum Press: New York, NY, USA, 1985; pp. 157–175.

17. Exposito, J.-Y.; Cluzel, C.; Garrone, R.; Lethias, C. Evolution of collagens. *Anat. Rec.* **2002**, *268*, 302–316. [CrossRef] [PubMed]

18. Mehbub, M.F.; Lei, J.; Franco, C.; Zhang, W. Marine sponge derived natural products between 2001 and 2010: Trends and opportunities for discovery of bioactives. *Mar. Drugs* **2014**, *12*, 4539–4577. [CrossRef] [PubMed]

19. Blunt, J.W.; Copp, B.R.; Keyzers, R.A.; Munro, M.H.G.; Prinsep, M.R. Marine natural products. *Nat. Prod. Rep.* **2016**, *33*, 382–431. [CrossRef] [PubMed]

20. Costantino, V.; Fattorusso, E.; Imperatore, C.; Mangoni, A.; Freigang, S.; Teyton, L. Glycolipids from sponges. 18. Corrugoside, a new immunostimulatory alpha-galactoglycosphingolipid from the marine sponge *Axinella corrugata*. *Bioorg. Med. Chem.* **2008**, *16*, 2077–2085. [CrossRef] [PubMed]

21. Lamoral-Theys, D.; Fattorusso, E.; Mangoni, A.; Perinu, C.; Kiss, R.; Costantino, V. An in vitro valuation of the anticancer activity of diterpene isonitriles from the sponge *Pseudoaxinella flava* in apoptosis-sensitive and apoptosis-resistant cancer cell lines. *J. Nat. Prod.* **2011**, *74*, 2299–2303. [CrossRef] [PubMed]

22. Gross, J.; Sokal, Z.; Rougvie, M. Structural and chemical studies on the connective tissue of marine sponges. *J. Histochem. Cytochem.* **1956**, *4*, 227–246. [CrossRef] [PubMed]

23. Junqua, S.; Robert, L.; Garrone, R.; Pavans de Ceccatty, M.; Vacelet, J. Biochemical and morphological studies on collagens of horny sponges. *Ircinia* filaments compared to sponginess. *Connect. Tissue Res.* **1974**, *2*, 193–203. [CrossRef] [PubMed]

24. Diehl-Seifert, B.; Kurelec, B.; Zahn, R.K.; Dorn, A.; Jeričevic, B.; Uhlenbruck, G.; Müller, W.E.G. Attachment of sponge cells to collagen substrata: Effect of a collagen assembly factor. *J. Cell Sci.* **1985**, *79*, 271–285. [PubMed]

25. Garrone, R.; Huc, A.; Junqua, S. Fine structure and physicochemical studies on the collagen of the marine sponge *Chondrosia reniformis* Nardo. *J. Ultrastruct. Res.* **1975**, *52*, 261–275. [CrossRef]

26. Swatschek, D.; Schatton, W.; Kellermann, J.; Müller, W.E.G.; Kreuter, J. Marine sponge collagen: Isolation, characterization and effects on the skin parameters surface-pH, moisture and sebum. *Eur. J. Pharm. Biopharm.* **2002**, *53*, 107–113. [CrossRef]

27. Pallela, R.; Bojja, S.; Janapala, V.R. Biochemical and biophysical characterization of collagens of marine sponge, *Ircinia fusca* (Porifera: Demospongiae: Irciniidae). *Int. J. Biol. Macromol.* **2011**, *49*, 85–92. [CrossRef] [PubMed]

28. Ferreira, A.M.; Gentile, P.; Chiono, V.; Ciardelli, G. Collagen for bone tissue regeneration. *Acta. Biomater.* **2012**, *8*, 3191–3200. [CrossRef] [PubMed]

29. Heinemann, S.; Ehrlich, H.; Douglas, T.; Heinemann, C.; Worch, H.; Schatton, W.; Hanke, T. Ultrastructural studies on the collagen of the marine sponge *Chondrosia reniformis* Nardo. *Biomacromolecules* **2007**, *8*, 3452–3457. [CrossRef] [PubMed]

30. Imhoff, J.M.; Garrone, R. Solubilization and characterization of *Chondrosia reniformis* sponge collagen. *Connect. Tissue Res.* **1983**, *11*, 193–197. [CrossRef] [PubMed]

31. Ehrlich, H.; Hanke, T.; Simon, P.; Goebel, C.; Heinemann, S.; Born, R.; Worch, H. Demineralisation von natürlichen Silikat-basierten Biomaterialien: Neue Strategie zur Isolation organischer Gerüststrukturen. *BIOmaterialien* **2005**, *6*, 297–302. [CrossRef]

32. Katzman, R.L.; Lisowska, E.; Jeanloz, R.W. Invertebrate connective tissue. Isolation of D-arabinose from sponge acidic polysaccharide. *Biochem. J.* **1970**, *119*, 17–19. [CrossRef] [PubMed]

33. Barzideh, Z.; Abd Latiff, A.; Gan, C.-Y.; Benjakul, S.; Abd Karim, A. Isolation and characterisation of collagen from the ribbon jellyfish (*Chrysaora* sp.). *Int. J. Food Sci. Technol.* **2014**, *49*, 1490–1499. [CrossRef]

34. Wang, L.; An, X.; Xin, Z.; Zhao, L.; Hu, Q. Isolation and characterization of collagen from the skin of deep-sea redfish (*Sebastes mentella*). *J. Food Sci.* **2007**, *72*, E450–E455. [CrossRef] [PubMed]

35. Duan, R.; Zhang, J.; Du, X.; Yao, X.; Konno, K. Properties of collagen from skin, scale and bone of carp (*Cyprinus carpio*). *Food Chem.* **2009**, *112*, 702–706. [CrossRef]

36. Kittiphattanabawon, P.; Benjakul, S.; Visessanguan, W.; Shahidi, F. Isolation and characterization of collagen from the cartilages of brownbanded bamboo shark (*Chiloscyllium punctatum*) and blacktip shark (*Carcharhinus limbatus*). *LWT-Food Sci. Technol.* **2010**, *43*, 792–800. [CrossRef]

37. Ahmad, M.; Benjakul, S. Extraction and characterization of pepsin-solubilised collagen from the skin of unicorn leatherjacket (*Aluterus monocerous*). *Food Chem.* **2010**, *120*, 817–824. [CrossRef]

38. Cao, H.; Xu, S.-Y. Purification and characterization of type II collagen from chick sternal cartilage. *Food Chem.* **2008**, *108*, 439–445. [CrossRef] [PubMed]

39. Kittiphattanabawon, P.; Benjakul, S.; Visessanguan, W.; Kishimura, H.; Shahidi, F. Isolation and characterisation of collagen from the skin of brownbanded bamboo shark (*Chiloscyllium punctatum*). *Food Chem.* **2010**, *119*, 1519–1526. [CrossRef]

40. Pati, F.; Adhikari, B.; Dhara, S. Isolation and characterization of fish scale collagen of higher thermal stability. *Bioresour. Technol.* **2010**, *101*, 3737–3742. [CrossRef] [PubMed]

41. Jackson, M.; Choo, L.; Watson, P.H.; Halliday, W.C.; Mantsch, H.H. Beware of connective tissue proteins: Assignment and implications of collagen absorptions in infrared spectra of human tissues. *Biochim. Biophys. Acta* **1995**, *1270*, 1–6. [CrossRef]

42. Plepis, A.M.D.G.; Goissis, G.; Das-Gupta, D.K. Dielectric and pyroelectric characterization of anionic and native collagen. *Polym. Eng. Sci.* **1996**, *36*, 2932–2938. [CrossRef]

43. Ikoma, T.; Kobayashi, H.; Tanaka, J.; Walsh, D.; Mann, S. Physical properties of type I collagen extracted from fish scales of *Pagrus major* and *Oreochromis niloticas*. *Int. J. Biol. Macromol.* **2003**, *32*, 199–204. [CrossRef]

44. Wong, D.W.S. *Mechanism and Theory in Food Chemistry*; Van Nostrand Reinhold: New York, NY, USA, 1989.

45. Na, G.C. UV spectroscopic characterization of type I collagen. *Collagen Relat. Res.* **1988**, *8*, 315–330. [CrossRef]

46. Junqua, S.; Lemonnier, M.; Robert, L. Glycoconjugates from "*Spongia officinalis*" (phylum porifera). Isolation, fractionation by affinity chromatography on lectins and partial characterization. *Comp. Biochem. Physiol. B* **1981**, *69*, 445–453. [CrossRef]

47. Bischof, J.C.; He, X. Thermal stability of proteins. *Ann. N. Y. Acad. Sci.* **2006**, *1066*, 12–33. [CrossRef] [PubMed]

48. Mu, C.; Li, D.; Lin, W.; Ding, Y.; Zhang, G. Temperature induced denaturation of collagen in acidic solution. *Biopolymers* **2007**, *86*, 282–287. [CrossRef] [PubMed]

49. Liu, W.; Li, G. Non-isothermal kinetic analysis of the thermal denaturation of type I collagen in solution using isoconversional and multivariate non-linear regression methods. *Polym. Degrad. Stab.* **2010**, *95*, 2233–2240. [CrossRef]

50. Miles, C.A. Kinetics of collagen denaturation in mammalian lens capsules studied by differential scanning calorimetry. *Int. J. Biol. Macromol.* **1993**, *15*, 265–271. [CrossRef]
51. Miles, C.A.; Burjanadze, T.V.; Bailey, A.J. The kinetics of the thermal denaturation of collagen in unrestrained rat tail tendon determined by differential scanning calorimetry. *J. Mol. Biol.* **1995**, *245*, 437–446. [CrossRef] [PubMed]
52. Vyazovkin, S.; Vincent, L.; Sbirrazzuoli, N. Thermal denaturation of collagen analyzed by isoconversional method. *Macromol. Biosci.* **2007**, *7*, 1181–1186. [CrossRef] [PubMed]
53. Nagai, T.; Ogawa, T.; Nakamura, T.; Ito, T.; Nakagawa, H.; Fujiki, K.; Nakao, M.; Yano, T. Collagen of edible jellyfish exumbrella. *J. Sci. Food Agric.* **1999**, *79*, 855–858. [CrossRef]
54. Kittiphattanabawon, P.; Benjakul, S.; Visessanguan, W.; Nagai, T.; Tanaka, M. Characterisation of acid-soluble collagen from skin and bone of bigeye snapper (*Priacanthus tayenus*). *Food Chem.* **2005**, *89*, 363–372. [CrossRef]
55. Nagai, T.; Araki, Y.; Suzuki, N. Collagen of the skin of ocellate puffer fish (*Takifugu rubripes*). *Food Chem.* **2002**, *78*, 173–177. [CrossRef]
56. Muyonga, J.H.; Cole, C.G.B.; Duodu, K.G. Characterisation of acid soluble collagen from skins of young and adult Nile perch (*Lates niloticus*). *Food Chem.* **2004**, *85*, 81–89. [CrossRef]
57. Bae, I.; Osatomi, K.; Yoshida, A.; Osako, K.; Yamaguchi, A.; Hara, K. Biochemical properties of acid-soluble collagens extracted from the skins of underutilised fishes. *Food Chem.* **2008**, *108*, 49–54. [CrossRef]
58. Venugopal, M.G.; Ramshaw, J.A.; Braswell, E.; Zhu, D.; Brodsky, B. Electrostatic interactions in collagen-like triple-helical peptides. *Biochemistry* **1994**, *33*, 7948–7956. [CrossRef] [PubMed]
59. Miki, A.; Inaba, S.; Baba, T.; Kihira, K.; Fukada, H.; Oda, M. Structural and physical properties of collagen extracted from moon jellyfish under neutral pH conditions. *Biosci. Biotechnol. Biochem.* **2015**, *79*, 1603–1607. [CrossRef] [PubMed]
60. Blumlein, A.; McManus, J.J. Reversible and non-reversible thermal denaturation of lysozyme with varying pH at low ionic strength. *Biochim. Biophys. Acta* **2013**, *1834*, 2064–2070. [CrossRef] [PubMed]
61. Harrington, W.F.; Josephs, R.; Segal, D.M. Physical chemical studies on proteins and polypeptides. *Ann. Rev. Biochem.* **1966**, *35*, 599–650. [CrossRef] [PubMed]
62. Engel, J.; Bächinger, H.P. Structure, stability and folding of the collagen triple helix. *Top. Curr. Chem.* **2005**, *247*, 7–33. [CrossRef]
63. Heidemann, E.; Roth, W. Synthesis and investigation of collagen model peptides. *Adv. Polym. Sci.* **1982**, *42*, 143–203. [CrossRef]
64. Burjanadze, T.V. New analysis of the phylogenetic change of collagen thermostability. *Biopolymers* **2000**, *53*, 523–528. [CrossRef]
65. Sadowska, M.; Kolodziejska, I.; Niecikowska, C. Isolation of collagen from the skins of Baltic cod (*Gadus morhua*). *Food Chem.* **2003**, *81*, 257–262. [CrossRef]
66. Brønsted, H.V.; Carlsen, F.E. A cortical cytoskeleton in expanded epithelium cells of sponge gemmules. *Exp. Cell Res.* **1951**, *2*, 90–96. [CrossRef]
67. Wiens, M.; Koziol, C.; Batel, R.; Müller, W.E.G. Prolidase in the marine sponge *Suberites domuncula*: Enzyme activity, molecular cloning, and phylogenetic relationship. *Mar. Biotechnol.* **1999**, *1*, 191–199. [CrossRef] [PubMed]
68. Sethmann, I.; Wörheide, G. Structure and composition of calcareous sponge spicules: A review and comparison to structurally related biominerals. *Micron* **2008**, *39*, 209–228. [CrossRef] [PubMed]
69. Sudharsan, S.; Seedevi, P.; Saravanan, R.; Ramasamy, P.; Vasanth Kumar, S.; Vairamani, S.; Srinivasan, A.; Shanmugam, A. Isolation, characterization and molecular weight determination of collagen from marine sponge *Spirastrella inconstans* (Dendy). *Afr. J. Biotechnol.* **2013**, *12*, 504–511. [CrossRef]
70. Wood, A.; Ogawa, M.; Portier, R.J.; Schexnayder, M.; Shirley, M.; Losso, J.N. Biochemical properties of alligator (*Alligator mississippiensis*) bone collagen. *Comp. Biochem. Physiol. Part B* **2008**, *151*, 246–249. [CrossRef] [PubMed]
71. Rigby, B.J. Amino acid composition and thermal stability of the skin collagen of the Antarctic ice fish. *Nature* **1968**, *219*, 166–167. [CrossRef] [PubMed]
72. Minh Thuy, L.T.; Okazaki, E.; Osako, K. Isolation and characterization of acid-soluble collagen from the scales of marine fishes from Japan and Vietnam. *Food Chem.* **2014**, *149*, 264–270. [CrossRef] [PubMed]

73. Shimizu, K.; Cha, J.; Stucky, G.D.; Morse, D.E. Silicatein a: Cathepsin L-like protein in sponge biosilica. *Proc. Natl. Acad. Sci. USA* **1998**, *95*, 6234–6238. [CrossRef] [PubMed]
74. Li, Y.; Douglas, E.P. Effects of various salts on structural polymorphism of reconstituted type I collagen fibrils. *Colloids Surf. B Biointerfaces* **2013**, *112*, 42–50. [CrossRef] [PubMed]

Article

Bioinspiring *Chondrosia reniformis* (Nardo, 1847) Collagen-Based Hydrogel: A New Extraction Method to Obtain a Sticky and Self-Healing Collagenous Material

Dario Fassini [1,2], Ana Rita C. Duarte [1,2], Rui L. Reis [1,2] and Tiago H. Silva [1,2,*]

[1] 3B's Research Group—Biomaterials, Biodegradables and Biomimetics, University of Minho, Headquarters of the European Institute of Excellence on Tissue Engineering and Regenerative Medicine, AvePark, Barco, Guimarães 4805-017, Portugal; dario.fassini@gmail.com (D.F.); aduarte@dep.uminho.pt (A.R.C.D.); rgreis@dep.uminho.pt (R.L.R.)
[2] ICVS/3B's PT Government Associated Laboratory, Braga/Guimarães 4710-057, Portugal
*   Correspondence: tiago.silva@dep.uminho.pt; Tel.: +351-253-510-900

Received: 30 September 2017; Accepted: 16 November 2017; Published: 4 December 2017

Abstract: Collagen is a natural and abundant polymer that serves multiple functions in both invertebrates and vertebrates. As collagen is the natural scaffolding for cells, collagen-based hydrogels are regarded as ideal materials for tissue engineering applications since they can mimic the natural cellular microenvironment. *Chondrosia reniformis* is a marine demosponge particularly rich in collagen, characterized by the presence of labile interfibrillar crosslinks similarly to those described in the mutable collagenous tissues (MCTs) of echinoderms. As a result single fibrils can be isolated using calcium-chelating and disulphide-reducing chemicals. In the present work we firstly describe a new extraction method that directly produces a highly hydrated hydrogel with interesting self-healing properties. The materials obtained were then biochemically and rheologically characterized. Our investigation has shown that the developed extraction procedure is able to extract collagen as well as other proteins and Glycosaminoglycans (GAG)-like molecules that give the collagenous hydrogel interesting and new rheological properties when compared to other described collagenous materials. The present work motivates further in-depth investigations towards the development of a new class of injectable collagenous hydrogels with tailored specifications.

Keywords: marine collagen; hydrogel; collagen rheology; marine sponge GAG; marine biomaterials; *Chondrosia reniformis*

1. Introduction

Collagen is the most abundant protein of the extra cellular matrix and can be found in all Phyla [1] with the remarkable exception of Placozoa [2] and Rotifera [3]. Collagen is the natural cell scaffolding and has several domains that bind proteoglycans [4,5], growth factors [6] and other cell signalling molecules [7,8]. In this view, it is regarded as an ideal material for many applications dealing with human health and wellbeing, including regenerative medicine [9–16]. Although collagens sources are still mainly derived by mammals (bovine and porcine), scientists and entrepreneurs are being challenged by specific concerns regarding zoonosis, potential immunogenic reactions [14,15] as well as ethical and religious concerns [17]. For those reasons marine organisms in the last years have gained much popularity as potential alternative source of safer and more acceptable source of collagen [17].

The potential applications of collagens extracted from marine organisms include Tissue Engineering and Regenerative Medicine (TERM); wound healing, drugs and gene delivery/carrier; cosmetic, food industry as well as nutraceutics [17]. Once sufficiently purified, collagen used in Tissue Engineering and Regenerative Medicine is generally processed in order to obtain solid scaffolds or highly hydrated scaffolds; also knew as hydrogels. One of the most common steps during collagen processing consists in the introduction of artificial crosslinks in order to stabilize the structure and to control its degradation rate once grafted. Crosslinks can be achieved by: physical [18,19]; chemical [20–27]; enzymatic [28] treatments or, less frequently, by combining two different approaches [29,30].

Chondrosia reniformis Nardo, 1847 is a common marine demosponge that lives in the shallow costs of the Mediterranean Sea and the South-West cost of the Atlantic Ocean [31]. The skeleton elements of the class Demospongiae are composed of an inorganic and an organic component. The inorganic component consists of a broad array of siliceous spicules [32] while spongin, collagen and chitin are the three constituents of the organic skeleton [33,34]. In contrast to most demosponges skeletons, which are constituted of spicules in association with one or more organic components, *C. reniformis* lacks both endogenous spicules and spongin/chitin elements. By contrast *C. reniformis* is particularly rich in collagen. This species has attracted the attention of scientists for the capability to modulate its mechanical properties by acting on the collagen crosslinks [35–37] and, since its high content, as an alternative source of collagen [11,38–41]. Intact fibrils, which resemble type I collagen [38], can be isolated using 4% EDTA [42], the alkali method described by Swaschtek and colleagues [11] or using a solution containing both EDTA and 2-mercaptoethanol based on a slightly modified protocol [37] firstly developed by Matsumura and coworkers for echinoderms [43]. *C. reniformis* collagen is particularly insoluble in acidic organic and inorganic media [38], and so far collagen/gelatin have been obtained only after trypsin digestion of sponge material [42]. Nonetheless, a recently reported study refers the use of water acidified with CO_2 for the successful extraction of collagen/gelatin from this sponge species [40,41], in a mixture with other unidentified compounds.

The aim of the present work was to establish a new method to increase the amount of collagen that can be extracted from the marine sponge *C. reniformis* preserving intact fibrils and able to directly produce a collagenous hydrogel. The extraction was performed separately for the two different sponge regions: the ectosome (Ec; the outer cortical layer), and the choanosome (Ch) that constitutes the main bulk of sponge. Independent collagen extractions from the two regions were run in parallel in order to evaluate potential differences in the characteristics of hydrogels given that type IV collagen is more abundant in the ectosome than in the choanosome [44].

2. Results

2.1. Collagen Extraction

In comparison with the method described by Fassini et al. [37] and herein referred as the reference procedure, a pre-treatment step in phosphate buffer saline/ethylenediaminetetraacetic acid (PBS/EDTA) followed by a significantly longer period of incubation in disaggregating solution (DS) was necessary to directly obtain the collagenous hydrogels (Scheme 1).

During the collagen extraction process, sponge pieces significantly shrink and take on a more whitish color during the treatment with EDTA. While still in DS the material has a homogeneous liquid consistency similar to that observed during the reference procedure (Figure 1a), during the dialysis the new material precipitates (Figure 1b).

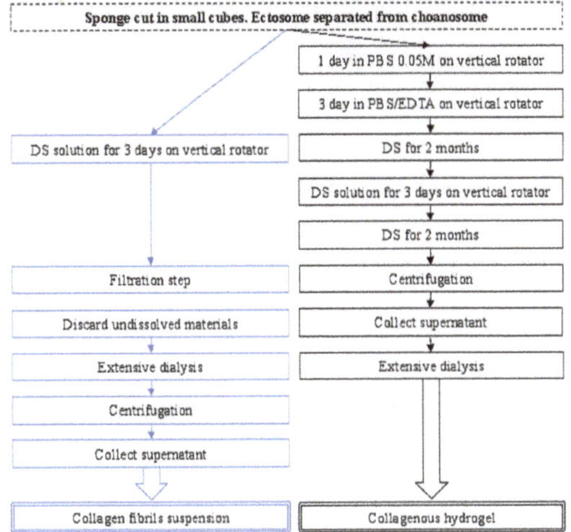

Scheme 1. Schematic representation of the extraction procedure as described in Fassini et al. 2014 [37] (blue squared text, left side) and the new extraction procedure (black squared text, right side). The initial step is the same for both the treatments (broken line); different treatments with similar aims are put on the same line in order to highlight the similarities and differences between the two extraction procedures. The final results of the two protocols is showed in the double bounded squares. PBS = phosphate buffer saline; EDTA = ethylenediaminetetraacetic acid; DS = disaggregating solution.

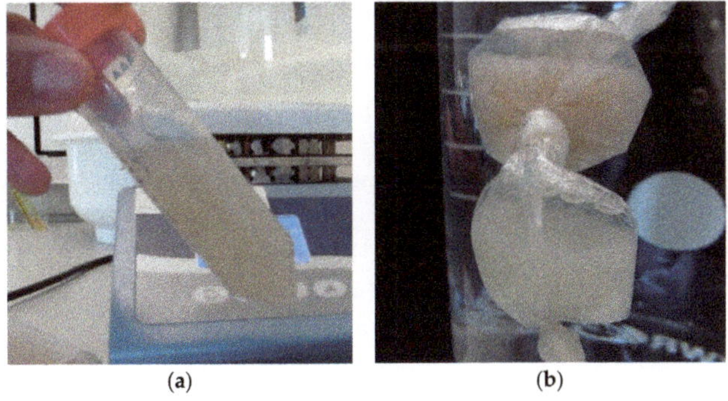

Figure 1. Image of a choanosome sample before (**a**) and during (**b**) the dialysis step.

At the end of the dialysis process, the collagenous materials obtained either from Ch or Ec consist of a sticky jelly-like body (see Video S1 in Supplementary Materials). The obtained materials were dramatically different from that where the pre-EDTA treatment was skipped. While the latter material was still liquid, the former material was almost solid and able to adhere moderately to plastic and nitrile materials (such as containers and gloves) and remarkably also onto wet surfaces. The materials showed also evident self-healing properties being able to regenerate its integrity after it has been disrupted. Following repeated manipulations aimed to stretch the collagenous gel-like body, it recovers the initial shape when the force is withdrawn (see Video S1 in Supplementary Materials). Furthermore, the material extracted could be concentrated by centrifugation or diluted by adding a water-based buffer and generous shaking.

A large variability in terms of concentrations (dry weight/mL) was observed. The range was comprised between 16.8 and 6.92 mg/mL. Higher concentrations were obtained from Ec samples given the fact that Ec is characterized by a higher abundance of both non fibrillar [44] and fibrillar [45,46] collagen than choanosome. In this view the higher yield of extracted material from Ec (126–109 mg/g of fresh sponge tissue) compared to Ch (77–50 mg/g of fresh sponge tissue) was not surprising. Remarkable differences in the concentrations were also observed among different hydrogels obtained from the same batch (Table 1, confront batch 2a and 3a). The variability is likely produced during the dialysis step. Indeed the use of dialysis membrane tubes does not allow to control the water uptake. Significant discrepancies were discovered between the total dry mass and the collagen/protein content (Table 1).

Table 1. Collagen and total proteins quantification of Ch batches determined by using Sircol and Bicinchoninic Acid (BCA) assays respectively. Batch 2a and 2b were obtained from an independent extraction starting from a different specimen.

Batch	Dry Weight (mg/mL)	Sircol (µg/mL)	BCA (µg/mL)
1	6.92	17.02	133.04
2	10.6	7.94	87.91
3	9.6	6.59	66.12

Collagen content of Ch batches extracted from different specimens was not directly related to the dry content and varied between 7 and 17 µg/mL while proteins range was comprised between 66 and 133 µg/mL (Table 1), being the latter quantified as smaller than 2% of the extract and collagen as roughly 10% of the total protein mass. This suggests that some other compounds like Glycosaminoglycans (GAG, see below) and, possibly, salts were present in the extracts. Moreover, it is highly plausible that sponge collagen did not react extensively with the chemicals in either the both Sircol and BCA assay due to the presence of non-solubilized collagen fibrils, the high HLys content and the presence of attached glycans [39].

2.2. Sodium Dodecyl Sulphate Polyacrylamide Gel Electrophoresis (SDS-PAGE)

The obtained extracts were submitted to electrophoretic analyses to better characterize their biochemical composition. Conventional collagen extracts and most of proteins are easily stained with Coomassie dye also at rather low concentrations. However, this was not the case of *C. reniformis* collagens, which were detected only after staining the gels with alcian blue (Figure 2).

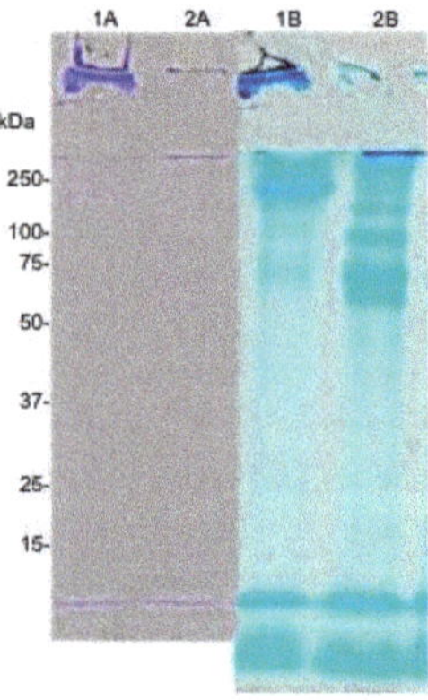

Figure 2. SDS-PAGE of ectosome (Ec) extract after one month (lane 1A, 1B) or two months (lane 2A, 2B) in DS stained with Coomassie R-250 (1A, 2A) or alcian blue (1B, 2B). A volume corresponding to 250 µg of material was added to each well.

Indeed, at 1 mg/mL, a concentration generally sufficient to well resolve the characteristic chain composition of fibrillar collagens, sponge collagen bands were rather evanescent when stained with R-250 Coomassie, but well visible when stained with alcian blue.

Comparing the protein profile one month after the start of the incubation (Figure 2—compare lanes 1A, 1B) and at the end of the second month (Figure 2—compare lanes 2A, 2B) in the DS, it is possible to see the effects of a prolonged incubation. Most of the collagen after one month in the DS is still in the form of large fibrils that cannot be solubilized and thus remain trapped in the stacking gel. However, after an additional month in DS, collagen fibrils are almost completely solubilized in their single components. In our gels we observed at least seven bands in the high molecular weight region plus a broad band that runs faster than the frontline. While some bands did not change over time (i.e., at one and two months of incubation), others appeared only after the complete incubation in DS solution, suggesting a significant increase in the solubility of collagen fibrils. Indeed, the fibrils trapped in the stacking gel (Figure 2—lane 1B), which are likely associated in small bundles as a result of interactions with other proteins, completely disappear between month 1 and 2 (Figure 2—lane 2B). On the other hand, we found a new band that was blocked at the beginning of the stacking gel. Considering its very high molecular weight, together with results detailed in Section 2.3 (below), i.e., the sensitivity to both Coomassie and alcian blue dyes as well as its resistance to a long pepsin digestion, this band was likely to have originated from the presence of isolated collagen fibrils.

2.3. Effect of Pepsin Digestion

Samples were treated with pepsin in order to evaluate whether some proteins or the collagen telopeptides were responsible for the gel consistency of the obtained materials and to confirm the collagenous materials of the bands observed in the SDS-PAGE.

We found that there were no qualitative macroscopic differences between untreated extracts and the ones treated with pepsin; on the other hand, the supernatant obtained after a rapid centrifugation, was significantly different. Supernatants of undigested samples formed a nearly 90° meniscus while the digestion caused a change in the meniscus angle that was >90° (Figure 3).

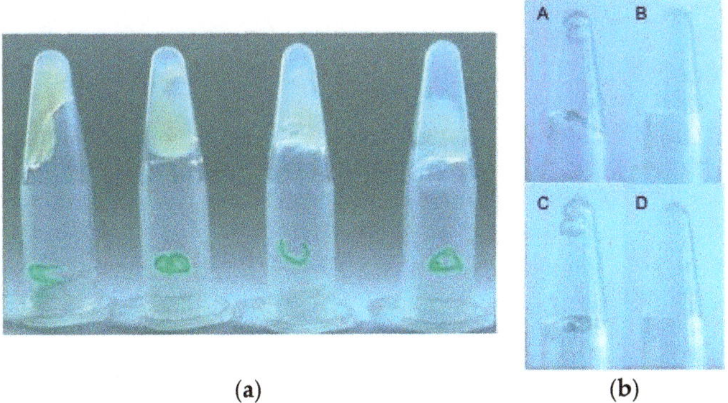

(a) (b)

Figure 3. Effects of pepsin digestion on choanosome (Ch) collagen extract. (**a**) Images of the materials soon after the digestion/control treatment; and (**b**) meniscus details of the supernatants obtained after the removal of the insoluble component. Tubes legend: A = Ec digested; B = Ec control; C = Ch digested; D = Ch control.

SDS-PAGE analysis was carried out on all the samples, including the supernatants obtained after the removal of the insoluble parts. In terms of composition, the material obtained from Ch was less complex than that extracted from Ec (Figure 4, compared undigested Ec and Ch).

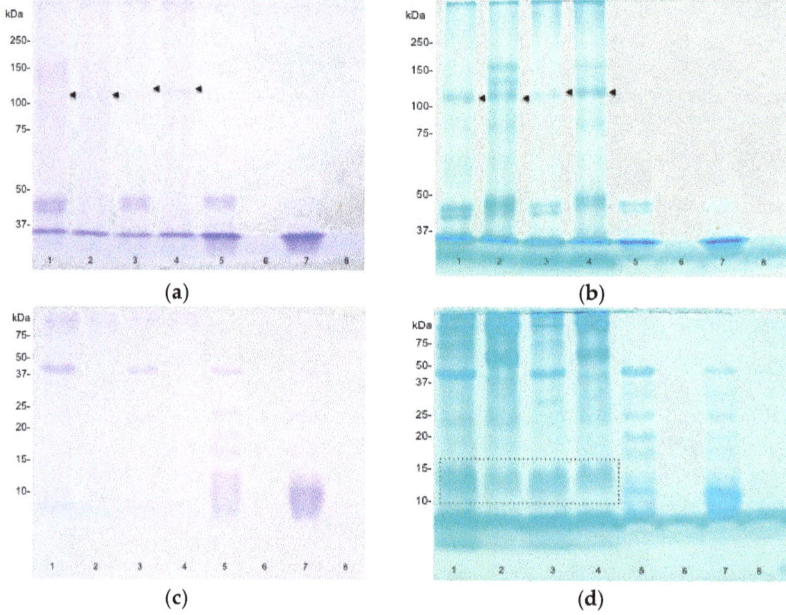

Figure 4. 7.5% (**a,b**) and 15% (**c,d**) sodium dodecyl sulphate-polyacrylamide gel electrophoresis (SDS-PAGE) stained with R-250 Coomassie (**a,c**) and alcian blue (**b,d**) of digested and undigested samples. Lanes 1/5 = digested Ec; 2/6 = undigested Ec; 3/7 = digested Ch; 4/8 = undigested Ch. Lanes 1–4 = pellet; 5–8 = supernatant. Pellets obtained after centrifugation were resuspended in 800 µL of loading buffer and freeze-dried supernatants in 80 µL; 40 µL of the resulting dispersions were added in each well. Arrowhead = presumptive collagen; broken line rectangle = possible glycan.

Moreover, the SDS-PAGE profile of collagens treated with pepsin revealed a significant difference with the untreated ones. In particular, several bands in the higher (>100 kDa) molecular weight appeared in the Ec (compare lane 12 s—tained with Coomassie of Figure 4) suggesting that pepsin treatment was able to solubilize different proteins. Both Ec and Ch-solubilized collagens had a lower molecular weight when treated with pepsin suggesting that the enzyme was able to remove collagens telopeptides. Our data suggests that a slight difference between Ec and Ch collagens does exist, while both seems to be homotrimeric. Pepsin treatment seems to increase the solubility and the Coomassie sensitivity of other kind of collagens present in the ectosome. Considering the evident effect of pepsin treatment on the supernatant (Figure 4) it is likely to be that some pepsin-sensitive molecules not stained with the employed dyes were present in the samples.

The combined analysis of Coomassie/alcian blue staining (Figure 4) revealed the presence of a glycan that, since it was not found in the supernatant, should be stably associated with the material itself.

The materials obtained from either Ch or Ec were completely dissolved after few hours when treated with papain at 50 °C. After the precipitation of GAGs and the following freeze-drying step, quite a large amount of material was obtained. About 3 mg per mL of starting materials (Ec 14.4 mg/mL and Ch 10.6 mg/mL) were recovered; however, when run trough Tris/borate/EDTA polyacrylamide gel electrophoresis (TBE-PAGE), only a small band was detected (Figure 5).

Neither of the bands were susceptible to an extensive incubation with Chondroitinase ABC (data not shown), and thus the nature of this molecules could not be assessed.

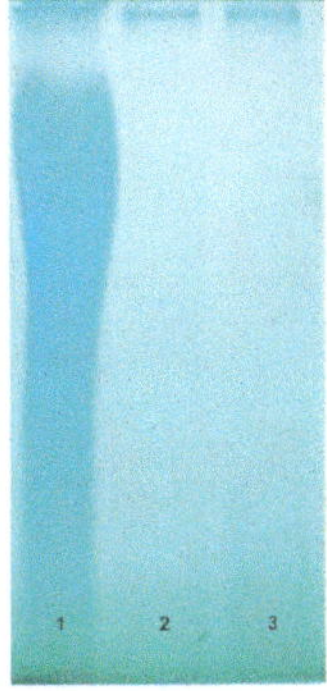

Figure 5. 10% Tris/borate/EDTA polyacrylamide gel electrophoresis (TBE-PAGE) of glycosaminoglycans (GAG) extracted from choanosome (Ch; lane 2) and ectosome (Ec; lane 3), after staining with alcian blue. A similar quantity (170 µg) of shark chondroitin sulfate was loaded into lane 1 as reference.

2.4. Fourier-Transformed Infrared Spectroscopy (FTIR)

Collagens extracted from both the Ec and Ch were characterized in their IR absorption spectra. Overall, the obtained spectra (Figure 6) are in accordance with the results described in literature [38,42].

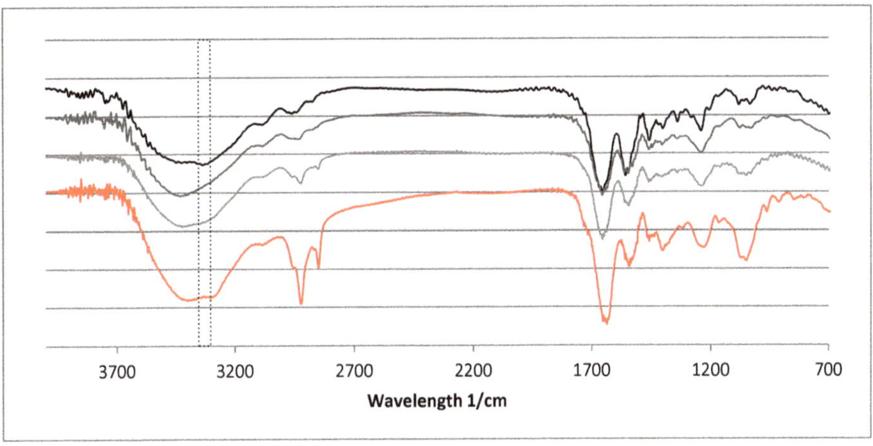

Figure 6. FTIR spectra of Sigma bovine type I collagen (black); Ec (dark grey); Ch (light grey); Ch supernatant (red). Broken line rectangle indicate Amide A region. Absorbance values have been shifted to facilitate the reading.

In more detail, the analysis of the obtained FTIR peaks indicated the presence of significant differences in the amide A region for both Ec and Ch. The asymmetric NH_2 stretching peak in the Ec samples was difficult to identify due to a nearby strong peak associated with the presence of hydrogen bonds. This result is again in accordance to the FTIR collagen spectra describe by Garrone and coworkers [42]. All other major peaks, listed in Table 2, correspond quite well type I collagen (Sigma Aldrich, Merck KGaA, Darmstadt, Germany), used as reference material.

The comparison of the FTIR spectra with Type I collagen (Sigma Aldrich, Merck KGaA, Darmstadt, Germany) revealed a substantial similitude in the peaks; finally the ratio of absorbance measured at $1235\ cm^{-1}$ and $1450\ cm^{-1}$, which were all >0.9, suggests the presence of native collagen triple helix in all the samples [49].

Table 2. List of representative FTIR peaks and the values obtained from FTIR analyses of Ch and Ec extracts. *S* = stretch; *B* = bend; *s* = symmetric; *as* = asymmetric; *HB* = hydrogen bonds.

Peak Wave Number/cm^{-1}			Assignment	Reference
Sigma Type I	**Ch**	**Ec**		
3414	3414	3421	NH *S*, coupled with H-bond	[47]
3325	3331		*as* NH	[38]
3080	3082	3082	NH *S*	[48]
2951	2956	2962	CHx	[38]
2926	2924	2935	*as* CH$_2$	[47]
	2852	2848	*s* CH$_2$	[47]
1666	1654	1660	C=O	[38]
1651	1647	1654	C=O *S*/*HB* coupled with COO-	[47]
1556	1560	1556	NH *B* coupled with CN *S*	[38]
1545	1540	1544		
1450	1448	1450	Pyrrolidine ring	[49]
1458	1458	1458	CH$_2$ *B*, CH$_2$ *S*	[38]
1400	1396	1404		
1339	1340	1340	CN *S*	[38]
1235	1240	1240	CH$_2$	[49]
1081	1082	1078	CO *S*	[38]
1033	1031	1033		

2.5. Rheology

Considering the unusual properties of the obtained collagenous materials and to better understand their mechanical performance the rheological properties of the collagenous materials isolated from both Ec and Ch were characterized using a rheometer. Both materials display a clear shear thinning behavior (Figure 7a,b respectively).

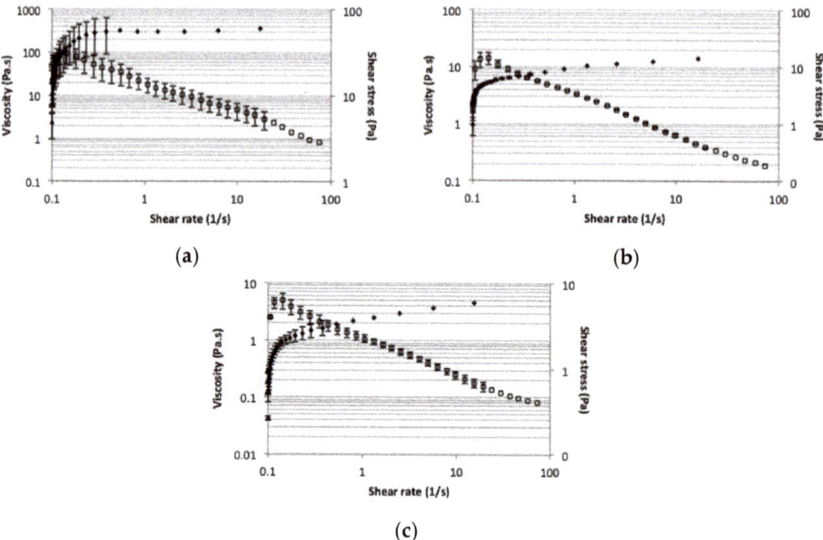

Figure 7. Flow curves of ectosome (Ec) and choanosome (Ch) subjected to increasing shear rates. (**a**) = Ec 14.4 mg/mL; (**b**) = Ch 9.6 mg/mL; (**c**) = Ch 8.2 mg/mL (Ch 9.6 mg/mL after centrifugation and resuspension). Square = viscosity; diamonds = shear stress. Values are expressed as the average of three repeated experiments; bars = standard deviation.

In all the samples the viscosity, which at low shear rates tends initially to increase, decreased almost constantly at higher shear rates. The sample extracted from the Ec was much more concentrated than the one obtained from the choanosome, which partially justifies the differences observed in absolutes values when viscosity is considered. Additionally, we cannot exclude that differences in the composition may also be responsible for the observed disparities in the viscosity values.

Figure 7b,c presents the results of the same Ch sample before and after a centrifugation/resuspension step. In this step, the sample was diluted in dH_2O 2:1, mixed thoroughly, centrifuged (6600 g for 1 min) to concentrate the collagenous part and then brought to the initial volume using dH_2O. Upon this procedure it is worth remarking that a significant amount of soluble material was extracted. Indeed, the dry weight decreased from 9.6 mg/mL to 8.2 mg/mL. This treatment revealed that the soluble molecules present in the supernatant play an important role in the rheological properties as their absence brings a significant change in the response of the material (Figure 7). In particular, the treatment decreased the viscosity at low shear strain and the overall shear thinning behavior.

The gel nature of the collagenous materials was confirmed by analyzing the storage (G′) and the loss modulus (G″) at frequency between 0.1 and 100 Hz. The oscillatory measures provide relevant information on the molecular arrangement of the network of the gel and allowed the determination of the solid-like (elastic) and liquid-like (viscous) regions of the material. In the frequency range studied, the gel formed by the choanosome-derived material is in the plateau zone of the viscoelastic region, i.e., for frequencies lower than 1 Hz both G′ and G″ are nearly constant (Figure 8).

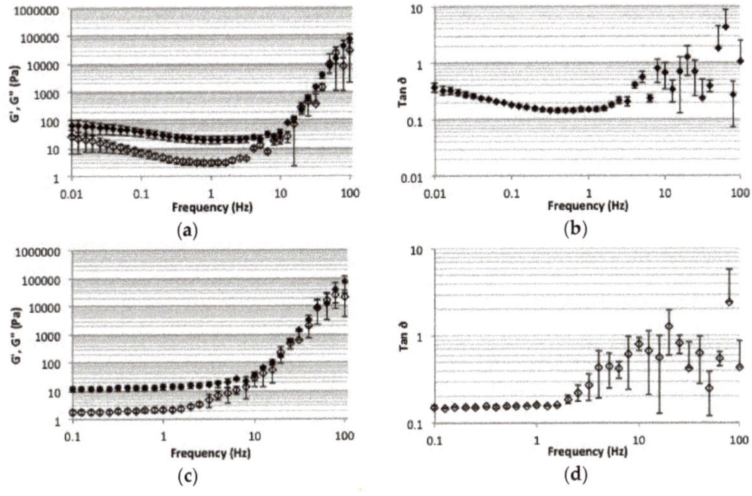

Figure 8. Storage and loss modulus measurements of choanosome (Ch) and ectosome (Ec) materials. Changes in the storage (G′, closed symbols), and loss (G″, open symbols) moduli at increasing frequency (**a,c**) and loss tangent (**b,d**) of Ec 14.4 mg/mL (**a,b**) and Ch 9.6 mg/mL (**c,d**). Values are expressed as the average of three repeated experiments; bars = standard deviation.

Furthermore, in this region the storage modulus is higher than the loss modulus, which is a characteristic behavior of a viscoelastic fluid. The G″/G′ ratio, called the loss tangent (*Tan ∂*), gives a clear indication of the solid/liquid responses. Values lower than 1 indicate a more solid behavior while at values >1 a liquid-like behavior is prevalent. The curve shown in Figure 8 revealed that at 1.5 Hz the material starts to have a more liquid response. At higher values we observe clear crossing points at 20 and 80 Hz. This phenomenon is related with the entanglements of collagen chains [50]. The entanglements limit the mobility of the chains contributing to a stiffer structure, and hence a higher loss tangent. From our observations the *Tan ∂* values suggest that the material retained a more solid response also at high frequencies. The storage modulus of Ec was also higher than the loss modulus, at low frequencies (Figure 8).

Here we noticed an apparent decrease in both G′ and G″, with the latter slightly more pronounced than the former as demonstrated by the *Tan ∂* values between 0.01 and 0.15 Hz. Ch samples (similarly to the Ec ones) at higher frequencies tend to remain at *Tan ∂* values lower than 1 although rapid changes in the *Tan ∂* values have been observed.

Following these observations, which suggest that the materials have a thixotropic behavior, specific rheological tests were further performed to confirm it. The thixotropic behavior of the Ch sample, less complex in terms of composition with respect to the Ec, was thus evaluated by the hysteresis loop obtained after a two-step experiment in which ramp-up and ramp-down experiments were performed. A large hysteresis area, as it can be observed in Figure 9, indicates the thixotropic properties of the gel, namely the ability of the gel to recover the initial structure after a certain stress is applied [51].

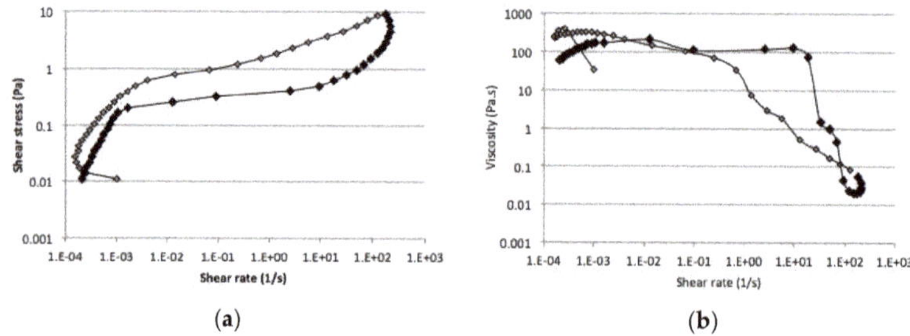

(a)   (b)

Figure 9. Ramp-up and ramp-down experiment showing the thixotropic property of the collagenous materials. Shear stress/rate (**a**) and viscosity/shear rate (**b**) curves of coanosome (Ch) 9.6 mg/mL in a ramp up/down experiment. Open grey symbols = ramp up; closed black symbols = ramp down. Lines connecting the points have been added, the points represent the average values of three different experiments. Standard deviations bars have been omitted to facilitate the reading of the graphs.

The thixotropy of a material is also characterized by the ability to restore its original viscosity after a stress has been applied. Figure 10 presents the thixotropic behavior of the Ch sample in terms of viscosity in a three-step experiment.

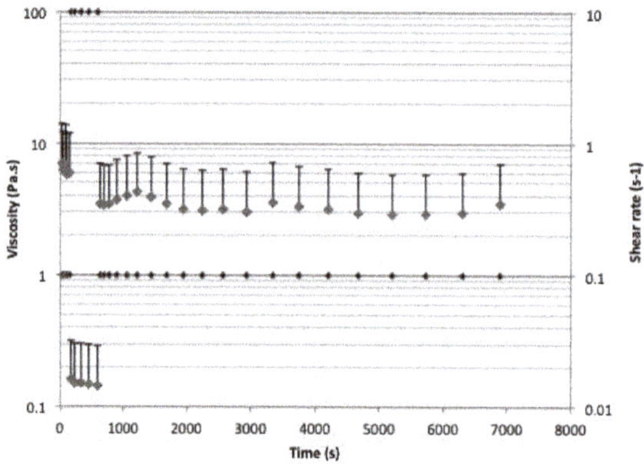

Figure 10. Viscosity values (grey dots) of coanosome (Ch) 9.6 mg/mL in a three-step experiment. Black dots = shear rate. Values are expressed as the average of three repeated experiments; bars = standard deviation.

We observed that during the ramp down phase, after a first rapid increase of viscosity, the values start to decrease and at very low shear rates are eventually much higher than the ones observed for higher shear rates, in which the viscosity should be around 10 Pa·s.

On the other hand, we found that after a steady increase of shear rate and a following decrease the material was not able to fully recovery its initial viscosity (Figure 10). However the evident initial loss of viscosity observed in the first phase of the experiments suggests that: small but prolonged shear stresses are able to modify the inner structure of the material suggesting that a much longer period to reach a viscosity equilibrium. In this view the viscosity values obtained after the stronger shear stress was removed are likely the real viscosity value of the material at the specific shear stress of $0.1~\text{s}^{-1}$.

3. Discussion

Collagen, which constitutes the bulk of the extracellular matrix of most of the animals, provides, among others features, the mechanical support of tissues and serves as well as anchoring points for cells and other molecules [52]. In this view, it is not surprising that it has long been proposed as a natural material for tissue engineering and regenerative medicine [53,54]. Among lower invertebrates, and sponges in particular, *C. reniformis* has been considered an interesting source of alternative and biocompatible collagen given its high collagen content [42] and for its interactions with silicic acids (Heinemann et al., 2007b). Moreover, among Porifera, *C. reniformis* is the reference model of dynamic collagenous tissues, which are characterized by the presence of labile and variable collagen crosslinks, and has been extensively investigated [35–37,55].

Standard acidic treatments, able to solubilize most of animal fibrillar collagens, have proven to be inefficient in the extraction of *C. reniformis* collagen and other dedicated protocols have been developed [11,37,41,42,56].

Our method is a substantial modification of the procedure described in Fassini et al. [34] and directly produced a collagenous hydrogel. In terms of FTIR spectra the collagenous hydrogels obtained were similar to data present in the literature. In particular, Ec collagen was very similar to the spectra provided by Garrone and coworkers [42], while Ch profiles were similar to what reported by Heinemann and coworkers [38]. Considering the fact that Garrone investigated the cortex (ectosome) while Heinemann investigated the whole sponge (where choanosome mass is predominant) our results are in line with both the previous publications. Our results provide evidence that some slow chemical modifications, able to increase the dissociation of collagen fibres, occur during the incubation in DS. Furthermore, we observed that the acidic pre-treatment with EDTA, which caused a significant bleaching and shrinkage of the mesohyl pieces, is necessary to obtain the sticky gel materials. Indeed the longest incubation period itself was not sufficient and the final product was a liquid collagen suspension identical to what described by Fassini et al. [37]. Given that acidic treatments do not influence the properties of *C. reniformis* collagen [38], the chelation activity on divalent ions of EDTA is most likely involved in the process. A solution of 4% EDTA is proven to influence the integrity of collagen bundles and to promote the formation of fibrils suspension [42], possibly by competing with the formation of the labile and calcium-dependent bonds that stabilize sponge collagen [36,37]. On the other side, our samples were exposed to significantly lower concentrations of EDTA, which apparently was not enough to dissociate the tissue. The significant increase in the solubilisation of the collagenous materials that was observed after two months is not clear. In the DS solution, EDTA and 2-mercaptoethanol are the active molecules that are mostly involved in the chelation of collagen stabilizing ions and in the reduction of disulphide bonds [43]. The fact that the solubilisation of the materials in reducing conditions occurred only after at least one month in DS, together with the relatively short half-life of 2-mercaptoethanol, signals that some other slow reaction occurred in the mixture during the incubation period. The nature of this possible reaction(s) is also still unclear.

The materials obtained from the two different regions of the sponge have quite similar compositions although material obtained from Ch region has a higher collagen/non-collagen ratio than the one obtained from the Ec. The SDS-PAGE profile of digested samples revealed the presence of hydrolysis products, one single putative band for collagen at around 110 kDa and another three main bands of other unidentified proteins that appears to be stably associated with collagen. Indeed, proteins such as pepsin and peptides originating from the enzymatic digestion were more abundant in the supernatant obtained after the centrifugation than in the pellets while, on another hand, no proteins were found in the supernatant of untreated samples. Collagen extracted and solubilized from sponge was composed of a homotrimeric chain of around 105/115 kDa, the small molecular weight differences found between digested and undigested collagens suggest that the telopeptides are maintained during the extraction while differences between Ec and Ch might result from a different glycosylation level, different transcriptions, or by the presence of two different collagen types. Indeed, while the presence of type I collagen has been demonstrated in the sponge *Ircina fusca* [57], recent transcriptomic data from several sponges revealed the presence of the fibrillar type XI collagen [58].

The gel structure of both Ch and Ec was also transferred to the surrounding media and can be disrupted by the enzymatic digestion suggesting that some pepsin-sensitive protein is involved in the gel structure formation. However our efforts to investigate the nature of the molecules present in the supernatant were not conclusive.

The rheological properties of the materials obtained are difficult to compare with other described collagenous gels due to the complex nature of the extracts obtained. Moreover, *C. reniformis* collagen fibrils are rather different from type I collagen having an aspect ratio of 1:5000 [35], thus significantly inhibiting a proper comparison with the shorter and bigger mammal type I collagens.

Our data demonstrates the shear thinning and pseudoplastic properties of the materials, which exhibited gel like behavior. Moreover, considering the thixotropic behavior and the dramatic increase of its viscosity once the shear rate is decreased (i.e., recovering the viscosity properties when the external stimulus—stress—is removed), the possibility of modulating the mechanical properties while keeping the collagenous matrix unaltered suggests that the materials might be further investigated in order to design an injectable collagenous hydrogel.

In the recent years much effort has been focussed on collagenous hydrogels and on the strategies to control their biochemical and physical properties [9,59,60]. Hydrogels are particularly interesting for several tissue-engineering applications given their capability to provide a suitable substrate for cell growth and encapsulation and their high water content [61,62]. Other interesting and regarded properties of new designed hydrogels concerns their injectability and the capability to strongly adhere to wet surfaces [60], as well as the possibility to control their stiffness with simple changes in the pH, temperature or ionic strength, remaining inside the physiological boundaries of human tissues. In this sense, the extracts obtained with the methodology proposed in this work may see exciting developments in the future, namely in the context of advanced tissue regeneration therapies.

4. Materials and Methods

4.1. Sponge Sampling

C. reniformis specimens were collected at Paraggi (Eastern Ligurian Sea, 44°18′40″ N, 9°12′46″ E), transferred in a thermic bag to a tank filled with artificial sea water (Instant Ocean®, Blacksburg, VA, USA) and left rest for at least one night. Specimens were then frozen at −20 °C until further processing.

4.2. Collagen Extraction

Sponges were thawed and cut in small (1 × 1 × 1 mm) cubes with a scalpel; the outer brownish layer—the ectosome (Ec)—was separated from the inner yellowish core—the choanosome (Ch). Briefly 5 g of dry material (Ec or Ch) were transferred into two 50 mL tubes and in each tube 40 mL of 5× phosphate buffer saline (PBS, Sigma Aldrich, Merck KGaA, Darmstadt, Germany) were added before putting the samples on a vertical rotator at room temperature (RT). After 24 h, 10 mL of supernatant were removed and 100 mg ethylenediaminetetraacetic acid (EDTA, Sigma Aldrich) per gram of sponge material were added in the each tube. After three days on the vertical rotator, the pieces were transferred into a new tube containing the disaggregating solution (DS) composed of 0.1 M Tris, 0.5 M NaCl, 0.05 M EDTA, and 0.2 M 2-mercaptoethanol. Ten milliliters of DS/g of fresh sponge material were used, and samples were left in DS for two months at RT. The resulting mixtures were then centrifuged (15 min at 500 g) to remove undissolved parts and the extracts obtained were then extensively dialyzed against dH_2O (ratio about 1:1000, two changes per day for 5 days) using a 14 kDa molecular weight cut off (MWCO) membrane tubing (Sigma Aldrich, Darmstadt, Germany) in order to remove all the 2-mercaptoethanol.

4.3. Collagen and Protein Quantification

Sircol Collagen Assay (Biocolor®, Carrick, UK) and BCA assay (MicroBCA protein detection kit, Pierce®, Thermo fisher Scientific, Waltam, MA, USA) were performed on the extracts from the choanosome according to the indications provided by the manufacturer, to quantify the amount of collagen and total proteins in the extracts, respectively. Readings were performed on a plate microreader (Synergy HT, Bio-Tek Instruments, Winooski, VT, USA), with calibration curve and concentrations of collagen/proteins being obtained by using the mean of at least three values for each standard/sample. Both Sircol and BCA assay were performed only on Ch hydrogels. The total amount of the extracted material was calculated by weighting standard amounts of freeze-dried material (mg/mL) and multiplying the values for the final volumes.

4.4. Enzymatic Digestions of Collagen and GAGs Extraction

In total 200 µL of extracts (in dH_2O) obtained either from Ch or Ec were mixed with 200 µL of pepsin (porcine pepsin, Sigma) solution (2.1 mg/mL in 6 mM HCl). The digestion was performed for 72 h at 37 °C. After the digestion, the samples were observed and photographed capsize-down to check the effect of the digestion. Samples were then centrifuged for 15 min at 17,000 g to better separate the collagenous part from the above supernatant. Both supernatants and the pellets were freeze dried and stored at −80 °C for further experiments. Control samples were obtained in parallel in the absence of pepsin in order to exclude possible changes induced by the temperature, acidity and dilution.

GAG extraction: 3 mL of the extracted collagen was mixed with 5 mL of digestion solution (5 mg/mL of papain in 1 M $NaBH_4$, 0.05 M NaOH) and placed at 45 °C overnight. Materials were completely dissolved after 2 h and GAG precipitation was performed slowly adding trichloroacetic acid (strong reaction) to reach a 50% vol/vol concentration according to the procedure reported by Nandini et al. [63]. The powders obtained were stored at RT for further characterization steps.

GAGs digestion: 10 mg of precipitated GAG were resuspended in 300 µL of PBS; 30 µL of 5 U/mL chondroitinase ABC (Sigma) were then added and enzymatic digestion was performed at 37 °C during 24 or 72 h. At each endpoint GAG were separated from the enzyme as described by Silva et al. [64]; briefly: 150 µL of the GAGs/enzyme mix were heated for 25 min at 70 °C and then centrifuged at 12,000 g for 25 min. The pellet was discarded while supernatant containing digested GAGs was recovered, freeze-dried and kept for further analysis.

4.5. Electrophoresis

4.5.1. SDS-PAGE

All used reagents were purchased from Fluka (SDS-PAGE Kit, Thermo fisher Scientific, Waltam, MA, USA). Collagen batches were analysed after one month in DS through a fast dialysis (2 step at 1:10,000) and at the end of the second month in DS after the extensive dialysis above described. Samples were mixed 1:1 with loading buffer, boiled for 5 min at 100 °C and run in 1.5 mm 7.5% or 15% gels in a MiniProtean system (Biorad, Hercules, CA, USA).

Coomassie R-250 staining: gel was stained according to Laemmli [65]. Alcian blue staining: gels previously stained with Coomassie were treated with 12.5% thrichloroacetic acid for 30 min.;

after 4 washes in dH_2O gels were treated with 1% sodium metaperiodate in 3% acetic acid for 1 h. After numerous washes (until the discarded water was not reacting with 1% silver nitrate) gels were immersed in 0.5% alcian blue in 3% acetic acid for 4 h; excess of alcian blue was removed with extensive washes in 7% acetic acid.

4.5.2. Tris/Borate/EDTA Polyacrylamide Gel Electrophoresis (TBE-PAGE)

TBE-PAGE was prepared using $10\times$ TBE buffer (Bioland) and reagents from Fluka (SDS-PAGE Kit). The 1.5 mm gels were run in a MiniProtean system (Biorad©), using 0.1 M Tris/borate, 1 mM NaEDTA, and pH 8.3 buffer. GAG (2 mg/mL) were mixed (1:5) with 2 M sucrose (Sigma) 0.02% bromophenol blue (Sigma) in $1\times$ TBE. Gels were run at constant 60 V for 15 min and then at 150 V for 60 min at RT. After, gel was stained with 0.5% alcian blue in 2% acetic acid for 30 min; excess of staining dye was removed by rinsing the gel in 2% acetic acid for 30 min.

All the staining/washes steps were performed on an orbital shaker at room temperature.

4.6. FTIR

FTIR analysis was performed on freeze-dried samples. Dried materials were mixed with potassium bromide (Sigma), crushed with a pestle and processed into a thin pellet using a hand press. Spectra were recorded from 32 scans with a resolution of 2 cm^{-1} from 4000 to 700 cm^{-1} using a Shimadzu IR Prestige 21 Infrared spectrometer (Shimatzu, Kyoto, Japan).

4.7. Rheology

The rheological properties of the collagenous materials obtained after the dialysis process were investigated by means of a Kinexus Prot rheometer (Malvern instruments, Malvern, UK). Steady-state flow measurements were carried out using a controlled-stress rheometer fitted with parallel plate geometry with a 10-mm diameter (PU20 SR1740 SS) and 1-mm gap. The torque amplitude was imposed by using a logarithmic ramp of shear rate, in a range of 0.1 to 100 s^{-1}. Rheological measurements in oscillatory frequency sweep strain controlled were performed in using the same parallel plate geometry with a 1 mm gap. Gel rheological behavior with storage/loss moduli and cross-over analysis was assessed at a constant 1% strain in a frequency region from 0.01 to 100 Hz.

All experiments were performed at a controlled temperature of 37 °C and results represent the average of three measurements.

5. Conclusions

We described a new method to extract collagen from the marine sponge *C. reniformis* resulting in sticky collagenous hydrogels. The materials obtained from two sponge regions, ectosome and choanosome, have singular rheological properties, with the observed shear thinning behavior seeming to depend on the presence of collagen fibrils and other associated proteins/GAGs. The weak interaction between those macromolecules have shown to be reversible and induce a fast recovery to the rest state of the viscosity after manipulation, suggesting their possible use as an injectable hydrogel medium for biomedical applications. The material obtained should be, however, further investigated to completely characterize its content and to fully address the role of the soluble compound(s). The data presented in this study represent a detailed starting point for further investigations aimed at producing tailored hydrogels with different interesting properties through a biomimetic approach.

Supplementary Materials: The following are available online at www.mdpi.com/1660-3397/15/12/380/s1, Video S1: demonstration of the adhesive and self-healing properties of the collagenous materials obtained from the choanosome.

Acknowledgments: The authors gratefully acknowledge the financial support from the European Union Seventh Framework Programme (FP7/2007–2013) under grant agreement ERC-2012-ADG 20120216-321266 (ERC Advanced Grant project ComplexiTE), as well as from the European Regional Development Fund (ERDF) under the projects "Accelerating tissue engineering and personalized medicine discoveries by the integration of key enabling nanotechonologies, marine-derived biomaterials and stem cells" (NORTE-01-0145-FEDER-000021), supported by Norte Portugal Regional Operational Program (NORTE 2020), under the PORTUGAL 2020 Partnership Agreement, and 0687_NOVOMAR_1_P, co-financed by transborder cooperation programme POCTEP. The authors are also thankful to the Area Marina Protetta Portofino (Italy) for permission to collect sponge specimens and to Daniela Candia (University of Milan, Italy) and Marco Giovine (University of Genoa, Italy) for the logistical support on the sponge sampling and immediate processing. We are grateful to the two anonymous referees for improving the quality of the present article.

Author Contributions: D.F., A.R.C.D. and T.H.S. conceived and designed the experiments; D.F. and A.R.C.D. performed the experiments; D.F., A.R.C.D. and T.H.S. analysed the data; R.L.R. and T.H.S. contributed in reagents/materials/analysis tools; D.F. wrote the draft of the manuscript and all the authors reviewed it to reach the final version.

Conflicts of Interest: The authors declare no conflict of interest.

References

1. Exposito, J.; Valcourt, U.; Cluzel, C.; Lethias, C. The Fibrillar Collagen Family. *Int. J. Mol. Sci.* **2010**, *11*, 407–426. [CrossRef] [PubMed]

2. Syed, T.; Schierwater, B. The Evolution of the Placozoa: A New Morphological Model. *Senckenberg. Lethaea* **2002**, *82*, 315–324. [CrossRef]

3. Clément, P. The Phylogeny of Rotifers: Molecular, Ultrastructural and Behavioural Data. *Hydrobiologia* **2003**, *255*, 527–544. [CrossRef]

4. Junqueira, L.C.; Montes, G.S. Biology of collagen-proteoglycan interaction. *Arch. Histol. Jpn.* **1983**, *46*, 589–629. [CrossRef] [PubMed]

5. Muller, L.J.; Pels, E.; Schurmans, L.R.; Vrensen, G.F. A new three-dimensional model of the organization of proteoglycans and collagen fibrils in the human corneal stroma. *Exp. Eye Res.* **2004**, *78*, 493–501. [CrossRef]

6. Schuppan, D.; Schmid, M.; Somasundaram, R.; Ackermann, R.; Ruehl, M.; Nakamura, T.; Riecken, E.O. Collagens in the liver extracellular matrix bind hepatocyte growth factor. *Gastroenterology* **1988**, *114*, 139–152. [CrossRef]

7. Czirok, A.; Rongish, B.; Little, C. Extracellular matrix dynamics during vertebrate axis formation. *Dev. Biol.* **2004**, *268*, 111–122. [CrossRef] [PubMed]

8. Di Lullo, G.; Sweeney, S.; Korkko, J.; Ala-Kokko, L.; San Antonio, J. Mapping the Ligand-binding Sites and Disease-associated Mutations on the Most Abundant Protein in the Human, Type I Collagen. *J. Biol. Chem.* **2001**, *277*, 4223–4231. [CrossRef] [PubMed]

9. Parenteau-Bareil, R.; Gauvin, R.; Berthod, F. Collagen-Based Biomaterials for Tissue Engineering Applications. *Materials* **2010**, *3*, 1863–1887. [CrossRef]

10. An, B.; Lin, Y.; Brodsky, B. Collagen interactions: Drug design and delivery. *Adv. Drug Deliv. Rev.* **2016**, *97*, 69–84. [CrossRef] [PubMed]

11. Swatschek, D.; Schatton, W.; Kellermann, J.; Muller, W.; Kreuter, J. Marine sponge collagen: Isolation, characterization and effects on the skin parameters surface-pH, moisture and sebum. *Eur. J. Pharm. Biopharm.* **2002**, *53*, 107–113. [CrossRef]

12. Kim, S.; Mendis, E. Bioactive compounds from marine processing byproducts—A review. *Food Res. Int.* **2006**, *39*, 383–393. [CrossRef]

13. Xinrong, P.; Ruiyue, Y.; Haifeng, Z.; Qiong, L.; Zhigang, L.; Junbo, W.; Yong, L. Preventive effect of marine collagen peptide on learning and memory impairment in SAMP8 Mice. *Food Ferment. Ind.* **2009**, *35*, 1–5.

14. Lynn, A.K.; Yannas, I.V.; Bonfield, W. Antigenicity and immunogenicity of collagen. *J. Biomed. Mater. Res. Part B Appl. Biomater.* **2004**, *71*, 343–354. [CrossRef] [PubMed]

15. Bentkover, S.H. The biology of facial fillers. *Facial Plast. Surg.* **2009**, *25*, 73–85. [CrossRef] [PubMed]

16. Xu, Y.J.; Han, X.L.; Li, Y. Effect of marine collagen peptides on long bone development in growing rats. *J. Sci. Food Agric.* **2010**, *90*, 1485–1491. [CrossRef] [PubMed]

17. Silva, T.; Moreira-Silva, J.; Marques, A.; Domingues, A.; Bayon, Y.; Reis, R. Marine Origin Collagens and Its Potential Applications. *Mar. Drugs* **2014**, *12*, 5881–5901. [CrossRef] [PubMed]

18. Weadock, K.; Miller, E.; Bellincampi, L.; Zawadsky, J.; Dunn, M. Physical crosslinking of collagen fibers: Comparison of ultraviolet irradiation and dehydrothermal treatment. *J. Biomed. Mater. Res.* **1995**, *29*, 1373–1379. [CrossRef] [PubMed]

19. Dewavrin, J.; Hamzavi, N.; Shim, V.; Raghunath, M. Tuning the architecture of three-dimensional collagen hydrogels by physiological macromolecular crowding. *Acta Biomater.* **2014**, *10*, 4351–4359. [CrossRef] [PubMed]

20. Berthod, F.; Hayek, D.; Damour, O.; Collombel, C. Collagen synthesis by fibroblasts cultured within a collagen sponge. *Biomaterials* **1993**, *14*, 749–754. [CrossRef]

21. Berthod, F.; Saintigny, G.; Chretien, F.; Hayek, D.; Collombel, C.; Damour, O. Optimization of thickness, pore size and mechanical properties of a biomaterial designed for deep burn coverage. *Clin. Mater.* **1994**, *15*, 259–265. [CrossRef]

22. Osborne, C.; Barbenel, J.; Smith, D.; Savakis, M.; Grant, M. Investigation into the tensile properties of collagen/chondroitin-6-sulphate gels: The effect of crosslinking agents and diamines. *Med. Biol. Eng. Comput.* **1998**, *36*, 129–134. [CrossRef] [PubMed]

23. Gough, J.; Scotchford, C.; Downes, S. Cytotoxicity of glutaraldehyde crosslinked collagen/poly(vinyl alcohol) films is by the mechanism of apoptosis. *J. Biomed. Mater. Res.* **2002**, *61*, 121–130. [CrossRef] [PubMed]

24. Zeugolis, D.; Paul, G.; Attenburrow, G. Cross-linking of extruded collagen fibers—A biomimetic three-dimensional scaffold for tissue engineering applications. *J. Biomed. Mater. Res. Part A* **2009**, *89*, 895–908. [CrossRef] [PubMed]

25. Sundararaghavan, H.; Monteiro, G.; Lapin, N.; Chabal, Y.; Miksan, J.; Shreiber, D. Genipin-induced changes in collagen gels: Correlation of mechanical properties to fluorescence. *J. Biomed. Mater. Res. Part A* **2008**, *87*, 308–320. [CrossRef] [PubMed]

26. Chandran, P.; Paik, D.; Holmes, J. Structural Mechanism for Alteration of Collagen Gel Mechanics by Glutaraldehyde Crosslinking. *Connect. Tissue Res.* **2012**, *53*, 285–297. [CrossRef] [PubMed]

27. Walters, B.; Stegemann, J. Strategies for directing the structure and function of three-dimensional collagen biomaterials across length scales. *Acta Biomater.* **2014**, *10*, 1488–1501. [CrossRef] [PubMed]

28. Orban, J.; Wilson, L.; Kofroth, J.; El-Kurdi, M.; Maul, T.; Vorp, D. Crosslinking of collagen gels by transglutaminase. *J. Biomed. Mater. Res. Part A* **2004**, *68*, 756–762. [CrossRef] [PubMed]

29. Wollensak, G.; Spoerl, E.; Seiler, T. Riboflavin/ultraviolet-a-induced collagen crosslinking for the treatment of keratoconus. *Am. J. Ophthalmol.* **2003**, *135*, 620–627. [CrossRef]

30. Fernandes-Silva, S.; Moreira-Silva, J.; Silva, T.; Perez-Martin, R.; Sotelo, C.; Mano, J.; Duarte, A.; Reis, R. Porous Hydrogels From Shark Skin Collagen Crosslinked under Dense Carbon Dioxide Atmosphere. *Macromol. Biosci.* **2013**, *13*, 1621–1631. [CrossRef] [PubMed]

31. Lazoski, C.; Solé-Cava, A.M.; Boury-Esnault, N.; Klautau, M.; Russo, C.A.M. Cryptic speciation in a high flow scenario in the oviparous marine sponge *Chondrosia reniformis*. *Mar. Biol.* **2001**, *139*, 421–429.

32. Uriz, M.; Turon, X.; Becerro, M.; Agell, G. Siliceous spicules and skeleton frameworks in sponges: Origin, diversity, ultrastructural patterns, and biological functions. *Microsc. Res. Tech.* **2003**, *62*, 279–299. [CrossRef] [PubMed]

33. Ehrlich, H.; Maldonado, M.; Spindler, K.; Eckert, C.; Hanke, T.; Born, R.; Goebel, C.; Simon, P.; Heinemann, S.; Worch, H. First evidence of chitin as a component of the skeletal fibers of marine sponges. Part I. Verongidae (demospongia: Porifera). *J. Exp. Zool. B Mol. Dev. Evol.* **2007**, *308*, 347–356. [CrossRef] [PubMed]

34. Aouacheria, A.; Geourjon, C.; Aghajari, N.; Navratil, V.; Deleage, G.; Lethias, C.; Exposito, J. Insights into Early Extracellular Matrix Evolution: Spongin Short Chain Collagen-Related Proteins Are Homologous to Basement Membrane Type IV Collagens and Form a Novel Family Widely Distributed in Invertebrates. *Mol. Biol. Evol.* **2006**, *23*, 2288–2302. [CrossRef] [PubMed]

35. Bonasoro, F.; Wilkie, I.C.; Bavestrello, G.; Cerrano, C.; Candia Carnevali, M.D. Dynamic structure of the mesohyl in the sponge *Chondrosia reniformis* (Porifera, Demospongiae). *Zoomorphology* **2001**, *121*, 109–121. [CrossRef]

36. Wilkie, I.C.; Parma, L.; Bonasoro, F.; Bavestrello, G.; Cerrano, C.; Candia Carnevali, M.D. Mechanical adaptability of a sponge extracellular matrix: Evidence for cellular control of mesohyl stiffness in *Chondrosia reniformis* Nardo. *J. Exp. Biol.* **2006**, *209*, 4436–4443. [CrossRef] [PubMed]

37. Fassini, D.; Parma, L.; Lembo, F.; Candia Carnevali, M.; Wilkie, I.; Bonasoro, F. The reaction of the sponge *Chondrosia reniformis* to mechanical stimulation is mediated by the outer epithelium and the release of stiffening factor(s). *Zoology* **2014**, *117*, 282–291. [CrossRef] [PubMed]

38. Heinemann, S.; Ehrlich, H.; Douglas, T.; Heinemann, C.; Worch, H.; Schatton, W.; Hanke, T. Ultrastructural Studies on the Collagen of the Marine Sponge *Chondrosia reniformis* Nardo. *Biomacromolecules* **2007**, *8*, 3452–3457. [CrossRef] [PubMed]

39. Heinemann, S.; Ehrlich, H.; Knieb, C.; Hanke, T. Biomimetically inspired hybrid materials based on silicified collagen. *Int. J. Mater. Res.* **2007**, *98*, 603–608. [CrossRef]

40. Barros, A.; Aroso, I.; Silva, T.; Mano, J.; Duarte, A.; Reis, R. Water and Carbon Dioxide: Green Solvents for the Extraction of Collagen/Gelatin from Marine Sponges. *ACS Sustain. Chem. Eng.* **2015**, *3*, 254–260. [CrossRef]

41. Silva, J.; Barros, A.; Aroso, I.; Fassini, D.; Silva, T.; Reis, R.; Duarte, A. Extraction of Collagen/Gelatin from the Marine Demosponge *Chondrosia reniformis* (Nardo, 1847) Using Water Acidified with Carbon Dioxide—Process Optimization. *Ind. Eng. Chem. Res.* **2016**, *55*, 6922–6930. [CrossRef]

42. Garrone, R.; Huc, A.; Junqua, S. Fine structure and physicochemical studies on the collagen of the marine sponge *Chondrosia reniformis* Nardo. *J. Ultrastruct. Res.* **1975**, *52*, 261–275. [CrossRef]

43. Matsumura, T.; Shinmei, M.; Nagai, Y. Disaggregation of connective tissue: Preparation of fibrous components from sea cucumber body wall and calf skin. *J. Biochem.* **1973**, *73*, 155–162. [PubMed]

44. Pozzolini, M.; Scarfì, S.; Mussino, F.; Ferrando, S.; Gallus, L.; Giovine, M. Molecular Cloning, Characterization, and Expression Analysis of a Prolyl 4-Hydroxylase from the Marine Sponge *Chondrosia reniformis*. *Mar. Biotechnol.* **2015**, *17*, 393–407. [CrossRef] [PubMed]

45. Bavestrello, G.; Cerrano, C.; Cattaneo-Vietti, R.; Sarà, M.; Calabria, F.; Cortesogno, L. Selective incorporation of foreign material in *Chondrosia reniformis* (Porifera, Demospongiae). *Ital. J. Zool.* **1996**, *63*, 215–220. [CrossRef]

46. Bavestrello, G.; Benatti, U.; Calcinai, B.; Cattaneo-Vietti, R.; Cerrano, C.; Favre, A.; Giovine, M.; Lanza, S.; Pronzato, R.; Sarà, M. Body polarity and mineral selectivity in the demosponge *Chondrosia reniformis*. *Biol. Bull.* **1998**, *195*, 120–125. [CrossRef] [PubMed]

47. Muyonga, J.; Cole, C.; Duodu, K. Characterisation of acid soluble collagen from skins of young and adult Nile perch (*Lates niloticus*). *Food Chem.* **2004**, *85*, 81–89. [CrossRef]

48. Tronci, G.; Doyle, A.; Russell, S.; Wood, D. Triple-helical collagen hydrogels via covalent aromatic functionalisation with 1,3-phenylenediacetic acid. *J. Mater. Chem. B* **2013**, *1*, 5478–5488. [CrossRef] [PubMed]

49. Silva-Júnior, Z.; Botta, S.; Ana, P.; França, C.; Fernandes, K.; Mesquita-Ferrari, R.; Deana, A.; Bussadori, S. Effect of papain-based gel on type I collagen—Spectroscopy applied for microstructural analysis. *Sci. Rep.* **2015**, *5*, 11448. [CrossRef] [PubMed]

50. Lai, G.; Li, Y.; Li, G. Effect of concentration and temperature on the rheological behavior of collagen solution. *Int. J. Biol. Macromol.* **2008**, *42*, 285–291. [CrossRef] [PubMed]

51. Ding, C.; Zhang, M.; Li, G. Rheological properties of collagen/hydroxypropyl methylcellulose (COL/HPMC) blended solutions. *J. Appl. Polym. Sci.* **2014**, *131*. [CrossRef]

52. Gelse, K. Collagens-structure, function, and biosynthesis. *Adv. Drug Deliv. Rev.* **2003**, *55*, 1531–1546. [CrossRef] [PubMed]

53. Lee, C.H.; Singla, A.; Lee, Y. Biomedical applications of collagen. *Int. J. Pharm.* **2001**, *221*, 1–22. [CrossRef]

54. Cen, L.; Liu, W.; Cui, L.; Zhang, W.; Cao, Y. Collagen Tissue Engineering: Development of Novel Biomaterials and Applications. *Pediatr. Res.* **2008**, *63*, 492–496. [CrossRef] [PubMed]

55. Fassini, D.; Parma, L.; Wilkie, I.C.; Bavestrello, G.; Bonasoro, F.; Carnevali, M.D.C. Ecophysiology of mesohyl creep in the demosponge *Chondrosia reniformis* (Porifera: Chondrosida). *J. Exp. Mar. Biol. Ecol.* **2012**, *428*, 24–31. [CrossRef]

56. Imhoff, J.M.; Garrone, R. Solubilization and Characterization of *Chondrosia reniformis* Sponge Collagen. *Connect. Tissue Res.* **1983**, *11*, 193–197. [CrossRef] [PubMed]

57. Pallela, R.; Boja, S.; Janapala, V. Biochemical and biophysical characterization of collagens of marine sponge, *Ircinia fusca* (Porifera: Demospongiae: Irciniidae). *Int. J. Biol. Macromol.* **2011**, *49*, 85–92. [CrossRef] [PubMed]

58. Riesgo, A.; Farrar, N.; Windsor, P.; Giribet, G.; Leys, S. The Analysis of Eight Transcriptomes from All Poriferan Classes Reveals Surprising Genetic Complexity in Sponges. *Mol. Biol. Evol.* **2014**, *31*, 1102–1120. [CrossRef] [PubMed]

59. Antoine, E.E.; Vlachos, P.P.; Rylander, M.N. Review of Collagen I Hydrogels for Bioengineered Tissue Microenvironments: Characterization of Mechanics, Structure, and Transport. *Tissue Eng. Part B Rev.* **2014**, *20*, 683–696. [CrossRef] [PubMed]

60. Ghobril, C.; Grinstaff, M.W. The chemistry and engineering of polymeric hydrogel adhesives for wound closure: A tutorial. *Chem. Soc. Rev.* **2015**, *44*, 1820–1835. [CrossRef] [PubMed]
61. Peppas, N.; Hilt, J.; Khademhosseini, A.; Langer, R. Hydrogels in Biology and Medicine: From Molecular Principles to Bionanotechnology. *Adv. Mater.* **2006**, *18*, 1345–1360. [CrossRef]
62. Zhu, J.; Marchant, E.R. Design properties of hydrogel tissue-engineering scaffolds. *Expert Rev. Med. Devices* **2011**, *8*, 607–626. [CrossRef] [PubMed]
63. Nandini, C.; Itoh, N.; Sugahara, K. Novel 70-kDa Chondroitin Sulfate/Dermatan Sulfate Hybrid Chains with a Unique Heterogenous Sulfation Pattern from Shark Skin, Which Exhibit Neuritogenic Activity and Binding Activities for Growth Factors and Neurotrophic Factors. *J. Biol. Chem.* **2004**, *280*, 4058–4069. [CrossRef] [PubMed]
64. Silva, C.; Novoa-Carballal, R.; Reis, R.; Pashkuleva, I. Following the enzymatic digestion of chondroitin sulfate by a simple GPC analysis. *Anal. Chim. Acta* **2015**, *885*, 207–213. [CrossRef] [PubMed]
65. Laemmli, U.K. Cleavage of structural proteins during the assembly of the head of bacteriophage T4. *Nature* **1970**, *227*, 680–685. [CrossRef] [PubMed]

 *marine drugs*

Article

Production, Characterization and Biocompatibility Evaluation of Collagen Membranes Derived from Marine Sponge *Chondrosia reniformis* Nardo, 1847

Marina Pozzolini [1,*], Sonia Scarfì [1], Lorenzo Gallus [1], Maila Castellano [2], Silvia Vicini [2], Katia Cortese [3], Maria Cristina Gagliani [3], Marco Bertolino [1], Gabriele Costa [1] and Marco Giovine [1]

[1] Department of Earth, Environment and Life Sciences (DISTAV), University of Genova, Via Pastore 3, 16132 Genova, Italy; soniascarfi@unige.it (S.S.); galluslorenzo@gmail.com (L.G.); marco.bertolino@edu.unige.it (M.B.); gabriele.costa@edu.unige.it (G.C.); mgiovine@unige.it (M.G.)

[2] Department of Chemistry and Industrial Chemistry (DCCI), University of Genova, Via Dodecaneso 31, 16146 Genova, Italy; maila@chimica.unige.it (M.C.); silvia.vicini@unige.it (S.V.)

[3] Department of Experimental Medicine (DIMES), Human Anatomy Section, University of Genova, Via De Toni 14, 16132 Genova, Italy; cortesek@unige.it (K.C.); gagliani@unige.it (M.C.G.)

* Correspondence: marina.pozzolini@unige.it; Tel.: +39-010-3533-8227

Received: 12 March 2018; Accepted: 27 March 2018; Published: 29 March 2018

Abstract: Collagen is involved in the formation of complex fibrillar networks, providing the structural integrity of tissues. Its low immunogenicity and mechanical properties make this molecule a biomaterial that is extremely suitable for tissue engineering and regenerative medicine (TERM) strategies in human health issues. Here, for the first time, we performed a thorough screening of four different methods to obtain sponge collagenous fibrillar suspensions (FSs) from *C. reniformis* demosponge, which were then chemically, physically, and biologically characterized, in terms of protein, collagen, and glycosaminoglycans content, viscous properties, biocompatibility, and antioxidant activity. These four FSs were then tested for their capability to generate crosslinked or not thin sponge collagenous membranes (SCMs) that are suitable for TERM purposes. Two types of FSs, of the four tested, were able to generate SCMs, either from crosslinking or not, and showed good mechanical properties, enzymatic degradation resistance, water binding capacity, antioxidant activity, and biocompatibility on both fibroblast and keratinocyte cell cultures. Finally, our results demonstrate that it is possible to adapt the extraction procedure in order to alternatively improve the mechanical properties or the antioxidant performances of the derived biomaterial, depending on the application requirements, thanks to the versatility of *C. reniformis* extracellular matrix extracts.

Keywords: collagen; Porifera; biomaterials; tissue engineering; membranes

1. Introduction

Collagen is the most abundant protein in animals; in particular, it is involved in the formation of a variety of fibrillar networks, providing the structural integrity of the tissues. This molecule, and its derived gelatin, is used as a biomaterial for tissue engineering and regenerative medicine (TERM) purposes, thanks to its low immunogenicity and mechanical properties. Collagen-derived biomaterials can be obtained in two ways: (i) by preserving as much as possible the shape of the original tissue via isolation of a de-cellularized extracellular matrix (ECM); or, (ii) by in vitro polymerization of soluble tropo-collagen, alone or combination with other ECM components, such as glycosaminoglycans (GAGs) [1,2], elastin [3,4] or chitosan [5,6]. These additions are meant to improve the enzymatic resistance and mechanical properties of the derived biomaterial. According to their application, collagen-based biomaterials can be used in the form of (i) collagen sponges, employed as three-dimensional (3-D) scaffolds in bone and cartilage repair; (ii) injectable hydrogels for drug delivery; and, (iii) two-dimensional (2-D) thin films or membranes suitable in wound dressing, dural closure, reinforcement of a compromised tissue and guided tissue regeneration [7]. The sources of collagen for these applications are mainly waste of bovine and porcine skin and bones. However, the risk of BSE (bovine spongiform encephalopathy), TSE (transmissible spongiform encephalopathy), and religious constraints (e.g., avoidance of porcine derivatives), has recently led to considering different sources of collagen. The most intriguing and promising ones come from the marine environment (for a review see [8]). In order to obtain economically sustainable amounts of these molecules, particular attention has been paid to collagens derived from fish or mollusks waste of food industry [9,10]. Besides, studies have also been done on collagens that were extracted from marine invertebrates, such as jellyfishes, generating real "blooms", in particular, environmental or seasonal conditions [11,12].

Marine sponges (Porifera) are the oldest metazoan group still extant on our planet [13]. Unlike other animal groups, they are characterized by a simple level of organization, they lack real tissues and organs, and are only formed by specialized cells types (e.g., choanocytes and pinacocytes [14]). These cells are embedded in a complex 3-D matrix network that is rich in collagen, which, in the Demospongiae class, is generically referred to as spongin [15]. It is formed by proteoglycans, glycoproteins, conventional striated collagen fibrils, and thin collagen microfilaments forming large structures that may be associated to siliceous spicules, or, in horny sponges, to foreign materials [16]. Although still being poorly characterized at the molecular level [17,18], spongin extracts [19], or sponge intact skeletons [20–22] provide a suitable framework for mammalian cell attachment and for migration and the proliferation of osteoprogenitor cells. These features reveal a promising biomaterial for bone repair. Furthermore, new fields of application of spongin are also arising, such as its use as supports for the immobilization of dyes [23–33]. Interestingly, in the Hexatinellida, a new unexpected structural role of collagen has been recently described. Indeed, this protein was identified within the basal siliceous spicules of the glass sponge *Hyalonema sieboldii* Gray, 1835 [34,35], stimulating further research on sponge collagen for a better comprehension of the skeletal structures of these deep-sea organisms.

Sponges are also organisms with a high biotechnological potential, since they are a relevant marine source of bioactive compounds. These animals, in fact, have developed a large variety of secondary metabolites for their defense from competitors, predators, and pathogens. The types of molecules produced are mainly nucleosides, toxins, and terpenoids with antibacterial, antifungal, or antiviral activity [36]. Notably, of the 18,000 marine natural bioactive products that have described until now in the pharmaceutical field, more than 30% have been isolated from sponges [37]. Unfortunately, wild sponge populations are frequently insufficient or inaccessible to produce sustainable commercial quantities of metabolites of interest. Thus, various mariculture [38,39] and in vitro cell culture [40,41] methods have been developed to fully exploit their interesting pharmaceutical features. In this perspective, the waste of sponge biomasses remaining after bioactive compound extraction could be used as a source of marine collagen.

One of the most studied sponge collagen is derived from the *Chondrosia reniformis* species, which is a member of the Demospongiae class. Its marine habitat includes a broad geographic (and bathymetric) range of marine environments, among which is the Mediterranean Sea. The body of *C. reniformis* is mainly made of tightly packed collagenous fibres that are lacking the typical needle-like silica spicules that strengthen the body of most other species of the same class. Its collagen is characterized by a unique dynamic plasticity due to the ability to reversibly alter its viscoelastic properties in an extraordinary short-time span. These features seem to be under the control of a calcium-dependent extracellular aggregating factor [42,43]. In the past, this collagen was partially biochemically characterized [44,45], and, more recently, some of its gene sequences were identified [46,47], as well as that of a collagen-maturation enzyme [48]. Moreover, an ultra-structural study was performed thanks to its partial solubilization [49]. Its biocompatibility on human skin has already been evaluated [50], and its use in the form of nanoparticles as carriers and coatings for drug preparations has been also described [51,52]. Notwithstanding, the use of *C. reniformis* collagen for membranous scaffold production in TERM applications has never been reported.

The aim of the present work was to evaluate the performance of various types of *C. reniformis* collagen extracts for the production of membranous scaffolds for TERM strategies. For this purpose, intact collagen fibrillar extracts were obtained, using four different methods that were previously described in literature [46,53–55]. Following their physico-chemical and biological characterization, the four extracts were used for the generation of collagenous membranes. Finally, the obtained membranes were characterized, and their biocompatibility evaluated, in order to establish the best extraction procedure to generate suitable tools for TERM applications.

2. Results

2.1. Fibrillar Collagen Suspensions Extraction and Characterization

Fibrillar collagen suspensions (FS) were extracted from *C. reniformis* tissue using four different methods, as indicated in Scheme 1, in order to evaluate the more suitable for collagenous membrane production. The total soluble proteins, collagen and GAGs content were then quantified in each FS. Table 1 reports the total yields obtained in each extract (from F1 to F4) in terms of FS, collagen, and GAGs, expressed as total mg/g of dry sponge tissue. Conversely, Table 2 shows a comparison of the composition by the quantification of total soluble proteins, collagen, and GAGs, which were calculated as weight percentage with respect to the total FS content of each extract. Moreover, a collagen/GAGs ratio ($R_{C/GAG}$) was calculated in order to establish the optimal collagen/sugar composition for the further membrane production.

Table 1. Yield value for each of the four extraction procedures (F1–F4) of the total FS (column 3), collagen (column 4) and glycosaminoglycan (GAGs) (column 5), expressed as mg/g dry sponge tissue. [1] From [53]; [2] from [54]; [3] from [46]; [4] from [55].

Sample #	Extraction Method	mg FS/g Dry Tissue	mg Collagen/g Dry Tissue	mg GAG/g Dry Tissue
F1	0.1% tryp, 2 rounds in H_2O [1]	338.27 ± 25.23	192.13 ± 22.15	12.85 ± 5.10
F2	8 M urea, tris 0.1 M, pH 9, βMe, EDTA [2]	422.83 ± 68.15	24.16 ± 1.30	20.44 ± 8.79
F3	NaCl 1 M, tris 50 mM pH 7.4, EDTA, βMe [3]	998.35 ± 16.12	355.68 ± 56.01	26.89 ± 7.52
F4	NaCl 0.5 M, tris 0.1 M pH 8, EDTA, βMe [4]	837.21 ± 75.87	139.02 ± 44.63	19.78 ± 3.67

Table 2. Percentage composition of total soluble proteins (column 2), GAGs (column 4) and collagen (column 5) in the four FSs expressed as mg/100 mg of total FS. Percentage composition of collagen (column 3) in the four FSs expressed as mg/100 mg of total soluble proteins. $R_{C/GAG}$ values (column 6) obtained from the ratio between the collagen content and GAG content in the four extracts.

Sample #	% Soluble Proteins/FS	% Collagen/Proteins	% GAG/FS	% Collagen/FS	$R_{C/GAG}$
F1	95.00	57.34	3.80	56.81	14.95
F2	34.00	16.76	4.83	5.73	1.18
F3	64.80	48.08	2.68	35.65	13.23
F4	27.39	60.62	2.36	16.63	6.99

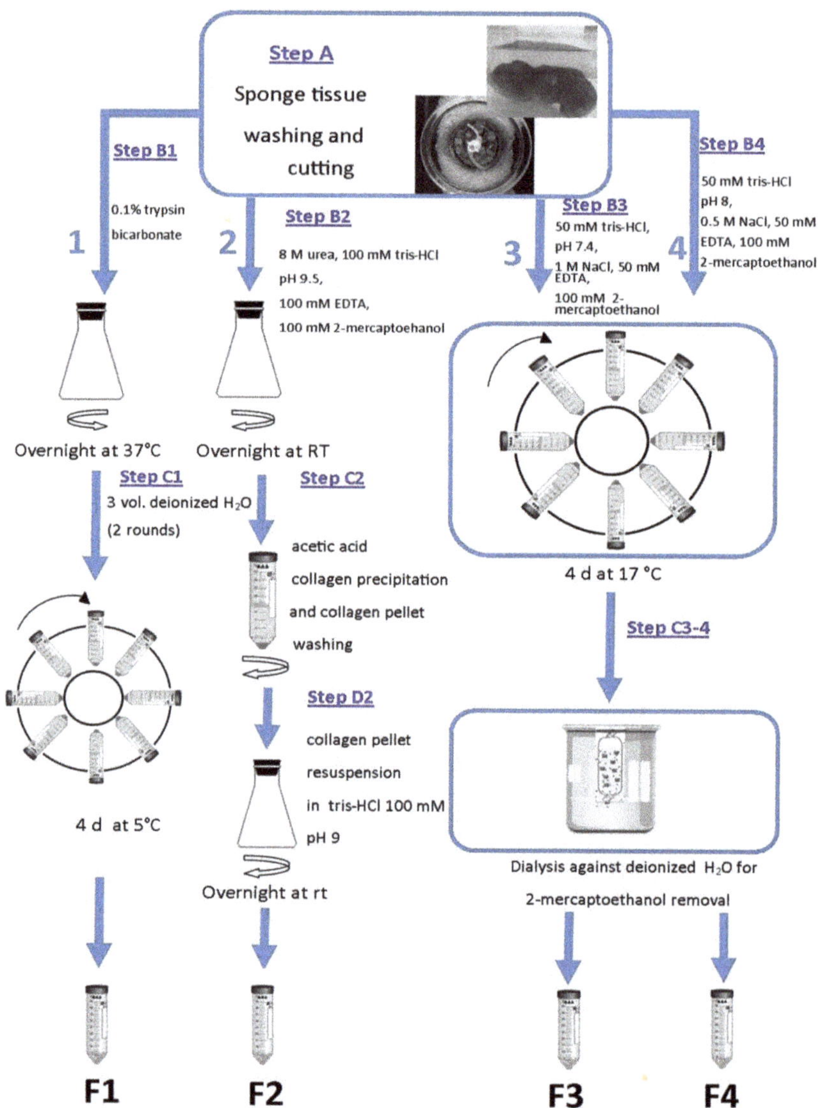

Scheme 1. Extraction methods used. Schematic representation of the four extraction procedures used to obtain four different sponge fibrillar suspensions (F1–F4) as detailed in Section 4.1. Method 1 from [53]; method 2 from [54]; method 3 from [46]; and, method 4 from [55].

F1 extract was obtained, as described by Gross et al., 1956 [53], by an overnight trypsin digestion at 37 °C in order to remove non collagenous proteins as much as possible. The digested sponge tissue was then subjected to two rounds of water extraction at 5 °C. After the second step of water extraction, the resulted sponge tissue completely dissolved. The collagen extract resulted viscous and light colored (Figure 1A). The yield of F1 suspension had the lowest values with respect to the yields of the other FS extracts (Table 1, column 3). The yield in collagen had intermediate values, while the GAGs had the lowest values of extraction with respect to the values that were obtained in the other extracts (Table 1, columns 4 and 5, respectively). Conversely, in terms of percentages, F1 suspension resulted composed almost all of the soluble proteins, half of which were collagen (Table 2, columns 2 and 3, respectively). Finally, the $R_{C/GAG}$ was of 14.95, the highest value observed with respect to the other FS extracts (Table 2).

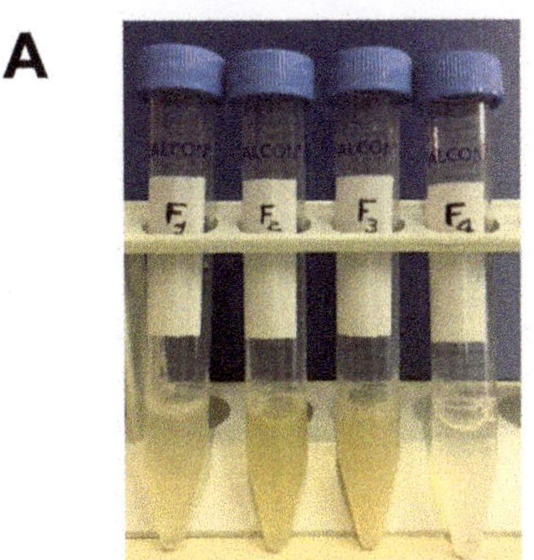

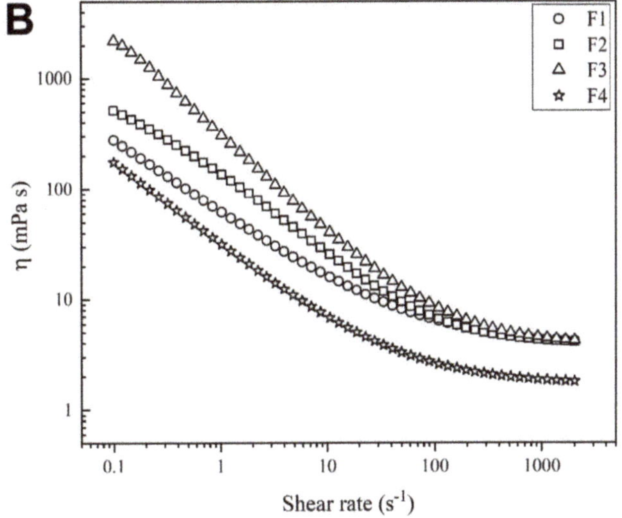

Figure 1. Fibrillar suspensions (FS) appearance and viscosity test. (**A**) Appearance of F1-F4 sponge FSs obtained through the four different extraction methods summarized in Scheme 1; (**B**) Flow sweep curves obtained in the viscosity tests of the four FSs by rheological measurements (for methods see Section 4.3.6). Curves were fitted by the Carreau-Gahleitner model (Mezger, 2006) as shown in Equation: $\frac{\eta - \eta_\infty}{\eta_0 - \eta_\infty} = \frac{1}{(1+(a \cdot \dot{\gamma})^b)^P}$ where η is the shear viscosity, η_∞ is the infinity-shear viscosity, η_0 is the zero-shear viscosity, a is the Carreau constant, b is the Gahleitner exponent, and P is the Carreau exponent. The experimental values of η_0 and η_∞ are shown in Table 3.

F2 extract obtained, as described previously by Diehl-Seifert et al. (1985) [54], was characterized by sponge tissue extraction under alkaline and reducing conditions in 8 M urea. The concentrated urea was used to improve fibre dissolving, whereas 2-mercapto-ethanol was used to ensure the intra and interchain disulfide collagen bridge break. At the end of the 24 h incubation at room temperature (RT), the sponge tissue was not completely dissolved and the insoluble fraction was removed by centrifugation. The collagen fibres were then precipitated by lowering the pH with the addition of acetic acid. The repeated centrifugation steps during the washings were necessary to neutralize the pH, but made the suspension less homogeneous than the others (Figure 1A), with several aggregates that necessitated a further homogenization step. It is to note that, in these extraction conditions the collagen retained in the F2 suspension had the lowest yield respect to the other extracts (Table 1). Furthermore, less than half of the F2 suspension was composed of soluble proteins, of which only the 16% was collagen (Table 2). Conversely, the percentage of GAGs extracted was the highest with respect to the other FSs, and consequently, the $R_{C/GAG}$ was the lowest (1.18), amounting to less than 1/10 of the F1 $R_{C/GAG}$ (Table 2, columns 4 and 6, respectively).

F3 and F4 extracts were obtained with a similar procedure [46,55], in both cases in fact, sponge fragments were incubated in an orbital shaker disk for four days at 17 °C in the presence of 2-mercapto-ethanol and EDTA. The only differences were that in F3 the pH was 7.4, whereas in F4, it was adjusted to 8.0, and that for F3 extraction, 1 M NaCl was used, while for F4, the NaCl concentration was only of 0.5 M. With the F3 extraction method, complete sponge tissue dissolution was observed at the end of the 4 day incubation. Conversely, using the F4 extraction method some insoluble tissue fragments were still present. F3 final extract resulted viscous and lightly colored, quite similar to F1, while F4 resulted clearer than the other FSs (Figure 1A). With these two extraction methods, the total yields of the two suspensions were the highest with respect to the other two methods (Table 1, column 3). Moreover, the F3 method showed the highest collagen and GAGs yields with respect to all of the other extraction procedures (Table 1, columns 4 and 5, respectively). For what concerns protein percentages, with the F3 method, the suspension obtained was composed of more than 60% of soluble proteins, of which nearly the 50% was collagen, while in the F4 method, less than 30% of the suspension was composed of soluble proteins, of which the collagen amounted to 60% (Table 2, column 2 and 3, respectively). Finally, the $R_{C/GAG}$ was higher in the F3 suspension, and was similar to the F1 value, and lower in the F4, pointedly nearly a half with respect to F1 and F3 (Table 2, column 6).

2.2. Viscosity Evaluation

Figure 1B reports the flow sweep curves fitted by the Carreau-Gahleitner model [56], while the values that were obtained by the experimental rheological measurements of η_0 and η_∞ are shown in Table 3.

Table 3. Rheological measurements values of η_0 and η_∞ of the four FSs from which were derived the flow sweep curves shown in Figure 1B.

Sample #	η_0 (mPa s)	η_∞ (mPa s)
F1	495	3.84
F2	911	3.93
F3	2412	3.90
F4	336	1.70

All of the tested solutions showed a low viscosity and typical gel behaviour, with a yield point in the low-shear range; moreover, an evident shear thinning behaviour was observable. Indeed, the samples were characterized by a shear-dependent viscosity (i.e., η decreases quickly with the increasing of the shear rate). F1, F2, and F3 samples showed similar values of η_∞, but significantly different values of zero-shear rate (η_0); in particular, in steady state, F3 and F4 appeared to be, respectively, the most and the lowest viscous samples, while F1 and F2 showed intermediate values of η.

2.3. Transmission Electron Microscopy Analysis of the FSs

To evaluate the state of integrity of the collagen fibrils that was extracted with the four different methods, TEM analyses were performed on the four different FSs.

At the ultrastructural level, all of the negatively-stained FS samples contained long unbranched single fibrils that was characterized by a periodicity pattern (Figure 2). F1, F2, and F4 samples (panel A, B and D, respectively) showed small clots (white arrows) that were closely apposed to the fibrils and thin curled filaments of *bona fide* mucopolysaccharides (MPS), which were completely absent in the F3 preparation (panel C). Of note, the F2 sample showed many curled fibrils, indicating a possible damage of the structural integrity of the fibrils (not shown). No curled collagen fibrils were observed in the other preparations.

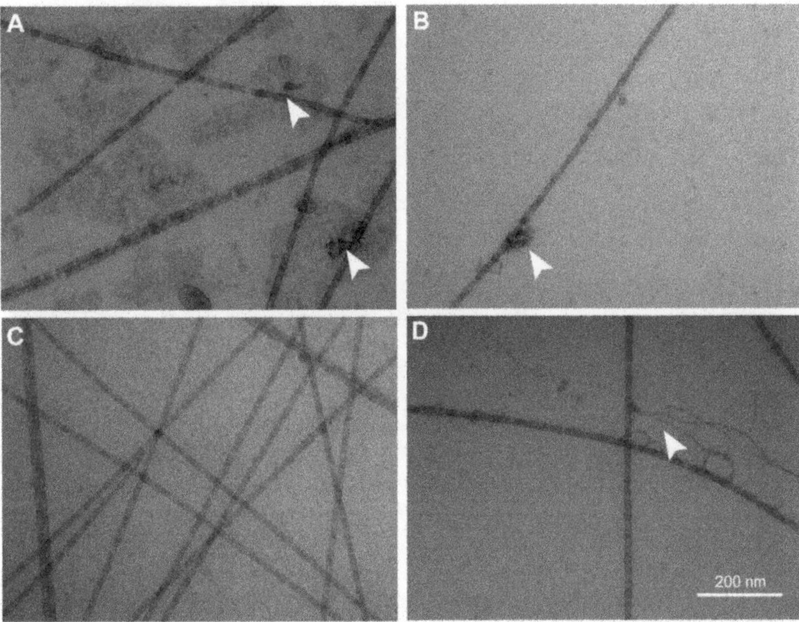

Figure 2. TEM analysis. Representative negatively stained transmission electron microscopy images of the four FSs (F1–F4) taken at 92,000× magnification (scale bar: 200 nm) with a CM10 Philips transmission electron microscope equipped with Megaview 3 camera and Olympus SIS iTEM software for digital image acquisition. (**A**) F1; (**B**) F2; (**C**) F3; and, (**D**) F4. All of the FSs show long unbranched banded fibrils of uniform size and periodicity band pattern. F1, F2, and F4 show small clots and filaments (white arrows) of putative mucopolysaccharides.

2.4. Qualitative Evaluation of the FSs by Histological Methods

FS composition was also qualitatively evaluated by standard histological methods. Different molecular components were highlighted on each FS, smeared and dried on histological slides, and observed in light microscopy (Figure 3).

Alcian staining (pH 2.5) allowed to highlight in light blue the presence of MPS by reacting with acidic groups. Alcian staining on the smeared FSs showed the following results: in F1 sample (panel A), small and uneven clots of MPS were observable; in the F2 (panel B), a strong coloration of uniformly distributed MPS clots was evidenced; in F3, only a weak MPS staining was detectable in some portions of the sample (panel C); in F4 (panel D), in addition to a weak and diffused coloration, some fibres with an irregular pattern were observable showing the presence of MPS.

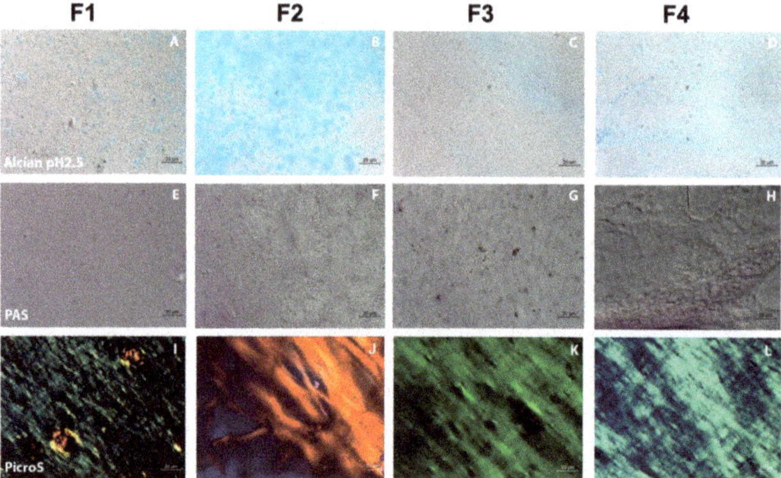

Figure 3. FS histological staining. The four FSs (F1–F4) smeared on histological slides and stained according to four different staining procedures, as described in Section **??**, and observed by optical/polarized microscopy. (**A–D**) Alcian pH 2.5 staining, highlighting mucopolysaccharides (MPS) and glycoproteins in blue, by reacting with acidic groups; (**E–H**) Periodic Acid Schiff (Hotchkiss–Mc Manus) (PAS)_staining highlighting MPS, glycoproteins, glycolipids, mucins, and polysaccharides such as glycogen in pink/red, by reacting with basic groups; (**I–L**) Picro-Sirius Red staining of collagen bundles in different shades of green, red, or yellow, depending on the thickness and the packing of fibres, from thinner to thicker, respectively. Each column of panels represents the various histological stainings of each FS. Scale bars in each panel span 20 micrometer.

PAS staining, on the other hand, highlights in pink/red, the presence of MPS by reacting with basic groups. In the F1 sample (panel E), a poorly concentrated staining of basic little clots was observable. In the F2 (panel F), basic MPS clots were still poorly concentrated. In F3 (panel G), a matrix of transparent/pink fibres was clearly visible by DIC (Differential Interference Contrast), with red clots indicating the presence of basic MPS, likely not being associated with fibres. Finally, in F4 samples (panel H) a weak PAS positive staining of fibrous material was observable, similarly to that observed with Alcian and Picro-Sirius Red (PicroS) in panels D and L, respectively, with the presence of MPS distributed in an irregular pattern.

With the PicroS collagen staining, and through the use of polarized microscopy, it is possible to distinguish small-caliber collagen fibres that result in highlighted in shades of green with respect to thicker fibres that appear to be yellow/red. PicroS staining of the F1 preparation (panel I), showed a discontinuous film of small-caliber fibres. Collagen appeared to be distributed non-uniformly, but in bundles with the presence of small lumps of about 20 μm in diameter showing larger caliber collagen fibres (in red/yellow). Conversely, in the F2 preparation, it was possible to observe an uneven distribution of thicker collagen fibres that are highlighted in red (showed in panel J). Finally, in the F3 and F4 preparations (panels K and L, respectively), the method was able to highlight two continuous fibre films, even if with different chromatic characteristics, pointing out smaller caliber fibres in F3, and thicker in F4.

2.5. Sponge Collagen Membrane Production

Sponge Collagen Membranes (SCMs) were produced using 2 mg/mL of F1, F2, F3, and F4 fibrillar suspensions in the presence of EDC/NHS crosslinking solution for 4 h at RT. As shown in Figure 4, the only FSs that were able to generate suitable SCMs were F1 and F3 extracts. In fact, F1- and

F3-derived SCMs looked like thin, clear, films (Figure 4A). Conversely, using F2 and F4 suspensions, even by increasing the FS concentration (data not shown), it was not possible to obtain any suitable membrane. In fact, when dried the derived films resulted lacking texture, extremely fragile and easy to break. Hence, all of the following characterization analyses were performed only on F1- and F3-derived SCMs named SCM-F1 and SCM-F3, respectively.

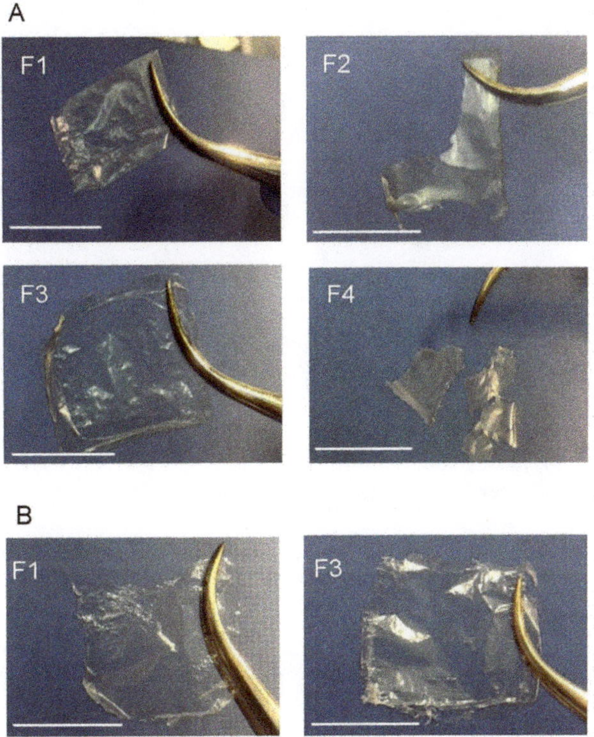

Figure 4. Sponge Collagen Membranes (SCMs). (**A**) SCMs derived from F1–F4 sponge extracts crosslinked with 1-ethyl-3-(3-dimethylaminopropyl) carbodiimide/*N*-hydroxysuccinimide (EDC/NHS) solution; (**B**) not crosslinked SCMs derived from F1 and F3 sponge extracts. Scale bars in each panel span 2.0 cm.

As shown in Figure 4B, the membranes were also prepared using not crosslinked F1 and F3 suspensions casting the collagen suspensions without adding the cross-linker. This procedure generated the nc-SCM-F1 and nc-SCM-F3 membranes, respectively, which were macroscopically similar to their crosslinked counterparts.

2.6. SCMs Characterization

2.6.1. Mechanical Tests

The elastic moduli G′ (filled symbols) and the viscous moduli G″ (open symbols) of the F1 and F3 SCMs and nc-SCMs are reported in Figure 5. G′ indicates the capability of the material to store energy and G″ refers to the capability of the material to dissipate energy. All of the samples that were examined showed relatively high moduli (in the range of 10^4–10^5 MPa for the elastic modulus).

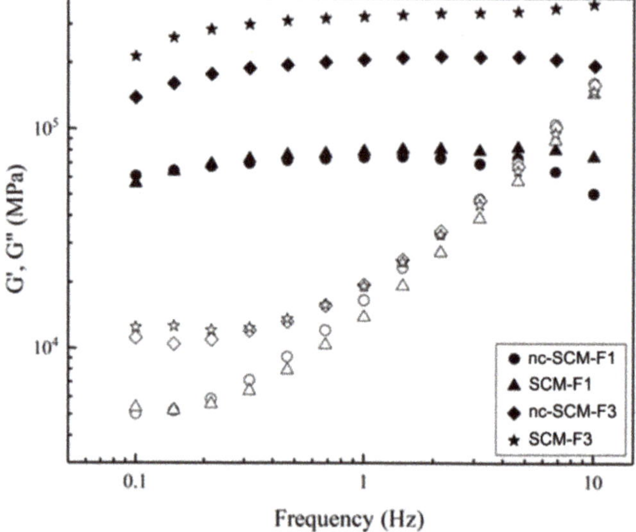

Figure 5. The elastic moduli G′ (filled symbols) and the viscous moduli G″ (open symbols) of the SCMs (SCM-F1, triangles; SCM-F3, stars) and not crosslinked SCMs (nc-SCM-F1, circles; nc-SCM-F3, diamonds) reported in function of the frequency applied by the rheometer in the dynamic mechanical analysis (for methods see Section 4.5.3).

As expected, for a gel system, a solid-like behaviour was observable. Indeed, it was possible to note the predominance of G′ upon G″ at low frequencies in all of the membranes, while G″ increased on increasing the frequency till reaching a crossover between the moduli. These data indicate that all of the tested samples show an evident elastic behaviour in the frequency range that was studied.

Both SCM-F1 (triangles) and nc-SCM-F1 (circles) showed lower moduli than SCM-F3 (stars) and nc-SCM-F3 (diamonds), indicating, in general, a lower mechanical stiffness. Moreover, as is clearly observable, the crosslinking did not affect the mechanical performance of F1-derived SCMs, with the two samples (nc-SCM-F1 and SCM-F1) showing comparable values of both G′ and G″. On the contrary, in F3-derived SCMs, a significant difference was observable between the nc-SCM-F3 and the SCM-F3, with an evident increase of the mechanical properties in the crosslinked sample, as evidenced by a higher G′ elastic modulus when compared to the nc-SCM-F3.

2.6.2. In Vitro Resistance to Enzymatic Degradation

In vitro biodegradation of F1 and F3 derived SCMs and nc-SCMs was evaluated by incubation at 37 °C with native fetal bovine serum (FBS) for 15 day and by collagenase digestion for 6 days. The results showed that all of the membranes were completely intact after both of the treatments. This indicates a strong resistance of the marine biomaterials to enzymatic degradation at physiological conditions, even for the not crosslinked versions nc-SCM-F1 and nc-SCM-F3 (not shown). Conversely, a parallel incubation of a commercial mammalian collagen membrane (Bio-Gide®, Geistlisch Pharma, Wolhusen, Switzerland), which is usually employed in dental surgery, showed a complete collagenase digestion after 6 days incubation. Furthermore, after FBS treatment, all of the SCMs (crosslinked or not) resulted visibly opaque (Figure 6A) as compared to the negative controls that were incubated in phosphate buffered saline (PBS). Other than opaque, FBS-treated membranes also appeared to be thicker than their controls in PBS, as observed by optical stereomicroscope (Figure 6B). Both features, i.e., opacity and thickness, were likely due to the serum protein adsorption onto the membrane surfaces.

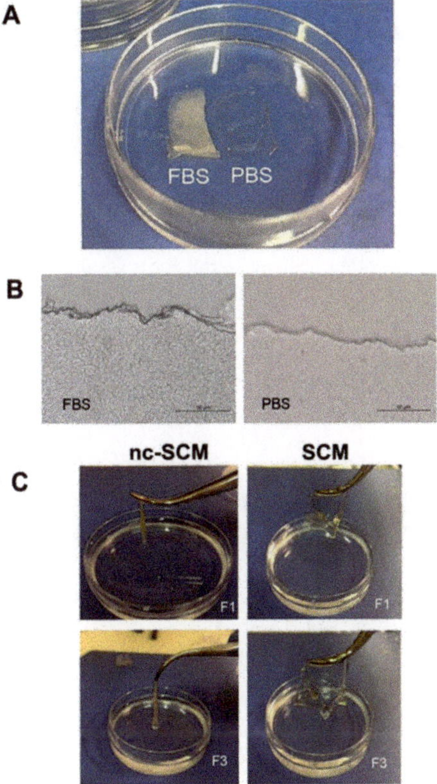

Figure 6. SCM serum interaction and water absorbing capacity. (**A**) SCM-F1 incubated in 1 mL of FBS (left) or PBS (right) for 15 day at 37 °C; (**B**) Stereo-microscope observation of SCM-F1after incubation for 15 days at 37 °C in FBS (left) or PBS (right); scale bar, 50 micrometer; (**C**) Not cross-linked (nc-SCM) and cross-linked SCM derived from F1 and F3 fibrillar extracts after soaking in PBS for 1 h.

2.6.3. Water Binding Capacity

The water binding capacity (WBC) of SCMs and nc-SCMs was assessed by the evaluation of their weight variation after 1 h soaking in PBS at RT. nc-SCM-F1 and SCM-F1 showed a WBC of 652 ± 35% and of 280 ± 33%, whereas the WBC of nc-SCM-F3 and of SCM-F3 were 701 ± 40% and 420 ± 55%, respectively. Thus, F3-derived membranes showed a slightly higher WBC than F1. Conversely, the commercial not crosslinked collagen membrane Bio-Gide® in the same soaking conditions showed a WBC of 442 ± 41%, which is a value that is significantly lower than the two nc-SCMs, but comparable to the SCMs.

Once hydrated, it was also clearly observable a significant different behavior of the SCMs with respect to the nc-SCMs. The latter, in fact, although much more hydrophilic than their crosslinked counterparts, completely loosed consistency and resulted in being very difficult to manipulate, as shown in Figure 6C.

2.6.4. Biocompatibility

To evaluate the biocompatibility of SCM-derived membranes, a fibroblast cell line and a keratinocyte cell line were tested for their ability to adhere and to grow on FS-coated plates using all of the extracts (F1–F4). Cell adhesion was evaluated 16 h after plating both qualitatively and quantitatively by the crystal violet staining and the MTT assay, respectively (Figure 7A,B). The crystal violet qualitative assessment of both fibroblast and keratinocyte cell adhesion revealed that the cell shape, and thus the interaction with the different matrices, was very similar in all of the conditions for both of the cell lines. A slight, qualitative difference could be observed only for the fibroblasts on F4 coating. In particular, a circular shape seemed to be predominant with respect to the physiological spindle/triangular shape typical of these cells, probably indicating a not preferred interaction with this type of coating (Figure 7A). Similarly, the MTT cell viability assay at 16 h of adhesion indicated a reduction of the attached cells only for fibroblasts (Figure 7B, black bars) and only on the F4 coating (21.75% cell reduction as compared to controls). These results confirm a poorer short-term F4 compatibility with respect to the other FSs and to the control, where cells were seeded onto rat tail collagen coated wells, namely stCol.

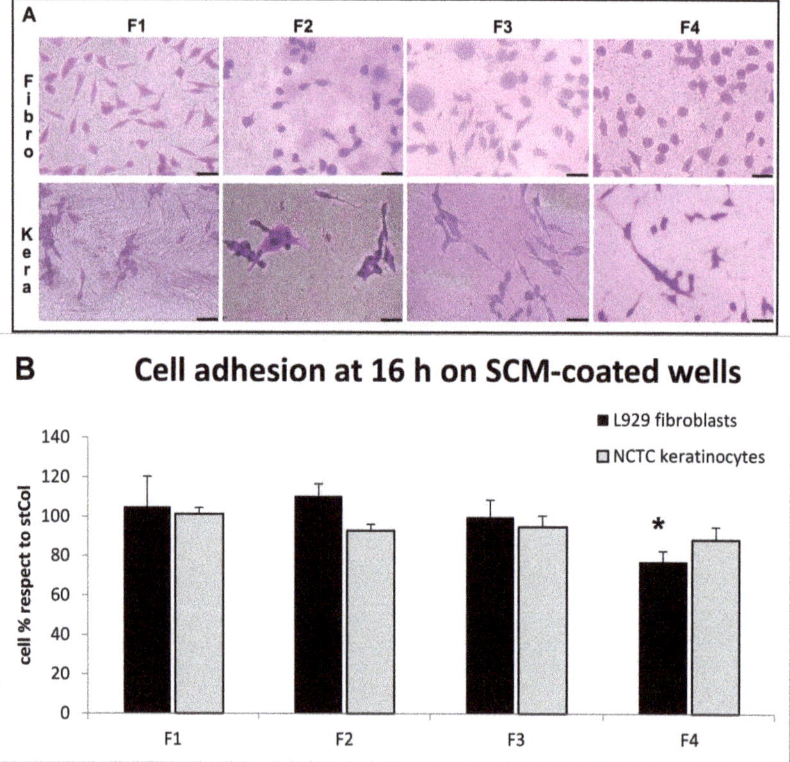

Figure 7. Cell adhesion evaluation. (**A**) Cell adhesion qualitative evaluation, by optical microscopy, of crystal violet stained L929 fibroblasts (first row) and National Collection of Type Cultures (NCTC) keratinocytes (second row) on the four different FS pre-coated plates (F1–F4) after 16 h incubation. Scale bars, 20 micrometer; (**B**) Cell adhesion quantitative evaluation, by MTT test, of L929 fibroblasts (black bars) and NCTC keratinocytes (grey bars) on the four different FS pre-coated plates after 16 h incubation. Results are expressed as cell percentages with respect to controls that were seeded on standard rat tail collagen-coated (stCol) wells and are the mean ± S.D. of two experiments that were performed in quadruplicate. Asterisks indicate significance in Tukey test (black bars ANOVA, $p < 0.05$; Tukey, F4 vs. stCol $p < 0.05$).

Cell viability and proliferation was also evaluated for longer periods of time on FS-coated plates by using the MTT assay. In particular, both fibroblasts and keratinocytes were evaluated after 3 days,

6 days, and 15 days of cultivation (Figure 8) on the FS-coated plates and compared to control cells that were grown onto rat tail collagen coated wells (stCol). In detail, fibroblasts showed a slight cell number reduction on F4 coating and a slight increase on F1 coating at 3 day with respect to the control (Figure 8A, black bars 19.4% decrease and 12.2% increase, respectively). Conversely, both at 6 day and 15 day of prolonged cell culture (Figure 8A, white bars and grey bars, respectively), no significant differences with respect to the controls grown on rat tail collagen were observed. This indicates a long-term good biocompatibility of the four FS coating for cells of fibroblastic nature. For what concerns keratinocytes, they showed good compatibility and reasonable cell growth, which was comparable to the controls, after 3 day of culture and after prolonged culture for 15 onto all of the FS-coated plates (Figure 8B, black bars and grey bars, respectively). Conversely, a slight, but significant, decrease of cell number at 6 day onto F3 and F4 coatings, as compared to the control, was observed (white bars, 36.0% and 37.2% decrease, respectively), probably indicating that keratinocyte cells undergo a period of adaptation, especially on F3 and F4 coatings, before restarting cell growth.

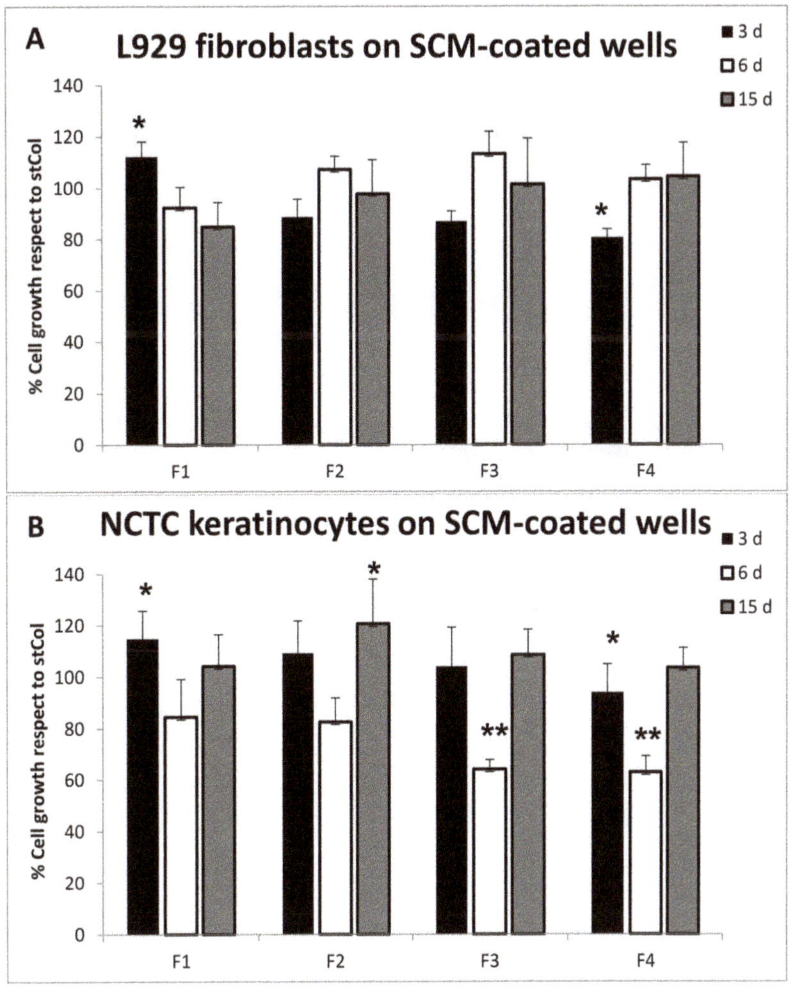

Figure 8. Cell growth evaluation. (**A**) L929 fibroblast cell growth evaluation, by MTT test, on the four

different FS pre-coated plates (F1–F4) after 3 day (black bars), 6 day (white bars) and 15 day (grey bars) incubation. Results are expressed as cell percentages respect to controls seeded on standard rat tail collagen-coated (stCol) wells and are the mean $\pm$ S.D. of two experiments performed in quadruplicate. Asterisks indicate significance in Tukey test (black bars ANOVA, $p < 0.0001$; white bars ANOVA, $p < 0.0001$; grey bars ANOVA, $p < 0.05$; Tukey vs. stCol: * $p < 0.05$, ** $p < 0.001$, respectively); (**B**) NCTC keratinocytes cell growth evaluation, by MTT test, on the four different FS pre-coated plates after 3 day (black bars), 6 day (white bars), and 15 day (grey bars) incubation. Results are expressed as cell percentages with respect to controls that are seeded on standard rat tail collagen-coated (stCol) wells and are the mean $\pm$ S.D. of two experiments that were performed in quadruplicate. Asterisks indicate significance in Tukey test (black bars ANOVA, $p < 0.00001$; Tukey vs. stCol: * $p < 0.05$).

2.6.5. Environmental Scanning Electron Microscope (ESEM) Analysis

ESEM analyses were performed on the SCMs that were derived from the four FSs, or fragments thereof for F2 and F4. Furthermore, the analyses were also performed on F1 and F3-derived membranes in the presence of fibroblasts and keratinocytes that were cultured on their sterilized surfaces for 3 day. The ultramicroscopic analysis of the four SCMs alone showed randomly distributed fibril patterns in F1, F3, and F4-derived SCMs (Figure 9A,C,D). Conversely, in F2-derived membranes, the surface that resulted was characterized by an unidentifiable clumped layer (Figure 9B). From F1, F3, and F4-derived SCMs, it was possible to calculate the average of the fibril diameter of *C. reniformis* collagen, which was of 21.08 ± 4.93 nm. Furthermore, in of all the three samples, it was possible to observe either the presence of free single fibrils or bundles of aggregated fibrils forming fibres of higher dimensions. No ultrastructural differences were observed between F1 and F3 crosslinked and not crosslinked membranes (data not shown). In SCM-F1 and F3, the random presence of pores on the surface of the membranes was also observed. SCM-F1 and SCM-F3 pores were measured and they spanned an average value of 6167 ± 2826 nm^2 and 3211 ± 1494 nm^2, respectively. The significant difference in the pore dimensions of the two membranes ($p < 0.005$) indicates a higher level of fibre cohesion in F3-derived membranes with respect to F1.

For what concerns the ultrastructural analysis of cells grown onto the SCM-F1 and SCM-F3 membranes, no significant differences could be observed in the degree of cell adherence and surface contact on both of the SCMs, either for fibroblasts (panel E for SCM-F1 and F for SCM-F3), either for keratinocytes (panel G for SCM-F1 and H for SCM-F3). In fact, all of the images that were obtained by the high resolution cell analysis showed a tight adherence to both SCM surfaces, especially in fibroblasts (E,F), and the presence of several cellular elongated processes taking contact with the SCM surface surrounding the cell body in keratinocytes (G,H). These results suggest that both membranes could have a good biocompatibility.

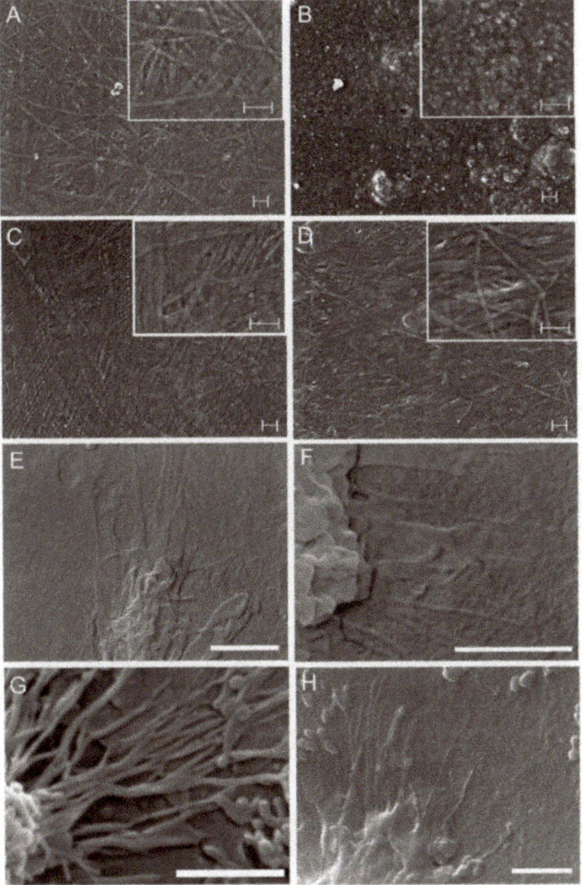

Figure 9. Environmental Scanning Electron Microscope (ESEM) analysis. All crosslinked SCMs (**A–D**), dehydrated at critical point and graphite covered, were observed with a FESEM Zeiss SUPRA 40 VP instrument, while only SCM-F1 and SCM-F3, in the presence of fibroblasts and keratinocytes (**E–H**) dehydrated at critical point and graphite covered, were observed with an ESEM Vega3–Tescan instrument. (**A**) SCM-F1; (**B**) SCM-F2; (**C**) SCM-F3; (**D**) SCM-F4; (**E,F**) Visualization of L929 fibroblasts adhesion to the two SCMs; (**E**) SCM-F1; (**F**) SCM-F3; (**G,H**) Visualization of NCTC keratinocytes adhesion to the two SCMs; (**G**) SCM-F1; (**H**) SCM-F3. In (**A–D**) scale bars span 200 nm; in (**E–H**) scale bars span 2 micrometer.

2.7. DPPH Radical Scavenging Activity

Since it is known that the amino acid residues that are present in collagens show a certain level of antioxidant activity [57], the radical scavenging activity of each FS was evaluated using the DPPH standard assay.

As indicated in Figure 10A, F1 showed the highest radical scavenging activity of 61.78 ± 2.84%, F2 of 14.61 ± 0.61%, whereas for F3 resulted of 26.97 ± 0.23%, and finally F4 showed a radical scavenging activity of 28.89 ± 2.28%. The radical scavenging activity resulted to persist also in F1- and F3-derived membranes (Figure 10B). In this case, the radical scavenging activity was expressed in the function of the membrane surface. In particular, the surface exerting the 50% of the total radical scavenging activity (SA_{50}) measured 242.81 mm^2 for SCM-F1, and 665.84 mm^2 for SCM-F3.

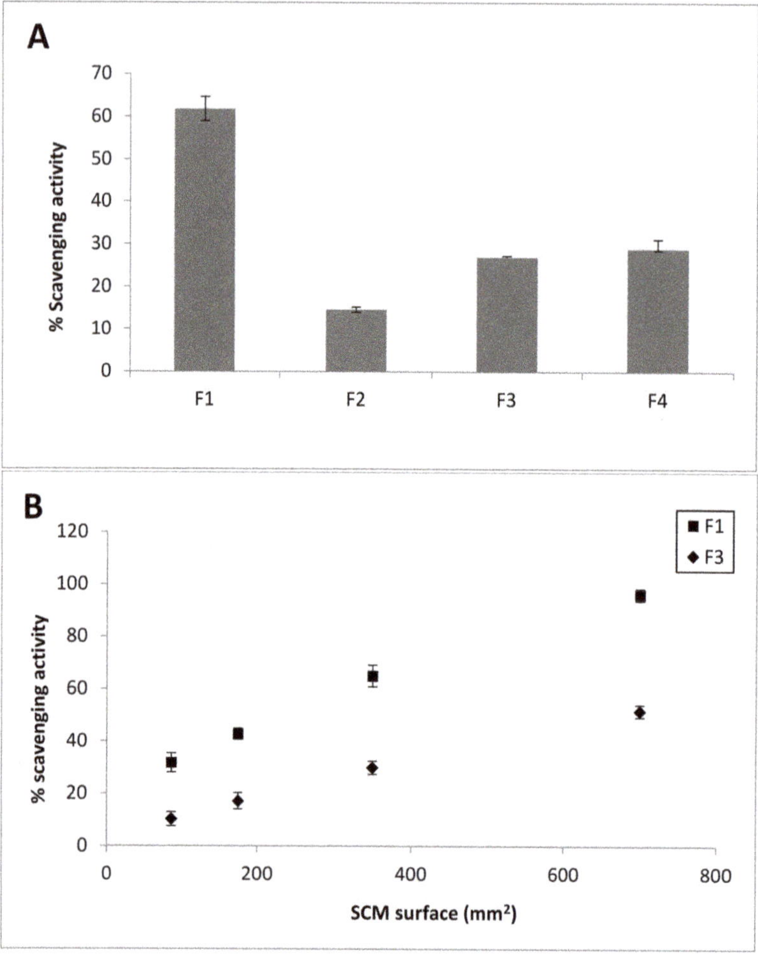

Figure 10. FS and SCM radical scavenging activity. (**A**) Antioxidant activity of the four different FSs (F1–F4) measured by the spectrophotometric DPPH oxidation assay using a concentration of 1 mg/mL of each FS. Results are expressed as percentages of radical scavenging activity based on the inhibition of DPPH oxidation calculated, as described in Section 4.7, and are the mean ± S.D. of two experiments performed in triplicate (ANOVA, $p < 0.000001$); (**B**) Antioxidant activity of SCM-F1 and SCM-F3 measured by the DPPH assay. Results are expressed as percentages of radical scavenging activity in function of the SCM surface area of the two membranes, and are the mean ± S.D. of two experiments performed in triplicate (ANOVA, $p < 0.000001$).

3. Discussion

The ability of the micro-environment and the adhesion substrate to affect various biological processes, i.e., differentiation and proliferation, has been shown in the evolution since the low metazoans [58]. The extracellular matrix (ECM) of multicellular animals is a complex, dynamic system. It is mainly formed by collagen fibrils structured in a highly organized, three-dimensional scaffold supporting cell adhesion and playing a crucial role in differentiation and tissue remodelling [59,60]. Indeed, collagen is considered to be extremely attractive to the manufacturing of biomaterials for TERM applications. The production of collagen-based biomaterials has intensively grown over the past decades. Besides, always new techniques are being developed to improve their performance in terms of mechanical properties and the resistance to enzymatic degradation. Collagenous biomaterials are used as composite materials [61], or as de-cellularized derivatives, obtaining scaffolds that closely mimic the ECM structure and its properties [62]. Marine collagen is considered as one of the most promising sources for these purposes. In particular, the collagen that was derived from the marine sponge *C. reniformis*, due to its dynamic plasticity [43], is particularly interesting for biotechnological applications. In fact, it is considered as a marine biopolymer that is suitable for the production of dynamic biomaterials thanks to its ability to rapidly change stiffness and viscosity similarly to what was observed in the mutable collagenous tissues of echinoderms [63]. Collagenous membranes that are made from ECM extracts, containing native collagen fibrils of echinoderms with good mechanical/biological properties, have been already produced [55]. Conversely, although *C. reniformis* collagen has been studied for a long time [44–46], collagenous membranes that are derived from this species have not yet been described.

The aim of the present work was to obtain marine collagen membranes using intact collagen fibres of this animal. In the production of de-cellularized tissue scaffolds, the purification procedures have a key role, since the removal or maintenance of certain residues may strongly affect their biocompatibility or mechanical properties [64]. In this study, intact sponge collagen fibres were isolated using four different extraction methods (see Scheme 1), and then SCMs were produced from each of them (Figure 4). Finally, the chemical and physical properties of the four extracts were related to the mechanical and biological performances of the respective membranes. These evaluations allowed for selecting the optimal procedures and obtaining the best compromise among yield, 3-D fibre integrity, resistance, and biocompatibility of said membranes.

When comparing the yield values that were obtained in the different FSs (Table 1), the highest total yield and the highest collagen/GAGs recovery was obtained in the F3 suspension. The strongly reducing conditions, due to the presence of 2-mercaptoethanol in the F3 and F4 method, likely contributed to the higher sponge tissue disaggregation and FS yields of these two extraction methods. These results could be explained by *C. reniformis* presence of peculiar short-chain collagen types, which were related to the mammalian type IV, which seem to be involved in the tissue stiffness. In particular, the five conserved cysteines at the C-terminal of these proteins [46], suggest the presence of a disulphide-bridge crosslinking system for a supramolecular arrangement, as already reported in type IV collagen organization in mammalian basal membranes [65]. Thus, the addition of the reducing agent is likely to act at this disulphide-bond level, promoting the complete sponge tissue disruption in the F3 and F4 extraction procedures.

Conversely, for what concerns collagen extraction total yield (Table 1), although the F3 and F4 methods were very similar, the total ionic strength of the extraction buffer seemed to be fundamental to enhancing collagen fibril release. Indeed, the presence of a double NaCl concentration in F3 determined a proportional double release of collagen fibrils from the sponge tissue with respect to the F4 extract.

In the F2 extraction method, where the chaotropic conditions were used in addition to the reducing agent, the total FS yield and low collagen content were significantly different than F3 and F4, which is probably due to the shorter time of incubation in this extraction buffer (i.e., 1 day vs. 4 day, respectively). Conversely, the F1 extraction method provided the highest collagen percentage with respect to the total FS content, indicating trypsin digestion as the best strategy to obtain highly collagen-enriched sponge FSs (Table 2). Furthermore, when comparing the total protein and the collagen percentages in the four FSs (Table 2), we observed that the four methods provided very different values, with F1 suspension showing the highest content of soluble proteins. Conversely, all of the other FSs showed lower and variable percentages of soluble proteins with respect to the total FS. This indicates that these methods (especially F2 and F4) also provided significant amounts of uncharacterized, microscopic materials that are retained in the final FS, despite centrifugation. Notwithstanding the variability of the soluble protein percentage, the collagen/protein ratio seems more homogeneous in the four FSs. Indeed, all of the extraction preparations showed a collagen/protein ratio in a range that was close to 1:2, except for

F2, in which the proportion was ≈1:6, indicating that the F2 extraction method retrieves the lowest amount of collagen fibrils from the sponge tissue. Overall, the four methods allowed for obtaining collagen fibrils in good state of integrity, as showed by the ultrastructural qualitative analysis by TEM (Figure 2). Although, in F2 suspension, the presence of many curled fibrils (not shown) likely indicate a slight damage of the extracted collagen. These observations strengthen the idea that the F2 extraction method is less efficient than the others.

The qualitative evaluation of the four FSs by histological staining indicated that the collagen component in F2 preparation had thicker collagen fibres than the other three extracts (Figure 3, PicroS series: panel J vs I-K-L, respectively). Since fibres of such dimensions are supposed to be removed in the centrifugation steps during the extraction, it is probable that the strong PicroS F2 coloration is due to the artificial clotting of the collagen fibres during the acetic acid precipitation step. This again suggests that F2 may not be optimal for subsequent scaffolding procedures. Conversely, a low-diameter fibrillar organization was observable in F1, F3, and F4 PicroS staining, with the bigger collagen bundles that are typically present in *C. reniformis* native tissues [66] being completely removed during the first centrifugation steps.

In *C. reniformis*, tissue collagen fibres are closely associated to complex carbohydrates [44]. Besides, it is known that this component improves the mechanical and the biocompatibility properties of the collagen scaffolds [67]. Thus, the GAG content was also evaluated quantitatively and qualitatively in the four FSs that were obtained from *C. reniformis*. Our data indicate that, even if GAGs are present in low percentages in the four FSs (Table 2), from a qualitative evaluation they seem associated to the collagen fibres extracted with the various methods (Figure3 Alcian and PAS panels). Although, the F2 suspension differs from the others, showing, respectively, the highest percentage of GAG and the lowest percentage of collagen extracted. In this case, the Alcian histological staining could highlight a significant concentration of GAGs (panel B) probably associated to non-fibrillar proteins. Interesting differences were also observed in the $R_{C/GAG}$, with F1 and F3 showing significantly higher ratios than the other two (Table 2). Considering that these two FSs were the only ones able to generate suitable SCMs and nc-SCMs, we could infer that the ratio between collagen and GAG content in the FS may be determinant for a successful reticulation during the membrane formation. Indeed, our results suggest that a $R_{C/GAG}$ <10 is inappropriate for SCM production.

The viscosity values that were measured in the four extracts (Figure 1B), do not seem to reflect the different collagen or GAG percentages, likely indicating that other factors (i.e., insoluble protein content, electrostatic interactions) could come into play, thus influencing this parameter. These results are in conflict with that observed visually in F1 and F3 suspensions, which appear as dark coloured viscous hydrogels respect to the other FSs. These features probably indicate a higher thixotropic behaviour and the presence of aggregates/micro-gels in solution. Conversely, the hydrogel consistency may be related to the collagen/FS percentage, which is the highest in these two samples, and where the fibrillar network could give a hydrogel aspect to the suspension [68].

SCMs were produced by EDC/NHS crosslinking of the four FSs; however, only F1 and F3 extracts could generate manageable membranes (Figure 4, SCM-F1 and SCM-F3, respectively). Conversely, F2 and F4 tentative membranes lacked structure and texture, were extremely fragile, and were difficult to recover from the silicon mould, also increasing the FS concentration (data not shown). These data indicate that *C. reniformis*-derived FSs with a collagen/FS percentage that is lower than 35% (Table 2) are unsuitable for membrane production.

ESEM analysis indicated that SCM-F1 and SCM-F3 (as well as also the fragmented SCM-F4) had a fibrillar organization. This further confirms that F1 and F3 extraction procedures were able to maintain an intact fibrillar structure with a significant difference in random pore dimensions between the two (Figure 9 A,C,D). Indeed, the higher mechanical resistance that was showed by SCM-F3 (Figure 5) could be explained by the presence of a smaller mesh of the fibrillar interlace with respect to SCM-F1.

Using the same FS concentration, F1 and F3 were also casted in silicon moulds without crosslinking. Surprisingly, both F1 and F3 formed collagenous membranes, named nc-SCM-F1 and nc-SCM-F3, respectively. These membranes showed macroscopic (Figure 4B) and ultrastructural (data not shown) features that were similar to their crosslinked counterparts. These data suggest that the collagen extracted from *C. reniformis* is naturally provided with a complex system of preformed fibrillar crosslinks, which may justify either their strong mechanical properties [44] or their insolubility in acidic solutions [49,50].

Another important consideration arising from the results of the membranes mechanical tests is about collagen/total protein content in F1 and F3 suspensions. Due to the higher resistance of SCM-F3 (Figure 5), we can also infer that the higher percentage of non-collagenous proteins present in F3 respect to F1 may have positive impact on the mechanical properties of the sponge FS-derived membranes. Although their exact nature is still unknown, it has been reported that these proteins constitute an amorphous inter-fibrillar matrix in the sponge ECM and are actively involved in the formation of the sponge fibre network [53]. These proteins are susceptible to trypsin treatment, being consequently digested in the F1 extraction procedure, and their absence could help to explain the significant reduction of the mechanical properties that were observed in SCM-F1. Moreover, while no differences were observed for SCM-F1 with respect to nc-SCM-F1 in terms of resistance improvement, a better performance was instead observed in SCM-F3 when compared to nc-SCM-F3. These data indicate that the higher non-collagenous protein content in F3 may participate to the formation of the crosslinks in the membrane, further strengthening this collagenous film.

Biomaterials originating from collagen still have some limitations to their use in human tissues due to inflammation arising from their biodegradation and relatively short durability [69]. It is known that, both living sponge tissues as well as isolated collagen fibres that are derived from *C. reniformis* are particularly resistant to bacterial collagenases [44]. In our strong experimental conditions (five times higher collagenase concentration than Garrone et al. [44], and 6 day incubation instead of 2 day), all of the membranes, crosslinked or not, maintained their intact structure. This evidence, in addition to confirming previous data on *C. reniformis* collagen, tell us that thin collagen membranes that are derived from this species possess a high stability in strong enzymatic degradation conditions. Conversely, in the same conditions, a commercial not crosslinked collagenous membrane, Bio-Gide®, was completely dissolved (not shown). Since this commercial product, which is mainly used in dental surgery, shows an in vivo resorption rate of 2–4 weeks [70], it is reasonable to believe that SMCs may show significantly longer resorption rates. If this was the case, another important requisite for implantable tissue engineering scaffolds would be met by these membranes. Also, extensive membrane treatment (15 day) with native FBS did not show any sign of disaggregation or enzymatic digestion. However, a higher opaqueness and thickness was observed with respect to control membranes that were incubated in PBS (Figure 6A,B). This was likely due to serum protein adsorption on the SCM surface. Indeed, also this feature is considered to be important in the long-term performance of implants [71]. In fact, once the proteins are adsorbed onto the material surface, cell adhesion and growth are facilitated. This adsorbed protein layer can also mediate the type of cells that adhere to the surface, which ultimately can determine the type of tissue that develops. The data collected in the present work indicate that F1 and F3 SCMs, either crosslinked or not, show strong in vitro enzymatic degradation resistance and are able to interact with serum proteins. Thus, they are suitable for the production of biomaterials needing long-term stability in guided tissue and bone regeneration applications.

The ability to bind water is another fundamental aspect of biomaterials. The WBC of SCMs and nc-SCMs indicated a slightly higher WBC in the F3-derived membranes than F1. Moreover, SCMs showed lower WBC than nc-SCMs, likely because most of the functional groups in the membranes are crosslinked and are less available for water interaction. Furthermore, the WBC values of the two nc-SCMs seem quite remarkable if compared to the value of the mammalian collagen not-crosslinked membrane Bio-Gide®. The latter, in fact, was 33% and 37% less hydrated than nc-SCM-F1 and nc-SCM-F3, respectively. The increased WBC in the nc-SCMs as compared to the commercial membrane could be explained by the presence, solely in the sponge membranes, of the ECM-derived GAGs. Indeed, their addition has been reported to proportionally improve the membrane's ability to bind water [72]. However, although nc-SCMs were much more hydrophilic than SCMs and commercial membranes, they appeared to be less easy to manipulate when hydrated (Figure 6C), which is a feature that has to be accurately considered when designing new biomaterials.

Preliminary biocompatibility assays showed good and encouraging results for a further employment of these biomaterials for medical purposes. Both fibroblasts and keratinocytes were able to adhere and grow on all coated plates when compared to control samples (Figures 7 and 8). In particular, for fibroblasts, the slight cell decrease and the rounded shape onto F4 coating at 16 h indicated a lower short-term biocompatibility with respect to the other FSs and to the controls that were grown onto standard rat tail collagen-coated plates. Conversely, for keratinocytes, a cell decrease was observed only at the 6 day mid-term for the F3 and F4 coatings, maybe indicating the necessity of a period of adaptation onto these two matrices with respect to the other FSs and to the controls. Anyway, this apparent difficulty of adhesion of fibroblasts at 16 h and the growth of keratinocytes at 6 day was only temporary. In fact, long-term analyses at 15 day showed similar growth rates for the two cell lines onto all of the FSs and all closely comparable to the controls (Figure 8). In addition to this, a good cell-SCM surface interaction was observed by ESEM analysis for both cell types on SCM-F1 and SCM-F3, as evidenced by the presence of a tight adherence and a multitude of cell processes interacting with the collagenous membranes (Figure 9E–H).

Marine sponges are also important sources of bioactive metabolites [36], including compounds with antioxidant properties [73,74], and it is also known that marine collagen-derived peptides possess radical scavenging activity. All of the four FSs that were extracted from *C. reniformis* showed antioxidant properties (Figure 10). These properties, however, did not reflect their collagen percentage; hence, the antioxidant activity may be determined by other molecular types, which were probably co-extracted during collagen fibril isolation. The significantly higher scavenging value of F1 suspension could be due to the absence of reducing conditions in this extraction method, as compared to the other three, which could have partially inactivated the antioxidant component. Surprisingly, the antioxidant properties were also retained in SCM-F1 and SCM-F3, even if reduced with respect to the radical scavenging values of the respective FS. This suggests that these marine membranes are suitable for wound healing applications, for skin repair after superficial cancer treatments, or for the prevention of skin photo-damage and photo-ageing.

In conclusion, as previously reported [20,21,75,76], collagen derived from marine sponges is an extremely performant biopolymer that is suitable for biomedical applications. Here, for the first time, a thorough analysis and chemical characterization of four different sponge collagenous extracts allowed for generating crosslinked thin collagenous membranes from *C. reniformis* demosponge that is suitable for TERM purposes. The two types of SCMs that were obtained showed good mechanical properties, enzymatic degradation resistance, water binding capacity, and biocompatibility. In addition to this, our results demonstrate that it is possible to adapt the extraction procedures in order to alternatively improve the mechanical properties or the antioxidant performances of the derived biomaterials thanks to the versatility of *C. reniformis*-derived extracts.

4. Materials and Methods

Chemicals
All reagents were acquired from SIGMA-ALDRICH (Milan, Italy), unless otherwise stated.

4.1. Sponge Sampling

Specimens of *C. reniformis* were collected in the area of the Portofino Promontory (Liguria, Italy) at depths of 10–20 m and were transferred in laboratory in a thermic bag. During transport, the temperature was maintained at 14–15 °C. A short-term stabulation was performed, as described in [27]. In particular, the sponges were stored at 14 °C in 200-L aquaria containing natural sea water that was collected in the same area of the Portofino Promontory with a salinity of 37‰ and was equipped with an aeration system. Finally, the sponge specimens were frozen at −20 °C until further processing.

4.2. Fibrillar Collagen Suspension Extracts

C. reniformis fibrillar collagen suspensions (FSs) were obtained using four different extraction procedures, obtaining, respectively, F1, F2, F3, and F4 extracts. As indicated in Scheme 1, for each procedure, about 25 g of frozen sponge tissue was thawed, extensively rinsed with cool deionized water, cut in small slices, and then minced in a blender in ice, with five volumes of the respective extraction buffer (step 1).

F1 collagen suspensions were obtained, as described in Gross et al. (1956) [53], with some modifications. Briefly, sponges tissue was minced in step 1 in five volumes of 100 mM ammonium bicarbonate pH 8.5, then 0.1% trypsin was added and the sample incubated overnight at 37 °C on a horizontal shaker. Afterwards, the fluid was removed by filtration with a metallic strainer and the solid material was suspended in three volumes of cool deionized water and incubated at 5 °C for three days in a rotary disk shaker aliquoted in 50 mL-tubes. The dark and viscous suspension was then filtered with a metallic strainer and the remained solid material was subjected at a second round of 3 day of water extraction. The viscous fluid that was obtained from the two rounds of water extraction was pooled and centrifuged at 1200× g, 10 min at 4 °C, in order to remove cell debris and sand particles. The supernatant fluid containing the collagen suspension was frozen at −20 °C for long-term storage.

F2 collagen suspensions were obtained using the protocol that was described by Diehl-Seifert et al. [52], with some modifications. Here, the sponge specimens were minced in step 1 in five volumes of 100 mM Tris–HCl buffer, pH 9.5, 10 mM EDTA, 8 M urea, and 100 mM 2-mercaptoethanol. The sample was incubated at room temperature (RT) continuously stirring for 24 h. Afterwards, the viscous extract was centrifuged at 5000× g, for 5 min at 4 °C. The pellet was discarded and the collagen precipitated from the supernatant by adding 1/3 of its volume of glacial acetic acid, and finally centrifuged at 20,000× g for 30 min at 4 °C. The collagen pellet was washed twice with distilled water until the pH was neutral and was finally suspended in 50 mL of 100 mM Tris–HCl buffer pH 9.0, homogenized for 30 s at 24,000 rpm with a T25 basic ULTRA-TURRAX® (IKA®-WERKE, Verke Staufen, Staufen im Breisgau, Germany) and stirred overnight at 4 °C. The collagen suspension that was obtained was frozen at −20 °C for long-term storage.

For F3 collagen suspension extraction, as already reported [46], sponge tissue was minced in step 1 in presence of 5 volumes of 50 mM Tris–HCl buffer pH 7.4, 1 M NaCl, 50 mM EDTA and 100 mM 2-mercaptoethanol. The sample was incubated at 17 °C for 4 days in a rotatory shaker disk aliquoted in 50 mL-tubes. The viscous extract was then centrifuged at 1200× g, for 10 min at 4 °C in order to remove cell debris and sand particles and the supernatant was extensively dialyzed against deionized water (ratio about 1:20, two changes per day for 5 days at 4 °C) using a 12 kDa molecular weight cutoff membrane tubing, in order to remove excess of 2-mercaptoethanol. The collagen suspension obtained was frozen at −20 °C for long-term storage.

F4 collagen suspensions were obtained as described by Di Benedetto et al. [55] with some modifications. Sponge tissue was minced in step 1 in five volumes of disaggregating solution that was composed of 100 mM Tris–HCl buffer pH 8.0, 0.5 M NaCl, 50 mM M EDTA, and 0.1 M 2-mercaptoethanol. The sample was incubated at 17 °C for 4 days in a rotatory shaker disk that was aliquoted in 50 mL-tubes. The viscous extract was then centrifuged at 1200× g, for 10 min at 4 °C in order to remove cell debris and sand particles, and the supernatant was extensively dialyzed against deionized water, as described for F3 collagen extraction, and at the end, it was frozen at −20 °C for long-term storage. Finally, in order to obtain the fibrillar concentration of each FS extract, 1 mL of suspension was lyophilized and weighted. All of the procedures were carried out two times in duplicate.

4.3. FS Characterization

4.3.1. BCA Total Protein Quantification

0.2 mL of each FS sample (F1, F2, F3, and F4) were centrifuged at 18,000× *g*, for 5 min at RT, the supernatant was discarded and the insoluble collagenous pellet was solubilized in 0.2 mL of 8 M urea pre-heated at 50 °C. The samples were then centrifuged at 18,000× *g*, for 2 min at RT in order to remove any insoluble residues. Total protein content was assayed in the soluble supernatant with Bicinchoninic Acid Protein Assay kit, following the manufacturer's instructions. Absorbance of each sample was read at 562 nm using a Beckman spectrophotometer (DU 640, Beckman Coulter SpA, Milan, Italy), in comparison to a skin porcine gelatin standard curve. The procedure was carried out in duplicate.

4.3.2. Collagen Quantification

The total collagen present in each FS was determined by the estimation of the hydroxyproline content using a modified method that was based on the Cloramine T reaction [77].

0.2 mL of each FS was hydrolyzed with 2 N NaOH by autoclaving at 120 °C for 20 min. Samples were neutralized by adding one volume of 2N HCl and were diluted fourfold in deionized water. The hydroxyproline concentration evaluation was obtained by adding Cloramine T and Ehrlich's reagent, as described previously [77]. Absorbance of each sample was read at 550 nm using a Beckman spectrophotometer (DU 640), in comparison to a cis-4-hydroxy-L-proline standard curve, and, finally, the content of hydroxylated proline residue was used to infer collagen content of each FS using the proportion factor of 1 g of hydroxyproline per 10 g of collagen [44]. The procedure was carried out in duplicate.

4.3.3. Alcian Blue GAG Assay

The glycosaminoglycan (GAGs) content was measured in each FS by the Alcian blue GAG assay, as described by [78]. To 20 µL of each FS extract, 20 µL of 0.027 M H_2SO_4, 0.375% Triton X-100, and 4 M guanidine-HCl were added, and then GAGs were stained with 0.2 mL of working dye solution containing 0.25% Triton X-100, 0.018M H_2SO_4 and 0.005% Alcian blue. All samples were incubated 10 min at RT in a horizontal shaker and then centrifuged at 18,000× *g*, for 10 min at 4 °C. The stained GAG pellet obtained was solubilized with 0.4 mL of 4 M guanidine-HCl and the absorbance of each sample was read at 620 nm using a Beckman spectrophotometer (DU 640), in comparison to shark cartilage chondroitin sulfate standard curve. The procedure was carried out in duplicate.

4.3.4. Transmission Electron Microscopy: Negative Staining

FS samples (F1, F2, F3, and F4) were fixed in 2% PFA in PBS for 20 min at RT and were washed out in PBS. 5 µL drops of fixed FS were placed on formwar-coated grids for 20 min. When the suspension was partially dried, grids were washed by touching them three times to the surface of a drop of distilled water. Grids adsorbed with FS samples were then stained with 2% uranyl acetate in 0.15 M oxalic acid for 5 min and an additional 5 min in a 9:1 mixture of 2% uranyl acetate and metylcellulose 25 ctp. FS samples were imaged with a CM10 Philips transmission electron microscope equipped with Megaview 3 camera and Olympus SIS iTEM software for digital image acquisition. Representative images of the four FS preparations were taken at 92,000× magnification (scale bar: 200 nm).

4.3.5. FS Qualitative Evaluation by Histological Methods

100 µL of each FS were smeared in triplicate on histological slides and were dried for 30 min at 37 °C. The slides were stained similarly to standard histological sections with various methods, (all products by Bio-Optica SpA, Milan, Italy): Periodic Acid Schiff (Hotchkiss-Mc Manus) (PAS) that produces a red staining reacting with glycol-containing cellular elements, e.g., glycogen or

neutral mucopolysaccharides; Alcian (pH 2.5), whih stains in blue acidic polysaccharides, such as glycosaminoglycans and some types of mucopolysaccharides, and finally, Picro-Sirius Red (PicroS), which stains specifically collagen fibres: in bright-field microscopy collagen appears red, when examined through crossed polarized light the larger collagen fibres are bright yellow or orange, and the thinner ones, including reticular fibres, are green [79]. The sections were observed through a Leica DMRB light and epifluorescence microscope equipped with DIC (Leica microsystems srl, Milan, Italy). Images were acquired using a Leica CCD camera DFC420C.

4.3.6. Rheological Characterization

The rheological measurements were performed with an Anton Paar Physica MCR 301 Rheometer (Anton Paar, GmbH, Ostfildern, Germany), which was equipped with a 50 mm cone/plate geometry (CP50). The viscosity curves were carried out using a shear rate range between 0.1 and 2500 s^{-1}, and each sample was tested twice to check for repeatability. The Rheometer was used with a Peltier heating system for an accurate control of the temperature. All of the measurements were performed at 20 °C.

4.4. SCM Production

For biocompatibility tests, all four FS were used to directly coat 24-well and 96-well plates. 300 µL (for 24-well plates) or 50 µL (for 96-well plates) of 2 mg/mL of each FS and of a standard rat tail collagen in the presence of 0.01% TritonX-100 were placed on the plates and were left to dry at 37 °C overnight. The coated plates were then incubated with 300 µL (24-well plates) or 50 µL (96-well plates) of EDC/NHS cross-linker solution: 30 mM (EDC 1-ethyl-3-(3-dimethylaminopropyl) carbodiimide/15 mM NHS N-hydroxysuccinimide in MES (N-morpholinoethanesulfonic acid buffer 100 mM, pH 5.5) for 4 h at RT in the dark. Crosslinked coated wells were then washed twice with 0.1 M Na_2HPO_4 for 30 min and twice with deionized H_2O for 15 min, and then finally dried/sterilized with 70% ethanol solution.

SCMs were produced using silicone molds as rectangular (25 × 28 mm) sheets for in vitro biodegradation, water binding capacity, and ultrastructural analyses that were filled with 3.3 mL of 2 mg/mL of each FS and as rectangular (10 × 45 mm) sheets for mechanical tests, filled with 2.25 mL of 2 mg/mL of each FS. The FSs were left to dry at 37 °C, successively incubated, as previously described with EDC/NHS cross-linker solution for 4 h, washed with Na_2HPO_4, and deionized H_2O and finally dried/sterilized with 70% ethanol solution.

Negative controls, lacking the cross-linker step, were prepared as well using 3.3 mL of 2 mg/mL F1 and F3 fibrillar suspensions casted in the same molds without adding the cross-linker solution.

4.5. SCM Characterization

4.5.1. In Vitro Enzymatic Resistance

In vitro enzymatic resistance of the F1- and F3-derived SCMs and nc-SCMs was determined by evaluating their stability in native fetal bovine serum (FBS, Euroclone, Milan, Italy) and in the presence of a commercial bacterial collagenase. For the FBS stability test, F1- and F3-derived SCMs (6 mg) were incubated with 1 mL of FBS at 37 °C in a humidified atmosphere for 15 day. The collagenase stability evaluation was determined, as already described [44]. Briefly, collagenase from *Clostridium histolyticum* in a ratio of 1:10 (enzyme:substrate) was used in 1 mL of PBS at 37 °C for 6 day. The enzymatic solution was refreshed daily to ensure continuous enzymatic activity on the SCMs. As a control, a commercial porcine collagen membrane (25 × 25 mm), called Bio-Gide® (Geistlich Pharma AG, Wolhusen, Switzerland), which is widely used in dental and bone surgery, was submitted to digestion as well. Experiments were performed three times in duplicate.

4.5.2. Water Binding Capacity

The water binding property of SCMs and nc-SCMs, was evaluated according to an already described method [80]. In brief, phosphate buffered saline (PBS, pH 7.4) was used as hydration medium, and the membranes were soaked for 1 h at RT by complete immersion. Then, the surface excess medium was removed by touching to a filter paper and then the SCMs were weighed (wet weight). As a control, the commercial porcine collagen membrane Bio-Gide®, with the same surface area of the SCMs and nc-SCMs, was submitted to hydration as well. The water binding capacity (WBC) was determined using the following Equation:

$$WBC\ (\%) = (Ww - Wd)/Wd \times 100.$$

Experiments were performed three times in duplicate.

4.5.3. Dynamic Mechanical Tests

Dynamic mechanical analysis was performed with an Anton Paar Physica MCR 301 Rheometer (Anton Paar, GmbH, Ostfildern, Germany), using a Solid Rectangular Fixtures (SFR) system. The temperature was set at 20 °C and each sample was tested at least twice. Tests were performed both in Amplitude Sweep and Frequency Sweep modes.

The values of the stress amplitude were checked by means of an amplitude sweep test, with a deformation range (γ) from 0.001 up to 0.1% at a fixed frequency of 0.1 Hz, in order to ensure that all of the measurements were performed within the linear viscoelastic region (LVER).

In order to obtain information about the storage (or elastic) modulus (G'), the loss (or viscous) modulus (G''), the complex viscosity (η^*) as a function of the frequency and the Frequency Sweep tests in the range 0.01–10 Hz, at a fixed deformation of 0.01% within LVER, were performed.

The data were collected and analysed using Rheoplus/32 Service V3.40 software.

4.6. SCM Biocompatibility Evaluation

4.6.1. Cell Cultures

The L929 mouse fibroblast cell line and the National Collection of Type Cultures (NCTC) human keratinocyte cell line were obtained from the American Type Culture Collection (LGC Standards srl, Milan, Italy). Cells were cultured at 37 °C in a humidified, 5% CO_2 atmosphere, in high glucose Dulbecco's modified Eagle's medium (D-MEM) with glutamax (Euroclone, Milan, Italy), which was supplemented with 10% FBS (Euroclone) with penicillin/streptomycin as antibiotics.

4.6.2. Cell Growth and Cell Adhesion

To evaluate cell growth on SCM-coated plates, experiments were performed in quadruplicate on 96-well plates. Both L929 and NCTC cell lines were plated at a density of 5000 cells/well on 96-well plates pre-coated with F1, F2, F3, and F4. *C. reniformis* FSs extracts, prepared, as described in Section 4.4. Conversely, controls were grown onto rat tail standard collagen coated plates, prepared, as already described in the same paragraph. Cells were incubated for 3 days, 6 days, and 15 days at 37 °C in complete medium. At the end of the experiments cell viability was assayed by the MTT test (0.5 mg/mL final concentration), as already reported [81]. For the evaluation of cell adhesion on FS-coated wells, experiments were performed in duplicate on 24-well plates. Both L929 and NCTC cell lines were plated at a density of 50,000 cells/well on 24-well plates that were pre-coated with F1, F2, F3, F4 extracts, or rat tail standard collagen as control. Cells were allowed to adhere for 16 h at 37 °C in complete medium and then the MTT test was performed as well to estimate the attached cells when compared to control cells on uncoated wells. Data are means ± S.D. of four independent experiments.

4.6.3. Light and ESEM Microscopy

For image acquisition in light microscopy cells were seeded at a density of 50,000 cells/well on 24-well plates pre-coated or not with F1, F2, F3, and F4 *C. reniformis* FSs, prepared as described in paragraph 4.4, and allowed to adhere for 16 h in complete medium. At the end of the experiment, the cells were washed with PBS to remove floating-unattached cells and stained with crystal violet by standard procedures (0.1% crystal violet in methanol for 30 min, followed by extensive washing with water). For image acquisition, an inverted optical microscope (IX53 Olympus, Tokyo, Japan) was used equipped with a CCD camera (U-LH100HG Olympus, Tokyo, Japan) and the relative software.

For ESEM observation of mammalian cells adhering to F1 and F3-derirved SCMs, 50,000 L929 fibroblasts or NCTC keratinocytes were seeded or not onto 0.25 cm^2 ethanol-sterilized membranes and incubated for 3 day at 37 °C in complete medium. At the end of the experiment SCMs were washed with PBS and fixed with with a mixture of 2% paraformaldehyde and 2.5 % glutaraldehyde 7.4 pH for 30 min, washed with PBS, extracted from the well plates, and mounted on a plastic support.

Specimens of F1, F2, F3, and F4 SCMs alone and of cells adhering to SCM-F1 and F3 were dehydrated by passing through a series of ethanol alcoholic solutions with an increasing concentration of up to 100%. The dehydrated membranes were further dehydrated at critical point, avoiding the use of acetone due to the presence of the plastic support, and then graphite was covered and observed. Observation and acquisition of the images of cells adhering to SCM-F1 and F3 were performed with an ESEM Vega3–Tescan, type LMU (Tescan Brno s.r.o., Brno, Czech Republic) equipped with a microanalyzer system EDS-Apollo_x and EDS Texture And Elemental Analytical Microscopy software (TEAM). Observation and acquisition of the four SCMs per se were performed with a FESEM Zeiss SUPRA 40 VP (Carl Zeiss AG, Oberkochen, Germany) and its associated software. The showed results are representative of three independent experiments.

Physical measurements of the fibrillar diameter and of the pore areas observed in the collagen membranes was performed on the images that were obtained by the FESEM analysis of the various membranes, using the ImageJ free software (Rasband, W.S., ImageJ, U. S. National Institutes of Health, Bethesda, MD, USA, https://imagej.nih.gov/ij/, 1997–2016). Means $\pm$ S.D. were calculated on at least 10 measurements of fibril diameter or pore areas performed on each membrane.

4.7. DPPH Radical Scavenging Activity

The radical scavenging activity was evaluated on each FS and on SCMs that were obtained from F1 and F3 suspensions. 500 µL of 1 mg/mL of each FS were added to 500 µL of methanol, and then to 250 µL of 0.1 mM DPPH in methanol solution (2,2-diphenyl-1-picrylhydrazyl, Calbiochem®, Millipore SpA, Milan, Italy). A negative control sample with deionized water was prepared in the same manner. All of the samples were incubated for 30 min at RT in the dark. Then, the samples were centrifuged at 18,000$\times$ *g*, for 3 min at RT, and finally the supernatant was read at 517 nm using a Beckman spectrophotometer (DU 640). In the blank sample, the DPPH solution was substituted with methanol. The antioxidant activity of the samples was evaluated by the inhibition percentage of DPPH radical using the following equation:

$$\text{DPPH radical scavenging activity (\%)} = (A0 - A)/A0 \times 100\% \tag{1}$$

where A was sample absorbance rate; A0 was the absorbance of the negative control. The procedure was carried out in duplicate.

For the evaluation of the radical scavenging activity of SCM-F1 and SCM-F3, fragments of 87.5, 175, 350, and 700 mm^2 were immersed in 500 µL of deionized water, and, after 15 min of incubation at RT, the samples were processed, as described above. The DPPH radical scavenging activity values were plotted in function of the SCM surface and the surface value of the 50% of scavenging activity (SA_{50}) was consequently calculated.

4.8. Statistical Analyses

Statistical analysis was performed using one-way ANOVA plus Tukey's post-test (GraphPad Software, Inc., San Diego, CA, USA). p values < 0.05 were considered to be significant.

Acknowledgments: The authors are indebted to Laura Negretti and Mauro Michetti for their precious technical support in ESEM analyses. This work was supported by University of Genova Funding to both S.S. and M.G. and to SIR funding by the Italian Ministry of University and Research (MIUR) to M.B.

Author Contributions: M.P.: conception and design of the work, manuscript writing, experimental design, data analysis and interpretation. S.S.: conception of the work and experimental design, manuscript writing, data analysis and interpretation, financial support. L.G., M.C., S.V., K.C. and M.C.G.: experimental design and data analysis. M.B. and G.C.: collection of specimens. M.G.: critical review of the manuscript and financial support.

Conflicts of Interest: The authors declare no conflict of interest.

References

1. Ellis, D.L.; Yannas, I.V. Recent advances in tissue synthesis in vivo by use of collagen glycosaminoglycan copolymers. *Biomaterials* **1996**, *17*, 291–299. [CrossRef]
2. Chen, P.; Marsilio, E.; Goldstein, R.H.; Yannas, I.V.; Spector, M. Formation of lung alveolar like structures in collagen-glycosaminoglycan scaffolds in vitro. *Tissue Eng.* **2005**, *11*, 1436–1448. [CrossRef] [PubMed]
3. Buijtenhuijs, P.; Buttafoco, L.; Poot, A.A.; Daamen, W.F.; van Kuppevelt, T.H.; Dijkstra, P.; de Vos, R.A.; Ster, L.M.; Geelkerken, B.R.; Feijen, J.; et al. Tissue engineering of bloodvessels: Characterization of smooth-muscle cells for culturing on collagen-and-elastin-based scaffolds. *Biotechnol. Appl. Biochem.* **2004**, *39*, 141–149. [CrossRef] [PubMed]
4. Garcia, Y.; Hemantkumar, N.; Collighan, R.; Griffin, M.; Rodriguez-Cabello, J.C.; Pandit, A. In vitro characterization of a collagen scaffold enzymatically cross-linked with a tailored elastin-like polymer. *Tissue Eng. Part A* **2009**, *15*, 887–899. [CrossRef] [PubMed]
5. Damour, O.; Gueugniaud, P.Y.; Berthin-Maghit, M.; Rousselle, P.; Berthod, F.; Sahuc, F.; Collombel, C. A dermal substrate made of collagen-GAG-chitosan for deep burn coverage: First clinical uses. *Clin. Mater.* **1994**, *15*, 273–276. [CrossRef]
6. Shahabeddin, L.; Berthod, F.; Damour, O.; Collombel, C. Characterization of skin reconstructed on a chitosan-cross-linked collagen-glycosaminoglycan matrix. *Skin Pharmacol.* **1990**, *3*, 107–114. [CrossRef] [PubMed]
7. Chattopadhyay, S.; Raines, R.T. Review collagen-based biomaterials for wound healing. *Biopolymers* **2014**, *101*, 821–833. [CrossRef] [PubMed]
8. Silva, T.; Moreira-Silva, J.; Marques, A.; Domingues, A.; Bayon, Y.; Reis, R. Marine origin collagens and its potential applications. *Mar. Drugs* **2014**, *12*, 5881–5901. [CrossRef] [PubMed]
9. Pati, F.; Adhikar, B.; Dhara, S. Isolation and characterization of fish scale collagen of higher thermal stability. *Biores. Technol.* **2010**, *101*, 3737–3742. [CrossRef] [PubMed]
10. Jridi, M.; Bardaa, S.; Moalla, D.; Rebaii, T.; Souissi, N.; Sahnoun, Z.; Nasri, M. Microstructure, rheological and wound healing properties of collagen-based gel from cuttlefish skin. *Int. J. Biol. Macromol.* **2015**, *77*, 369–374. [CrossRef] [PubMed]
11. Boero, F.; Bouillon, J.; Gravili, C.; Miglietta, M.P.; Parsons, T.; Piraino, S. Gelatinous plankton: Irregularities rule the world (sometimes). *Mar. Ecol. Prog. Ser.* **2008**, *356*, 299–310. [CrossRef]
12. Song, E.; Kim, S.Y.; Chun, T.; Byun, H.J.; Lee, Y.M. Collagen scaffolds derived from a marine source and their biocompatibility. *Biomaterials* **2006**, *27*, 2951–2961. [CrossRef] [PubMed]
13. Feuda, R.; Dohrmann, M.; Pett, W.; Philippe, H.; Rota-Stabelli, O.; Lartillot, N.; Wörheide, G.; Pisani, D. Improved modeling of compositional heterogeneity supports sponges as sister to all other animals. *Curr. Biol.* **2017**, *27*, 3864–3870. [CrossRef] [PubMed]
14. Simpson, T.L. Collagen fibrils, spongin, matrix substances. In *The Cell Biology of Sponges*; Springer: New York, NY, USA, 1984; ISBN 978-1-46-129740-6.
15. Garrone, R. *Phylogenesis of Connective Tissue: Morphological Aspects and Biosynthesis of Sponge Intercellular Matrix*; Karger, S., Ed.; University of Michigan: Ann Arbor, MI, USA, 1978; pp. 1–250, ISBN 978-3-80-552767-5.

16. Junqua, S.; Robert, L.; Garrone, R.; Pavans de Ceccatty, M.; Vacelet, J. Biochemical and morphological studies on collagens of horny sponges. Ircinia filaments compared to spongines. *Connect. Tissue Res.* **1974**, *2*, 193–203. [CrossRef] [PubMed]

17. Exposito, J.Y.; Garrone, R. Characterization of a fibrillar collagen gene in sponges reveals the early evolutionary appearance of two collagen gene families. *Proc. Natl. Acad. Sci. USA* **1990**, *87*, 6669–6673. [CrossRef] [PubMed]

18. Exposito, J.Y.; Le Guellec, D.; Lu, Q.; Garrone, R. Short chain collagens in sponges are encoded by a family of closely related genes. *J. Biol. Chem.* **1991**, *266*, 21923–21928. [PubMed]

19. Kim, M.M.; Mendis, E.; Rajapakse, N.; Lee, S.H.; Kim, S.K. Effect of spongin derived from *Hymeniacidon sinapium* on bone mineralization. *J. Biomed. Mater. Res. B Appl. Biomater.* **2009**, *90*, 540–546. [CrossRef] [PubMed]

20. Green, D.; Howard, D.; Yang, X.; Kelly, M.; Oreffo, R.O. Natural marine sponge fibre skeleton: A biomimetic scaffold for human osteoprogenitor cell attachment, growth, and differentiation. *Tissue Eng.* **2003**, *9*, 1159–1166. [CrossRef] [PubMed]

21. Lin, Z.; Solomon, K.L.; Zhang, X.; Pavlos, N.J.; Abel, T.; Willers, C.; Dai, K.; Xu, J.; Zheng, Q.; Zheng, M. In vitro evaluation of natural marine sponge collagen as a scaffold for bone tissue engineering. *Int. J. Biol. Sci.* **2011**, *7*, 968–977. [CrossRef] [PubMed]

22. Nandi, S.K.; Kundu, B.; Mahato, A.; Thakur, N.L.; Joardar, S.N.; Mandal, B.B. In vitro and in vivo evaluation of the marine sponge skeleton as a bone mimicking biomaterial. *Integr. Biol.* **2015**, *7*, 250–262. [CrossRef] [PubMed]

23. Norman, M.; Bartczak, P.; Zdarta, J.; Tylus, W.; Szatkowski, T.; Stelling, A.L.; Ehrlich, H.; Jesionowski, T. Adsorption of C.I. Natural Red 4 onto spongin skeleton of marine demosponge. *Materials* **2014**, *8*, 96–116. [CrossRef] [PubMed]

24. Norman, M.; Bartczak, P.; Zdarta, J.; Ehrlich, H.; Jesionowski, T. Anthocyanin dye conjugated with *Hippospongia communis* marine demosponge skeleton and its antiradical activity. *Dyes Pigment.* **2016**, *134*, 541–552. [CrossRef]

25. Norman, M.; Zdarta, J.; Bartczak, P.; Piasecki, A.; Petrenko, I.; Ehrlich, H. Marine sponge skeleton photosensitized by copper phthalocyanine: A catalyst for Rhodamine B degradation. *Open Chem.* **2016**, *14*, 243–254. [CrossRef]

26. Norman, M.; Bartczak, P.; Zdarta, J.; Tomala, W.; Żurańska, B.; Dobrowolska, A.; Piasecki, A.; Czaczyk, K.; Ehrlich, H.; Jesionowsk, T. Sodium copper chlorophyllin immobilization onto *Hippospongia communis* marine demosponge skeleton and its antibacterial activity. *Int. J. Mol. Sci.* **2016**, *17*, 1564. [CrossRef] [PubMed]

27. Norman, M.; Żółtowska-Aksamitowska, S.; Zgoła-Grześkowiak, A.; Ehrlich, H.; Jesionowski, T. Iron(III) phthalocyanine supported on a spongin scaffold as an advanced photocatalyst in a highly efficient removal process of halophenols and bisphenol A. *J. Hazard. Mater.* **2018**, *347*, 78–88. [CrossRef] [PubMed]

28. Zdarta, J.; Norman, M.; Smułek, W.; Moszynski, D.; Kaczorek, E.; Stelling, A.L.; Ehrlich, H.; Jesionowski, T. Spongin-based scaffolds from *Hippospongia communis* demosponge as an effective support for lipase immobilization. *Catalysts* **2017**, *7*, 147. [CrossRef]

29. Szatkowski, T.; Wysokowski, M.; Lota, G.; Pęziak, D.; Bazhenov, V.V.; Nowaczyk, G.; Walter, J.; Molodtsov, S.L.; Stöcker, H.; Himcinschi, C.; et al. Novel nanostructured hematite-spongin composite developed using extreme biomimetic approach. *RSC Adv.* **2015**, *5*, 79031–79040. [CrossRef]

30. Szatkowski, T.; Stefańska, K.S.; Wysokowski, M.; Stelling, A.L.; Joseph, Y.; Ehrlich, H.; Jesionowski, T. Immobilization of titanium(IV) oxide onto 3D spongin scaffolds of marine sponge origin according to extreme biomimetics principles for removal of C.I. Basic Blue 9. *Biomimetics* **2017**, *2*, 4. [CrossRef]

31. Szatkowski, T.; Kopczyński, K.; Motylenko, M.; Borrmann, H.; Mania, B.; Graś, M.; Lota, G.; Bazhenov, V.V.; Rafaja, D.; Roth, F.; et al. Extreme Biomimetics: Carbonized 3D spongin scaffold as a novel support for nanostructured manganese oxide(IV) and its electrochemical applications. *Nano Res.* **2018**. [CrossRef]

32. Heinemann, S.; Ehrlich, H.; Knieb, C.; Hanke, T. Biomimetically inspired hybrid materials based on silicified collagen. *Int. J. Mater. Res.* **2007**, *98*, 603–608. [CrossRef]

33. Heinemann, S.; Heinemann, C.; Ehrlich, H.; Meyer, M.; Baltzer, H.; Worch, H.; Hanke, T. A novel biomimetic hybrid material made of silicified collagen: Perspectives for bone replacement. *Adv. Eng. Mater.* **2007**, *9*, 1061–1068. [CrossRef]

34. Ehrlich, H.; Heinemann, S.; Heinemann, C.; Simon, P.; Bazhenov, V.V.; Shapkin, N.P.; Born, R.; Zabachnick, K.R.; Hanke, T.; Worch, H. Nanostructural organisation of naturally occuring composites: Part I. Silica-Collagen-Based Biocomposites. *J. Nanomater.* **2008**. [CrossRef]

35. Ehrlich, H.; Deutzmann, R.; Brunner, E.; Cappellini, E.; Koon, H.; Solazzo, C.; Yang, Y.; Ashford, D.; Thomas-Oates, J.; Lubeck, M.; et al. Mineralization of the Meter-long Biosilica Structures of Glass Sponges is template on Hydroxylated Collagen. *Nat. Chem.* **2010**, *2*, 1084–1088. [CrossRef] [PubMed]

36. Mehbub, M.F.; Lei, J.; Franco, C.; Zhang, W. Marine Sponge Derived Natural Products between 2001 and 2010: Trends and Opportunities for Discovery of Bioactives. *Mar. Drugs* **2014**, *12*, 4539–4577. [CrossRef] [PubMed]

37. Leal, M.C.; Puga, J.; Serôdio, J.; Gomes, N.C.; Calado, R. Trends in the discovery of new marine natural products from invertebrates over the last two decades—Where and what are we bioprospecting? *PLoS ONE* **2012**, *7*, e30580. [CrossRef] [PubMed]

38. Sipkema, D.; Osinga, R.; Schatton, W.; Mendola, D.; Tramper, J.; Wijffels, R.H. Large-scale production of pharmaceuticals by marine sponges: Sea, cell, or synthesis? *Biotechnol. Bioeng.* **2005**, *90*, 201–222. [CrossRef] [PubMed]

39. Ruiz, C.; Valderrama, K.; Zea, S.; Castellanos, L. Mariculture and natural production of the antitumoural (+)-discodermolide by the Caribbean marine sponge *Discodermia dissoluta*. *Mar Biotechnol.* **2013**, *15*, 571–583. [CrossRef] [PubMed]

40. Custodio, M.R.; Prokic, I.; Steffen, R.; Koziol, C.; Borojevic, R.; Brümmer, F.; Nickel, M.; Müller, W.E. Primmorphs generated from dissociated cells of the sponge *Suberites domuncula*: A model system for studies of cell proliferation and cell death. *Mech. Ageing Dev.* **1998**, *105*, 45–59. [CrossRef]

41. Pozzolini, M.; Mussino, F.; Cerrano, C.; Scarfi, S.; Giovine, M. Sponge cell cultivation: Optimization of the model *Petrosia ficiformis* (Poiret 1789). *J. Exp. Mar. Biol. Ecol.* **2014**, *454*, 70–77. [CrossRef]

42. Wilkie, I.C.; Parma, L.; Bonasoro, F.; Bavestrello, G.; Cerrano, C.; Carnevali, M.D. Mechanical adaptability of a sponge extracellular matrix: Evidence for cellular control of mesohyl stiffness in *Chondrosia reniformis* Nardo. *J. Exp. Biol.* **2006**, *209*, 4436–4443. [CrossRef] [PubMed]

43. Fassini, D.; Parma, L.; Lembo, F.; Candia Carnevali, M.D.; Wilkie, I.C.; Bonasoro, F. The reaction of the sponge *Chondrosia reniformis* to mechanical stimulation is mediated by the outer epithelium and the release of stiffening factor(s). *Zoology* **2014**, *117*, 282–291. [CrossRef] [PubMed]

44. Garrone, R.; Huc, A.; Junqua, S. Fine structure and physicochemical studies on the collagen of the marine sponge *Chondrosia reniformis* Nardo. *J. Ultrastruct. Res.* **1975**, *52*, 261–275. [CrossRef]

45. Imhoff, J.M.; Garrone, R. Solubilization and Characterization of *Chondrosia reniformis* Sponge Collagen. *Connect. Tissue Res.* **1983**, *11*, 193–197. [CrossRef] [PubMed]

46. Pozzolini, M.; Bruzzone, F.; Berilli, V.; Mussino, F.; Cerrano, C.; Benatti, U.; Giovine, M. Molecular characterization of a nonfibrillar collagen from the marine sponge *Chondrosia reniformis* Nardo 1847 and positive effects of soluble silicates on its expression. *Mar. Biotechnol.* **2012**, *14*, 281–293. [CrossRef] [PubMed]

47. Pozzolini, M.; Scarfi, S.; Mussino, F.; Ghignone, S.; Vezzulli, L.; Giovine, M. Molecular characterization and expression analysis of the first Porifera tumor necrosis factor superfamily member and of its putative receptor in the marine sponge *C. reniformis*. *Dev. Comp. Immunol.* **2016**, *57*, 88–98. [CrossRef] [PubMed]

48. Pozzolini, M.; Scarfi, S.; Mussino, F.; Ferrando, S.; Gallus, L.; Giovine, M. Molecular cloning, characterization, and expression analysis of a Prolyl 4-Hydroxylase from the marine sponge *Chondrosia reniformis*. *Mar. Biotechnol.* **2015**, *17*, 393–407. [CrossRef] [PubMed]

49. Heinemann, S.; Ehrlich, H.; Douglas, T.; Heinemann, C.; Worch, H.; Schatton, W.; Hanke, T. Ultrastructural studies on the collagen of the marine sponge *Chondrosia reniformis* Nardo. *Biomacromolecules* **2007**, *8*, 3452–3457. [CrossRef] [PubMed]

50. Swatschek, D.; Schatton, W.; Kellermann, J.; Muller, W.; Kreuter, J. Marine sponge collagen: Isolation, characterization and effects on the skin parameters surface-pH, moisture and sebum. *Eur. J. Pharm. Biopharm.* **2002**, *53*, 107–113. [CrossRef]

51. Nicklas, M.; Schatton, W.; Heinemann, S.; Hanke, T.; Kreuter, J. Enteric coating derived from marine sponge collagen. *Drug Dev. Ind. Pharm.* **2009**, *35*, 1384–1388. [CrossRef] [PubMed]

52. Kreuter, J.; Muller, W.; Swatschek, D.; Schatton, W.; Schatton, M. Method for Isolating Sponge Collagen and Producing Nanoparticulate Collagen, and the Use Thereof. U.S. Patent 20030032601 A1, 3 March 2000.

53. Gross, J.; Sokal, Z.; Rougvie, M. Structural and chemical studies on the connective tissue of marine sponges. *J. Histochem. Cytochem.* **1956**, *4*, 227–246. [CrossRef] [PubMed]

54. Diehl-Seifert, B.; Kurelec, B.; Zahn, R.K.; Dorn, A.; Jericevic, B.; Uhlenbruck, G.; Müller, W.E. Attachment of sponge cells to collagen substrata: Effect of a collagen assembly factor. *J. Cell Sci.* **1985**, *79*, 271–285. [PubMed]

55. Di Benedetto, C.; Barbaglio, A.; Martinello, T.; Alongi, V.; Fassini, D.; Cullorà, E.; Patruno, M.; Bonasoro, F.; Barbosa, M.A.; Carnevali, M.D.; et al. Production, characterization and biocompatibility of marine collagen matrices from an alternative and sustainable source: The sea urchin *Paracentrotus lividus*. *Mar. Drugs* **2014**, *12*, 4912–4933. [CrossRef] [PubMed]

56. Mezger, T.G. *The Rheology Handbook: For Users of Rotational and Oscillatory Rheometers*, 2nd ed.; Vincentz Network: Hannover, Germany, 2006; ISBN 978-3-86-630842-8.

57. Nam, K.A.; You, S.G.; Kim, S.M. Molecular and physical characteristics of squid (*Todarodes pacificus*) skin collagens and biological properties of their enzymatic hydrolysates. *J. Food Sci.* **2008**, *73*, 249–255. [CrossRef] [PubMed]

58. Pozzolini, M.; Valisano, L.; Cerrano, C.; Menta, M.; Schiaparelli, S.; Bavestrello, G.; Benatti, U.; Giovine, M. Influence of rocky substrata on three-dimensional sponge cells model development. *In Vitro Cell. Dev. Biol. Anim.* **2010**, *46*, 140–147. [CrossRef] [PubMed]

59. Mauch, C.; Hatamochi, A.; Scharffetter, K.; Krieg, T. Regulation of collagen synthesis in fibroblasts within a three-dimensional collagen gel. *Exp. Cell Res.* **1988**, *178*, 1508–1515. [CrossRef]

60. Nusgens, B.; Merrill, C.; Lapiere, C.; Bell, E. Collagen biosynthesis by cells in a tissue equivalent matrix in vitro. *Coll. Relat. Res.* **1984**, *4*, 351–361. [CrossRef]

61. Parenteau-Bareil, R.; Gauvin, R.; Berthod, F. Collagen-based biomaterials for tissue engineering applications. *Materials* **2010**, *3*, 1863–1887. [CrossRef]

62. Badylak, S.F. The extracellular matrix as a biologic scaffold material. *Biomaterials* **2007**, *28*, 3587–3593. [CrossRef] [PubMed]

63. Wilkie, I.C. Mutable collagenous tissue: Overview and biotechnological perspective. *Prog. Mol. Subcell. Biol.* **2005**, *39*, 221–250. [PubMed]

64. Bondioli, E.; Fini, M.; Veronesi, F.; Giavaresi, G.; Tschon, M.; Cenacchi, G.; Cerasoli, S.; Giardino, R.; Melandri, D. Development and evaluation of a decellularized membrane from human dermis. *J. Tissue Eng. Regen. Med.* **2014**, *8*, 325–336. [CrossRef] [PubMed]

65. Wu, J.J.; Eyre, D.R.; Slayter, H.S. Type VI collagen of the intervertebral disc. Biochemical and electron-microscopic characterization of the native protein. *Biochem. J.* **1987**, *248*, 373–381. [CrossRef] [PubMed]

66. Pozzolini, M.; Scarfi, S.; Gallus, L.; Ferrando, S.; Cerrano, C.; Giovine, M. Silica-induced fibrosis: An ancient response from the early metazoans. *J. Exp. Biol.* **2017**, *220*, 4007–4015. [CrossRef] [PubMed]

67. Haugh, M.G.; Murphy, C.M.; McKiernan, R.C.; Altenbuchner, C.; O'Brien, F.J. Crosslinking and mechanical properties significantly influence cell attachment, proliferation, and migration within collagen glycosaminoglycan scaffolds. *Tissue Eng. Part A* **2011**, *17*, 1201–1208. [CrossRef] [PubMed]

68. Li, C.; Duan, L.; Tian, Z.; Liu, W.; Li, G.; Huang, X. Rheological behavior of acylated pepsin-solubilized collagen solutions: Effects of concentration. *Korea-Aust. Rheol. J.* **2015**, *27*, 287–295. [CrossRef]

69. Goo, H.C.; Hwangb, Y.-S.; Choib, Y.R.; Choc, H.N.; Suha, S. Development of collagenase-resistant collagen and its interaction with adult human dermal fibroblasts. *Biomaterials* **2003**, *24*, 5099–5113. [CrossRef]

70. Wang, J.; Wang, L.; Zhou, Z.; Lai, H.; Xu, P.; Liao, L.; Wei, J. Biodegradable polymer membranes applied in guided bone/tissue regeneration: A review. *Polymers* **2016**, *8*, 115. [CrossRef]

71. Collier, T.O.; Jenney, C.R.; DeFife, K.M.; Anderson, J.M. Protein adsorption on chemically modified surfaces. *Biomed. Sci. Instrum.* **1997**, *33*, 178–183. [PubMed]

72. Pieper, J.S.; Oosterhof, A.; Dijkstra, P.J.; Veerkamp, J.H.; van Kuppevelt, T.H. Preparation and characterization of porous crosslinked collagenous matrices containing bioavailable chondroitin sulphate. *Biomaterials* **1999**, *20*, 847–858. [CrossRef]

73. Liu, Y.; Ji, H.; Dong, J.; Zhang, S.; Lee, K.J.; Matthew, S. Antioxidant alkaloid from the South China Sea marine sponge *Iotrochota* sp. *Z. Naturforsch. C* **2008**, *63*, 636–638. [CrossRef] [PubMed]

74. Alonso, E.; Alvariño, R.; Leirós, M.; Tabudravu, J.N.; Feussner, K.; Dam, M.A.; Rateb, M.E.; Jaspars, M.; Botana, L.M. Evaluation of the antioxidant activity of the marine pyrroloiminoquinone makaluvamines. *Mar. Drugs* **2016**, *14*, 197. [CrossRef] [PubMed]

75. Pallela, R.; Ehrlich, H.; Bhatnagar, I. Biomedical applications of marine sponge collagens, marine sponges: Chemicobiological and biomedical applications. In *Marine Sponges: Chemicobiological and Biomedical Application*; Pallela, R., Ehrlich, H., Eds.; Springer: New Delhi, India, 2016; pp. 373–381.

76. Fassini, D.; Duarte, A.R.C.; Reis, R.L.; Silva, T.H. Bioinspiring *Chondrosia reniformis* (Nardo, 1847) collagen-based hydrogel: A new extraction method to obtain a sticky and self-healing collagenous material. *Mar. Drugs* **2017**, *15*, 380. [CrossRef] [PubMed]

77. Reddy, G.K.; Enwemeka, C.S. A simplified method for the analysis of hydroxyproline in biological tissues. *Clin. Biochem.* **1996**, *29*, 225–239. [CrossRef]

78. Frazier, S.B.; Roodhouse, K.A.; Hourcade, D.E.; Zhang, L. The quantification of glycosaminoglycans: A comparison of HPLC, Carbazole, and Alcian Blue methods. *Open Glycosci.* **2008**, *1*, 31–39. [CrossRef] [PubMed]

79. Junqueira, L.C.; Bignolas, G.; Brentani, R.R. Picrosirius staining plus polarization microscopy, a specific method for collagen detection in tissue sections. *Histochem. J.* **1979**, *11*, 447–455. [CrossRef] [PubMed]

80. Pan, Y.; Dong, S.; Hao, Y.; Chu, T.; Li, C.; Zhang, Z.; Zhou, Y. Demineralized bone matrix gelatin as scaffold for tissue engineering. *Afr. J. Microbiol. Res.* **2010**, *4*, 865–870. [CrossRef]

81. Pozzolini, M.; Scarfi, S.; Benatti, U.; Giovine, M. Interference in MTT cell viability assay in activated macrophage cell line. *Anal. Biochem.* **2003**, *313*, 338–341. [CrossRef]

 *marine drugs*

Article

Development of an Integrated Mariculture for the Collagen-Rich Sponge *Chondrosia reniformis*

Mert Gökalp [1,2,*], **Tim Wijgerde** [2], **Antonio Sarà** [3], **Jasper M. de Goeij** [1,4] **and Ronald Osinga** [1,2]

1   Porifarma BV, Poelbos 3, 6718 HT Ede, The Netherlands; J.M.deGoeij@uva.nl (J.M.d.G.);
     ronald.osinga@wur.nl (R.O.)
2   Marine Animal Ecology, Wageningen University, P.O. Box 338, 6700 AH Wageningen, The Netherlands;
     tim.wijgerde@wur.nl
3   Studio Associato Gaia, Piazza della Vittoria 15/23, 16121 Genova, Italy; a.sara@studioassociatogaia.com
4   Department of Freshwater and Marine Ecology, Institute for Biodiversity and Ecosystem Dynamics,
     University of Amsterdam, P.O. Box 94248, 1090 GE Amsterdam, The Netherlands
*   Correspondence: mert.gokalp@gmail.com; Tel.: +90-5377086534

Received: 29 November 2018; Accepted: 27 December 2018; Published: 5 January 2019

Abstract: In this study, novel methods were tested to culture the collagen-rich sponge *Chondrosia reniformis* Nardo, 1847 (Demospongiae, Chondrosiida, Chondrosiidae) in the proximity of floating fish cages. In a trial series, survival and growth of cultured explants were monitored near a polluted fish farm and a pristine control site. Attachment methods, plate materials, and plate orientation were compared. In a first trial, chicken wire-covered polyvinyl chloride (PVC) was found to be the most suitable substrate for *C. reniformis* (100% survival). During a second trial, survival on chicken wire-covered PVC, after six months, was 79% and 63% for polluted and pristine environments, respectively. Net growth was obtained only on culture plates that were oriented away from direct sunlight (39% increase in six months), whereas sponges decreased in size when sun-exposed. Chicken wire caused pressure on explants and it resulted in unwanted epibiont growth and was therefore considered to be unsuitable for long-term culture. In a final trial, sponges were glued to PVC plates and cultured for 13 months oriented away from direct sunlight. Both survival and growth were higher at the polluted site (86% survival and 170% growth) than at the pristine site (39% survival and 79% growth). These results represent a first successful step towards production of sponge collagen in integrated aquacultures.

Keywords: mariculture; sponge; *Chondrosia reniformis*; fishfarm; integrated multitrophic aquaculture

---

1. Introduction

The first attempts to farm sponges date back to the 19th century, presumably as a consequence of periodical depletion of "bath-sponge" stocks [1,2], or—in more recent times—in pursuit of a safer and economically more attractive alternative to wild collection [3,4]. Overfishing and repeated outbreaks of mass mortality events halted the ancient tradition of Mediterranean fishing of commercially important "bath sponge" species, such as *Spongia officinalis* (Linnaeus) and *Hippospongia communis* (Lamarck) [3–6]. Sponge mariculture has received increased attention over the last two decades (e.g., [7–9]; see also reviews or comparative studies by [10–12]), particularly driven by the discovery of biologically active metabolites in many sponges (e.g., [13,14]). Sponge mariculture could potentially provide for a sustainable supply of sponge-derived bioactive compounds and biomaterials.

Sponges can be co-cultured with other organisms in so-called integrated mariculture systems, in which sponges take up metabolic wastes from other system components, including bacterioplankton growing on these metabolic wastes [15–19]. This way, sponges can effectively reduce waste streams from fish farms [2,5,20], since they have been shown to exhibit retention efficiencies of up to

99% for nano- and picoplankton (e.g., [21–23]), while processing large volumes of water, up to 0.6 cm^3 cm^{-3} sponge s^{-1} (e.g., [24–26]). Hence, a large-scale sponge culture facility that is constructed near a fish farm may positively affect the quality of the surrounding water. Conversely, the additional nutrition originating from the farmed fish may enhance the growth of the sponges in culture, thus providing a more efficient and profitable business.

In 2006–2007, an integrated mariculture approach using sponges was tested in the coastal waters around the Bodrum Peninsula, Turkey [27]. Two Mediterranean demosponge species with possible commercial interest, *Dysidea avara* (Schmidt, 1862) and *Chondrosia reniformis* Nardo, 1847 (Demospongiae, Chondrosiida, Chondrosiidae), were cultured at a pristine site (i.e., no fish farms within the nearest 30 km) and an organically polluted fish farm site, the latter sponges being directly cultured underneath an open cage fish farm. *D. avara* was chosen since it produces the bioactive compound avarol, a potential anti-psoriasis agent [14,28]. *Chondrosia reniformis* synthesizes large amounts of collagen, which is suitable for cosmetic and medical applications [29–31]. Type I & IV mammalian-like collagens can be effectively extracted from *C. reniformis* [32,33] and they can be used to promote the regeneration of human tissue and bone tissue engineering scaffolds [31]. *Chondrosia reniformis* showed better growth and survival rates at the pristine site, whereas *D. avara* grew and survived better at the polluted site [27]. The low survival rates of *C. reniformis* at the polluted site were largely due to the farming protocol used. *Chondrosia reniformis* is a highly plastic sponge, being able to de-attach and move around [34], a phenomenon that was frequently encountered using common culturing structures, such as pins, lines, plaques or metal/net grids [5,8,35,36]. To avoid displacement, explants of *C. reniformis* were put inside cages on the seafloor [27]. However, due to the high particle load in the water around the fish farm, the explants in these cages were suffocated by sediment.

This study describes progress towards the development of a raw collagen production pipeline using the sponge *C. reniformis* in an integrated multi-trophic aquaculture approach, i.e., by culturing the sponges in the vicinity of offshore floating fish cages. Using thin plastic plates as substratum, a series of consecutive trials were executed, aimed at developing an optimal, species-specific culture method. We monitored survival and growth rates of cultured explants of *C. reniformis*, thereby comparing a polluted fish farm site to a pristine site. Variables studied included methods for attaching explants to plates, plate materials and plate orientation. The culture methods (glue, cable-ties on plaques, net/mesh cover) were applied previously on other sponge species by several authors; for detailed information, see review by Duckworth et al. [10].

2. Materials and Methods

2.1. Mariculture Sites and Monitoring of Water Quality

All of the studies were carried out in the coastal waters around the Bodrum Peninsula, Southwest Turkey (Figure 1). Meteoroloji Bay (Figure 1 Pr.1), a shallow area with an abundant population of *C. reniformis* was selected as a pre-culture and initial testing site (Trial 1). Based on water visibility (Secchi disk, cf. [37]) and organic loading (total organic carbon (TOC) measurements), two additional sites were selected for subsequent testing (Trials 2 and 3): Kargi Island (Figure 1 Pr. 2) at the Southern side of the Bodrum Peninsula was selected as a pristine site, whereas Guvercinlik Bay (Figure 1 Po. 1), located at the Northern side of the peninsula, was selected as a polluted site.

Water temperature (Uwatec Aladin Air X Nitrox dive computer, calibrated with mercury thermometer) and visibility were measured 17 times during periodic visits at the polluted site throughout the experimental period from April 2011 to December 2013. To determine organic loading, three replicate water samples (50 mL) were taken within 10 m from the culture platforms by SCUBA diving from each location for TOC analysis using the wet oxidation method [38]). TOC samples were stored in pre-combusted (450 °C for 6 h) 50 mL glass bottles with glass caps at −20 °C until analysis. Prior to analysis, sulphuric acid was added to the samples (end concentration 2 mmol L^{-1})

to remove dissolved inorganic carbon species. The acidified samples were supplemented with sodium tetraborate and potassium persulphate and processed using segmented flow analysis (SFA) on a Continuous Flow Analyser (Skalar, Breda, The Netherlands). In SFA, TOC is first oxidized using UV light and then measured as CO_2 while using infrared detection. The TOC detection limit of the method, as intercalibrated with other labs, is 25 μmol L^{-1}, the average TOC variation among replicate measurements is 10 μmol L^{-1}. The internal standards used were 3.3 μmol L^{-1} EDTA, 3.3 μmol L^{-1} of a humic acid, and 3.6 μmol L^{-1} phenylalanine.

Figure 1. Map of the Bodrum Peninsula. Small white circle (Pr.1: Pre-culture site—Meteoroloji Bay). Large white circles (Pr.2: Pristine—Kargı island) and large black circles (Po.1: Polluted—Guvercinlik Bay) circles point the approximate locations of the sites used for growing sponge explants in this study. GPS coordinates 36.9444444, 27.27611111; 36.95166667, 27.30694444; 36.96861111, 27.45083333, respectively. (Source: Google Earth, 2018).

2.2. Sponge Collection and Seeding

Sponge specimens were collected by SCUBA at 5–10 m water depth in the Bay of Meteoroloji (Figure 1 Pr.1). Explants were cut with sharp razor blades and detached from rock surfaces with a spatula [8,39], leaving the majority (at least 75%) of the donor sponge unaffected. The explants received maximally two cut surfaces and more than 50% of their surface was always covered with intact pinacoderm (i.e., the sponge outer tissue layer). The explant size ranged between 10–15 cm^2 with an average thickness of 2 cm, and all of the explants had at least one osculum (i.e., outflow opening). Explants were stored in perforated plastic containers that allowed water flow and they were left underwater until the seeding operations started. To enable sponges to attach and acclimatize after seeding, the seeding plates with the explants were left horizontally next to the culture platforms for three days before the plates were secured to their spots on the culture frame [14].

2.3. Mariculture Trials

Within the period between April 2011 and December 2013, three subsequent mariculture trials were executed in order to develop and improve a culture method for *C. reniformis*.

1st mariculture trial, April 2011–June 2011: testing materials and attachment procedures—Sponge explants (*n* = 5 specimens per plate) were attached to four types of supports (autoclaved aerated concrete, white polyvinyl chloride (PVC), black PVC, and cemented PVC) using six different combinations of attachment methods and substrates (Table 1 (a)). The cementation of plates may improve attachment of the sponges to the support and enhance growth, since quartz and silica are found to promote collagen formation in sponges [34,40]. Accordingly, coarse sea-sand was used to make cement to cover the cemented-PVC supports. All of the supports were positioned in Meteoroloji Bay (Figure 1 Pr. 1) at 2–3 m water depth under overhangs or in crevices (i.e., not in direct sunlight) and fixed with diving weights.

2nd mariculture trial, June 2011–June 2012: testing culture plate orientation and site—Based on the results of the first trial, PVC plates were chosen for the second mariculture experiment. Explants (250 in total per site) were positioned on both sides of five 50 × 50 cm plates, with 25 explants on each side of every plate. In order to find the optimal positioning of the sponges, explants were cultured at nine different angles, under light (exposed side of plate) or dark (underside of plate) conditions, resulting in 10 different conditions (n = 1 plate per treatment) (0°, 30°, 45°, 60°, 90° light, 90°, 120°, 135°, 150°, and 180° dark; Supplementary Table S1, Figure 2a,b). In order to keep the sponges attached to the plates until natural attachment took place, the sponges were covered with chicken wire and left on the seabed for two days. After the attachment of the explants to the plates, the plates were mounted on a metal frame. Frames were installed in July 2011 at the two selected sites (Kargı Island (Figure 1 Pr. 2)—pristine and Guvercinlik Bay (Figure 1 Po. 1)—polluted) at a water depth of 10 m.

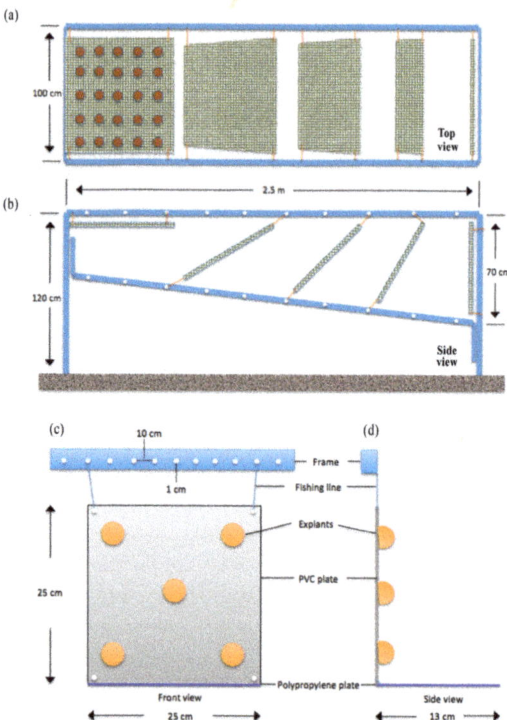

Figure 2. Schematic drawings of the culture platform designs. (**a**) 2nd trial—Top view; aluminum culture frame and attached 50 × 50 cm polyvinyl chloride (PVC) plates covered with chicken wire, each carrying 25 sponge explants on one side, totaling 50 sponges per plate. (**b**) 2nd trial—Side view; plates positioned at 9 different testing angles 0°, 30°, 45°, 60°, 90°, 120°, 135°, 150°, 180°. 3rd trial, (**c**) 3rd trial—front view, and (**d**) 3rd trial—side view showing positioning of glued *C. reniformis* explants on 25 × 25 cm PVC plates. PVC plates were secured tightly to the aluminum frame from four corners with 6-mm thick fishing lines, blue PP plates were attached to the bottom sides of the PVC plates.

Table 1. Overview of the results of the three culture trials executed in between May 2011 and November 2013. (a) 1st trial, material test and attachment procedures, pristine site (b) 2nd trial, testing culture orientation and site (c) 3rd trial, assessment of productivity at the optimal culture orientation.

	Material	Attachment Method	Advantage	Disadvantage	Result
a. 1st trial Material test and attachment procedures	Air-concrete	Iron screw		No attachment	Not suitable
	PVC—white	Cable-ties Superglue Chicken wire	Survival (80%) Ease of operation High survival (100%)	Dispersion of explants Lower survival (60%) handling time	Not selected for 2nd trial Selected for 3rd trial Selected for 2nd trial
	PVC—black	Cable-ties	Survival (80%)	Dispersion of explants	No preference; the color of the plate did not affect the result
	Cemented PVC	Cable-ties	High survival (100%)	Cost, handling time and weight	Not selected

	Site	Material	Disadvantage	Survival Rate	Average Growth	Result	Orientation (°)
b. 2nd trial Testing culture orientation and site	Pristine	PVC chicken wire	Squeezed explants, resulted in unwanted epibiont growth, time & cost	63% survival after 6 months of culture	Culture frame demolished by an anchor	Chicken wire method is not suitable	90° was selected for the next trial
	Polluted			79% survival after 6 months & 1 year of culture	39.2 ± 36.2% in 12 months for dark angles −40.9 ± 37.7% in 12 months for light angles		

	Site	Species	Survival Rate	Average Growth	Growth per Culture Interval		Range of Growth for Individuals
					0–6 Months	7–13 Months	
c. 3rd trial Assessment of productivity at the optimal culture orientation	Pristine	*C. reniformis* (*n* = 15 plates)	39%	79.0 ± 37.4% in 13 months	69.8 ± 33.6%	5.4 ± 5.7%	−3.6–135.6%
	Polluted	*C. reniformis* (*n* = 16 plates)	86%	170.4 ± 109.1% in 13 months	114.0 ± 94.6%	30.1 ± 27.9%	0.8–322.9%

3rd mariculture trial, June 2012–July 2013: assessment of growth at the optimal culture orientation—Based upon the observations of the 2nd mariculture trial, it was decided to choose an angle of 90° for primary upscaling of the cultures. Two new frames were installed in early June 2012, one at the pristine site and one at the polluted site, each carrying 20 white PVC plates of 25 × 25 cm with a total of 100 explants of *C. reniformis* (5 per plate), attached using gel-based polyacrylate superglue. A gel-based polyacrylate superglue method was preferred over chicken wire to improve the handling time and reduce the weight and cost of the culture materials. In addition, horizontal blue polypropylene PP plates were placed underneath the 90° PVC plates to provide extra surface for the explants to attach and grow, should they fall (Figure 2c,d). Explants were grown for 13 months, photographed, and measured in June, July, August, September, October, and November 2012 (both sites), in March and May 2013 (polluted site only, due to weather restrictions), and in July 2013 (both sites).

2.4. Survival Rate Analysis and Sponge Explant Growth

Survival rates in Trial 1 were assessed by visual observation. For Trials 2 and 3, survival was monitored by underwater photography. Explants were photographed using a Nikon D300s digital single lens reflex camera (Nikon Corporation, Tokyo, Japan) and a Sigma 10–20 mm wide-angle lens set (Sigma Corporation, Ronkonkoma, NY, USA), coupled with dedicated Ikelite housing and an DS160 substrobe (Ikelite, Indianapolis, IN, USA). Survival was calculated from the initial and final number (N) of explants residing on the PVC and/or PP plates, as follows:

$$\text{Survival} = (N_{\text{final}}/N_{\text{initial}}) \times 100 \tag{1}$$

For Trial 3, as described above, blue PP plates were used to collect detached sponges. To calculate survival data, sponges that had fallen onto the PP plates were pooled with the sponges that remained on the PVC plates. However, fallen sponges were excluded from the growth analysis, as PP may affect sponge growth differently from PVC.

For Trial 2, explant growth rates were calculated by measuring wet weights by using a scale (Sinbo SKS 4514) at the start and end of the experiment. To reduce measurement error, the sponges were briefly wiped with clean paper to remove seawater for a duration of approximately 10 s. Growth was calculated from initial and final explant wet weights (WW) as follows:

$$\text{Growth (\%)} = ((\text{WW}_{final} - \text{WW}_{initial})/\text{WW}_{initial}) \times 100 \tag{2}$$

For Trial 3, growth was monitored by underwater photography using the same setup, as described above, for the monitoring of survival. Following each dive, recorded images were transferred to Photoshop CS5 software (Adobe Systems Incorporated, San Jose, CA, USA) and lens distortion was corrected using a Camera Raw 6.7.1 plug-in (Adobe Systems Incorporated, San Jose, CA, USA). The images were calibrated using known plate dimensions, peripheries of explants were marked, and surface area was calculated from pixel counts of the marked areas [39]. Growth was expressed as the increase in the number of pixels, calculated with the pixel counter function of the image editing software. At each time point, the growth in percentage increase in projected surface area was calculated from initial (at start of the time point) and final explant surface areas (A) as follows:

$$\text{Growth (\%)} = (\text{A}_{final} - \text{A}_{initial})/\text{A}_{initial}) \times 100 \tag{3}$$

To assess the correlation between surface area growth to both biomass and volumetric growth, an additional 20 sponges of random size were collected from a neighboring site. For all 20 sponges, the wet weights were determined, as described above. In addition, volume was determined by measuring displaced seawater in a graduated cylinder. Finally, surface area was determined by using photography and a ruler as a reference, and photographs were processed, as described above.

2.5. Statistical Analysis

The normality of data was tested by plotting the residuals of each dataset versus the predicted values, and by performing a Shapiro-Wilk test. Homogeneity of variances was determined using Levene's test. All data were found to be normally distributed and showed homogeneity of variance after a log10 transformation ($p > 0.05$). Student's independent t-test was used to determine growth differences between the light and dark group in the second trial, with $n = 5$ plates for both groups. A two-way mixed factorial ANOVA was used to determine the main and interactive effects of culture site and time on *C. reniformis* growth in the third trial, with culture site as a between-subjects factor, and time as a within-subjects factor. In all analyses, the culture plate was considered as an experimental unit, i.e., data of explants growing on a single plate was pooled. To correlate surface area to mass and volume, Pearson's product-moment correlation was used. A p-value of less than 0.05 was considered to be statistically significant. Statistical analysis was performed with SPSS Statistics 22.0 (IBM, Somers, NY, USA). Graphs were plotted with SigmaPlot 12 (Systat software, San Jose, CA, USA).

3. Results

3.1. Polluted versus Pristine Site: Water Temperature, Visibility and TOC

Visibility was on average 3.8 times lower at the polluted site (6.5 ± 1.7 m; mean ± SD throughout text unless stated otherwise) when compared with pristine site (25 ± 1.1 m; Figure 3). The water temperatures that were recorded at the polluted site ranged between 19–26 °C (Figure 3). During summer periods, especially in August, as a result of intensive fish feeding activities, Secchi disk water depth dropped to 4–6 m at the polluted site (Figure 3). TOC levels at the polluted site (280 ± 0.07 µmol L^{-1}) were 2.4 times higher than the TOC levels at the pristine site (110 ± 0.01 µmol L^{-1}).

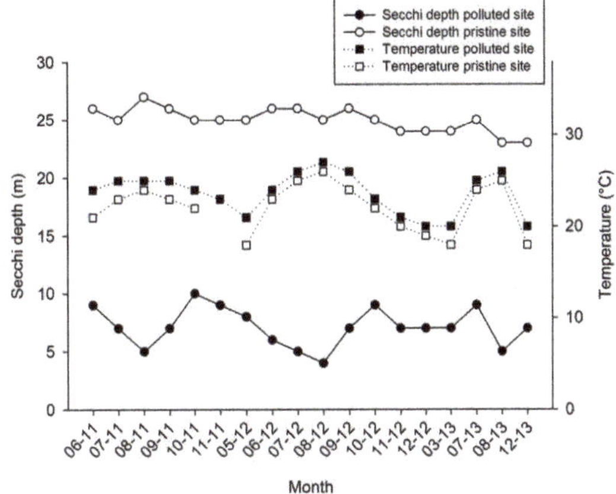

Figure 3. Water temperature (in °C, black squares with dottled line) and Secchi water depth (in m, black and white circles with continuous line) measurements for the pristine and polluted site over a 31-month time frame (June 2011–December 2013).

3.2. 1st Mariculture Trial: Testing Materials and Attachment Procedures

The air concrete material was found to be unsuitable for further experimentation, as none of the explants attached to it (Table 1 (a)). In addition, the material was positively buoyant in seawater, which hampered easy handling. There was no difference in preference between white and black PVC (80% survival for both plates), as sponges attached equally well to both substrates without showing any signs of disparity (Supplementary Figure S1a,b, Table 1 (a)). Cable-ties gave a better recovery percentage than super glue (80% vs. 60%), but, in addition to increasing handling time, cable-ties also triggered the dispersion of *C. renifomis* explants into two parts for both black and white PVC's (fission; Supplementary Figure S1b). The combinations PVC/chicken wire and cemented PVC/cable-tie were the most successful methods in terms of recovery percentage (all sponges survived on plate). However, the cost of material, plate weight, and handling time were factors favoring the PVC/chicken wire method (Table 1 (a)). Accordingly, PVC/chicken wire method was selected for the 2nd trial.

3.3. 2nd Mariculture Trial: Testing Culture Plate Orientation and Site

The explants at the polluted site (Supplementary Figure S1c) and pristine site (Supplementary Figure S1d) showed signs of bacterial infections and decay within a week after initiation of the cultures, causing initial losses at both sites (4.8% and 2% of explants deteriorated in the polluted and pristine sites, respectively). Among the explants that survived the initial deterioration, overall survival after six months was slightly better at the polluted site (79% of 238 explants survived at the polluted site and 63% of 245 explants survived at the pristine site; Table 1 (b)). The culture frame at the pristine site was found to be demolished, when revisited in May 2012. It was found 50 m away from the culture site. As a consequence, it was not possible to deduce annual survival and growth rates for the culture at the pristine site. At the polluted site, survival was highest among sponges that were put at an angle of 90° or higher (Figure 4). Growth rates were highly variable among treatments (Figure 5), but the average growth at "light" angles of 0–90° (-41 ± 38%; negative values points to loss of WW biomass) was significantly lower than the average growth at "shade" angles of 90–180° (39 ± 36%; Student's *t*-Test $z = -3.4$, $p < 0.01$, $n = 5$). The 90° plate was selected as the preferred culture orientation in Trial 3, based on the survival rate and the ease of operation (photography, measurements, and handling; see Table 1 (b) for details). Photographic measurement of growth was found to be impossible with the

PVC-chicken wire method as a result of continuous movement, splitting, and fusing of *C. reniformis* explants, and epibiont growth. In addition, chicken wire compressed the explants, which may not be beneficial for their development. Also, installing the large 50 × 50 cm PVC plates was time consuming. Therefore, smaller (25 × 25 cm) PVC plates were used in Trial 3 and superglue was selected as the attachment method.

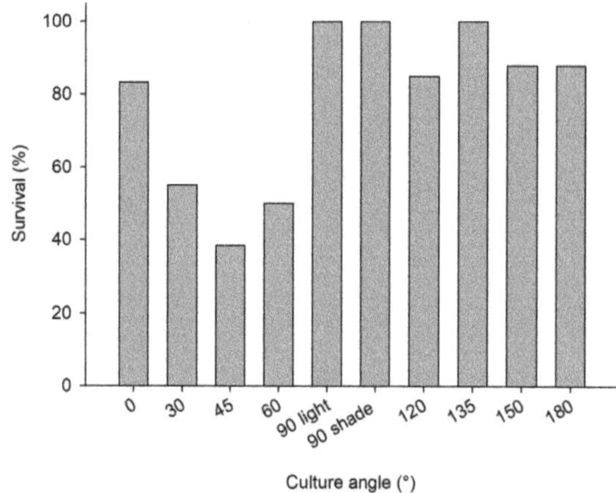

Figure 4. 2nd mariculture trial, polluted site, June 2011–June 2012; survival percentage of *C. reniformis* explants on PVC plates with various angles (0–90° light represents PVC plates with greater light exposure and 90° shade −180° plates receiving less light exposure).

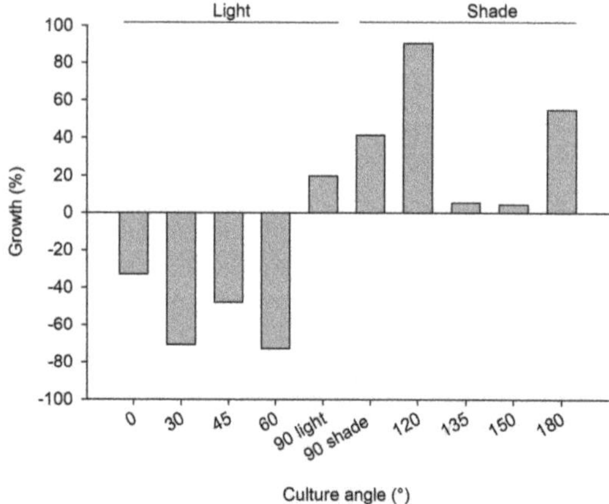

Figure 5. 2nd mariculture trial, polluted site June 2011–June 2012; growth rate as percentage wet weight increase of *C. reniformis* explants on PVC plates with various angles (0–90 light represents PVC plates with greater light exposure and 90 shade −180 plates receiving less light exposure; 25 explants for each plate).

3.4. 3rd Mariculture Trial: Assessment of Sponge Culture Productivity Polluted vs. Pristine Site

During the first week of the experiment, 69 (polluted) and 70 (pristine) out of 200 explants dropped off the PVC plates. Fortunately, the PP plates that were placed under the PVC plates were able to catch 55 (at polluted site) and nine (at pristine site) of these explants, which attached onto the PP plates and continued to increase their surface area. Because they could not be related anymore to their original size, explants that were attached on the PP plates were left out of the surface area increase analysis.

However, they were included in calculation of survival rates, which were highly different between the pristine and polluted sites (39–86%, respectively—Table 1 (c)). A total of 61 explants survived on the vertical PVC plates (30 explants on 15 plates at the pristine site, 31 explants on 16 plates at the polluted site), which were used for surface area increase analysis. The average increase in surface area over time of these *C. reniformis* explants is presented in Figure 6. After being cultured for 13 months, the average surface area increase was 79.0 ± 37.4% at the pristine site and 170.4 ± 109.1% at the polluted site (Table 1 (c)). Both culture site and time had a significant main effect on sponge surface area increase rates (Table 2). At both sites, explant surface area increased significantly, but it slowed down after the first six months at both sites (two-way factorial ANOVA, $F_{1,27} = 55.550$; $p < 0.001$), and for the pristine site even stalled after six months. Surface increase was significantly higher at the polluted site as compared to the pristine site (two-way factorial ANOVA, $F_{1,27} = 14.439$; $p = 0.001$), irrespective of time.

For *C. reniformis*, a highly significant correlation between surface area and wet weight was found (Pearson correlation, $r = 0.92$, $n = 20$, $p = 0.000$, two-tailed, Supplementary Figure S2), as well as between surface area and volume ($r = 0.92$, $n = 20$, $p = 0.000$, two-tailed, Supplementary Figure S3). The relationships are size-independent, leading to fixed conversion factors of 1.2 g wet mass per cm^2 of surface area and 1.1 cm^3 sponge volume per cm^2 surface area, respectively.

Table 2. Two-way mixed factorial ANOVA, demonstrating main and interactive effects of culture site and time on *C. reniformis* growth rates ($n = 15$–16).

Factor	F	df	Error	p
Culture site	14.439	1	27	0.001 **
Time	55.550	1	27	0.000 **
Culture site x Time	2.686	1	27	0.113

** Indicates significant effect ($p < 0.01$).

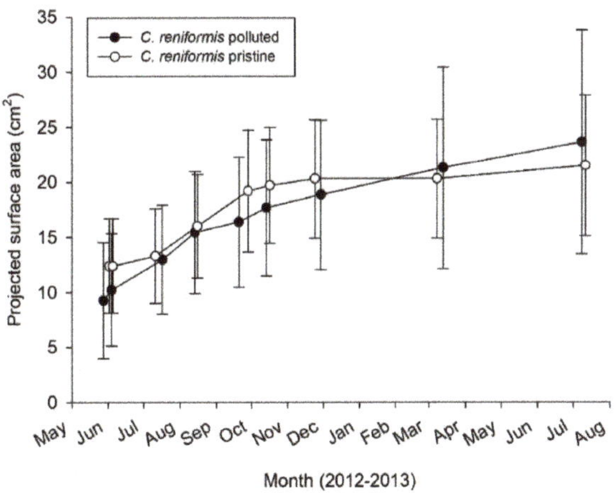

Figure 6. 3rd mariculture trial. Annual growth rate as surface area increase for *C. reniformis* ($n = 15$–16 plates) explants in polluted and pristine sites.

4. Discussion

This study explored the feasibility to integrate fish culture with a biomedically promising Mediterranean sponge species, *C. reniformis*. The main aim of the study was to derive the best mariculture practices of *C. reniformis* from a series of subsequent culture trials.

4.1. Explant Survival Rates

Survival of explants can be compromised by detachment and by disease. In terms of initial survival, cable-ties and chicken wire were the most effective means of attaching explants onto PVC substrates, with glue giving a slightly lower survival. In the long term, however, the use of chicken wire (mesh culture) gave ambiguous results. Sandwiched mesh structures were designed to promote the explants to grow out of the pocket and to ease harvesting [8]. Mesh culture that is used in turbid waters might reduce water flow and subsequently decrease available food for the explants if mesh size is too small [10]. Although the mesh size used in the second trial was sufficiently large (5 × 5 cm) as recommended in [41], after some time the space between meshes and the PVC plate was covered by epibionts, and the mesh did not prevent some explants from moving or even dropping themselves off the plate. Despite these drawbacks, the survival rate at the polluted site after one year was 79%, which is higher than in the study by [8], who reported 55% survival after seven months and who lost entire *C. reniformis* explants with the sandwiched mesh method. However, the mesh method is labor intensive, especially when considering that increased cleaning of biofouling on the mesh is recommended. By attaching the explants with glue in the third trial, it was anticipated to reduce both handling time and fouling. Despite the predicted improvements regarding initiation time (May vs. June) and culture angle (all at 90°), the third trial showed low survival for the pristine site. This was probably due to occasional strong currents that prevail at this site, which may make the explants more prone to dropping of the plates and physical removal from the site. During the whole month of September 2013, flow velocities above 20 cm/s were recorded at this site by analyzing the velocity of neutrally buoyant particles (video clips of laterally moving natural particles, data not shown). *Chondrosia reniformis* inhabits both nearly stagnant to occasional high flow waters (M. Gokalp; personal observation), however the attachment of explants to PVC plates is probably less firm than attachment to natural substrates, especially during the acclimatization time after wounding them to explant the parent sponges. At the polluted site, the use of glue instead of chicken wire did slightly improve long-term survival rate, which shows that gluing is a suitable method to attach explants of *C. reniformis*. It is also the fastest and easiest method. A future recommendation is to perform the initial acclimatization (of 7–10 days; see [42]) at a more secluded site, after which the attached explants are placed at the study site.

During culture Trial 2, initial sponge survival was compromised by disease-like phenomena. Bacterial infections, which were possibly due to late seeding of explants in Mid-June with relatively higher water temperatures, might have been responsible for the initial losses at both sites. High water temperatures in summer have been reported to be a risk for sponge mariculture in temperate and subtropical climates [11,39,41,43], as it makes cuttings more vulnerable to bacterial attack, although such increased vulnerability had not been observed in our earlier studies on this species in this area [27].

Culture angle directly affected explant survival, mainly in association with prevailing light levels. Lower light levels at the more turbid polluted site may, therefore, also explain the higher explant survival at the light-exposed angles at the polluted site. These results corroborate the findings of [35], which purport *C. reniformis* prefers shaded habitats.

4.2. Explant Growth

Since surface area of *C. reniformis* showed a size-independent relationship with wet weight and volume, surface area can be used as a proxy for growth. This enables a direct comparison of growth data obtained for this species using different methods.

Culture of *C. reniformis* has been considered to be difficult, to even unsuitable with the methods applied [5,8]. Wilkinson and Vacelet [35] reported moderate growth rates of 95% per year (55 weeks doubling time in volume, measured using volume displacement) when *C. reniformis* was cultured under shaded conditions. Ref. [27] obtained grow rates of 100 to 200% per year when growing *C. reniformis* on the bottom of metal wire cages under pristine conditions, but this study failed to achieve such results at a fish farm site as the explants cultured were smothered by effluents from the fish farm. Conversely, the current study demonstrates that if cultured using an appropriate method, *C. reniformis* will survive and grow (up to 170% in 13 months), even in a fish farm environment with a considerable particle load. These growth rates are considerably higher than those reported for naturally growing specimen. Garrabou and Zabala [36] reported an in situ growth rate of 2.3% per year (deduced from two-dimensional (2D) areal growth) for *C. reniformis*, which was an order of magnitude lower than the growth rate of three other Mediterranean sponge species in their study *Hemimycale Ccolumella* (Bowerbank), *Oscarella lobularis*, and *Crambe Crambe* (Schmidt). They ascribed the slow growth rate of *C. reniformis* to a greater energy investment in tissue production per unit area as a result of its thick collagenous cortex. However, the data found by Osinga et al. [27] and those from the current study indicate that in aquaculture, *C. reniformis* exhibits growth rates that are nearly two orders of magnitude higher than the in situ rates reported by Garrabou and Zabala [35]. Under optimal circumstances, the production of collagen is apparently not hampered by energy input. The current results show a clear potential for collagen production through the aquaculture of *C. reniformis*. The highly variable growth of *C. reniformis* under different conditions and the high variability within treatments highlight the need for further optimization studies.

During Trial 3, *C. reniformis* surface area increase rates were significantly different between culture sites, with an approximate two-fold higher growth at the polluted site. This may relate to the higher food availability—i.e., higher TOC concentration as a result of fish farm activities—and, as mentioned earlier, correspondingly lower light levels at the polluted culture site. Hence, the combination with fish farming is potentially beneficial for culture success of this sponge species. The surface area increase of *C. reniformis* was clearly higher in the first six months after initiation of the cultures, regardless of culture site. Although this may partially be explained by seasonal effects (growth might cease in autumn and winter [11,44]), it is possible that the sponges exhibit lower specific growth rates when being in culture for a longer period [39]. This could be due to initial enhanced surface area increase due to explant cutting [45], which could hamper growth at later stages, due to high costs of wound healing and regeneration [42]. Fast initial surface area increase was also found in a side experiment where the explants were cultured starting in autumn 2013 (data not shown), pointing towards a wound healing and regeneration effect rather than a seasonal effect, but this observation needs to be further investigated.

4.3. Culture of Chondrosia reniformis—Best Practices

As stated in Schippers et al. [11], initial mariculture trials should span a complete annual cycle in order to perceive effects of seasonality, substrate preference, and growth physiology of the sponge, and possible external impacts to the culture site, such as the occurrence of fouling and specific sponge predators, boat traffic and anchoring, and the presence of fishermen and divers. Accordingly, in this study, valuable information was acquired regarding the preferences for attachment, survival, and growth of *C. reniformis* during the first two trials. *Chondrosia reniformis* explants attached to PVC plates tend to move on the plate, thus obfuscating multiple genotypic comparisons on one plate (our study was initially designed to investigate genotype effects, but this part of the study could not be completed due to random movement of the explants over the plates). In addition, fusion and fission of explants makes the proper assessment of survival difficult. Even though attachment to the PVC plates has succeeded, some *C. reniformis* still found ways to divide their body into several parts, moved around the PVC plate (possibly in pursuit of shaded areas), or dropped themselves to the ground possibly in search for better living conditions. Survival and growth is best at culture angles of 90° and above, where explants are not being exposed to direct sunlight, as *C. reniformis* performs better at low illumination levels.

In experiments 2 & 3, the initial losses and/or droppings of explants were slightly high and unpredictable, despite the variety of methods applied. Once attached for a longer time, the explants would remain attached. Therefore, initial losses and/or droppings are the main problem to be solved to secure better culture performance. Restraining bacterial attack on freshly cut explants by initiation of cultures early in the season (spring) and preventing exposure to high currents during the first months should be practiced together with the best performing methods regarding attachment.

Based upon the three mariculture trials described above, the following best practices have been deduced for culturing *C. reniformis* in sea-based aquaculture under turbid conditions:

1. Culture method: Sponge explants cut from parent sponges are glued to PVC plates using gel-based polyacrylate superglue. PVC plates are best positioned vertically onto frames and they should be extended with a basket on the bottom site to recover explants falling off the plates. Chicken wire may be applied during the first few weeks after explanting to prevent early losses but should be removed once the attachment is stable. Prolonged use of chicken wire cover tends to hold sediments and promotes epibiont growth and hence undesired space competition with the cultured sponge.

2. Site selection: Sites should not be prone to strong fluctuations in weather. The area should be secured and should be clear of boat traffic and anchoring [10,11]. Sites should be carefully assessed for (e.g., seasonal) strong currents. High water turbidity and increased load of organic content associated with the presence of fish farms does not appear to hamper growth of *C. reniformis* on vertical plates, making this sponge an interesting candidate for integrated multitrophic mariculture. Daily fish feeding activities and occasional net replacing hinders the use of culture platforms inside the fish farm area. Thus, sponge culture platforms have to be placed outside boat traffic area. To eliminate this problem, one method that we consider for future applications is using layered scallop lanterns placed in between an anchor and a submerged buoy system (just outside the fish farm culture area), a method that was successfully applied by both Duckworth et al. [46] and Kelly et al. [12], for *Latrunculia wellingtonensis*, *Polymastia croceus*, and *(Heterofibria) manipulatus*, respectively.

3. Seasonality: Initiating a culture of *C. reniformis* in the Mediterranean is best done in either spring (April-May) or autumn (October-November) to prevent bacterial infections following cutting of explants from parent sponges.

4.4. Recommendations for Future Research

Culture success can be further improved by optimizing the period of culture. Optimal culture time can be determined by observing sponge growth rates over a period of two or more subsequent years. Page et al. [39], found reduced growth rates for *Mycale (Carmia) hentscheli* (Bergquist & Fromont) [47] over time. The growth rates in their study dropped from 2437% year^{-1} to 1355% year^{-1} from the first to the third culture period. Moreover, the growth rates of cloned sponges harvested from cultured explants should also be followed, as Page et al. [39], found reduced growth rates and even negative growth through repeated cloning (F0 to F2).

Other important aspects to include in future studies are seasonality (e.g., is the fast initial growth observed in this study season-influenced or a wound-healing response that is irrespective of season) and genotypic variability.

Supplementary Materials: The following are available online at http://www.mdpi.com/1660-3397/17/1/29/s1, Figure S1: Overview of culture methods; (a) 1st trial; explants tie-wrapped to black PVC, experiment start (b) 1st trial; sponges tie-wrapped to white PVC, 1 year later explants splitting into two fragments and relocating position. (c) 2nd trial, polluted site; image of 50 × 50 cm plates, sponge explants secured with chicken wire (0–30° plates can be seen, 25 additional explants are on the other side of each plate). (d) 2nd trial, pristine site; infected explants shortly after seeding. (e) 3rd trial, polluted site; first upscaling of aquaculture of Chondrosia reniformis on vertical PVC plates. Explants are attached with gel-based polyacrylate superglue to PVC plates (10 plates per site). (f) 3rd trial, pristine site; explants after 4 months of culture. Some of the explants tend to travel towards the polypropylene plate and grow on it, but others prefer to stay on the PVC plates. Figure S2: Correlation between surface area (cm^2) and wet weight (g) for Chondrosia reniformis (Pearson correlation, $r = 0.92$, $n = 20$, $p = 0.000$, two-tailed). Figure S3: Correlation between surface area (cm^2) and volume (mL) for Chondrosia reniformis (Pearson correlation, $r = 0.92$, $n = 20$, $p = 0.000$, two-tailed). Table S1: Experimental design of aquaculture trials 2 and 3. In Trial 3, 10 explants were taken per parent sponge, five explants were attached to each plate.

Author Contributions: Conceptualization, M.G., A.S., J.M.d.G. and R.O.; Data curation, M.G.; Formal analysis, M.G., T.W. and R.O.; Funding acquisition, J.M.d.G. and R.O.; Investigation, M.G.; Methodology, M.G., A.S., J.M.d.G. and R.O.; Resources, R.O.; Supervision, R.O.; Validation, M.G., T.W. and R.O.; Visualization, M.G. and T.W.; Writing—original draft, M.G.; Writing—review & editing, T.W., A.S., J.M.d.G. and R.O.

Funding: This study was funded by the European Union Seventh Framework Programme (FP7/2007-2013) under grant agreement no. KBBE-2010-266033 (Project SPECIAL). Funding was also received from The Innovational Research Incentives Scheme of the Netherlands Organization for Scientific Research (NWO-VENI; 863.10.009; personal grant to Jasper M. de Goeij). The funders had no role in study design, data collection and analysis, decision to publish, or preparation of the manuscript.

Acknowledgments: Intergrup is acknowledged for providing the PVC plates. Special thanks to Laura Valderrama Ballesteros, Holger Kuehnhold, Tjitske Kooistra, Marretje Adriaanse and Mustafa Gökalp for dive support, Suha and Alev Gökalp, Yasemin and Hakan Akyuz for logistics, and Erdal Betin for continuous support.

Conflicts of Interest: The authors declare no conflicts of interest.

References

1. Storr, J.F. *Ecology of the Gulf of Mexico Commercial Sponges and its Relation to the Fishery*; Special Scientific Report; US Fish and Wildlife Service: Washington, DC, USA, 1964; Volume 466, p. 73.

2. Manconi, R.; Cubeddu, T.; Corriero, G.; Pronzato, R. Commercial sponges farming as natural control of floating cages pollution. In *New Species for Mediterranean Aquaculture*; Enne, G., Greppi, G.F., Eds.; Elsevier: Amsterdam, The Netherlands, 1999; pp. 269–274.

3. Pronzato, R.; Manconi, R. Mediterranean commercial sponges: Over 5000 years of natural history and cultural heritage. *Mar. Ecol.* **2008**, *29*, 146–166. [CrossRef]

4. Voultsiadou, E.; Dailianis, T.; Antoniadou, C.; Vafidis, D.; Dounas, C.; Chintiroglou, C.C. Aegean Bath Sponges: Historical Data and Current Status. *Rev. Fish. Sci.* **2011**, *19*, 34–51. [CrossRef]

5. Pronzato, R.; Bavestrello, G.; Cerrano, C.; Magnino, G.; Manconi, R.; Pantelis, J.; Sarà, A.; Sidri, M. Sponge farming in the Mediterranean Sea: New perspectives. *Mem. Qld. Mus.* **1999**, *44*, 485–491.

6. Milanese, M.; Sarà, M.; Manconi, R.; Ben Abdalla, A.; Pronzato, R. Commercial sponge fishing in Libya: Historical records, present status and perspectives. *Fish. Res.* **2008**, *89*, 90–96. [CrossRef]

7. Müller, W.E.G.; Wimmer, W.; Schatton, W.; Böhm, M.; Batel, R.; Filic, Z. Initiation of an aquaculture of sponges for the sustainable production of bioactive metabolites in open systems: Example, *Geodia cydonium*. *Mar. Biotechnol.* **1999**, *1*, 569–579. [CrossRef]

8. Van Treeck, P.; Eisinger, M.; Muller, J.; Paster, M.; Schuhmacher, H. Mariculture trials with Mediterranean sponge species: The exploitation of an old natural resource with sustainable and novel methods. *Aquaculture* **2003**, *218*, 439–455. [CrossRef]

9. De Voogd, N.J. The mariculture potential of the Indonesian reefdwelling sponge *Callyspongia (Euplacella) biru*: Growth, survival and bioactive compounds. *Aquaculture* **2007**, *262*, 54–64. [CrossRef]

10. Duckworth, A. Farming sponges to supply bioactive metabolites and bath sponges: A review. *Mar. Biotechnol.* **2009**, *11*, 669–679. [CrossRef]

11. Schippers, K.J.; Sipkema, D.; Osinga, R.; Smid, H.; Pomponi, S.A.; Martens, D.E.; Wijffels, R.H. Cultivation of Sponges, Sponge Cells and Symbionts: Achievements and Future Prospects. *Adv. Mar. Biol.* **2012**, *62*, 273–337. [CrossRef]

12. Kelly, M.; Handley, S.; Page, M.; Butterfield, P.; Hartill, B.; Kelly, S. Aquaculture trials of the New Zealand bath-sponge *Spongia (Heterofibria) manipulatus* using lanterns. *N. Z. J. Mar. Freshw. Res.* **2010**, *38*, 231–241. [CrossRef]

13. Pomponi, S.A. The oceans and human health: The discovery and development of marine-derived drugs. *Oceanography* **2001**, *14*, 28–42. [CrossRef]

14. Sipkema, D.; Osinga, R.; Schatton, W.; Mendola, D.; Tramper, J.; Wijffels, R.H. Large-scale production of pharmaceuticals by marine sponges: Sea, cell, or synthesis. *Biotechnol. Bioeng.* **2005**, *90*, 201–222. [CrossRef] [PubMed]

15. Milanese, M.; Chelossi, E.; Manconi, R.; Sarà, A.; Sidri, M.; Pronzato, R. The marine sponge *Chondrilla nucula* Schmidt, 1862 as an elective candidate for bioremediation in integrated aquaculture. *Biomol. Eng.* **2003**, *20*, 363–368. [CrossRef]

16. Fu, W.; Sun, L.; Zhang, X.; Zhang, W. Potential of the marine sponge *Hymeniacidon perleve* as a bioremediator of pathogenic bacteria in integrated aquaculture ecosystems. *Biotechnol. Bioeng.* **2006**, *93*, 1112–1122. [CrossRef]

17. Zhang, X.; Zhang, W.; Xue, L.; Zhang, B.; Jin, M.; Fu, W. Bioremediation of bacteria pollution using the marine sponge *Hymeniacidon perlevis* in the intensive mariculture water system of turbot *Scophthalmus maximus*. *Biotechnol. Bioeng.* **2009**, *105*, 59–68. [CrossRef] [PubMed]

18. Ledda, F.D.; Pronzato, R.; Manconi, R. Mariculture for bacterial and organic waste removal: A field study of sponge filtering activity in experimental farming. *Aquac. Res.* **2014**, *45*, 1389–1401. [CrossRef]

19. Longo, C.; Cardone, F.; Corriero, G. The co-occurrence of the demosponge *Hymeniacidon perlevis* and the edible mussel *Mytilus galloprovincialis* as a new tool for bacterial load mitigation in aquaculture. *Environ. Sci. Pollut. Res.* **2016**, *23*, 3736–3746. [CrossRef]

20. Pronzato, R.; Cerrano, C.; Cubeddu, T. Sustainable development in coastal areas: Role of sponge farming in integrated aquaculture. In *Aquaculture and Water: Fish Culture, Shellfish Culture and Water Usage*; Special Publication No. 26; Grizel, H., Kesmont, P., Eds.; European Aquaculture Society: Bordeaux, France, 1998; pp. 231–232. [CrossRef]

21. Reiswig, H.M. In-situ pumping activities of tropical Demospongiae. *Mar. Biol.* **1971**, *9*, 38–50. [CrossRef]

22. Pile, A.J.; Witman, J. In situ grazing on plankton <10 μm by the boreal sponge. *Mycale lingua. Mar. Ecol. Prog. Ser.* **1996**, *141*, 95–102. [CrossRef]

23. McMurray, S.E.; Johnson, Z.I.; Hunt, D.E.; Pawlik, J.R.; Finelli, C.M. Selective feeding by the giant barrel sponge enhances foraging efficiency. *Limnol. Oceanogr.* **2016**, *61*, 1271–1286. [CrossRef]

24. Vogel, S. Current-Induced Flow through Living Sponges in Nature. *Proc. Natl. Acad. Sci. USA* **1977**, *74*, 2069–2071. [CrossRef] [PubMed]

25. Savarese, M.; Patterson, M.R.; Chernykh, V.I.; Fialkov, V.A. Trophic effects of sponge feeding within Lake Baikal's littoral zone. 1. In situ pumping rates. *Limnol. Oceanogr.* **1997**, *42*, 171–178. [CrossRef]

26. Weisz, J.B.; Lindquist, N.; Martens, C.S. Do associated microbial abundances impact marine demosponge pumping rates and tissue densities? *Oecologia* **2008**, *155*, 367–376. [CrossRef] [PubMed]

27. Osinga, R.; Sidri, M.; Cerig, E.; Gokalp, S.Z.; Gokalp, M. Sponge Aquaculture Trials in the East-Mediterranean Sea: New Approaches to Earlier Ideas. *Open Mar. Biol. J.* **2010**, *4*, 74–81. [CrossRef]

28. Tommonaro, G.; Iodice, C.; AbdEl-Hady, F.K.; Guerriero, G.; Pejin, B. The Mediterranean Sponge *Dysidea avara* as a 40 Year Inspiration of Marine Natural Product Chemists. *J. Homeopath. Ayurvedic Med.* **2014**. [CrossRef]

29. Swatschek, D.; Schatton, W.; Kellerman, J.; Muller, W.E.G.; Kreuterc, J. Marine sponge collagen: Isolation, characterization and effects on the skin parameters surface-pH, moisture and sebum. *Eur. J. Pharm. Biopharm.* **2002**, *53*, 107–113. [CrossRef]

30. Nickel, M.; Brummer, F. In vitro sponge fragment culture of *Chondrosia reniformis* (Nardo, 1847). *J. Biotechnol.* **2003**, *100*, 147–159. [CrossRef]

31. Silva, T.H.; Moreira-Silva, J.; Marques, A.L.; Domingues, A.; Bayon, Y.; Reis, R.L. Marine origin collagens and its potential applications. *Mar. Drugs* **2014**, *12*, 5881–5901. [CrossRef] [PubMed]

32. Dos Reis, R.L. Method to Obtain Collagen/Gelatin from Marine Sponges. WO Patent WO2015151030A1, 8 October 2015. Available online: https://patentimages.storage.googleapis.com/8c/50/55/0a910f35b77782/WO2015151030A1.pdf (accessed on 26 November 2018).

33. Silva, J.; Barros, A.; Aroso, I.; Fassini, D.; Silva, T.H.; Reis, R.L.; Duarte, A. Extraction of Collagen/Gelatin from the Marine Demosponge Chondrosia reniformis (Nardo, 1847) Using Water Acidified with Carbon Dioxide—Process Optimization. *Ind. Eng. Chem. Res.* **2016**, *55*, 6922–6930. [CrossRef]

34. Bavestrello, G.; Benatti, U.; Calcinai, B.; Cattaneo-Vietti, R.; Cerrano, C.; Favre, A.; Giovine, M.; Lanza, S.; Pronzato, R.; Sara, M. Body Polarity and Mineral Selectivity in the Demosponge *Chondrosia reniformis*. *Biol. Bull.* **1998**, *195*, 120–125. [CrossRef]

35. Wilkinson, C.; Vacelet, J. Transplantation of marine sponges to different conditions of light and current. *J. Exp. Mar. Biol. Ecol.* **1979**, *37*, 91–104. [CrossRef]

36. Garrabou, J.; Zabala, M. Growth dynamics in four Mediterranean Demosponges. *Estuar. Coast. Shelf Sci.* **2001**, *52*, 293–303. [CrossRef]

37. Hannah, L.; Pearce, C.M.; Cross, S.F. Growth and survival of California sea cucumbers (*Parastichopus californicus*) cultivated with sablefish (*Anoplopoma fimbria*) at an integrated multi-trophic aquaculture site. *Aquaculture* **2013**, *406–407*, 34–42. [CrossRef]

38. Menzel, D.W.; Vaccaro, R.F. The measurement of dissolved organic and particulate carbon in seawater. *Limnol. Oceanogr.* **2003**, *9*, 138–142. [CrossRef]

39. Page, M.J.; Handley, S.J.; Northcote, P.T.; Cairney, D.; Willan, R.C. Successes and pitfalls of the aquaculture of the sponge *Mycale hentscheli*. *Aquaculture* **2011**, *312*, 52–61. [CrossRef]

40. Le Pennec, G.; Perovic, S.; Ammar, M.S.A.; Grebenjuk, V.A.; Steffen, R.; Brummer, F.; Mueller, W.E.G. Cultivation of primmorphs from the marine sponge *Suberites domuncula*: Morphogenetic potential of silicon and iron a review. *J. Biotechnol.* **2003**, *100*, 93–108. [CrossRef]

41. Duckworth, A.R.; Battershill, C.N.; Bergquist, P.R. Influence of explant procedures and environmental factors on culture success of three sponges. *Aquaculture* **1997**, *156*, 251–267. [CrossRef]

42. Alexander, B.E.; Achlatis, M.; Osinga, R.; van der Geest, H.G.; Cleutjens, J.P.M.; Schutte, B.; de Goeij, J.M. Cell kinetics during regeneration in the sponge *Halisarca caerulea*; how local is the response to tissue damage? *PeerJ* **2015**, *3*, e820. [CrossRef] [PubMed]

43. Maldonado, M.; Zhang, X.C.; Cao, X.P.; Xue, L.Y.; Cao, H.; Zhang, W. Selective feeding by sponges on pathogenic microbes: A reassessment of potential for abatement of microbial pollution. *Mar. Ecol. Prog. Ser.* **2010**, *403*, 75–89. [CrossRef]

44. Corriero, G.; Longo, C.; Mercurio, M.; Marzano, C.N.; Lembo, G.; Spedicato, M.T. Rearing performance of *Spongia officinalis* on suspended ropes off the Southern Italian Coast (Central Mediterranean Sea). *Aquaculture* **2004**, *238*, 195–205. [CrossRef]

45. Wulff, J.L. Trade-Offs in Resistance to Competitors and Predators, and Their Effects on the Diversity of Tropical Marine Sponges. *J. Anim. Ecol.* **2005**, *74*, 313–321. [CrossRef]

46. Duckworth, A.; Battershill, C.; Schiel, D.R. Effects of depth and water flow on growth, survival and bioactivity of two temperate sponges cultured in different seasons. *Aquaculture* **2004**, *242*, 237–250. [CrossRef]

47. Lerner, C.; Hajdu, E. Two new Mycale (Naviculina) Gray (Mycalidae, Poecilosclerida, Demospongiae) from the Paulista Biogeographic Province (Southwestern Atlantic). *Rev. Bras. Zool.* **2002**, *19*, 109–122. [CrossRef]

 *marine drugs*

Article

Biphasic Scaffolds from Marine Collagens for Regeneration of Osteochondral Defects

Anne Bernhardt *, Birgit Paul and Michael Gelinsky

Centre for Translational Bone, Joint and Soft Tissue Research, University Hospital Carl Gustav Carus and Faculty of Medicine of Technische Universität Dresden, Fetscherstraße 74, 01307 Dresden, Germany; birgit.paul@tu-dresden.de (B.P.); michael.gelinsky@tu-dresden.de (M.G.)
* Correspondence: anne.bernhardt@tu-dresden.de; Tel.: +49-351-458-6692

Received: 10 January 2018; Accepted: 10 March 2018; Published: 13 March 2018

Abstract: Background: Collagens of marine origin are applied increasingly as alternatives to mammalian collagens in tissue engineering. The aim of the present study was to develop a biphasic scaffold from exclusively marine collagens supporting both osteogenic and chondrogenic differentiation and to find a suitable setup for in vitro chondrogenic and osteogenic differentiation of human mesenchymal stroma cells (hMSC). Methods: Biphasic scaffolds from biomimetically mineralized salmon collagen and fibrillized jellyfish collagen were fabricated by joint freeze-drying and crosslinking. Different experiments were performed to analyze the influence of cell density and TGF-β on osteogenic differentiation of the cells in the scaffolds. Gene expression analysis and analysis of cartilage extracellular matrix components were performed and activity of alkaline phosphatase was determined. Furthermore, histological sections of differentiated cells in the biphasic scaffolds were analyzed. Results: Stable biphasic scaffolds from two different marine collagens were prepared. An in vitro setup for osteochondral differentiation was developed involving (1) different seeding densities in the phases; (2) additional application of alginate hydrogel in the chondral part; (3) pre-differentiation and sequential seeding of the scaffolds and (4) osteochondral medium. Spatially separated osteogenic and chondrogenic differentiation of hMSC was achieved in this setup, while osteochondral medium in combination with the biphasic scaffolds alone was not sufficient to reach this ambition. Conclusions: Biphasic, but monolithic scaffolds from exclusively marine collagens are suitable for the development of osteochondral constructs.

Keywords: jellyfish collagen; mineralized salmon collagen; osteochondral tissue engineering; biphasic scaffold; osteochondral medium; alginate

---

1. Introduction

Collagen is one of the most frequently applied biomaterials for biomedical research as well as clinical applications [1,2]. The main industrial sources of collagen are bovine and porcine tissues; however, there is increasing demand for alternative sources. Marine collagens which can be obtained from both invertebrates and vertebrates [3,4] show promising features and have the potential to overrule mammalian collagens in biomedical applications for several reasons. Marine collagens do not bear the risk of disease translation and are not allergy-causing; they are not subjected to ethical or religious concerns, show low inflammatory response and can be obtained with high yield [1]. From the beginning of the 21st century research on marine collagens has continuously emerged [5,6]. Scaffolds for tissue engineering applications are increasingly developed from collagens of marine origin, such as from fish collagen, collagen of marine sponges, jellyfish collagen and collagen from marine gastropods [5]. In two own studies we applied marine collagens for the fabrication of porous scaffolds: first we adapted the procedure of biomimetic mineralization of bovine collagen to collagen of the Atlantic salmon *Salmo salar* and prepared mineralized porous scaffolds from salmon collagen for the application in bone tissue engineering [7]. A second study applied collagen of the jellyfish *Rhopilema esculentum* which is structurally similar to human collagen II [8] for the preparation of porous scaffolds to be used for chondral tissue engineering [9]. In the present study, we combined fibrillized jellyfish collagen with biomimetically mineralized salmon collagen to a biphasic scaffold suitable for osteochondral defect regeneration. The applied technique was already described in 2007 for the generation of biphasic, but monolithic scaffolds from mineralized bovine tendon collagen and fibrillized collagen from calf skin/hyaluronic acid composite [10]. Joint freeze-drying and chemical crosslinking of the two different phases resulted in a scaffold material which overcame the risk of delamination of the mineralized and non-mineralized phases, since the scaffolds consisted of a unified whole [10]. Challenge in the fabrication of biphasic scaffolds for the regeneration of both cartilage and the subchondral bone layer is the mechanically stable conjunction of the different phases, which have different mechanical properties to mimic the chemical nature of elastic, water-rich chondral ECM, and the stiff, mineralized bone ECM. The main challenge, however, is the simultaneous chondrogenic and osteogenic differentiation, guided by the scaffold properties which should recapitulate the native milieu of bone and cartilage development [11–14]. Furthermore, when osteochondral tissue engineering constructs are differentiated in vitro, a suitable osteochondral medium should be developed [15,16]. The aim of the present study was to generate biphasic, but monolithic porous scaffolds from fibrillized jellyfish collagen for the chondral part and biomimetically mineralized salmon collagen for the bony part. Multipotent human mesenchymal stromal cells (hMSC) were differentiated simultaneously into chondrocytes and osteoblasts, respectively, in this biphasic scaffold in vitro. To generate a microenvironment, which guides the cells to the two differentiation lineages, we inserted the cells into the chondral part with a tenfold higher cell density compared to the bony part and, furthermore, used an alginate hydrogel for embedding the cells in the porous chondral part. Furthermore, an osteochondral medium was developed and sequential seeding of the scaffold phases with pre-differentiated MSC was performed.

2. Results

2.1. Preparation and Characterization of Biphasic Scaffolds from Marine Collagen

Biphasic, but monolithic scaffolds, composed of biomimetically mineralized salmon collagen [7] and fibrillized jellyfish collagen [9] were successfully obtained by overlaying the two different phases as liquid suspensions, joint freeze-drying and crosslinking. Figure 1 shows the morphology of these scaffolds, both in dry and wet state. Although the stronger swelling of the jellyfish collagen phase is clearly visible, the stability of the whole scaffold is not affected by the different swelling behavior of the phases. This becomes especially obvious, when the microstructure of the scaffolds is analyzed. Interconnecting pores were verified in the transition zone between the two phases (Figure 2). Scanning electron microscopy (SEM) images show the smooth surface of the non-mineralized jellyfish collagen pore walls, as well as the rough morphology of the mineralized salmon collagen, which originates in the presence of hydroxyapatite nanocrystals, decorating the surface (Figure 2).

2.2. Evaluation of Optimal Seeding Density for the Osteogenic Differentiation of hMSC

Single scaffolds from mineralized salmon collagen were seeded with hMSC at three different densities, 2.4×10^5 cells/cm^3, 6×10^5 cells/cm^3, and 1.2×10^6 cells/cm^3 and cultivated under osteogenic stimulation for 14 and 28 days. Number of viable cells, visualized by MTT staining (Figure 3a) was still highest in the scaffolds with the highest initial seeding density after 28 days of osteogenic stimulation. Quantitative analysis of cell number (Figure 3b) confirmed these findings. Furthermore, it was shown, that cell number did not increase between d14 and d28 of cultivation. Osteogenic differentiation was evaluated by quantification of specific alkaline phosphatase (ALP) activity (Figure 3c). Highest specific ALP activities were obtained for the constructs with the lowest seeding density; however, due to high variations between single samples the effect was not statistically significant. Furthermore, the ALP activity in scaffolds seeded with the lowest cell density was close to the detection limit of the colorimetric test. Therefore, for the seeding of biphasic scaffolds, the intermediate cell density (6×10^5 cells/cm^3) was chosen, since specific ALP activity was still higher compared to the highest seeding density.

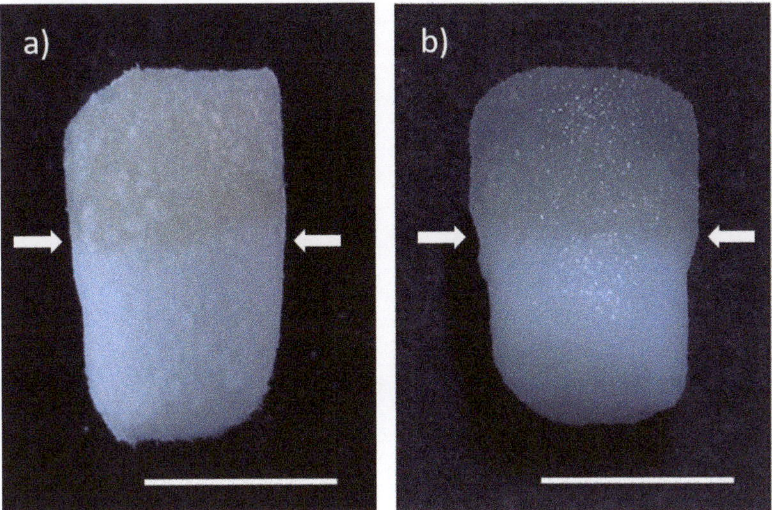

Figure 1. Biphasic scaffolds from mineralized salmon collagen (**lower layer**) and fibrillized jellyfish collagen (**upper layer**) in dry state (**a**) as well in wet state (**b**). Arrows indicate the transition zone between the layers. Scale bar represents 5 mm.

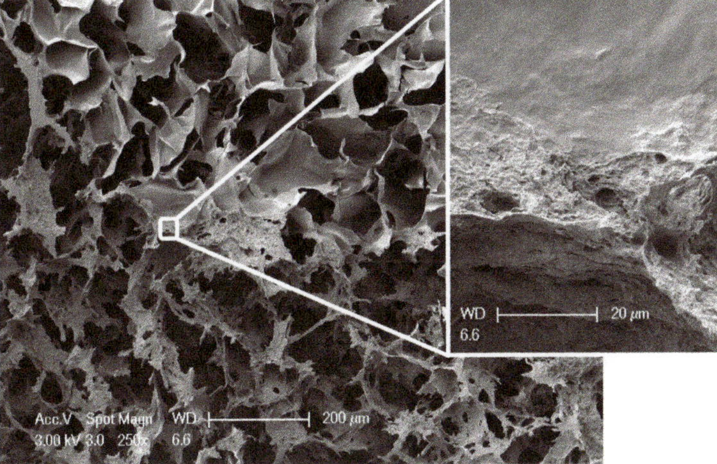

Figure 2. SEM image of a cross section of a biphasic scaffold at the transition zone between jellyfish collagen (**upper layer**) and mineralized salmon collagen (**lower layer**). Scale bar represents 200 µm. Insert showing the transition zone with higher magnification; scale bar represents here 20 µm.

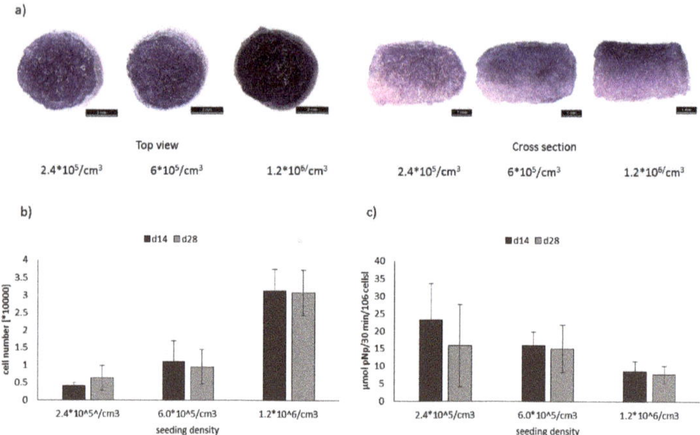

Figure 3. (**a**) MTT staining of osteogenically differentiated hMSC seeded in mineralized salmon collagen scaffolds at different densities and cultivated for 28 days; (**b**) cell number, calculated from DNA content and (**c**) specific ALP activity (ALP activity related to cell number) of osteogenically differentiated hMSC in salmon collagen scaffolds with different seeding densities after 14 and 28 days of cultivation, n = 3, mean +/− standard deviation. Significances from 2-way ANOVA were as follows: Cell number: d14 ↔ d28 n.s.; 2.4×10^5 ↔ 6×10^5 n.s., 2.4×10^5 ↔ 1.2×10^6 $p < 0.05$; 6×10^5 ↔ 1.2×10^6 $p < 0.05$, specific ALP activity: d14 ↔ d28 n.s.; $2.4 \times * 10^5$ ↔ $6 \times * 10^5$ n.s., 2.4×10^5 ↔ 1.2×10^6 $p < 0.05$; 6×10^5 ↔ 1.2×10^6 n.s.

2.3. Influence of TGF-β on Osteogenic Differentiation

We tested the influence of TGF-β3 on the osteogenic differentiation of hMSC, seeded in monophasic scaffolds from mineralized salmon collagen. Specific ALP activity after 14 days of osteogenic stimulation showed a significant ($p < 0.05$) decrease in the presence of TGF-β3 both for hMSC and osteogenically prestimulated hMSC (Figure 4). The experiment was also performed to test the effectivity of osteogenic pre-stimulation in the monolayer, and a significantly ($p < 0.001$) higher specific ALP activity was detected both at day 1 and day 14.

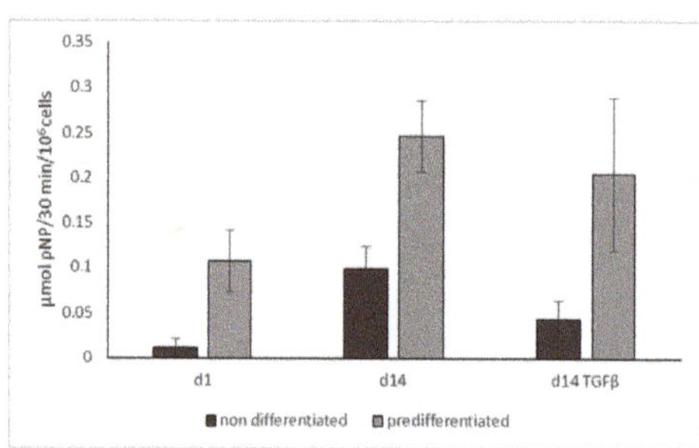

Figure 4. Specific ALP activity of hMSC which were cultivated in monophasic scaffolds from mineralized salmon collagen under osteogenic stimulation (10^{-7} M Dex, 10 mM β-GP, 12.5 µg/mL AAP = osteogenic supplements, OS+) in the presence and absence of 10 ng/mL TGF-β3. Dark bars: cells were prestimulated with OS+ in monolayer culture for 9 days before seeding the scaffolds. Cells of two different donors were used, each n = 3 per condition. ALP activity was related to cell number. Values are presented as mean (n = 6) +/− standard deviation. Significances from 2-way ANOVA were as follows: d1 ↔ d14 $p < 0.001$; d1 ↔ d14TGFβ $p < 0.05$; d14 ↔ d14TGFβ $p < 0.05$; non-differentiated ↔ pre-differentiated $p < 0.001$.

2.4. Chondrogenic and Osteogenic Differentiation of hMSC in Biphasic Marine Scaffolds

A sequential seeding procedure for the biphasic collagen scaffolds was performed (Figure 5). In the first step of sequential cultivation, hMSC were suspended in alginate solution, the jellyfish collagen phase of the biphasic scaffolds was infiltrated with this mixture with a cell density of 6×10^6 cells/cm^3 and the constructs were cultivated with complete chondrogenic medium for 9 to 12 days. At the same time, hMSC from the same batch were seeded into flasks and cultivated in the presence of osteogenic medium for 9 to 12 days. After 9 to 12 days of pre-stimulation, the osteogenically induced cells were seeded into the mineralized salmon collagen layer of the biphasic scaffolds with an initial cell density of 6×10^5 cells/cm^3. The constructs were cultivated until d21 from the initial seeding with osteochondral medium, containing 5 ng/mL TGF-β3, ITS, 10^{-7} M Dex and 50 µg/mL AAP.

The biphasic constructs were stable during the whole cultivation period. MTT staining of viable cells after 1 and 9 days of cultivation demonstrated that the chondrogenically induced cells did not migrate out of the jellyfish collagen/alginate phase to the mineralized salmon collagen phase below (Figure 6). This was confirmed by confocal laser scanning microscopic (cSM) investigations at the area between the two phases (Figure 7).

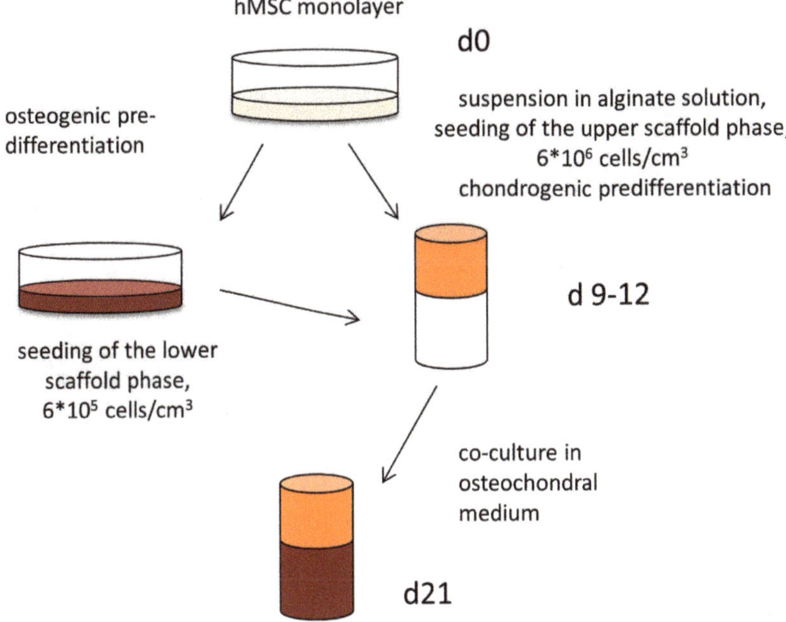

Figure 5. Pre-differentiation and sequential seeding of hMSC onto biphasic scaffolds from jellyfish collagen and mineralized salmon collagen.

At the end of the cultivation, a contraction of the biphasic scaffolds was visible. This contraction has been observed before in monophasic collagen scaffolds and it has been shown to be diminished with the application of alginate as cell carrier. Histological sections of biphasic scaffolds after 21 days of cultivation showed the presence of cells in both phases (Figure 8A–C). Furthermore, toluidine blue staining of histological sections revealed the production of cartilage extracellular matrix in the chondrogenic part of the scaffold (Figure 8D–F). The cellularity of the constructs is considerably lower compared to pellet cultures of chondrogenically stimulated cells. However, we have shown in a previous study, that higher cell densities in porous collagen scaffolds did not increase extracellular matrix production per cell [9].

Figure 6. MTT Staining of viable cells in biphasic collagen scaffolds, freshly seeded with alginate-embedded hMSC (d1), after 9 days of chondrogenic differentiation (d9) and after 21 days of cultivation, seeded with osteogenically pre-differentiated hMSC at day 9. Scale bars represent 2 mm.

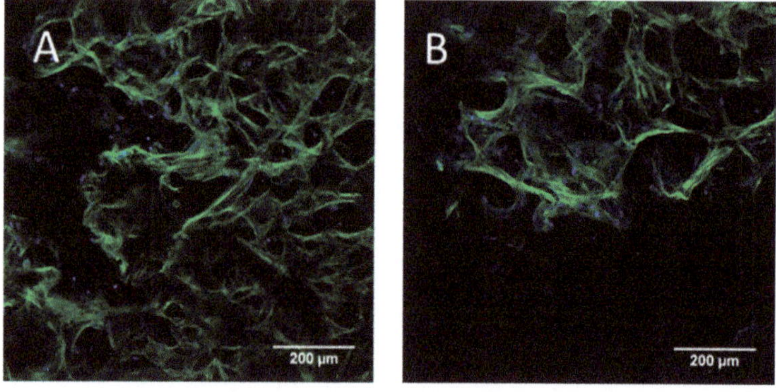

Figure 7. (**A**) cLSM images of a cross section of biphasic collagen scaffold seeded with hMSC after 9 days of chondrogenic differentiation; (**B**) transition area between jellyfish collagen (upper) and salmon collagen (lower) phase, after 9 days of chondrogenic differentiation before seeding of osteogenic cells, cytoskeleton stained with Alexa Fluor 488 phalloidin (green), nuclei stained with DAPI (blue). Scale bars represent 200 µm.

Chondrogenic differentiation of the cells in the jellyfish collagen/alginate phase was verified by gene expression analysis of collagen II, which was also detected at protein level (Figure 9).

Gene expression of collagen II increased from d1 to d12, but decreased slightly during the cultivation of the biphasic constructs in the osteochondral medium. Similar results were obtained for the quantification of collagen II by ELISA; however, there were always high variations between the different samples (Figure 9b). Additionally, the production of sulfated glycosaminoglycans increased during chondrogenic differentiation of jellyfish collagen/alginate embedded hMSC, but it did not further increase during cultivation in the osteochondral medium (Figure 9c).

Osteogenic differentiation of hMSC in the mineralized salmon collagen phase was demonstrated by ALP gene expression, which was strongest immediately after seeding of the biphasic scaffolds with osteogenically pre-differentiated hMSC. During further cultivation in osteochondral medium the ALP gene expression decreased (Figure 9a). Furthermore, gene expression of osteocalcin was analyzed in both scaffold parts, which was relatively low in all examined samples, however, cells in the chondrogenic layer showed down-regulation of osteocalcin (Figure 9).

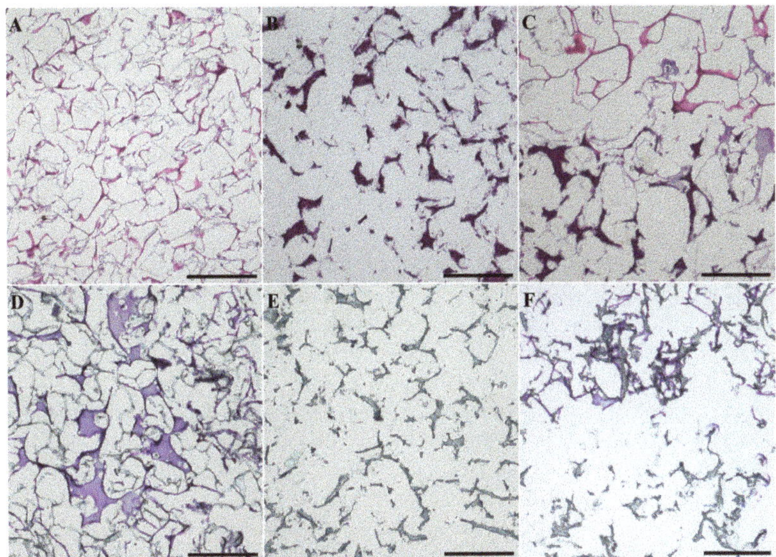

Figure 8. Histological sections of biphasic scaffolds after 21 days of osteochondral cultivation: Haematoxylin/Eosin staining of (**A**) jellyfish collagen phase, (**B**) mineralized salmon collagen phase and (**C**) transition zone between jellyfish collagen part (**top**) and mineralized salmon collagen part (**bottom**). Deeply purple stained regions in (**B,C**) represent mineralized salmon collagen which is stained by haematoxylin too. Toluidine blue staining of (**D**) jellyfish collagen phase, (**E**) mineralized salmon collagen phase and (**F**) transition zone between jellyfish collagen part (**top**) and mineralized salmon collagen part (**bottom**).

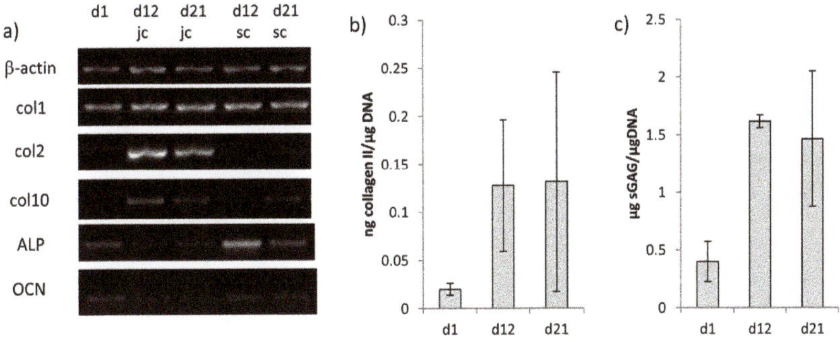

Figure 9. (**a**) RT-PCR products of different osteogenic and chondrogenic marker genes jc = jellyfish collagen, sc = mineralized salmon collagen; (**b**) Collagen II content in the jellyfish collagen phase of biphasic scaffolds, detected by ELISA; (**c**) concentration of sulfated glycosaminoglycanes content in the jellyfish collagen phase of biphasic scaffolds. $n = 3$, mean $+/-$ standard deviation.

3. Discussion

Biphasic, but monolithic scaffolds, exclusively from the marine biopolymers jellyfish collagen, biomimetically mineralized salmon collagen and alginate were fabricated for the first time. Due to the concerted freeze drying of the two phases before crosslinking the layers were tightly connected showing interconnecting pores through the area of the different layers. In the wet state, the two layers showed different swelling behavior. While the biomimetically mineralized salmon collagen phase did not swell, the volume of the fibrillized jellyfish collagen phase increased somewhat in wet state. The reason for the different swelling behavior could be the difference between the collagen types. Collagen II was shown to have higher swelling capacity compared to collagen I [17]. Furthermore, the presence of nanocrystalline hydroxyapatite connected to the collagen fibers of the salmon collagen could be responsible to the reduced swelling in this phase. Nevertheless, the different swelling behavior did not impair the union between the two phases, which have been cross-linked with higher EDC concentrations compared to the monophasic scaffolds. Despite the high concentration of carbodiimide for crosslinking to allow the tight union between the phases, we did not detect any negative effects on cytocompatibility. In a similar approach to our method mineralized and non-mineralized collagen (from equine origin) were combined to a triphasic scaffold for osteochondral regeneration [18]. In contrast to our study, crosslinking of the single layers was performed before freeze-drying, which required an additional knitting procedure to anchor the layers. Biphasic scaffolds from collagen I in the chondrogenic part and collagen I mineralized with Mg^{2+} substituted hydroxyapatite in the osteogenic part were recently prepared by Sartori and co-workers [19]. Also in this study, the collagen layers were cross-linked before freeze-drying. Nevertheless, the resulting biphasic scaffolds were stable enough to withstand subcutaneous implantation in mice for up to 8 weeks. In vitro investigations for MSC differentiation, however, were only performed with monophasic scaffolds in this study.

For the generation of osteochondral tissue constructs in vitro progenitor cells like human mesenchymal stroma cells need to be differentiated into two different lineages in spatially separated phases of the same scaffold. Optimally, scaffold structure and chemical composition provide the necessary stimuli for differentiation into the osteogenic and chondrogenic cell line. Jellyfish collagen from R. esculentum has a similar structure to human collagen II, since it consists of α-chain homotrimers and shows a degree of glycosylation similar to that of vertebrate collagen type II [20]. Chondrogenic differentiation is triggered by clustering of hMSC, which can be realized by 3D pellet formation or seeding of scaffolds with high cell densities [21]. Additionally, embedding of cells into alginate hydrogel, infiltrated into porous collagen scaffolds, induced a chondrogenic phenotype and increased collagen II expression of the cells [22]. Therefore, the upper (chondral) phase of the marine biphasic scaffolds was seeded with hMSC in high density (6×10^6 cells/mL) which were incorporated into alginate hydrogel completely filling the pores of the jellyfish collagen part of the scaffolds. Osteogenic differentiation is favored by the presence of hydroxyapatite, and it has been shown in porous collagen scaffolds in vitro, that specific ALP activity which is the main osteogenic marker increases with decreasing seeding density [23]. Similar results were also obtained in the present study for monophasic scaffolds from mineralized salmon collagen (Figure 3c). The lower phase of the biphasic marine scaffolds was therefore seeded with hMSC in a tenfold lower density compared to the chondral part, and the cells were allowed to attach directly at the pore walls of the mineralized salmon collagen without addition of alginate. Nevertheless, as shown in our previous studies with monophasic scaffolds from jellyfish collagen, scaffold structure and composition as well as cell density alone are not sufficient to induce chondrogenic differentiation of hMSC. Likewise, biomimetically mineralized collagen scaffolds, despite their bone-like composition of collagen and nanocrystalline hydroxyapatite, are not osteoinductive. The addition of osteogenic stimuli is necessary to induce osteogenic differentiation of MSC seeded in the scaffolds. For simultaneous osteogenic and chondrogenic differentiation of hMSC in the marine biphasic scaffolds in vitro, a suitable osteochondral medium needed to be developed. Cell culture media for the chondrogenic and osteogenic differentiation of MSC have equal and distinct components [24]. While dexamethasone and ascorbate are included in both chondrogenic and osteogenic differentiation medium, both media have exclusive components (Table 1).

Table 1. Main differences in the composition of commonly used osteogenic and chondrogenic differentiation media (FCS = fetal calf serum, ITS = insulin, selen, transferrin mix, AAP = ascorbic acid-2-phosphate, Dex = dexamethasone, β-GP = β-glycerophosphate, TGF-β = transforming growth factor β).

Osteogenic medium	FCS	-	Low glucose	AAP	Dex	β-GP	
Chondrogenic medium	-	ITS	High glucose	AAP	Dex	-	TGF-β

To develop an osteochondral medium supporting both osteogenic and chondrogenic differentiation of MSC (additionally to the stimuli, which are exerted by the scaffold matrix) it is not sufficient just to combine the single media 1:1. In a previous study we have analyzed the impact of FCS on the chondrogenic differentiation of hMSC in monophasic scaffolds of jellyfish collagen [25] and demonstrated, that even small amounts of FCS (2%) in the culture medium significantly decrease the mRNA expression of chondrogenic markers. In addition, reduction of glucose content caused a decreased mRNA expression of chondrogenic markers as well as a decreased extracellular matrix production of the chondrogenically differentiated cells. In contrast, osteogenic differentiation of hMSC in mineralized salmon collagen scaffolds was not affected when the glucose concentration of the medium was increased (data not shown). Furthermore, osteogenic differentiation of hMSC is even favored by low serum conditions (1% and 5% compared to 10%) [26]. B-glycerophosphate which is an integral component of osteogenic differentiation media to provide a phosphate source for the mineralizing osteoblasts, is regarded as hypertrophy promoting reagent for chondrocytes [27]. The main component of chondrogenic differentiation medium, TGF-β, was shown to cause a downregulation of ALP expression [28], which was also found in our study for MSC seeded in scaffolds from mineralized salmon collagen (Figure 4). Based on these results we propose an osteochondral medium containing a reduced amount of TGF-β (5 ng/mL), no FCS, high glucose content, bovine serum albumin, as well as the factors stimulating both osteogenic and chondrogenic differentiation: dexamethasone and ascorbic acid-2-phosphate. However, even the combination of (1) different seeding densities in the osteogenic and chondrogenic part (2) different scaffold layer composition (3) alginate infiltration in the chondrogenic part preventing cell adhesion to the scaffolds pores and therefore supporting chondrogenic cell phenotype and (4) osteochondral medium was not sufficient to stimulate osteogenic and chondrogenic differentiation spatially separated in the respective scaffold parts. Similar results were obtained by Gupta et al. who tried to simultaneously differentiate rat MSC into chondrogenic and osteogenic lineage in gradient PLGA scaffolds with encapsulated chondroitin sulfate for chondrogenic priming as well as tricalcium phosphate for osteogenic priming [29]. The authors admittedly reported a better interaction of cells with the materials and a greater cellularity, but it was not possible to drive differentiation specifically into either of the planned directions. Caliari and Harley investigated the impact of scaffold local biophysical properties like mineral content and density on hMSC differentiation in the presence of mixed soluble signals for osteogenic and chondrogenic differentiation [30]. Unexpectedly, the authors observed an increased osteogenic response just in nonmineralized scaffolds with low density, which were intended to induce chondrogenic differentiation. Therefore, in the present study, a sequential seeding and pre-differentiation approach was developed. It has been shown before, that temporal stimulation of chondrogenic cells with TGF-β is sufficient to induce the chondrogenic phenotype in MSC. Buxton and co-workers stimulated hMSC in hydrogels with TGF-β1 and demonstrated, that total production of collagen II after three weeks of cultivation was not decreased in comparison to the controls, when TGF-β was withdrawn after 7 days of prestimulation [31]. Likewise Fensky and co-workers demonstrated, that a 10 day stimulation of hMSC embedded in a collagen I hydrogel is sufficient to induce upregulation of the chondrogenic marker genes for collagen II and aggrecan after three weeks of cultivation [32]. Short delivery of high TGF-β doses (100 ng/mL) for 7 days on bovine MSC embedded in hyaluronic acid hydrogels was sufficient to induce and maintain the chondrogenic phenotype over a period of 9 weeks [33]. Chondrogenically pre-differentiated hMSC maintained their chondrogenic potential after embedding in methacrylated hyaluronic acid gels [34]. Osteogenic pre-differentiation of MSC before application in bone regeneration has successfully been applied by several groups. Peters et al. demonstrated that the injection of osteogenically pre-differentiated MSC enhanced healing of a critical bone defect in rats [35]. It is hypothesized that the application of osteogenically pre-differentiated instead of undifferentiated hMSC may prevent the transplanted cells from neoplasia and tumor formation [36]. Osteogenic pre-differentiation of hMSC has been successfully applied in our study. Osteogenically pre-differentiated hMSC seeded into monophasic scaffolds from mineralized salmon collagen showed significantly increased ALP activity both at the start of the 3D cultivation and after 14 days. Furthermore, the negative effect of TGF-β on ALP activity is less pronounced for osteogenically pre-differentiated cells compared to non-differentiated MSC, which is a further point to apply osteogenic pre-differentiation for osteochondral constructs. Osteocalcin gene expression, which is a further marker of osteogenic differentiation, was quite low in the biphasic scaffolds. This is in accordance to a fundamental study of Jaiswal and co-workers, demonstrating that vitamin D3 is necessary to induce adequate osteocalcin expression in osteogenically differentiated hMSC in vitro [37]. We refrained from adding vitamin D3 to the osteochondral medium to preserve the chondrogenic phenotype in the chondral layer. However, it might be beneficial to include vitamin D3 into the osteogenic pre-differentiation medium in future. In the present study, osteogenic and chondrogenic pre-differentiation was performed simultaneously. While chondrogenic pre-differentiation was realized already during cultivation of alginate-embedded hMSC in the jellyfish collagen part of the biphasic scaffold, osteogenic pre-differentiation of hMSC of the same batch was performed in the monolayer. Gene expression analysis revealed collagen II expression exclusively in the chondral part of the scaffolds; however, there was a slight decrease in collagen II expression during the cultivation with osteochondral medium in the presence of osteogenically pre-differentiated cells. Likewise, collagen II production on protein level and production of sGAG were somewhat decreased during co-culture (Figure 9). It has been demonstrated that continuous treatment of hMSC pellet cultures with TGF-β provided significantly higher production of extracellular matrix and expression of chondrogenic genes compared to short-time TGF-β supplementation of 3 and 10 days [38]. Possibly, the reduced TGF-β supplementation during osteochondral cultivation was not sufficient to stabilize the chondrogenic phenotype at the starting level. Similar observations were made for the osteogenic part. Highest ALP expression was detected immediately after seeding of the biphasic scaffolds, followed by a decrease of ALP expression during further osteochondral stimulation. Since ALP is an early osteogenic marker, the decrease could also relate to further differentiation along the osteogenic lineage. However, it has to be noted, that the applied osteochondral medium does not provide optimal conditions for osteogenic differentiation. Nevertheless, the RNA amount isolated from the mineral part of the biphasic scaffolds after 3 weeks of cultivation was not reduced compared to the start of 3D cultivation, suggesting that the osteogenically pre-differentiated cells survived the osteochondral conditions with serum deprivation. After three weeks of cultivation osteochondral constructs were obtained in vitro with spatial separated expression of chondrogenic ECM and osteogenic differentiation.

4. Materials and Methods

4.1. Preparation of Biphasic Scaffolds

Biphasic but monolithic collagen scaffolds were prepared by overlaying suspensions of biomimetically mineralized salmon collagen and fibrillized jellyfish collagen in the cavities of a 96-well plate. Biomimetically mineralized collagen and fibrillized jellyfish collagen suspensions were prepared as already published [7,9]. After freezing of the overlayed samples at 1 K/min to a final temperature of −20 °C lyophilization was conducted for about 24 h (Alpha 1–2, Christ). For stabilization the scaffolds were cross-linked with a 30 g/L solution of *N*-(3-dimethylaminopropyl)-*N'*-ethylcarbodiimide (EDC) hydrochloride in 80% *v/v* ethanol for 12 h. Subsequently, the scaffolds were rinsed thoroughly in deionized water, 1% glycine solution, once again in water and finally freeze-dried.

4.2. Cultivation of hMSC

Bone marrow derived hMSC harvested from the iliac crest of two healthy donors were kindly provided by the group of Prof. Martin Bornhäuser (Medical Clinic I, University Hospital Dresden). Written informed consent from the donors was obtained for the use of these samples in research. Cells were characterized as hMSC according to the criteria of the International Society of Cellular Therapy [39]. All procedures were approved by the Ethical Commission of the Medical Faculty of Technische Universität Dresden. Cells were expanded in Dulbecco's modified Eagle's Medium (DMEM, Gibco) supplemented with 10% fetal calf serum, 2 mM L-glutamine, 100 U/mL penicillin and 100 µg/mL streptomycin (all from Biochrom, Berlin, Germany) (expansion medium) until passage five.

General seeding preparation for porous collagen scaffolds: Scaffolds were immersed with expansion medium for 24 h, which was removed before seeding wet scaffolds were placed on sterile filter paper to remove excess immersion medium from the scaffold pores.

4.3. Osteogenic Differentiation of hMSC in Mineralized Salmon Collagen Scaffolds

Monophasic scaffolds from mineralized salmon collagen (d = 6 mm, h = 3 mm) were prepared as already published [7] and sterilized by γ-irradiation before use in cell culture. Scaffolds were seeded with 2×10^4, 5×10^4, and 1×10^5 cells in 50 µL of expansion medium. After 30 min of initial adhesion, further 500 µL of expansion medium were added to each scaffold. Cell-seeded scaffolds were cultivated for up to 4 weeks, with medium change every 3–4 days. Osteogenic differentiation medium contained Minimal essential medium α-modification (α-MEM) (Biochrom Berlin, Germany), supplemented with 10% fetal calf serum, 2 mM L-glutamine, 100 U/mL penicillin and 100 µg/mL streptomycin (all from Biochrom), 10^{-7} M Dex, 12.5 µg/mL AAP and 10 mM β-GP (all from Sigma, Taufkirchen, Germany). Since we wanted to analyze the effect of TGF-β on the osteogenic differentiation additionally 10 ng/mL TGF-β3 (Miltenyi, Bergisch Gladbach, Germany) were added to one experimental group. After 1, 14 and 28 days samples for DNA and ALP quantification were washed twice and frozen at −80 °C.

4.4. Cultivation of Osteochondral Constructs

Biphasic scaffolds from jellyfish collagen and mineralized salmon collagen (d = 6 mm, h = 8 mm) were sterilized by γ-irradiation before use in cell culture. Sodium alginate (Sigma) was dissolved in Ca^{2+} free DMEM (Sigma) without any further supplements at 12 mg/mL and the solution was filtered through a syringe filter with 0.45 µm pore size. 5×10^5 cells were suspended in 50 µL of alginate solution and the suspension was applied to the top of each biphasic scaffold. With 6 mm diameter and 3 mm height the chondral phase of the biphasic scaffold has an approximate volume of 85 mm^3 resulting in a seeding density of 6×10^6 cells/cm^3. After 15 min of incubation at 37 °C, 1 mL of sterile $CaCl_2$ (100 mM) solution was added and the constructs were incubated for further 15 min for the formation of alginate gel. The constructs were washed with expansion medium and cultivated further with chondrogenic differentiation medium (DMEM high glucose (Gibco, Dublin, Ireland, distributed by Thermo Fisher, Waltham, MA, USA), 100 U/mL penicillin, 100 µg/mL streptomycin, 1% ITS-X-Mix (Gibco; results in concentrations of 10 µg/mL insulin, 5.5 µg/mL transferrin, 6.7 ng/mL sodium selenite and 2 µg/mL ethanolamine in the medium), 0.35 µM proline, 50 µg/mL AAP, 10^{-7} M Dex, 0.15% bovine serum albumin (BSA) (all Sigma) and 10 ng/mL TGF-β3 (Miltenyi)) for 9–12 days with medium changes every 3–4 days. At the same time, hMSC from the same batch were cultivated in T-flasks with α-MEM supplemented with 10% fetal calf serum, 2 mM L-glutamine, 100 U/mL penicillin and 100 µg/mL streptomycin, 10^{-7} M Dex, 12.5 µg/mL AAP and 10 mM βGP. After 9 to 12 days of separate cultivation, biphasic scaffolds seeded with chondrogenically pre-differentiated hMSC were gently dried on sterile filter paper, flipped, that the mineralized layer was on top of the scaffold and were seeded in the mineralized salmon collagen layer with 5×10^4 osteogenically pre-differentiated hMSC in 50 µL cell culture medium, resulting in a seeding density of 6×10^5 cells/cm^3. The constructs were cultivated up to 21 days of total cultivation time with osteochondral medium consisting of DMEM high glucose supplemented with 100 U/mL penicillin, 100 µg/mL streptomycin, 1% ITS-X-Mix, 5 ng/mL TGF-β3, 10^{-7} M Dex, 0.35 µM proline, 0.15% BSA and 50 µg/mL AAP. For each time point of the experiment 9 biphasic scaffolds were used (3 for gene expression analysis, 3 for measurement of collagenII and sGAG, 1 for MTT staining, 1 for fluorescence staining and 1 for histological staining).

4.5. MTT Staining

To detect the distribution and amount of viable cells, medium of cell-seeded monophasic and biphasic scaffolds was supplemented with 1.2 mM 3-(4,5-dimethylthiazol-2-yl)-2,5-diphenyltetrazolium bromide (MTT; Sigma), followed by further incubation at 37 °C for 4 h. Cell seeded scaffolds were imaged using a Leica stereomicroscope.

4.6. Fluorescence Staining and Confocal Laser Scanning Microscopy

Cell-seeded scaffolds were fixed with 4% buffered formaldehyde and permeabilized with 0.2% Triton X-100 (Sigma) in Hank's balanced salt solution (HBS) (Gibco). Autofluorescence was blocked with, 30 min incubation in 3% BSA in HBS. Staining of cytoskeleton was performed with Alexa Fluor 488 phalloidin and staining of the nuclei with 0.3 µM DAPI (4′,6-diamidino-2-phenylindole dihydrochloride; both Invitrogen) in HBS. Samples were imaged using a confocal laser scanning microscope LSM 510 (Zeiss, Jena, Germany) applying an excitation/emission wavelength of 405/461 nm (diode laser) for DAPI and 488/519 nm (argon laser) for Alexa Fluor 488.

4.7. Gene Expression Analysis

Biphasic scaffolds were divided with a scalpel into the two phases prior to RNA extraction, for which three samples of each experimental group were used. During the RNA isolation procedure, cell lysates of the three samples of each group were pooled.

To dissolve the alginate in alginate-containing jellyfish collagen constructs, samples were mixed thoroughly with 55 mM sodium citrate (Fluka, distributed by Sigma-Aldrich, Taufenkirchen, Germany) with 0.9% sodium chloride (Sigma), in DEPC-water (Gibco), incubated at 37 °C for 45 min and afterwards mixed thoroughly again. Cells in the supernatant were centrifuged at 3600 rpm for 10 min, resuspended in PBS, and centrifuged again. Both the pellet and the collagen scaffold were treated with lysis buffer from peqGOLD MicroSpin total RNA Kit (Peqlab, Erlangen, Germany). Lysates from the pellets and scaffolds were pooled and RNA was extracted according to the manufacturer's instructions of the kit. To isolate RNA from the mineralized salmon collagen phase, cell seeded scaffolds were treated with lysis buffer from the RNA kit directly.

For polymerase chain reaction (PCR) 200 ng per experimental condition were transcribed into cDNA in a 20 µL reaction mixture containing 200 U of superscript II reverse transcriptase, 0.5 mM dNTPs (both Invitrogen), 12.5 ng/µL random hexamers (Eurofins MWG Operon, Ebersberg, Germany) and 40 U of RNase inhibitor RNase OUT (Invitrogen, Carlsbad, CA, USA). 1 µL cDNA in 20 µL reaction mixtures containing specific primer pairs were used for amplification in PCR analysis to detect transcripts of collagen I, collagen IIa, collagen X, ALPL and β-actin, respectively. Primer sequences (Eurofins MWG Operon), annealing temperatures and amplicon sizes for each gene are summarized in Table 2.

For analysis, PCR products were visualized in 2% agarose gels (Ultra PureTMAgarose, Invitrogen).

Table 2. Primer and conditions for reverse transcriptase PCR.

Marker	Bp	Primer (Forward/Reverse)	Buffer/±Enhancer	$T_{annealing}$	Amplification Cycles
β-Act	234	5′-GGACTTCGAGCAAGAGATGG-3′ 5′-AGCACTGTGTTGGCGTACAG-3′	buffer S/−	55 °C	30x
Col 1	331	5′-GGATGAGGAGACTGGCAAC-3′ 5′-GAAGAAGAAATGGCAAAGAGAAAG-3′	buffer S/−	55 °C	25x
Col 2	388	5′-GAACATCACCTACCACTGCAAG-3′ 5′-GCAGAGTCCTAGAGTGACTGAG-3′	buffer Y/+	60 °C	35x
Col 10	196	5′-GCCCACTACCCAACACCAAGAC-3′ 5′-CCTGGCAACCCTGGCTCTC-3′	buffer S/−	50 °C	30x
ALP	162	5′-ACCATTCCCACGTCTTCACATTTG-3′ 5′-ATTCTCTCGTTCACCGCCCAC-3′	buffer S/−	55 °C	30x
OCN	177	5′-CAA AGG TGC AGC CTT TGT GTC-3′ 5′-TCA CAG TCC GGA TTG AGC TCA-3′	buffer S/−	55 °C	35x

4.8. Analysis of DNA Content, ALP Activity in Monophasic Scaffolds from Mineralized Salmon Collagen sGAG Content and Collagen II Content

Frozen cell-seeded scaffolds were homogenized in ice-cold PBS (2 × 10 s at 5900 rpm) using a Precellys24 apparatus (Peqlab, Erlangen, Germany). After homogenization, 10% Triton X-100 in PBS were added to a final concentration of 1% Triton-X-100 and the samples were incubated on ice for 50 min. ALP activity and DNA content were analyzed from the same lysate. ALP activity was analyzed by conversion of p-nitrophenyl phosphate (Sigma, 1 mg/mL), in 0.1 M diethanolamine pH 9.8, 1% Triton X-100 and 1 mM $MgCl_2$, to p-nitrophenol after 30 min of incubation at 37 °C and absorption measurement at 405 nm (Infinite® M200 Pro, Tecan, Männedorf, Switzerland). DNA content was quantified from the same lysate with Quantifluor dye (Promega, Madison, WI, USA) at an excitation/emission wavelength of 485/535 nm. Cell number was calculated from DNA content of defined cell numbers. ALP activity was related to the cell number of the respective sample.

4.9. sGAG Content and Collagen II Content in Biphasic Scaffolds

Biphasic scaffolds were divided with a scalpel into the two phases prior to freezing for subsequent biochemical analyzes which included always three biphasic scaffolds per group. After thawing of alginate-containing jellyfish collagen constructs, samples were covered with 500 μL of 55 mM sodium citrate (Fluka) with 0.9% sodium chloride (Sigma), in deionized water, homogenized (1 × 10 s at 5900 rpm, Precellys24, Peqlab) incubated at 37 °C for 30 min and afterwards mixed thoroughly again. After thawing mineralized salmon collagen constructs, samples were covered with 450 μL PBS, homogenized (2 × 10 s at 5900 rpm), 50 μL of 10% Triton X100 were added, and the mixture was incubated for 30 min at 37 °C. All samples were centrifuged at 3600 rpm for 10 min. The supernatants were used for collagen II ELISA, while the pellet was further processed for sGAG determination. Collagen II ELISA was performed as already published [25]. Briefly, each 50 μL of the supernatants was added to wells which were precoated with primary antibody (Mouse Anti-Chick Collagen II Capture Antibody (clone 35; Chondrex, Redmont, WA, USA) diluted 1:500 in PBS). A calibration line was established using human collagen II (Millipore) diluted in PBS with 3% normal goat serum (NGS; Life Technologies, Carlsbad, CA, USA). Afterwards 50 μL of secondary antibody (Biotin-labeled Detection Antibody (Mouse monoclonal anti-Type 2 Collagen (Chondrex)) diluted 1:100 in PBS with 3% NGS) were added. Detection was performed with streptavidin-horseradish-peroxidase (R&D Systems, Minneapolis, MN, USA) and 3,3′,5,5′-tetramethylbenzidin substrate-solution (Sigma). Absorbance was assessed at 450 nm (reference 570 nm) in a microplate reader (Infinite® M200 Pro, Tecan). Absorbance values from scaffolds without cells carried along during the experiments were subtracted as correction factors.

sGAG content from the pellets was quantified as previously described [25]. Briefly, 1 mL of papain digestion solution (containing 125 μg/mL papain, 5 mM EDTA (both from Sigma), 100 mM Na_2HPO_4, and 5 mM cystein (both from Carl Roth, Karlsruhe, Germany) in deionized water) were added to each pellet. After incubation at 60 °C for 24 h. 50 µL of the digested solution were subjected to a commercially available sGAG assay (Kamiya, Seattle, WA, USA) according to manufacturer's instructions. Absorbance was assessed at 610 nm (Infinite® M200 Pro, Tecan).

4.10. Histological Investigations on Biphasic Scaffolds

Biphasic scaffolds cultivated for 21 days under osteochondral stimulation were fixed with 4% buffered formaldehyde, dehydrated and embedded in paraffin. 5 µm sections were cut and mounted to cover slides. After deparaffinization, sections were stained with hematoxylin/eosin (H/E) to visualize cell distribution and toluidine blue to visualize the production of cartilage proteoglycans. Stained samples were images using a BZ-9000 (Biorevo) (Keyence, Neu-Isenburg, Germany) microscope.

4.11. Statistical Analysis

Statistical analyses for cell number and ALP activity were performed by two-way analysis of variance (ANOVA). Post-hoc analysis was performed in all cases to determine multiple comparisons using the Tukey method (Origin 9.1, OriginLab). Significance levels were set as $p < 0.05$, $p < 0.01$ and $p < 0.001$.

5. Conclusions

Biphasic, but monolithic scaffolds exclusively from marine collagens are stable under cell culture conditions for up to three weeks without any delamination of the phases. We have tried to simultaneously differentiate hMSC into osteogenic and chondrogenic lineage spatially separated in the bone and cartilage layer of the marine scaffolds. However, the different chemical nature and mineralization of the layers, as well as different seeding densities, and the application of alginate hydrogel to embed the cells into the jellyfish collagen layer of the scaffold were not sufficient to trigger the differentiation of hMSC adequately into the respective direction. We therefore propose a sequential seeding of the biphasic scaffolds and pre-differentiation of the cells into both osteogenic and chondrogenic lineage to obtain functional osteochondral constructs.

Acknowledgments: We are grateful to the German Research Foundation (DFG) for funding the present study (BE 5139/1-1) and acknowledge Sophie Brüggemeier for excellent technical assistance. Thanks to Diana Jünger for preparing histological sections.

Author Contributions: A.B. performed the cell culture experiments, analyzed the data, and wrote the manuscript. B.P. prepared the monophasic mineralized salmon collagen scaffolds, developed the method for the preparation of the biphasic marine scaffolds, prepared all biphasic scaffolds, and did the SEM investigation. M.G. developed the original method of formation of biphasic but monolithic collagen scaffolds and was engaged in writing the manuscript.

References

1. Silvipriya, K.; Kumar, K.; Bhat, A.; Kumar, B.; John, A.; Lakshmanan, P. Collagen: Animal Sources and Biomedical Application. *J. Appl. Pharm. Sci.* **2015**, *5*, 123–127. [CrossRef]
2. Ramshaw, J.A.M. Biomedical applications of collagens. *J. Biomed. Mater. Res. B Appl. Biomater.* **2016**, *104*, 665–675. [CrossRef] [PubMed]
3. Ehrlich, H. *Biological Materials of Marine Origin: Invertebrates*; Biologically-Inspired Systems; Springer: Dordrecht, The Netherlands; New York, NY, USA, 2010; ISBN 978-90-481-9129-1.
4. Ehrlich, H. *Biological Materials of Marine Origin Vertebrates*; Springer: Dordrecht, The Netherlands, 2015; ISBN 978-94-007-5730-1.
5. Silva, T.; Moreira-Silva, J.; Marques, A.; Domingues, A.; Bayon, Y.; Reis, R. Marine Origin Collagens and Its Potential Applications. *Mar. Drugs* **2014**, *12*, 5881–5901. [CrossRef] [PubMed]
6. Subhan, F.; Ikram, M.; Shehzad, A.; Ghafoor, A. Marine Collagen: An Emerging Player in Biomedical applications. *J. Food Sci. Technol.* **2015**, *52*, 4703–4707. [CrossRef] [PubMed]

7. Hoyer, B.; Bernhardt, A.; Heinemann, S.; Stachel, I.; Meyer, M.; Gelinsky, M. Biomimetically mineralized salmon collagen scaffolds for application in bone tissue engineering. *Biomacromolecules* **2012**, *13*, 1059–1066. [CrossRef] [PubMed]
8. Sewing, J.; Klinger, M.; Notbohm, H. Jellyfish collagen matrices conserve the chondrogenic phenotype in two- and three-dimensional collagen matrices. *J. Tissue Eng. Regen. Med.* **2017**, *11*, 916–925. [CrossRef] [PubMed]
9. Hoyer, B.; Bernhardt, A.; Lode, A.; Heinemann, S.; Sewing, J.; Klinger, M.; Notbohm, H.; Gelinsky, M. Jellyfish collagen scaffolds for cartilage tissue engineering. *Acta Biomater.* **2014**, *10*, 883–892. [CrossRef] [PubMed]
10. Gelinsky, M.; Eckert, M.; Despang, F. Biphasic, but monolithic scaffolds for the therapy of osteochondral defects. *Int. J. Mater. Res.* **2007**, *98*, 749–755. [CrossRef]
11. Jeon, J.E.; Vaquette, C.; Klein, T.J.; Hutmacher, D.W. Perspectives in Multiphasic Osteochondral Tissue Engineering: Perspectives in Multiphasic Osteochondral Tissue Engineering. *Anat. Rec.* **2014**, *297*, 26–35. [CrossRef] [PubMed]
12. Shimomura, K.; Moriguchi, Y.; Murawski, C.D.; Yoshikawa, H.; Nakamura, N. Osteochondral Tissue Engineering with Biphasic Scaffold: Current Strategies and Techniques. *Tissue Eng. Part B Rev.* **2014**, *20*, 468–476. [CrossRef] [PubMed]
13. Yousefi, A.-M.; Hoque, M.E.; Prasad, R.G.S.V.; Uth, N. Current strategies in multiphasic scaffold design for osteochondral tissue engineering: A review: Current Strategies in Multiphasic Scaffold Design. *J. Biomed. Mater. Res. A* **2015**, *103*, 2460–2481. [CrossRef] [PubMed]
14. Gadjanski, I.; Vunjak-Novakovic, G. Challenges in engineering osteochondral tissue grafts with hierarchical structures. *Expert Opin. Biol. Ther.* **2015**, *15*, 1583–1599. [CrossRef] [PubMed]
15. Mano, J.F.; Reis, R.L. Osteochondral defects: Present situation and tissue engineering approaches. *J. Tissue Eng. Regen. Med.* **2007**, *1*, 261–273. [CrossRef] [PubMed]
16. Li, J.; Mareddy, S.; Tan, D.M.; Crawford, R.; Long, X.; Miao, X.; Xiao, Y. A minimal common osteochondrocytic differentiation medium for the osteogenic and chondrogenic differentiation of bone marrow stromal cells in the construction of osteochondral graft. *Tissue Eng. Part A* **2009**, *15*, 2481–2490. [CrossRef] [PubMed]
17. Pieper, J.S.; van der Kraan, P.M.; Hafmans, T.; Kamp, J.; Buma, P.; van Susante, J.L.C.; van den Berg, W.B.; Veerkamp, J.H.; van Kuppevelt, T.H. Crosslinked type II collagen matrices: Preparation, characterization, and potential for cartilage engineering. *Biomaterials* **2002**, *23*, 3183–3192. [CrossRef]
18. Nicoletti, A.; Fiorini, M.; Paolillo, J.; Dolcini, L.; Sandri, M.; Pressato, D. Effects of different crosslinking conditions on the chemical–physical properties of a novel bio-inspired composite scaffold stabilised with 1,4-butanediol diglycidyl ether (BDDGE). *J. Mater. Sci. Mater. Med.* **2013**, *24*, 17–35. [CrossRef] [PubMed]
19. Sartori, M.; Pagani, S.; Ferrari, A.; Costa, V.; Carina, V.; Figallo, E.; Maltarello, M.C.; Martini, L.; Fini, M.; Giavaresi, G. A new bi-layered scaffold for osteochondral tissue regeneration: In vitro and in vivo preclinical investigations. *Mater. Sci. Eng. C* **2017**, *70*, 101–111. [CrossRef] [PubMed]
20. Bermueller, C.; Schwarz, S.; Elsaesser, A.F.; Sewing, J.; Baur, N.; von Bomhard, A.; Scheithauer, M.; Notbohm, H.; Rotter, N. Marine collagen scaffolds for nasal cartilage repair: Prevention of nasal septal perforations in a new orthotopic rat model using tissue engineering techniques. *Tissue Eng. Part A* **2013**, *19*, 2201–2214. [CrossRef] [PubMed]
21. Yoo, J.U.; Barthel, T.S.; Nishimura, K.; Solchaga, L.; Caplan, A.I.; Goldberg, V.M.; Johnstone, B. The chondrogenic potential of human bone-marrow-derived mesenchymal progenitor cells. *J. Bone Jt. Surg. Am.* **1998**, *80*, 1745–1757. [CrossRef]
22. Pustlauk, W.; Paul, B.; Gelinsky, M.; Bernhardt, A. Jellyfish collagen and alginate: Combined marine materials for superior chondrogenesis of hMSC. *Mater. Sci. Eng. C* **2016**, *64*, 190–198. [CrossRef] [PubMed]
23. Lode, A.; Bernhardt, A.; Gelinsky, M. Cultivation of human bone marrow stromal cells on three-dimensional scaffolds of mineralized collagen: Influence of seeding density on colonization, proliferation and osteogenic differentiation. *J. Tissue Eng. Regen. Med.* **2008**, *2*, 400–407. [CrossRef] [PubMed]
24. Vater, C.; Kasten, P.; Stiehler, M. Culture media for the differentiation of mesenchymal stromal cells. *Acta Biomater.* **2011**, *7*, 463–477. [CrossRef] [PubMed]
25. Pustlauk, W.; Paul, B.; Brueggemeier, S.; Gelinsky, M.; Bernhardt, A. Modulation of chondrogenic differentiation of human mesenchymal stem cells in jellyfish collagen scaffolds by cell density and culture medium: Chondrogenic differentiation of hMSCs in jellyfish collagen scaffolds. *J. Tissue Eng. Regen. Med.* **2017**, *11*, 1710–1722. [CrossRef] [PubMed]

26. Binder, B.Y.K.; Sagun, J.E.; Leach, J.K. Reduced Serum and Hypoxic Culture Conditions Enhance the Osteogenic Potential of Human Mesenchymal Stem Cells. *Stem Cell Rev. Rep.* **2015**, *11*, 387–393. [CrossRef] [PubMed]

27. Mackay, A.M.; Beck, S.C.; Murphy, J.M.; Barry, F.P.; Chichester, C.O.; Pittenger, M.F. Chondrogenic differentiation of cultured human mesenchymal stem cells from marrow. *Tissue Eng.* **1998**, *4*, 415–428. [CrossRef] [PubMed]

28. Glueck, M.; Gardner, O.; Czekanska, E.; Alini, M.; Stoddart, M.J.; Salzmann, G.M.; Schmal, H. Induction of Osteogenic Differentiation in Human Mesenchymal Stem Cells by Crosstalk with Osteoblasts. *BioRes. Open Access* **2015**, *4*, 121–130. [CrossRef] [PubMed]

29. Gupta, V.; Mohan, N.; Berkland, C.J.; Detamore, M.S. Microsphere-Based Scaffolds Carrying Opposing Gradients of Chondroitin Sulfate and Tricalcium Phosphate. *Front. Bioeng. Biotechnol.* **2015**, *3*, 96. [CrossRef] [PubMed]

30. Caliari, S.R.; Harley, B.A.C. Collagen-GAG Scaffold Biophysical Properties Bias MSC Lineage Choice in the Presence of Mixed Soluble Signals. *Tissue Eng. Part A* **2014**, *20*, 2463–2472. [CrossRef] [PubMed]

31. Buxton, A.N.; Bahney, C.S.; Yoo, J.U.; Johnstone, B. Temporal Exposure to Chondrogenic Factors Modulates Human Mesenchymal Stem Cell Chondrogenesis in Hydrogels. *Tissue Eng. Part A* **2011**, *17*, 371–380. [CrossRef] [PubMed]

32. Fensky, F.; Reichert, J.C.; Traube, A.; Rackwitz, L.; Siebenlist, S.; Nöth, U. Chondrogenic predifferentiation of human mesenchymal stem cells in collagen type I hydrogels. *Biomed. Tech.* **2014**, *59*, 375–383. [CrossRef] [PubMed]

33. Kim, M.; Erickson, I.E.; Choudhury, M.; Pleshko, N.; Mauck, R.L. Transient exposure to TGF-β3 improves the functional chondrogenesis of MSC-laden hyaluronic acid hydrogels. *J. Mech. Behav. Biomed. Mater.* **2012**, *11*, 92–101. [CrossRef] [PubMed]

34. Lin, S.; Lee, W.Y.W.; Feng, Q.; Xu, L.; Wang, B.; Man, G.C.W.; Chen, Y.; Jiang, X.; Bian, L.; Cui, L.; et al. Synergistic effects on mesenchymal stem cell-based cartilage regeneration by chondrogenic preconditioning and mechanical stimulation. *Stem Cell Res. Ther.* **2017**, *8*, 221. [CrossRef] [PubMed]

35. Peters, A.; Toben, D.; Lienau, J.; Schell, H.; Bail, H.J.; Matziolis, G.; Duda, G.N.; Kaspar, K. Locally applied osteogenic predifferentiated progenitor cells are more effective than undifferentiated mesenchymal stem cells in the treatment of delayed bone healing. *Tissue Eng. Part A* **2009**, *15*, 2947–2954. [CrossRef] [PubMed]

36. Han, D.; Han, N.; Zhang, P.; Jiang, B. Local transplantation of osteogenic pre-differentiated autologous adipose-derived mesenchymal stem cells may accelerate non-union fracture healing with limited pro-metastatic potency. *Int. J. Clin. Exp. Med.* **2015**, *8*, 1406–1410. [PubMed]

37. Jaiswal, N.; Haynesworth, S.E.; Caplan, A.I.; Bruder, S.P. Osteogenic differentiation of purified, culture-expanded human mesenchymal stem cells in vitro. *J. Cell. Biochem.* **1997**, *64*, 295–312. [CrossRef]

38. Kim, H.-J.; Kim, Y.-J.; Im, G.-I. Is continuous treatment with transforming growth factor-beta necessary to induce chondrogenic differentiation in mesenchymal stem cells? *Cells Tissues Organs* **2009**, *190*, 1–10. [CrossRef] [PubMed]

39. Dominici, M.; Le Blanc, K.; Mueller, I.; Slaper-Cortenbach, I.; Marini, F.; Krause, D.; Deans, R.; Keating, A.; Prockop, D.; Horwitz, E. Minimal criteria for defining multipotent mesenchymal stromal cells. The International Society for Cellular Therapy position statement. *Cytotherapy* **2006**, *8*, 315–317. [CrossRef] [PubMed]

 *marine drugs* 

Article

Unique Collagen Fibers for Biomedical Applications

Dafna Benayahu [1,*], **Mirit Sharabi** [2], **Leslie Pomeraniec** [1], **Lama Awad** [1], **Rami Haj-Ali** [2] **and Yehuda Benayahu** [3,*]

1   Department of Cell and Developmental Biology, Sackler School of Medicine, Tel Aviv University,
     Tel Aviv 69978, Israel; leslieyael.p@gmail.com (L.P.); lamaawad@mail.tau.ac.il (L.A.)
2   The Fleischman Faculty of Engineering, Tel Aviv University, Tel Aviv 69978, Israel;
     miritsharabi@gmail.com (M.S.); rami98@tau.ac.il (R.H.-A.)
3   School of Zoology, George S. Wise Faculty of Life Sciences, Tel Aviv University, Tel Aviv 69978, Israel
*   Correspondence: dafnab@tauex.tau.ac.il (D.B.); yehudab@tauex.tau.ac.il (Y.B.);
     Tel.: +972-3-640-7432 (D.B.); +972-3-640-9090 (Y.B.)

Received: 15 February 2018; Accepted: 17 March 2018; Published: 23 March 2018

Abstract: The challenge to develop grafts for tissue regeneration lies in the need to obtain a scaffold that will promote cell growth in order to form new tissue at a trauma-damaged site. Scaffolds also need to provide compatible mechanical properties that will support the new tissue and facilitate the desired physiological activity. Here, we used natural materials to develop a bio-composite made of unique collagen embedded in an alginate hydrogel material. The collagen fibers used to create the building blocks exhibited a unique hyper-elastic behavior similar to that of natural human tissue. The prominent mechanical properties, along with the support of cell adhesion affects cell shape and supports their proliferation, consequently facilitating the formation of a new tissue-like structure. The current study elaborates on these unique collagen fibers, focusing on their structure and biocompatibility, in an in vitro model. The findings suggest it as a highly appropriate material for biomedical applications. The promising in vitro results indicate that the distinctive collagen fibers could serve as a scaffold that can be adapted for tissue regeneration, in support of healing processes, along with maintaining tissue mechanical properties for the new regenerate tissue formation.

Keywords: marine biomaterials; medical device; scaffold; soft corals; tissue regeneration

---

1. Introduction

Acute and chronic injury can cause temporary or permanent damage to tissue. When the body's natural tissue repair mechanisms are inefficient there is a need to facilitate tissue regeneration. Autologous grafts are limited by the patient's own tissue and also involve additional surgical procedures. Allografts and xenografts are hard to acquire and require a high compatibility in order to avoid transplant rejection by the recipient as an immune response to the foreign body [1–4]. There is a growing demand for alternative and improved biomedical devices that are able not only to replace the damaged tissue but also to enable its regeneration and the bio-integration of the implant. Cell function during regeneration relies on the scaffold in order to form new tissue to produce the extracellular matrix (ECM) that will support the cells' neo-tissue formation. The ECM plays a pivotal role during tissue regeneration, by directing the cell arrangement and modifying their function. Collagen is the most abundant structural protein in the ECM and defines the three-dimensional (3D) structure and biomechanical properties of tissues.

The challenges in the development of grafts lie not only in promoting cell proliferation and differentiation, but also in providing the grafts with compatible mechanical properties to support the new tissue and facilitate the desired physiological activity [5]. This represents a continuous challenge for researchers and much effort has been focused on developing new materials for tissue repair.

Scaffolds can be manufactured from synthetic polymers or natural products [6–8]. The advantage of synthetic materials derives from their reproducibility and ability to tailor their physical properties and degradation rate. The degradation rate and porosity allow the tailoring of their use to specific applications, while their limitation is that they cannot be absorbed or integrated with the host tissue, and frequently trigger an immune response [9]. Natural materials that possess a 3D structure are bio-compatible with cells, resulting in better cell attachment and proliferation and consequently facilitating the formation of new tissue and improving regeneration and healing processes [9–16]. The main drawbacks of such materials may relate to their poor mechanical properties or their potential to elicit an immuno-pathological reaction when derived from other mammalian sources (such as bovine, pig, or rat). The use of natural biodegradable bio-polymers for scaffold engineering has several advantages over synthetic materials, such as their similarity to biological macro-molecules, which minimizes immunological reactions and chronic inflammation [5,17,18]. In addition, the natural structures of biomaterials (biopolymers) are compatible and possess properties that support cell attachment. Marine source for bio-polymers such as chitin [14,15], silicified collagen [19–21], collagen [12,16,22], or mineralized skeletons [11,13] have been purified from marine organisms such as sponges, jellyfish, and corals and investigated for tissue engineering applications [10]. Corals have also been studied previously as biomaterials for tissue engineering, especially bone tissue engineering, but those studies were mainly focused on mineralized biomaterials [10,11,19–21,23]. Here we present a study on unique collagen fibers isolated from the soft coral *Sarcophyton ehrenbergi*, which reveal a 3D structure and superior mechanical properties. The collagen fibers reported were used to create bio-composite materials that demonstrated a unique hyper-elastic behavior similar to that of natural human tissues [24–26]. The prominent mechanical properties make these collagen fibers a highly suitable material for biomedical applications. In addition the collagen fibers are endowed by sequence motifs that facilitate and encourage cell adhesion. The current study elaborates upon these collagen fibers embedded with a hydrogel matrix, and these composites were analyzed for their mechanical behavior and biocompatibility in an in vitro model.

2. Results and Discussion

A recent study on the soft coral *Sarcophyton ehrenbergi* (Figure 1A) revealed ultra-long collagen fibers. These fibers when mechanically pulled from the soft-coral colony (torn parts, Figure 1B) retain their natural physical properties and structure, as opposed to collagen harvested by other methods that destroy its natural structure [6]. These are cord-like fibers that when detached from the mesenteries of the soft-coral tissue reveal a coiled and wavy feature (Figure 1C,D). In earlier studies we revealed the microanatomy of *Sarcophyton auritum* and *Sarcophyton ehrenbergi* colonies, in which the mesoglea between the polyps was shown to contain sparse and short collagen fibers following Masson Trichrome histological staining [27,28]. Additionally, the histological sections revealed eight mesenteries radiating from the inner polyp body-wall across the gastro-vascular cavity and connecting to the pharynx, which contain a mass of collagen fibers. Based on protein MS/MS analysis, these fibers were identified as collagen [27,28].

The fibers are produced and maintained as cord-like bundles in the soft-coral mesenteries. Under a light microscope they demonstrate an undulating appearance. Under fluorescence microscopy an intrinsic auto-fluorescence was detected in the range of 305–450 nm, which is typical for collagen (Figure 1C). Scanning electron microscopy images of these long fibers revealed a coiled spring-like organization (Figure 1D). The coiling presented a pitch-range of 6–40 μm, most probably related to the force applied during their extraction from the polyps. The diameter of the fibers averaged 8.70 ± 1.27 μm (n = 57). Such a natural arrangement of the fibers indicates their potential elasticity and suggests that they could serve as a biomaterial with potential for scaffolding applications. The use of high-resolution imaging of scanning and transmission electron microscopy analysis confirmed the striated arrangement and revealed the prominent fibrillary collagenous structure of these fibers [28]. The collagenous type was also confirmed by X-ray diffraction analysis [27].

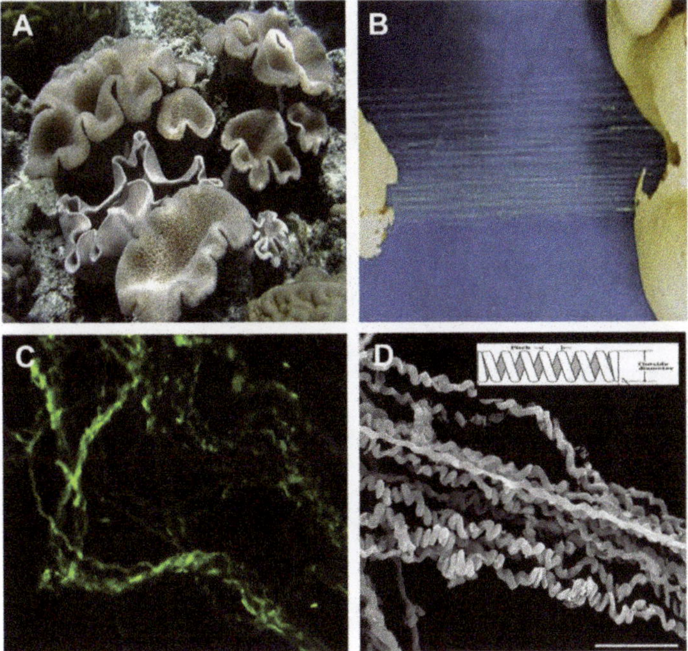

Figure 1. (**A**) Macro image of *Sarcophyton* and (**B**) its torn-apart polypary, revealing collagen fibers pulling from these sections; (**C**) Auto-fluorescence of collagen fibers observed by fluorescence microscopy; (**D**) E-SEM micrographs of collagen fibers that feature a coiled structure.

For potential use of the cord-like fibers for scaffold applications, their mechanical properties were measured (Figure 2) and they were analyzed for cell growth support (Figures 3 and 4). The collagen fibers were pulled out from the soft coral and aligned on a frame (as demonstrated in Figure 2A). In order to create a composite material, the aligned fibers were embedded in an alginate solution (a biocompatible polysaccharide) (Figure 2B) [24–26], then immersed in a calcium solution that created ionic bridges and cross-link the alginate hydrogel to create a collagen-alginate bio-composite (Figure 2C). The alginate hydrogel binds the collagen fibers together, creating both an aqueous surrounding and mechanical integrity [26], similar to the mechanical function of proteoglycans in the extracellular matrix (ECM) [29]. The high content of alginic acid with the addition of calcium ions become hydrogel that mechanically resembles the ECM structure of the tissue. The uniqueness of the collagen fibers is manifested in their macro-fibrous structure, ultra-long length, simple isolation, and superior mechanical properties which are rare for a non-mineralized bio-polymer.

Biomechanics of the Collagen Fibers and Collagen-Alginate Bio-Composites

The collagen fibers were analyzed by differential scanning calorimetry (DSC) to determine the melting point of 68 °C [30], suggesting that these fibers are naturally cross-linked and providing a possible explanation for their exceptional mechanical properties [14,16]. A detailed mechanical analysis of the collagen fibers revealed a non-linear and hyper-elastic mechanical behavior with large deformations similar to that of collagenous soft tissues [31]. The stress-strain curve shown in Figure 2E can be divided into three main regions as reported in the literature for soft tissues [32]: (1) toe region, (2) heel region, and (3) linear region. These regions are also presented for the collagen fiber stress-strain curve in Figure 2E. In the toe region only, a low stress is necessary to achieve a large deformation.

This region is a result of the 3D coiling of the fibers (Figure 1D) where the collagen fiber coiling becomes straightened [32]. In the heel region there is non-linear stiffening in the stress-strain curve (Figure 2E), in which the load increases and the collagen fibers straighten with the load direction [32]. The synchrotron by X-ray revealed that a lateral arrangement of the collagen fibrils increased linearly with the strain [33]. The linear region follows the heel region, in which the stress-strain relation is linear, the molecular crimps disappear, and the collagen fibers become straighter at high tensile stresses. The collagen fibers are aligned together in the load direction, becoming straight and strongly resistant to loading, which makes the tissue stiffer and the stress-strain relation returns to linear. Beyond the third phase the ultimate tensile strength is reached and fibers begin to break until tissue failure [32]. For the collagen fibers, the ultimate tensile strength (UTS) was 39–59 MPa (for ~15% strain) and the Young's modulus in the linear region was 0.34–0.54 GPa [30]. The 3% alginate hydrogel also demonstrated non-linear behavior (n = 12), with high deformations, a tensile modulus of 0.91 ± 0.26 MPa and an ultimate tensile strength of 0.19 ± 0.05 MPa, and a failure strain of 0.29 ± 0.08, three orders of magnitude less than that of coral collagen fibers (Figure 2B,E). The collagen fibers retained their coiled structure through post-processing and alignment inside the alginate matrix. The collagen/alginate bio-composite (n = 8) with a fiber fraction of 0.33 ± 0.09 demonstrated a similar hyper-elastic behavior to that of the collagen fibers with a UTS of 0.77 ± 0.21 MPa, tensile modulus of 5.54 ± 1.89 MPa, and failure strain of 0.181 ± 0.024 (Figure 2C–E). The alginate hydrogel was found to influence the toe region mechanical behavior of the bio-composite stress-strain curve, while the collagen fibers created the stiffening effect at the heel and linear regions [24,25].

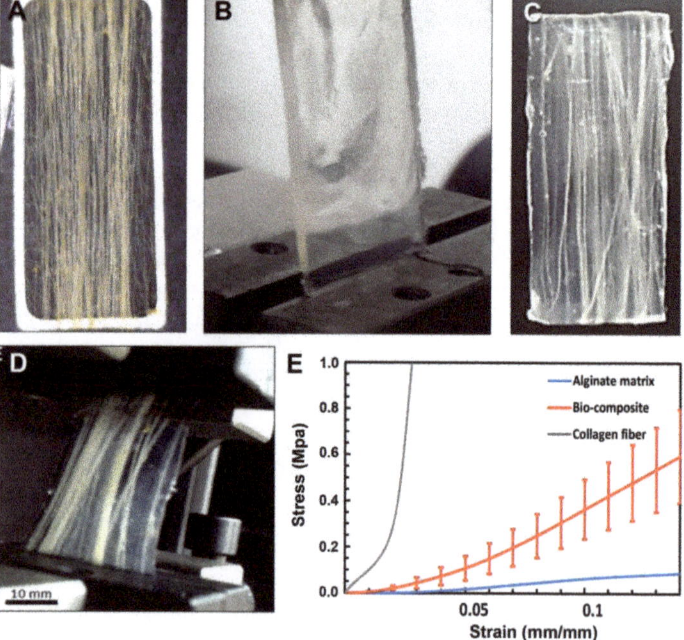

Figure 2. Bio-composite fabrication and mechanical behavior. (**A**) *Sarcophyton* collagen fibers aligned on a metal frame; (**B**) 3% Alginate hydrogel; (**C**) Fabricated uniaxial bio-composite; (**D**) Uniaxial bio-composite under tensile test; (**E**) Mechanical behavior of collagen fibers, alginate matrix, and uniaxial bio-composite. The toe, heel, and linear regions are demonstrated on the *Sarcophyton* collagen fiber stress-strain curve.

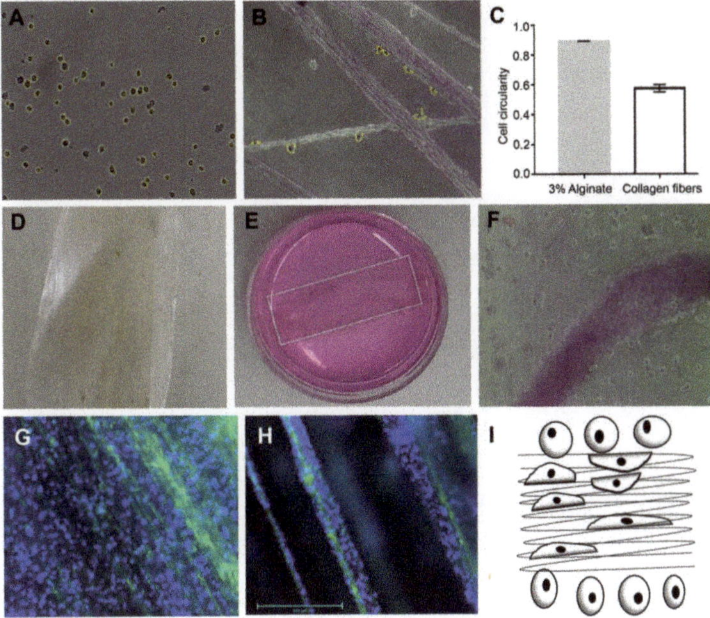

Figure 3. Live imaging and quantification for cells seeded on 3% hydrogel alginate (**A**) on *Sarcophyton* collagen fibers; (**B**) 24 h after seeding cells. Images are at ×100 magnification. (**C**) Cell circularity was analyzed by ImageJ software. Bio-composite was made of collagen fibers embedded in 3% hydrogel alginate. (**D,E**) Phase microscopic images of cells seeded on the bio-composite in a tissue culture dish (**F**), were observed to have a mixed morphology of cells: elongated cells are visualized on the collagen fibers along their orientation. On alginate the cells maintained a rounded shape. (**G,H**) Cells grown on collagen fibers up to nine weeks formed a tissue-like structure (collagen fibers are shown in green and cell nuclei are DAPI-stained blue). (**I**) Model of substrate stiffness and cells' morphology. Cells on the fibers are attached and become elongated with the substrate orientation, while cells on the alginate are rounded as a result of no adhesion and the pressure applied by the alginate matrix.

The bio-composite material enables the production of 3D structures that allow cell growth while retaining the natural mechanical features of the fibers. In the literature, some composite materials have been presented based on synthetic electro-spun fibers such as poly lactic-co-glycolic acid (PLGA) or on natural materials such as jellyfish collagen [12,18]. The hybrid scaffold demonstrates an effective mechanical stability and biocompatibility. Considering its marine origin and the evolutionary distance from vertebrates, this minimizes the risk normally associated with a mammalian collagen [12]. The hybrid combination affects the biocompatibility, proliferation, degradation rate, and mechanical strength and its duration [12]. The challenge is to design a 3D scaffold constructed of natural materials that is also biocompatible with human cells. The biocompatibility of the fibers allows cells to better attach and proliferate, subsequently leading to new tissue formation and better healing and regeneration rates. To date, the main drawback related to natural scaffolding concerns its poor mechanical properties. Therefore, collagen is considered a good source of biomaterials as it allows the preservation of the original tissue shape and the ECM structure of the matrix. When collagen is subjected to the chemical processes of extraction, isolation, purification, and polymerization, its natural properties are reduced (such as the cross-linking that gives its tensile strength and proteolytic resistance to collagenase) [6]. The particular collagen source is an important parameter for consideration since at present it is derived from bovine or porcine source and thus constitutes a risk of being a vector of pathogens. Recently, it has been reported that collagen extracted from jellyfish is biocompatible and supports cell viability [16,22]. The jellyfish collagen used as a scaffold is, however, known to possess poor mechanical properties, and is thus of limited applicative use [34].

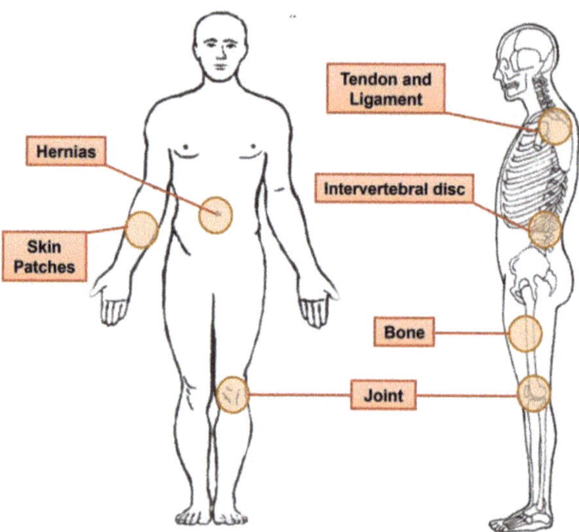

Figure 4. Illustration of potential use of collagen fibers embedded in a hydrogel bio-composite for medical devices with adjusted mechanical properties that provide support and allow motion and flexibility of the tissue under repair.

The soft-coral collagen fibers presented here have previously been studied by us at the molecular level using a proteomic analysis that enabled us to characterize their organic matrices. A molecular amino-acid sequence based on mass spectrometry (MS/MS) revealed a series of peptides [15,16] that correspond to fibrillary collagen. The sequences were analyzed, revealing that the collagen family of proteins has been well conserved through evolution, with their molecular sequence already documented in some metazoans [35]. Regarding vertebrates, 29 collagens have been characterized thus far, while for cnidarians the data available on collagens are limited (taxonomy browser NCBI—https://www.ncbi.nlm.nih.gov/protein/). Among this latter group, most of the recognized references are from sea anemones and jellyfish, and most of the reports provide only partial sequences (RefSeq Protein Database, NCBI). Moreover, the majority of collagen and collagen-like proteins described in invertebrates belong to the non-fibrillary and soluble structure subtype IV. Only a few are associated with collagen Type I, the constituent with the taxonomically closest source. A comparison of the peptides sequences obtained from the MS/MS assay on *Sarcophyton ehrenbergi* and *Sarcophyton auritum* collagen fibers to the UniProt database identify a remarkable homology to fibrillary mammalian collagen [15,16].

It is well recognized that fibrillary collagen harbors well-conserved motifs such as integrin-binding, the von Willebrand factor type A (vWFA) domain, the vWFA subfamily, the metal ion-dependent adhesion site (MIDAS) and fibronectin type 3 domain [36], and the immunoglobulin-like and cytokine receptor motif. Integrins are the main family of cell adhesion molecules, functioning to transduce signaling in and out of the cell by linking the cytoskeleton to ECM proteins, and their essential role in cell adhesion is well characterized. In addition, a laminin G domain motif recognized in the collagen fiber sequence also has a role in cell adhesion, signaling, migration, and differentiation [37].

Thus, fibrillar collagen possesses motifs that support cell adhesion and make it an essential component for scaffold formation in providing the conditions for cell attachment and spread, thereby supporting cell growth. In addition to the numerous cell adhesion motifs, high molecular weight glutenin subunits are also present in collagen fibers. This is attributed to the elastomeric proteins, which are characterized by their ability to withstand significant physical deformations without breaking, and to return to their original conformation when the stress is removed [38]. The outstanding mechanical properties of the cord-like long collagen fibers revealed in our studies [24–26] may thus be in part attributed to their elastomeric nature. We also identified a proteoglycan-rich organization that might underlay their structure and mechanical strength [27]. The biomolecular data on these collagen fibers isolated from *Sarcophyton ehrenbergi*, compiled from their molecular and protein features along with their mechanical properties, make these fibers a unique candidate for use as a scaffold material.

Our experiments were based on (i) cells seeded on 3% alginate (Figure 3A,C), (ii) cells seeded on collagen fibers (Figure 3B,C), and (iii) cells seeded on collagen-alginate composite (Figure 3D–F). In these cultures, we observed different cell morphologies (Figure 3) according to their substrate stiffness (Figure 2). The cells' different morphologies were examined using live imaging, which allowed us to quantify cell shape parameters as circularity [defined as (4πArea/[Perimeter]^2)] by ImageJ software). Cells seeded on 3% alginate hydrogel displayed a circularity of 0.89 ± 0.0014 (n = 1222) (Figure 3A,C), while for cells seeded on collagen fibers the measured cell circularity was 0.575 ± 0.025 (n = 79) (Figure 3B,C), exhibiting a significant difference of $p < 0.0001$. This is explained by the stiffer properties of the collagen fibers (Figure 2), which bring to transduction of forces that lead cells to spread on the collagen fibers via adhesion molecules that bound the domains on the collagen (as described above). The transduced forces activate the cytoskeleton response to the cells' tensional forces, demonstrating a more elongated shape. The activated signaling pathways affect cell proliferation and growth, leading to the tissue-like formation observed in Figure 3F–H.

The bio-composite was tested as a potential support for cell growth support (Figure 3D–F), by seeding cells on a collagen-alginate composite, in which the alginate hydrogel surrounded the stiffer collagen fibers. We noticed that cells sensed the different rigidities of the components, which affected their morphology (Figure 3A–C). When cells were plated on collagen fibers only they proliferated and grew for several weeks into a tissue-like structure as seen following nine weeks in culture (Figure 3G,H). This phenomenon is explained at the molecular level as tethering that underlies the cellular contractile forces, creating a connection with molecules that sense the substrate stiffness. ECM proteins anchorage and tethering play a role in regulating mechano-responsive cellular behaviors. The cellular sensing of an effective environmental rigidity combined with molecular tethering transduce through cytoskeleton backbone deformation [39]. By applying various densities of the collagen fibers in a bio-composite material, we produced a range of stiffness levels for different purposes of scaffolding. Collagen fibers able to provide different degrees of substrate stiffness, enabling different cues for cell activation, growth, and fate, can be adapted for specific applications as suggested in our previous study [7] and as related to the various applications suggested in Figure 4. The presented collagen-alginate composite, which demonstrated superior mechanical compatibility (e.g., strength and elasticity) alongside the biocompatibility properties as presented in Figure 3, therefore holds great potential for use in tissue grafts in different tissues e.g., Figure 4.

Concerning future biomedical devices, the production of soft tissue grafts that support biocompatibility and cell function together with an appropriate mechanical compatibility is still a challenge [5]. Although many efforts have been focused on the aspects of biological compatibility, the mechanical behavior of the graft is no less important. A functioning graft will require compatible mechanical properties in order to support the new tissue formation and produce the appropriate physiological activity. The challenge for researchers is thus to develop new materials for tissue repair that demonstrate biocompatibility and integration with the native tissue, while also maintaining the mechanical properties of the new tissue formed. The risk in using a mechanically unsuitable implant lies in the formation of stress concentration occurring at the interphase with the native tissue, which can lead to hyperplasia or even graft failure [5,17].

Here we used a bio-composite material that mimics the components of the natural tissue: the collagen fibers provide the load-bearing ability while the hydrogel provides the ECM-like properties for the cells' surroundings. The effective mechanical behavior of the bio-composite lies between that of the fibers and the matrix, where the local mechanical behavior is soft enough for the cells and stiff enough to support new tissue formation, as demonstrated in the studied macroscopic mixture combined from the two constituent materials. The current study on composite materials with the potential to create a hybrid material was inspired by both industrial composites and the structure of natural tissues. Tailoring the mechanical behavior of the bio-composite to the targeted tissue can be achieved by changing the collagen fiber fraction and orientation [6–8].

3. Materials and Methods

3.1. Coral Used for Isolation of Collagen Fibers

Sarcophyton soft corals (Figure 1A) were frozen following harvesting and defrosted prior to fiber extraction. A piece of the colony was torn to expose the fibers, which were then physically pulled out from the soft coral (Figure 1B). The fibers were manually spun around a thin rectangular-shaped metal frame to create unidirectional, straight, and organized array of fiber bundles (Figure 2A). The aligned fibers were carefully washed several times in water, PBS, and then with 70% ethanol.

3.2. Microscopy Fiber Analysis

Fluorescence microscopy—Collagen fibers isolated from the coral were visualized under a fluorescence microscope at 305–450 nm (Optiphot, Nikon, Tokyo, Japan).

Phase contrast microscopy—Live cell cultures were observed under phase contrast microscopy and digital photography (Optiphot, Nikon, Tokyo, Japan).

Scanning electron microscopy (SEM)—For scanning electron microscopy, subsamples were fixed in 4% glutaraldehyde in filtered seawater (0.22 μm FSW), decalcified as described above, dehydrated through a graded series of ethanol up to 100%, and critical point dried with liquid CO_2. The preparations were fractured under a compound microscope, using the tips of fine forceps, and the gastrovascular cavities were then carefully exposed. Next, they were gold-coated and examined under a scanning electron microscope (SEM Jeol-840a, Jeol LTD., Tokyo, Japan). In order to obtain fiber preparations, fibers were isolated as explained above and stored in 70% ethanol. They were then processed, coated with gold-palladium alloy, and examined at high vacuum under an environmental scanning electron microscope (SEM and ESEM, JSM-6700 Field Emission Scanning Electron Microscope, Jeol LTD., Tokyo, Japan). The diameters of collagen fibers and fibrils were obtained from the images, using the ImageJ software.

3.3. Bio-Composite Fabrication

The extracted fiber bundles were sterilized (Figure 2A), arranged on a metal frame, and then inserted into a dialysis membrane (6000–8000 MWCO, Spectra Por, Spectrum Labs Inc., Rancho Dominguez, CA, USA) together with 3 mL sodium alginate solution (3% *w/v* in DDW (Protanal LF 10/60, FMC BioPolymer, Philadelphia, PA, USA)). The membrane was sealed, flattened, and soaked in $CaCl_2$ (0.02 M Ca for Tissue culture (TC) experiments or 0.1 M for mechanical testing) solution used as a cross-linker through diffusion for 48 h at room temperature to allow the gelation of the alginate to hydrogel. The bio-composite was then removed from the membrane and the frame was sterilized by immersion in 70% ethanol and further used for cell seeding. Alginate hydrogel samples were fabricated by the same protocol, excluding the embedding of fibers.

3.4. Fiber Volume Fraction in the Bio-Composite

The aligned fiber images were placed against a dark background and their images were acquired using a digital microscope (AM311S, BigCatch, Torrance, CA, USA). The images were processed into binary numerical arrays, and the percentage of white pixels (representing the fibers) was calculated in order to determine the fiber fraction. The fraction was normalized to the final bio-composite thickness. The calculations were done using Matlab code.

3.5. Mechanical Testing of the Bio-Composites and Alginate Hydrogels

Tensile testing was performed on an Instron machine model 5582 with Blue hill 2 operating software and a 100-N load cell at a rate of 0.05 mm s^{-1}. The tensile measurements included three cycles of preconditioning up to 10%, followed by stretching to failure parallel to the fibers' direction.

3.6. Tissue Culture

Mouse mesenchymal 3T3-L1 cells were seeded in a growth medium consisting of Dulbecco's modified Eagle's medium (450 mg/dL; Biological Industries, Beit Haemek Ltd., Beit Haemek, Israel), 10% fetal bovine serum (Biological Industries), 1% L-glutamine (Biological Industries), and 0.1% penicillin-streptomycin (Sigma, St. Loius, MO, USA). Cell morphology was analyzed using ImageJ software (National Institute of Health, Bethesda, MD, USA).

Acknowledgments: The Israeli Ministry of Science (53441 grant to DB and RHA) supported this study. We thank The Interuniversity Institute for Marine Sciences in Eilat (IUI) for assistance and use of facilities. We acknowledge M. Weis for assistance, Z. Barkay for ESEM work, V. Wexler for digital editing, and N. Paz for editorial assistance. We are grateful to the anonymous reviewers for their constructive comments that greatly contributed to the manuscript.

Author Contributions: M.S., D.B., and R.H.-A. conceived and designed the experiments; M.S., L.P., and L.A. performed the experiments; M.S. and L.A. analyzed the data; D.B., Y.B., and R.H.-A. contributed reagents/materials/analysis tools; D.B., M.S., L.P., and Y.B. wrote the paper. Authorship is limited to those who contributed substantially to the work reported.

References

1. Langer, R.; Vacanti, J.P. Tissue Engineering. *Science* **1993**, *260*, 920–926. [CrossRef] [PubMed]
2. Vangsness, C.T.; Wagner, P.P.; Moore, T.M.; Roberts, M.R. Overview of safety issues concerning the preparation and processing of soft-tissue allografts. *Arthrosc. J. Arthrosc. Relat. Surg.* **2006**, *22*, 1351–1358. [CrossRef] [PubMed]
3. Chen, J.; Xu, J.; Wang, A.; Zheng, M. Scaffolds for tendon and ligament repair: Review of the efficacy of commercial products. *Expert Rev. Med. Devices* **2009**, *6*, 61–73. [CrossRef] [PubMed]
4. Hodde, J.; Hiles, M. Constructive soft tissue remodelling with a biologic extracellular matrix graft: Overview and review of the clinical literature. *Acta Chir. Belg.* **2007**, *107*, 641–647. [CrossRef] [PubMed]
5. Mazza, E.; Ehret, A.E. Mechanical biocompatibility of highly deformable biomedical materials. *J. Mech. Behav. Biomed. Mater.* **2015**, *48*, 100–124. [CrossRef] [PubMed]
6. Parenteau-Bareil, R.; Gauvin, R.; Berthod, F. Collagen-Based Biomaterials for Tissue Engineering Applications. *Materials* **2010**, *3*, 1863–1887. [CrossRef]
7. Vats, A.; Tolley, N.; Polak, J.; Gough, J. Scaffolds and biomaterials for tissue engineering: A review of clinical applications. *Clin. Otolaryngol.* **2003**, *28*, 165–172. [CrossRef] [PubMed]
8. Drury, J.L.; Mooney, D.J. Hydrogels for tissue engineering: Scaffold design variables and applications. *Biomaterials* **2003**, *24*, 4337–4351. [CrossRef]
9. Dhandayuthapani, B.; Yoshida, Y.; Maekawa, T.; Kumar, D.S. Polymeric scaffolds in tissue engineering application: A review. *Int. J. Polym. Sci.* **2011**, *2011*, 290602. [CrossRef]
10. Clarke, S.; Walsh, P.; Maggs, C.; Buchanan, F. Designs from the deep: Marine organisms for bone tissue engineering. *Biotechnol. Adv.* **2011**, *29*, 610–617. [CrossRef] [PubMed]

11. Ehrlich, H.; Etnoyer, P.; Litvinov, S.; Olennikova, M.; Domaschke, H.; Hanke, T.; Born, R.; Meissner, H.; Worch, H. Biomaterial structure in deep-sea bamboo coral (Anthozoa: Gorgonacea: Isididae): Perspectives for the development of bone implants and templates for tissue engineering. *Mater. Werkst.* **2006**, *37*, 552–557. [CrossRef]

12. Jeong, S.I.; Kim, S.Y.; Cho, S.K.; Chong, M.S.; Kim, K.S.; Kim, H.; Lee, S.B.; Lee, Y.M. Tissue-engineered vascular grafts composed of marine collagen and PLGA fibers using pulsatile perfusion bioreactors. *Biomaterials* **2007**, *28*, 1115–1122. [CrossRef] [PubMed]

13. Martina, M.; Subramanyam, G.; Weaver, J.C.; Hutmacher, D.W.; Morse, D.E.; Valiyaveettil, S. Developing macroporous bicontinuous materials as scaffolds for tissue engineering. *Biomaterials* **2005**, *26*, 5609–5616. [CrossRef] [PubMed]

14. Mutsenko, V.V.; Bazhenov, V.V.; Rogulska, O.; Tarusin, D.N.; Schütz, K.; Brüggemeier, S.; Gossla, E.; Akkineni, A.R.; Meißner, H.; Lode, A. 3D chitinous scaffolds derived from cultivated marine demosponge Aplysina aerophoba for tissue engineering approaches based on human mesenchymal stromal cells. *Int. J. Biol. Macromol.* **2017**, *104*, 1966–1974. [CrossRef] [PubMed]

15. Mutsenko, V.V.; Gryshkov, O.; Lauterboeck, L.; Rogulska, O.; Tarusin, D.N.; Bazhenov, V.V.; Schütz, K.; Brüggemeier, S.; Gossla, E.; Akkineni, A.R. Novel chitin scaffolds derived from marine sponge Ianthella basta for tissue engineering approaches based on human mesenchymal stromal cells: Biocompatibility and cryopreservation. *Int. J. Biol. Macromol.* **2017**, *104*, 1955–1965. [CrossRef] [PubMed]

16. Song, E.; Yeon Kim, S.; Chun, T.; Byun, H.-J.; Lee, Y.M. Collagen scaffolds derived from a marine source and their biocompatibility. *Biomaterials* **2006**, *27*, 2951–2961. [CrossRef] [PubMed]

17. Tremblay, D.; Zigras, T.; Cartier, R.; Leduc, L.; Butany, J.; Mongrain, R.; Leask, R.L. A Comparison of Mechanical Properties of Materials Used in Aortic Arch Reconstruction. *Ann. Thorac. Surg.* **2009**, *88*, 1484–1491. [CrossRef] [PubMed]

18. Mauck, R.L.; Baker, B.M.; Nerurkar, N.L.; Burdick, J.A.; Li, W.-J.; Tuan, R.S.; Elliott, D.M. Engineering on the straight and narrow: The mechanics of nanofibrous assemblies for fiber-reinforced tissue regeneration. *Tissue Eng. Part B Rev.* **2009**, *15*, 171–193. [CrossRef] [PubMed]

19. Heinemann, S.; Ehrlich, H.; Knieb, C.; Hanke, T. Biomimetically inspired hybrid materials based on silicified collagen. *Int. J. Mater. Res.* **2007**, *98*, 603–608. [CrossRef]

20. Heinemann, S.; Heinemann, C.; Ehrlich, H.; Meyer, M.; Baltzer, H.; Worch, H.; Hanke, T. A novel biomimetic hybrid material made of silicified collagen: Perspectives for bone replacement. *Adv. Eng. Mater.* **2007**, *9*, 1061–1068. [CrossRef]

21. Heinemann, S.; Ehrlich, H.; Douglas, T.; Heinemann, C.; Worch, H.; Schatton, W.; Hanke, T. Ultrastructural studies on the collagen of the marine sponge Chondrosia reniformis Nardo. *Biomacromolecules* **2007**, *8*, 3452–3457. [CrossRef] [PubMed]

22. Widdowson, J.P.; Picton, A.J.; Vince, V.; Wright, C.J.; Mearns-Spragg, A. In vivo comparison of jellyfish and bovine collagen sponges as prototype medical devices. *J. Biomed. Mater. Res. Part B Appl. Biomater.* **2017**. [CrossRef] [PubMed]

23. Ehrlich, H.; Deutzmann, R.; Brunner, E.; Cappellini, E.; Koon, H.; Solazzo, C.; Yang, Y.; Ashford, D.; Thomas-Oates, J.; Lubeck, M. Mineralization of the metre-long biosilica structures of glass sponges is templated on hydroxylated collagen. *Nat. Chem.* **2010**, *2*, 1084–1088. [CrossRef] [PubMed]

24. Sharabi, M.; Varssano, D.; Eliasy, R.; Benayahu, Y.; Benayahu, D.; Haj-Ali, R. Mechanical flexure behavior of bio-inspired collagen-reinforced thin composites. *Compos. Struct.* **2016**, *153*, 392–400. [CrossRef]

25. Sharabi, M.; Benayahu, D.; Benayahu, Y.; Isaacs, J.; Haj-Ali, R. Laminated collagen-fiber bio-composites for soft-tissue bio-mimetics. *Compos. Sci. Technol.* **2015**, *117*, 268–276. [CrossRef]

26. Sharabi, M.; Mandelberg, Y.; Benayahu, D.; Benayahu, Y.; Azem, A.; Haj-Ali, R. A new class of bio-composite materials of unique collagen fibers. *J. Mech. Behav. Biomed. Mater.* **2014**, *36*, 71–81. [CrossRef] [PubMed]

27. Orgel, J.P.; Sella, I.; Madhurapantula, R.S.; Antipova, O.; Mandelberg, Y.; Kashman, Y.; Benayahu, D.; Benayahu, Y. Molecular and ultrastructural studies of a fibrillar collagen from octocoral (Cnidaria). *J. Exp. Biol.* **2017**, *220*, 3327–3335. [CrossRef] [PubMed]

28. Mandelberg, Y.; Benayahu, D.; Benayahu, Y. Octocoral Sarcophyton auritum Verseveldt & Benayahu, 1978: Microanatomy and Presence of Collagen Fibers. *Biol. Bull.* **2016**, *230*, 68–77. [PubMed]

29. Yanagishita, M. Function of proteoglycans in the extracellular matrix. *Pathol. Int.* **1993**, *43*, 283–293. [CrossRef]

30. Benayahu, Y.; Benayahu, D.; Kashman, Y.; Rudi, A.; Lanir, Y.; Sela, I.; Raz, E. Coral Derived Collagen and Methods of Farming Same. U.S. Patent US20110038914A1, 17 February 2011.

31. Holzapfel, G.A. Biomechanics of soft tissue. *Handb. Mater. Behav. Models* **2001**, *3*, 1049–1063.

32. Fratzl, P.; Weinkamer, R. Nature's hierarchical materials. *Prog. Mater.Sci.* **2007**, *52*, 1263–1334. [CrossRef]

33. Fratzl, P.; Misof, K.; Zizak, I.; Rapp, G.; Amenitsch, H.; Bernstorff, S. Fibrillar structure and mechanical properties of collagen. *J. Struct. Biol.* **1998**, *122*, 119–122. [CrossRef] [PubMed]

34. Addad, S.; Exposito, J.-Y.; Faye, C.; Ricard-Blum, S.; Lethias, C. Isolation, characterization and biological evaluation of jellyfish collagen for use in biomedical applications. *Mar. Drugs* **2011**, *9*, 967–983. [CrossRef] [PubMed]

35. Exposito, J.Y.; Cluzel, C.; Garrone, R.; Lethias, C. Evolution of collagens. *Anat. Rec.* **2002**, *268*, 302–316. [CrossRef] [PubMed]

36. Chi-Rosso, G.; Gotwals, P.J.; Yang, J.; Ling, L.; Jiang, K.; Chao, B.; Baker, D.P.; Burkly, L.C.; Fawell, S.E.; Koteliansky, V.E. Fibronectin type III repeats mediate RGD-independent adhesion and signaling through activated β1 integrins. *J. Biol. Chem.* **1997**, *272*, 31447–31452. [CrossRef] [PubMed]

37. Guicheney, P.; Vignier, N.; Zhang, X.; He, Y.; Cruaud, C.; Frey, V.; Helbling-Leclerc, A.; Richard, P.; Estournet, B.; Merlini, L. PCR based mutation screening of the laminin alpha2 chain gene (LAMA2): Application to prenatal diagnosis and search for founder effects in congenital muscular dystrophy. *J. Med. Genet.* **1998**, *35*, 211–217. [CrossRef] [PubMed]

38. Tatham, A.S.; Shewry, P.R. Elastomeric proteins: Biological roles, structures and mechanisms. *Trends Biochem. Sci.* **2000**, *25*, 567–571. [CrossRef]

39. Shao, Y.; Fu, J. Integrated micro/nanoengineered functional biomaterials for cell mechanics and mechanobiology: A materials perspective. *Adv. Mater.* **2014**, *26*, 1494–1533. [CrossRef] [PubMed]

 *marine drugs*

Article

Optimization of Extraction Conditions and Characterization of Pepsin-Solubilised Collagen from Skin of Giant Croaker (*Nibea japonica*)

Fangmiao Yu, Chuhong Zong [†], Shujie Jin [†], Jiawen Zheng, Nan Chen, Ju Huang, Yan Chen, Fangfang Huang, Zuisu Yang, Yunping Tang * and Guofang Ding

Zhejiang Provincial Engineering Technology Research Center of Marine Biomedical Products, School of Food and Pharmacy, Zhejiang Ocean University, Zhoushan 316022, China; fmyu@zjou.edu.cn (F.Y.); zongchuhong1997@163.com (C.Z.); m18868006087@163.com (S.J.); jwzheng1996@163.com (J.Z.); chennan_0576@163.com (N.C.); qiuqiu20130621@163.com (J.H.); cyancy@zjou.edu.cn (Y.C.); gracegang@126.com (F.H.); abc1967@126.com (Z.Y.); dinggf2007@163.com (G.D.)
* Correspondence: tangyunping1985@163.com; Tel.: +86-580-229-9809
† The authors contributed equally to this study and share first authorship.

Received: 27 December 2017; Accepted: 10 January 2018; Published: 14 January 2018

Abstract: In the present study, response surface methodology was performed to investigate the effects of extraction parameters on pepsin-solubilised collagen (PSC) from the skin of the giant croaker *Nibea japonica*. The optimum extraction conditions of PSC were as follows: concentration of pepsin was 1389 U/g, solid-liquid ratio was 1:57 and hydrolysis time was 8.67 h. Under these conditions, the extraction yield of PSC was up to 84.85%, which is well agreement with the predict value of 85.03%. The PSC from *Nibea japonica* skin was then characterized as type I collagen by using sodium dodecyl sulfate polyacrylamide gel electrophoresis (SDS-PAGE). The fourier transforms infrared spetroscopy (FTIR) analysis revealed that PSC maintains its triple-helical structure by the hydrogen bond. All PSCs were soluble in the pH range of 1.0–4.0 and decreases in solubility were observed at neutral or alkaline conditions. All PSCs had a decrease in solubility in the presence of sodium chloride, especially with a concentration above 2%. So, the *Nibea japonica* skin could serve as another potential source of collagen.

Keywords: marine collagen; *Nibea japonica*; response surface methodology; optimization; characterization

---

1. Introduction

Collagen is the predominant structural protein in the extracellular matrix of animals, making up about 30% of the total protein content [1,2]. Nowadays, collagen has been widely used in biomedical fields, such as sponges for wound healing [3,4], cornea for ophthalmology [5], hydrogels for articular cartilage [6], scaffolds for bone regeneration [7], and so on. Collagen is also a very attractive ingredient in cosmetics [8]. Furthermore, its hydrolysate (collagen peptide) has also been widely used as function foods or cosmetic additive with its antioxidant activity [9,10]. Collagens used in these fields are commonly extracted from skins or bones from bovine and porcine, while porcine collagens are unacceptable for some religions and bovine collagens are at risk of contamination with prion diseases [11]. Taking into account these limitations, there is the need for preparing safe, high quality collagens from alternative resources.

Recently, recombinant technology has been used to produce human collagen or collagen-like protein, especially expression of hydroxylated collagen [12,13]. Due to its high cost of production and low yield (no more than 2.0 g/L of hydroxylated collagen), it seems not to be a suitable method for industrial production of collagen. Nowadays, collagen extracted from marine fish byproducts has gathered more attention due to non-religious restrictions and safety when compared to other animals [14]. Various marine fish by-products have been used for extracting collagen, such as the skin of *Aluterus monocerous* [2], scales of *Pseudosciaena crocea* [9], skin and bone of *Scomberomorous niphonius* [15], skin and swim bladder of *Lates calcarifer* [16], and so on. The biochemical and functional characteristics of the collagen from different by-products will be different. In addition, the extracted collagen is also used for enzymatic hydrolysis to obtain the bioactive collagen peptides [4,17,18].

Giant croaker, *Nibea japonica* is a carnivorous fish which is cultured and considered as a promising species for marine aquaculture in East Asia because of its high value, fast growth speed, easy receptivity to captivity and the availability of production technology [19,20]. However, collagen from *Nibea japonica* has not been reported and its characterization is also unknown. In this study, pepsin-solubilised collagen (PSC) from *Nibea japonica* skin was extracted and characterized for the first time. So far there is no published work on studies on various extraction conditions on the yield of PSC from *Nibea japonica* skin. As many factors may affect the extraction yield of collagen, response surface methodology (RSM) and Box-Behnken design (BBD) was performed in this study to optimize the extraction conditions for extracting higher yield of PSC. Furthermore, the properties of PSC were also characterized by determining its protein patterns, amino acid composition, fourier transforms infrared spetroscopy (FTIR) spectra and so on.

2. Results and Discussion

2.1. Single Factor Results

2.1.1. Effect of Enzyme Concentration on the Extraction Yield of PSC

In the previous studies, collagens were often extracted by using acid extraction and enzymatic extraction [15,16]. However, the acid solubilisation process gives a low yield of collagen. Since pepsin or papain is able to cleave peptides in the telopeptide region of collagen and the helical arrangement can exist in the PSC or papain digested collagen. However, there are some other proteins (above 97.4 kDa) in the papain digested collagen when compared to the PSC [21]. So, pepsin was chosen for extracting collagen in this study. Then, different pepsin concentrations (800, 1200, 1600, 2000 and 2400 U/g) were used to investigate the effect of pepsin concentration on the extracting yield of PSC. The other two extraction parameters were set as follows: solid-liquid ratio was 1:45 and hydrolysis time was 8 h in 0.5 M acetic acid buffer. As shown in Figure 1a, the PSC extraction yield significantly increased from 66.35% to 79.93% when pepsin concentration varied from 800 to 1200 U/g and then slightly increased when pepsin concentration exceeded 1200 U/g. Considering the higher-cost industrial extraction process, the amount of 1200 U/g pepsin was used for further optimization.

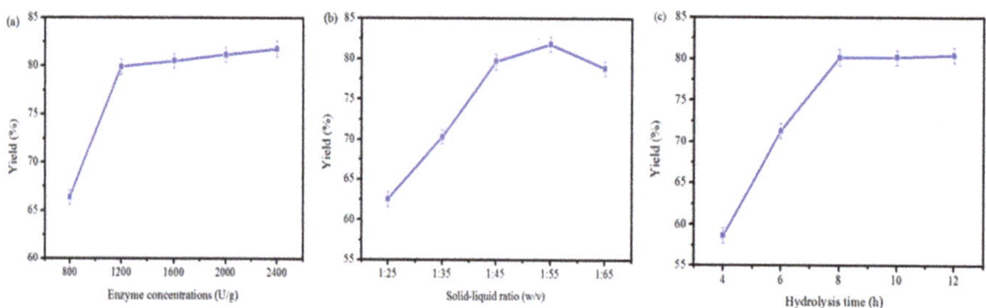

Figure 1. Effects of enzyme concentration (**a**), liquid-solid ratio (**b**) and hydrolysis time (**c**) on extraction yield of collagen from *Nibea japonica* skin.

2.1.2. Effect of Solid-Liquid Ratio on the Extraction Yield of PSC

Different solid-liquid ratio (1:25, 1:35, 1:45, 1:55 and 1:65) was used to study the effect of solid-liquid ratio on the extraction yield of PSC. The other two extraction parameters were set as follows: pepsin concentration was 1200 U/g and hydrolysis time was 8 h in 0.5 M acetic acid buffer. As shown in Figure 1b, the extraction yield of PSC was significantly increased with increasing liquid-solid ratio between 1:25 and 1:55. So, the solid-liquid ratio of 1:55 was selected for next optimization.

2.1.3. Effect of Hydrolysis Time on the Extraction Yield of PSC

To check the effect of hydrolysis time on the extraction yield of PSC, different hydrolysis time (4, 6, 8, 10 and 12 h) was carried out in this study. The other two extraction parameters were set as follows: pepsin concentration was 1200 U/g and solid-liquid ratio was 1:55 in 0.5 M acetic acid buffer. As shown in Figure 1c, the extraction yield of PSC was significantly increased with hydrolysis time between 4 and 8 h. Therefore, the hydrolysis time of 8 h was selected for the next optimization.

2.2. Optimization of Extraction Parameters of PSC Using RSM

2.2.1. Response Surface Analysis

After screening, RSM using BBD was used to obtain the optimal levels of the three above factors that significantly affected the yield of PSC. The experimental design and results are shown in Table 1 On the basis of the regression analysis of the data in Table 1, the effects of these three factors on the extraction yield of PSC were predicted by using a second-order polynomial function as follows:

$$Y = 83.73 + 3.53X_1 + 1.47X_2 + 1.78X_3 - 0.11X_1X_2 - 0.59X_1X_3 - 0.80X_2X_3 - 3.68X_1^2 - 2.03X_2^2 - 1.89X_3^2$$

(where Y was the extraction yield of PSC, and X_1, X_2, X_3 were the pepsin concentration, solid-liquid ratio and hydrolysis time, respectively).

Table 1. The Box-Behnken design and the response for the extraction yield of pepsin-solubilised collagen (PSC).

Runs	Enzyme Concentration (X_1)	Solid-Liquid Ratio (X_2)	Hydrolysis Time (X_3)	PSC Yield (%) (Y)
1	0	0	0	83.88
2	0	1	−1	81.38
3	−1	1	0	76.01
4	0	1	1	82.78
5	1	1	0	82.36
6	0	0	0	83.39
7	−1	0	−1	75.97
8	0	−1	−1	75.24
9	0	−1	1	80.85
10	1	0	−1	80.95
11	1	0	1	82.91
12	1	−1	0	80.25
13	0	0	0	83.91
14	−1	0	1	76.53
15	−1	−1	0	74.44

In order to determine the significance of the quadratic model, the analysis of variance (ANOVA) was performed and the results are shown in Table 2. As suggested by the model F value and a low probability value (p = 0.0001), it was obvious that the model was highly significant. The lack of fit F value in this model was about 6.13 and it suggested that the lack of fit was not significant relative to the pure error. The determination coefficient (R^2 = 0.9920) by ANOVA of this model and the adjusted determination coefficient (Adj R^2 = 0.9777) also indicated that the model was highly significant. So, this model was selected in this study for optimizing.

Table 2. Analysis of variance of regression model.

Source	Sum of Squares	df	Mean Square	F Value	p Value
Model	216.13	9	24.01	69.10	0.0001
X_1	99.83	1	99.83	287.24	<0.0001
X_2	17.26	1	17.26	49.66	0.0009
X_3	25.45	1	25.45	73.24	0.0004
X_1X_2	0.053	1	0.053	0.15	0.7125
X_1X_3	1.37	1	1.37	3.94	0.1040
X_2X_3	2.58	1	2.58	7.41	0.0417
$X_1{}^2$	50.13	1	50.13	144.23	<0.0001
$X_2{}^2$	15.17	1	15.17	43.65	0.0012
$X_3{}^2$	13.15	1	13.15	37.83	0.0017
Residual	1.74	5	0.35		
Lack of fit	1.57	3	0.52	6.13	0.1435
Pure Error	0.17	2	0.085		
Cor Total	217.86	14			
R^2					0.9526
Adj R^2					0.9777

Furthermore, three-dimensional response surfaces and contour plots were generated from the model equation to visualize the relationship between the extraction yield of PSC and extraction factors (Figure 2). It also show the optimal levels of each component required for the extraction of PSC (Figure 3). These three-dimensional response surfaces and contour plots provided a visual interpretation of the mutual interactions between two factors. The maximum predicted yield of PSC was 85.03% under the following conditions: concentration of enzyme was 1389 U/g, solid-liquid ratio was 1:57 and hydrolysis time was 8.67 h.

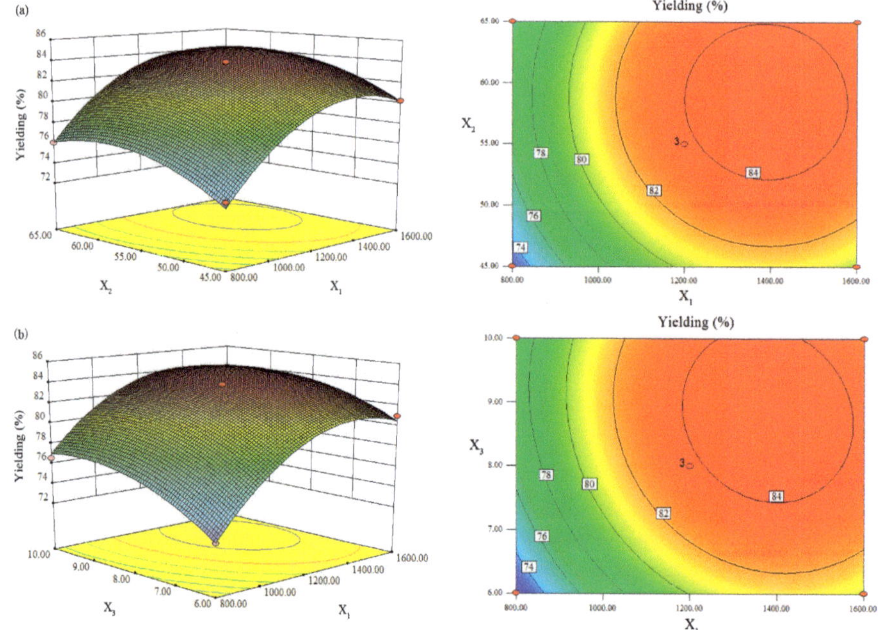

Figure 2. *Cont.*

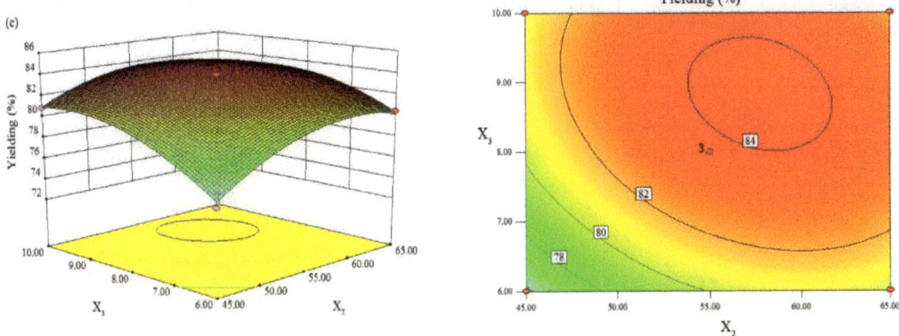

Figure 2. Three-dimensional response surface plots (**left**) and two-dimensional contour plots (**right**) showing the effects of (**a**) enzyme concentration (X_1) vs. liquid-solid ratio (X_2), (**b**) enzyme concentration (X_1) vs. hydrolysis time (X_3) and (**c**) liquid-solid ratio (X_2) vs. hydrolysis time (X_3) on extraction yield of collagen from *Nibea japonica* skin.

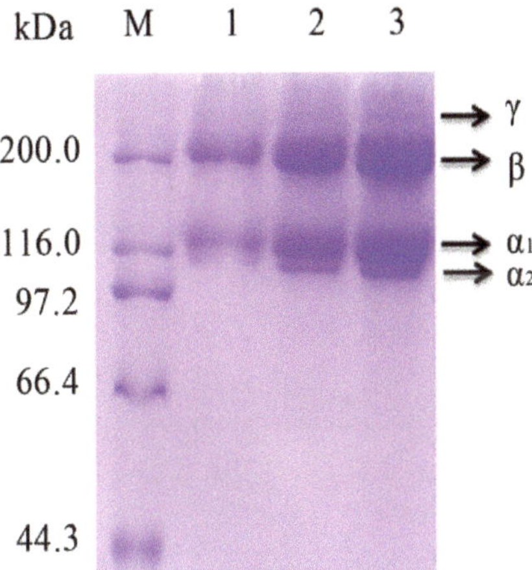

Figure 3. Sodium dodecyl sulfate polyacrylamide gel electrophoresis (SDS-PAGE) analysis of PSC from *Nibea japonica* skin. M: Protein molecular weight marker; Lane 1–3: Purified PSC from *Nibea japonica* skin.

2.2.2. Validation of the Models

Three additional experiments were performed in order to verify the predicted yield under the optimal extraction conditions. The mean value of PSC yield was 84.85%, which was in excellent agreement with the predicted value, under the similar conditions.

2.3. SDS-PAGE Analysis

The protein patterns of PSC from *Nibea japonica* skin were analyzed by SDS-PAGE (Figure 3). As shown in Figure 3, PSC from *Nibea japonica* skin consisted of two α_1-chains and one α_2-chain. The β and γ chains as well as the cross-linked constituents were also observed in this study (Figure 3). PSC extracted from *Nibea japonica* skin may have the structure of $(\alpha_1)_2\alpha_2$, which was classified as Type I collagen. Our results were consistent with the collagens from other marine fish skins, such as

PSC from *Aluterus monocerous* [2], PSC from *Scomberomorous niphonius* [15] and PSC from *Istiophorus platypterus* [22].

2.4. Amino Acid Composition of PSC

The amino acid composition of PSC from *Nibea japonica* skin was determined and the results are shown in Table 3 and compared with collagen from calf skin, type I collagen from porcine skin and human [23–25]. The most abundant amino acids found in PSC from *Nibea japonica* skin were glycine (Gly), alanine (Ala), proline (Pro) and hydroxyproline (Hyp). In this study, Gly was found to be the major amino acid in PSC (348 residues/1000 residues), the result is accordance with the (Gly-Xaa-Yaa) n repeat structure in all collagen molecules. It is known that the Xaa and Yaa positions can be occupied by any other amino acid, but the most common residue for Xaa is Pro and for Yaa is Hyp [26], forming the most common triplet repeats that found in most collagens (Gly-Pro-Hyp) n [22]. The Pro and Hyp contents of the PSC was 116 residues/1000 residues and 75 residues/1000 residues, respectively, which is similar to that of PSC from skin of *Aluterus monocerous* [2]. The rate of proline hydroxylation was about 39.3% for PSC from *Nibea japonica* skin. There were no tryptophan and cysteine residues in the PSC from *Nibea japonica* skin.

Table 3. Amino acid compositions of PSC from *Nibea japonica* skin (results are expressed as residues/1000 residues).

Amino Acid	*Nibea japonica* Skin PSC	Calf Skin Collagen [23]	Type I Collagen of Porcine Skin [24]	Type I Collagen of Human [25]
Aspartic acid	43	45	44	43
Threonine	20	18	16	17
Serine	29	33	33	33
Glutamic acid	73	75	72	71
Glycine	348	330	341	335
Alanine	128	119	115	111
Cysteine	0	0	0	0
Valine	19	21	22	26
Methionine	10	6	6	6
Isoleucine	9	11	10	9
Leucine	25	23	22	23
Tyrosine	3	3	1	2
Phenylalanine	6	3	12	12
Histidine	8	5	5	6
Lysine	30	26	27	23
Arginine	51	50	48	50
Proline	116	121	123	120
Hydroxyproline	75	94	97	103
Imino acid	191	215	220	223

2.5. UV-Visible Spectroscopy

It is known that collagen has a single absorption peak at 230 nm because of its triple helical structure, so UV-visible spectroscopy of collagen can be used to evaluate its purity [14,27]. As shown in Figure 4, PSC extracted from *Nibea japonica* skin showed a single absorption peak at 230 nm. Our result was similar to the collagen that has been isolated from other fish species. It is also necessary to point out that no any other obvious peaks were found at 280–300 nm while other proteins usually have absorption peaks at 280 nm. This is because the tyrosine content in collagen was very low. Finally, the UV-visible spectroscopy of PSC indicated that the extracted proteins using pepsin extraction was collagen and it also shown that pepsin extraction was the efficient methods to obtain purity collagens.

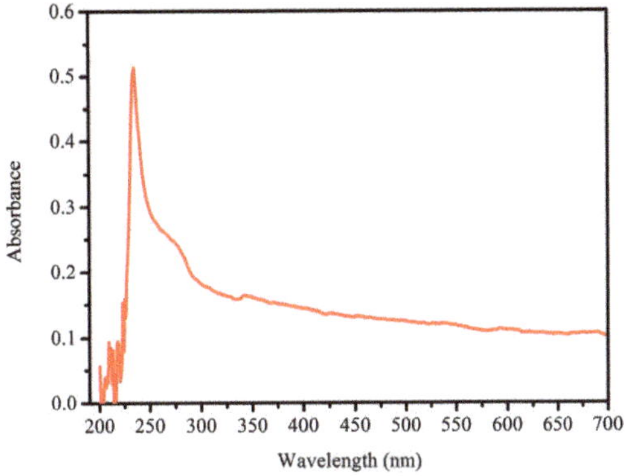

Figure 4. UV-visible spectroscopy of PSC from *Nibea japonica* skin.

2.6. Fourier Transforms Infrared Spetroscopy (FTIR) Analysis

The FTIR spectra of PSC from *Nibea japonica* skin is shown in Figure 5. These peaks correspond to five main amide bonds (amide A, B, I, II and III). The amide A bands of PSC was measured at 3305.90 cm^{-1}. The value is associated with N-H stretching frequency and indicate the presence of hydrogen bonds. The free N-H frequency vibration occurs at 3400–3440 cm^{-1} and shifts lower to 3300 cm^{-1} [28]. The amide B band of PSC was measured at 2928.38 cm^{-1}, which was consistent with asymmetrical stretch of CH$_2$ [29]. The amide I band of PSC was detected at 1641.35 cm^{-1}, fitting well with the range of 1600–1700 cm^{-1} for general amide I band position. The amide II band of PSC was measured at 1550.26 cm^{-1}, fitting well with the range of amide II band position (1550–1600 cm^{-1}). Finally, the amide III band of PSC was measured at 1240.47 cm^{-1}, which indicated the helical arrangement existed in the PSC from *Nibea japonica* skin [29,30].

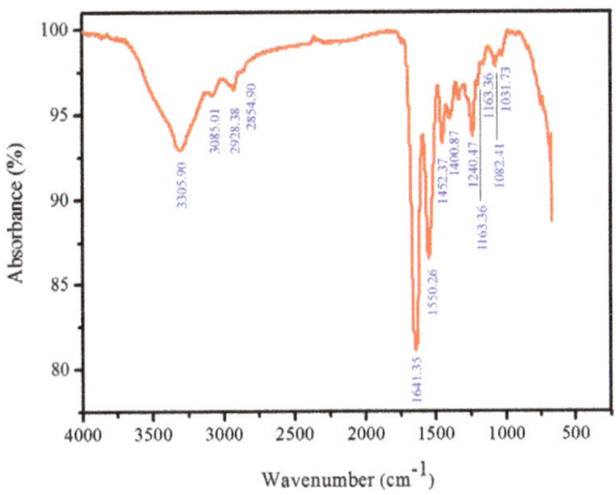

Figure 5. FTIR analysis of PSC from *Nibea japonica* skin.

2.7. Effects of pH and Sodium Chloride on PSC Solubility

The effects of pH and sodium chloride on the solubility of PSC from *Nibea japonica* skin were also investigated in the present study. As shown in Figure 6a, the PSC was dissolved in the acidic pH range of 1.0–4.0. The decrease in solubility was observed in the pH range of 5.0–7.0 and the dissolved protein was found to be deposited in this pH range. However, the slight increase in solubility was shown in the pH range of 8.0–10.0. Our results were consistent with the collagens from the skin of *Spanish mackerel* [15], bone and skin of *Hypophthalmichthys molitrix* [14] and skin of *Ictalurus punctatus* [31]. As shown in Figure 6b, all PSCs had a slight decrease in solubility with the concentrations of sodium chloride lower than 2% and drastic decrease was observed with the concentrations of sodium chloride higher than 2%. Similar reports were reported from the skin of *Ictalurus punctatus* [31] and skin of *Aluterus monocerous* [2].

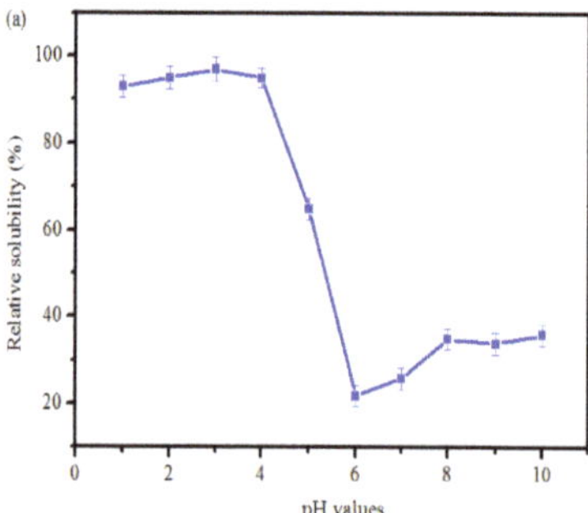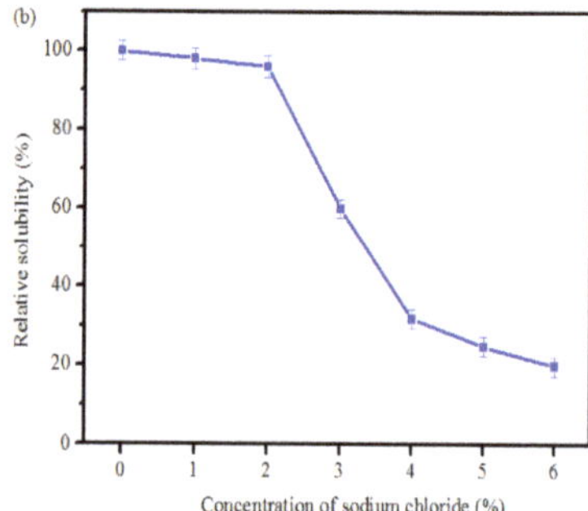

Figure 6. Effects of pH (**a**) and sodium chloride (**b**) on PSC solubility.

3. Materials and Methods

3.1. Materials and Chemical Reagent

Nibea japonica skins were provided by Zhejiang Marine Fisheries Research Institution (Zhoushan, China). The age of these fish was about one year and the average weight is about 0.4–0.5 kg. The fish skins were thawed at 4 °C and the residues under skins were removed. The cleaned fish skins were cut into small pieces and then stored at −20 °C until used for extracting of collagen. Pepsin was purchased from YTHX Biotechnology Co., Ltd. (Beijing, China). L-hydroxyproline and chloramine T trihydrate were purchased from Aladdin (Shanghai, China). All other reagents used were analytical grade.

3.2. Extraction of PSC from Nibea japonica Skin

The PSC from *Nibea japonica* skin was extracted according to the previous methods with slight modification [11,32]. Ten grams of fish skins were weighed precisely and in order to remove other proteins, the fish skins were then treated with 10 volumes (*v*/*w*) of 0.1 M NaOH for 24 h at 4 °C with continuous stirring. The alkali-treated fish skins were neutralized by washing repeatedly with pre-cooling distilled water and extracted by 1200 U/g pepsin in a ratio of 1:55 (*w*/*v*) for 10 h at 4 °C in 0.5 M acetic acid. The extractions were then centrifuged at 12,000 rpm for 10 min at 4 °C and the supernatants were collected. The supernatants were dialyzed against cold distilled water until the neutral pH was reached by using a dialysis bag with molecular weight cut-off of 25 kDa at 4 °C with a gentle stirring. The final solution was then lyophilized using d freeze dryer (ALPHA 1-2 LD plus, Christ, Germany). The Hyp content in the PSC or fish skins was calculated according to the previous study [33] and the extraction yield of PSC was calculated using the equation as follows:

$$\text{PSC extraction yield (\%)} = \frac{\text{Hydroxyproline content in PSC}}{\text{Hydroxyproline content in fish skin}} \times 100\% \tag{1}$$

3.3. Experimental Design and Statistical Analysis

Single factor experiments were carried out for establishing the preliminary range of the extraction variables, such as pepsin concentration, solid-liquid ratio and hydrolysis time. Then, RSM and BBD were applied to optimize the three extraction parameters for improving the yield of PSC from

Nibea japonica skin. The range and levels of the variables investigated in the present study were given in Table 4.

Table 4. Independent factors and their levels used in the response surface design.

Independent Factors	Symbol	Level of Factor		
		−1	0	1
Enzyme concentration (U/g)	X_1	800	1200	1600
Solid-liquid ratio (v/w)	X_2	1:45	1:55	1:65
Hydrolysis time (h)	X_3	6	8	10

RSM with BBD was performed to obtain the optimum conditions for PSC extraction. For statistical calculations, the factors were coded according the equation as follows:

$$\chi_i = \frac{X_i - X_o}{\Delta X} \tag{2}$$

where, χ_i is the coded value of the independent factor, X_i is the actual value of the independent factor, X_o is the actual value of X_i at the center point and ΔX is the step change value. As shown in Table 2, the Box-Behnken design in the experiment design consists of 15 experimental points and the data from experiment design were explained by multiple regressions to fit the second-order polynomial equation as follows:

$$Y = \beta_o + \sum \beta_i X_i + \sum \beta_{ij} X_i X_j + \sum \beta_{ii} X_i X_i \tag{3}$$

where, Y is the dependent variable (PSC yield, %); β_o is the intercept term, β_i is the linear regression coefficient, β_{ii} is the aquared coefficient and β_{ij} is the interaction coefficient. X_i and X_j are levels of the independent variables. Each experiment design was determined in triplicate and the data were analyzed using the software Design-Expert 8.0.5 (State-Ease Inc., Minneapolis, MN, USA).

3.4. SDS-PAGE Analysis

The PSC samples from *Nibea japonica* skin were then analyzed by using SDS-PAGE according to the method described by Tang et al. [34]. The PSC samples were firstly dissolved in 0.5 M acetic acid and mixed with the loading buffer. Electrophoresis was performed on 7.5% gels and high protein molecular weight marker (Takara, Dalian, China) was used to estimate the molecular weight of PSC.

3.5. Amino Acid Composition of PSC

The lyophilized PSC samples were hydrolyzed with 6 M HCl at 110 °C for 24 h without oxygen and then vaporized. The hydrolysates were then analyzed by using a Hitachi amino acid analyser L-8800 (Hitachi, Tokyo, Japan). The content of Hyp was accurately measured according to the protocol described by Tang et al. [1].

3.6. UV-Visible Spectroscopy of PSC

The UV-visible spectrum of PSC from *Nibea japonica* skin was determined by a Shimadzu spectrophotometer (Shimadzu, Kyoto, Japan). The lyophilized PSC samples were dissolved in 0.5 M acetic acid and then centrifuged at 12,000 rpm for 10 min at 4 °C. The absorbance of the supernatant was measured at different wavelengths (from 200 nm to 700 nm) to get its UV-visible spectrum.

3.7. FTIR Spectra of PSC

FTIR spectra of lyophilized PSC samples were measured on a Bruker Tensor 27 FTIR spectrometer (Bruker, Rheinstetten, Germany) using the method described by previous studies. The spectra were produced with a wavelengths range from 4000 to 450 cm^{-1} at a resolution of 1 cm^{-1} for a single scan.

3.8. Effects of pH and Sodium Chloride on PSC Solubility

Lyophilized PSC samples (3 mg/mL) were first dissolved in 0.5 M acetic acid. Then, PSC solution (8 mL) was transferred to a centrifuge tube (15 mL) and the pH was adjusted with 6 N HCl or 6 N NaOH to get the final pH ranging from 1.0 to 10.0 and constant-volumed to 10 mL by deionized water. The mixtures were centrifuged at 12,000 rpm for 10 min at 4 °C and the protein content in the supernatant was measured by the Bradford method.

The effect of sodium chloride on PSC solubility was determined as follows: PSC (6 mg/mL) was dissolved in 0.5 M acetic acid and 5 mL of this solution was added with 5 mL sodium chloride (in 0.5 M acetic acid) with a series of concentrations (0, 2, 4, 6, 8, 10 and 12%) to the final concentrations of 0, 1, 2, 3, 4, 5 and 6%. The mixtures were then stirred at 4 °C for 30 min and then centrifuged at 12,000 rpm for 10 min at 4 °C. The protein content in the supernatant was measured as described above.

4. Conclusions

In the present study, RSM was used to optimize the extraction process of PSC from *Nibea japonica* skin. The pepsin concentration of 1389 U/g, solid-liquid ratio of 1:57 and hydrolysis time of 8.67 h was found to be optimal for PSC extraction; giving a yield of 84.85%. The extracted PSC was then characterized as type I collagen using SDS-PAGE electrophoresis, and the FTIR analysis also revealed that PSC maintains its triple-helical structure. All PSCs were soluble in the pH range of 1.0–4.0 and decreases in solubility were observed at neutral or alkaline conditions. All PSCs had a decrease in solubility in the presence of sodium chloride, especially with a concentration above 2%. Further study will be performed to investigate whether this collagen can be used in biomedical applications and other fields.

Acknowledgments: This work was financially supported by the Natural Science Foundation of Zhejiang Province (grant No. LQ16H300001 and No. LQ15C200009), the National Natural Science Foundation of China (grant No. 81773629 and No. 21502170), the National Spark Program of China (grant No. 2015GA700044), the National Undergraduate Training Program for Innovation and Entrepreneurship (grant No. 201710340011), the Foundation of Zhejiang Educational Committee (grant No. Y201534400), the Public Welfare Program of Zhoushan (grant No. 2015C31012 and No.2016C41003) and the Zhejiang Xinmiao Talents Program (grant No. 2017R411009).

Author Contributions: Yunping Tang conceived and designed the experiments. Fangmiao Yu, Chuhong Zong, Shujie Jin, Jiawen Zheng performed the experiments. Nan Chen, Ju Huang, Yan Chen, Fangfang Huang and Zuisu Yang and Guofang Ding carried out statistical analysis of the date. All authors contributed to the writing of the manuscript.

Conflicts of Interest: The authors declare no conflict of interest.

References

1. Tang, Y.P.; Yang, X.L.; Hang, B.J.; Li, J.T.; Huang, L.; Huang, F.; Xu, Z.N. Efficient production of hydroxylated human-like collagen via the co-expression of three key genes in *Escherichia coli* Origami (DE3). *Appl. Biochem. Biotechnol.* **2016**, *178*, 1458–1470. [CrossRef] [PubMed]

2. Ahmad, M.; Benjakul, S. Extraction and characterisation of pepsin-solubilised collagen from the skin of unicorn leatherjacket (*Aluterus monocerous*). *Food Chem.* **2010**, *120*, 817–824. [CrossRef]

3. Jeon, E.Y.; Choi, B.-H.; Jung, D.; Hwang, B.H.; Cha, H.J. Natural healing-inspired collagen-targeting surgical protein glue for accelerated scarless skin regeneration. *Biomaterials* **2017**, *134*, 154–165. [CrossRef] [PubMed]

4. Hu, Z.; Yang, P.; Zhou, C.X.; Li, S.D.; Hong, P.Z. Marine collagen peptides from the skin of Nile Tilapia (*Oreochromis niloticus*): Characterization and wound healing evaluation. *Mar. Drugs* **2017**, *15*, 102–112. [CrossRef] [PubMed]

5. Sachdev, G.S.; Sachdev, M. Recent advances in corneal collagen cross-linking. *Indian J. Ophthalmol.* **2017**, *65*, 787–796. [CrossRef] [PubMed]

6. Vazquez-Portalatin, N.; Kilmer, C.E.; Panitch, A.; Liu, J.C. Characterization of collagen type I and II blended hydrogels for articular cartilage tissue engineering. *Biomacromolecules* **2016**, *17*, 3145–3152. [CrossRef] [PubMed]

7.   Lee, J.C.; Volpicelli, E.J. Bioinspired collagen scaffolds in cranial bone regeneration: From bedside to bench. *Adv. Healthc. Mater.* **2017**, *6*, 1700232–1700252. [CrossRef] [PubMed]

8.   Peng, Y.Y.; Stoichevska, V.; Vashi, A.; Howell, L.; Fehr, F.; Dumsday, G.J.; Werkmeister, J.A.; Ramshaw, J.A.M. Non-animal collagens as new options for cosmetic formulation. *Int. J. Cosmet. Sci.* **2015**, *37*, 636–641. [CrossRef] [PubMed]

9.   Wang, B.; Wang, Y.-M.; Chi, C.-F.; Luo, H.-Y.; Deng, S.-G.; Ma, J.-Y. Isolation and characterization of collagen and antioxidant collagen peptides from scales of croceine croaker (*Pseudosciaena crocea*). *Mar. Drugs* **2013**, *11*, 4641–4661. [CrossRef] [PubMed]

10. Li, X.-R.; Chi, C.-F.; Li, L.; Wang, B. Purification and identification of antioxidant peptides from protein hydrolysate of scalloped hammerhead (*Sphyrna lewini*) Cartilage. *Mar. Drugs* **2017**, *15*, 61–76. [CrossRef] [PubMed]

11. Reza, M.; Amin Mohammadifar, M.; Mohammad Mortazavian, A.; Milad, R.; Jahan, B.G.; Zohre, D. Extraction optimization of pepsin-soluble collagen from eggshell membrane by response surface methodology (RSM). *Food Chem.* **2016**, *190*, 186–193.

12. Pakkanen, O.; Hamalainen, E.R.; Kivirikko, K.I.; Myllyharju, J. Assembly of stable human type I and III collagen molecules from hydroxylated recombinant chains in the yeast Pichia pastoris—Effect of an engineered C-terminal oligomerization domain foldon. *J. Biol. Chem.* **2003**, *278*, 32478–32483. [CrossRef] [PubMed]

13. Baez, J.; Olsen, D.; Polarek, J.W. Recombinant microbial systems for the production of human collagen and gelatin. *Appl. Microbiol. Biotechnol.* **2005**, *69*, 245–252. [CrossRef] [PubMed]

14. Abdollahi, M.; Rezaei, M.; Jafarpour, A.; Undeland, I. Sequential extraction of gel-forming proteins, collagen and collagen hydrolysate from gutted silver carp (*Hypophthalmichthys molitrix*), a biorefinery approach. *Food Chem.* **2018**, *242*, 568–578. [CrossRef] [PubMed]

15. Li, Z.R.; Wang, B.; Chi, C.F.; Zhang, Q.H.; Gong, Y.D.; Tang, J.J.; Luo, H.Y.; Ding, G.F. Isolation and characterization of acid soluble collagens and pepsin soluble collagens from the skin and bone of Spanish mackerel (*Scomberomorous niphonius*). *Food Hydrocoll.* **2013**, *31*, 103–113. [CrossRef]

16. Sittichoke, S.; Soottawat, B.; Kishimura, H. Comparative study on molecular characteristics of acid soluble collagens from skin and swim bladder of seabass (*Lates calcarifer*). *Food Chem.* **2013**, *138*, 2435–2441.

17. Zhao, Y.-Q.; Zeng, L.; Yang, Z.-S.; Huang, F.-F.; Ding, G.-F.; Wang, B. Anti-fatigue effect by peptide fraction from protein hydrolysate of croceine croaker (*Pseudosciaena crocea*) swim bladder through inhibiting the oxidative reactions including DNA damage. *Mar. Drugs* **2016**, *14*, 221. [CrossRef] [PubMed]

18. Zhuang, Y.; Sun, L.; Zhang, Y.; Liu, G. Antihypertensive Effect of long-term oral administration of jellyfish (*Rhopilema esculentum*) collagen peptides on renovascular hypertension. *Mar. Drugs* **2012**, *10*, 417–426. [CrossRef] [PubMed]

19. Han, T.; Li, X.Y.; Wang, J.T.; Hu, S.X.; Jiang, Y.D.; Zhong, X.D. Effect of dietary lipid level on growth, feed utilization and body composition of juvenile giant croaker *Nibea japonica*. *Aquaculture* **2014**, *434*, 145–150. [CrossRef]

20. Li, X.Y.; Wang, J.T.; Han, T.; Hu, S.X.; Jiang, Y.D. Effects of dietary carbohydrate level on growth and body composition of juvenile giant croaker *Nibea japonica*. *Aquac. Res.* **2015**, *46*, 2851–2858. [CrossRef]

21. Bakar, J.; Mohamad, R.U.H.; Mat, H.D.; Qurni, S.A.; Harvinder, K. Collagen Extraction from Aquatic Animals. WIPO Patent WO2010074552 A1, 23 December 2008.

22. Somasundaram, T.; Anguchamy, V.; Muthuvel, A. Isolation and characterization of acid and pepsin-solubilized collagen from the skin of sailfish (*Istiophorus platypterus*). *Food Res. Int.* **2013**, *54*, 1499–1505.

23. El-Rashidy, A.A.; Gad, A.; Abu-Hussein, A.G.; Habib, S.I.; Badr, N.A.; Hashem, A.A. Chemical and biological evaluation of Egyptian Nile Tilapia (*Oreochromis niloticas*) fish scale collagen. *Int. J. Biol. Macromol.* **2015**, *79*, 618–626. [CrossRef] [PubMed]

24. Zhang, J.J.; Rui Duan, R. Characterisation of acid-soluble and pepsin-solubilised collagen from frog (*Rana nigromaculata*) skin. *Int. J. Biol. Macromol.* **2017**, *101*, 638–642. [CrossRef] [PubMed]

25. Nokelainen, M.; Tu, H.; Vuorela, A.; Notbohm, H.; Kivirikko, K.I.; Myllyharju, J. High-level production of human type I collagen in the yeast *Pichia pastoris*. *Yeast* **2001**, *18*, 797–806. [CrossRef] [PubMed]

26. Ramshaw, J.A.M.; Shah, N.K.; Brodsky, B. Gly-X-Y tripeptide frequencies in collagen: A context for host-guest triple-helical peptides. *J. Struct. Biol.* **1998**, *122*, 86–91. [CrossRef] [PubMed]

27. Kumar, P.G.; Nidheesh, T.; Suresh, P.V. Comparative study on characteristics and in vitro fibril formation ability of acid and pepsin soluble collagen from the skin of catla (*Catla catla*) and rohu (*Labeo rohita*). *Food Res. Int.* **2015**, *76*, 804–812.

28. Doyle, B.B.; Bendit, E.G.; Blout, E.R. Infrared spectroscopy of collagen and collagen-like polypeptides. *Biopolymers* **1975**, *14*, 937–957. [CrossRef] [PubMed]

29. Muyonga, J.H.; Cole, C.G.B.; Duodu, K.G. Fourier transform infrared (FTIR) spectroscopic study of acid soluble collagen and gelatin from skins and bones of young and adult Nile perch (*Lates niloticus*). *Food Chem.* **2004**, *86*, 325–332. [CrossRef]

30. Liu, H.Y.; Han, J.; Guo, S.D. Characteristics of the gelatin extracted from Channel Catfish (*Ictalurus Punctatus*) head bones. *LWT-Food Sci. Technol.* **2009**, *42*, 540–544. [CrossRef]

31. Yuqing, T.; Chang, S.K.C. Isolation and characterization of collagen extracted from channel catfish (*Ictalurus punctatus*) skin. *Food Chem.* **2018**, *242*, 147–155.

32. Dasong, L.; Li, L.; Regenstein, J.M.; Peng, Z. Extraction and characterisation of pepsin-solubilised collagen from fins, scales, skins, bones and swim bladders of bighead carp (*Hypophthalmichthys nobilis*). *Food Chem.* **2012**, *133*, 1441–1448.

33. Jamall, I.S.; Finelli, V.N.; Hee, S.S.Q. A simple method to determine nanogram levels of 4-hydroxyproline in biological tissues. *Anal. Biochem.* **1981**, *112*, 70–75. [CrossRef]

34. Tang, Y.P.; Zhang, G.M.; Wang, Z.; Liu, D.; Zhang, L.L.; Zhou, Y.F.; Huang, J.; Yu, F.M.; Yang, Z.S.; Ding, G.F. Efficient synthesis of a (S)-fluoxetine intermediate using carbonyl reductase coupled with glucose dehydrogenase. *Bioresour. Technol.* **2017**, *250*, 457–463. [CrossRef] [PubMed]

 *marine drugs*

Article

Evaluation of the Potential of Collagen from Codfish Skin as a Biomaterial for Biomedical Applications

Ana M. Carvalho [1,2]**, Alexandra P. Marques** [1,2]**, Tiago H. Silva** [1,2,*]** and Rui L. Reis** [1,2,3]

[1] 3B's Research Group, I3Bs–Research Institute on Biomaterials, Biodegradables and Biomimetics, University of Minho, Headquarters of the European Institute of Excellence in Tissue Engineering and Regenerative Medicine Avepark–Parque de Ciência e Tecnologia, Zona Industrial da Gandra, 4805-017 Barco, Guimarães, Portugal; ammpfc@i3bs.uminho.pt (A.M.C.); apmarques@i3bs.uminho.pt (A.P.M.); rgreis@i3bs.uminho.pt (R.L.R.)
[2] ICVS/3B's–PT Government Associate Laboratory, 4805-017 Braga/Guimarães, Portugal
[3] The Discoveries Centre for Regenerative and Precision Medicine, Headquarters at University of Minho, Avepark, 4805-017 Barco, Guimarães, Portugal
* Correspondence: tiago.silva@i3bs.uminho.pt; Tel.: +351-253-510-931

Received: 6 November 2018; Accepted: 5 December 2018; Published: 8 December 2018

Abstract: Collagen is one of the most widely used biomaterials, not only due its biocompatibility, biodegradability and weak antigenic potential, but also due to its role in the structure and function of tissues. Searching for alternative collagen sources, the aim of this study was to extract collagen from the skin of codfish, previously obtained as a by-product of fish industrial plants, and characterize it regarding its use as a biomaterial for biomedical application, according to American Society for Testing and Materials (ASTM) Guidelines. Collagen type I with a high degree of purity was obtained through acid-extraction, as confirmed by colorimetric assays, SDS-PAGE and amino acid composition. Thermal analysis revealed a denaturing temperature around 16 °C. Moreover, collagen showed a concentration-dependent effect in metabolism and on cell adhesion of lung fibroblast MRC-5 cells. In conclusion, this study shows that collagen can be obtained from marine-origin sources, while preserving its bioactivity, supporting its use in biomedical applications.

Keywords: marine-origin collagen; codfish; biophysical characterization; biologic activity; ASTM guidelines; biomedical application; marine biomaterials

---

1. Introduction

Collagen is the most abundant protein in vertebrates, playing a dominant role in the maintenance of the biological and structural integrity of tissues, and contributing to the molecular architecture, shape, and mechanical features [1]. It is a trimeric molecule consisting of three polypeptide α-chains, which are woven in a triple helix forming homotrimers or heterotrimers, depending on the collagen type. Currently, about 28 types of collagen are identified in human tissues. About one-half of the total human body collagen is present in the skin, being mostly collagen type I [2,3], with each tropocollagen (individualized) molecule composed of two equal α_1 chains and one α_2 chain.

Collagen has revealed excellent biocompatibility, biodegradability and biorenewal, and weak antigenicity [4]. Moreover, collagen-cell interaction is well understood. Such properties and knowledge led to the development of a plethora of collagen-based biomedical devices, including drug delivery systems, surgical sutures, hemostatic agents and tissue-engineering applications [2,5]. In addition, collagen has a well-preserved structure and amino acid composition between species, which contributes to the mentioned properties when considering the use of non-human collagens in the biomedical context.

Considering commercial exploitation, the available collagen is mainly obtained from terrestrial animals, namely bovine, porcine, caprine and rat species. Despite the fine screening of transmissible diseases, the risk of transmission of bovine spongiform encephalopathy, ovine and caprine scrape, foot-and-mouth disease and swine influenza from animals to humans is nonetheless a concern [6,7]. Collagen obtained from fish may be a safer alternative, since there is no known risk of disease transmission. Moreover, fish processing by-products, such as skin, bones and scales, are abundantly available, assuring the sustainable exploitation of marine resources through the valorization of residues, while addressing associated environmental pollution issues [8,9]. Besides fish, many other marine organisms from different taxa have been studied as origins of biomass for the extraction of collagen, from a fundamental perspective or as raw-material for biopolymer production. Examples can be found on marine sponges [10–15], jellyfish [16–18], anemones [19,20], corals [21], squids [22–24] and different echinoderms [25–27].

To allow the use of such marine-origin collagen in pharmaceutical, cosmetics or biomedical fields, adequate processing of raw material is required, following highly demanding and strict regulations, which can be particularly challenging when using fish by-products as skins and hence associated with high cost. Although those efforts would be fully covered by the high added value of the envisaged applications, one should also consider the expected differences between marine origin collagens and mammal collagens, particularly regarding denaturation temperature, which is commonly lower for marine collagens [9], but also the rheological properties exhibited by aqueous solutions of this molecule, which we can infer in a rough way by the normally less-viscous character of marine collagen solutions or dispersions when compared with those produced with mammal collagens. These differences are commonly associated with alterations in the hydroxyproline contents and/or hydroxylation degree of proline amino acids, which are reduced in marine collagens. This has been recently confirmed by a comparative study by Bao et al. [28], which also raised the possibility of a synergistic effect of hydroxyproline and cysteine, affecting not only the thermal properties of collagen, but also its mechanical properties. Other possibilities to overcome the limited mechanical properties may be the establishment of extraction protocols that render more native macromolecular entities, namely, preserving interaction with glycosaminoglycans [29], or addressing processing features, namely, chemical crosslinking [30,31].

Fish collagen has been obtained from different species, including rohu, catla, megrim, dover sole, codfish, hake, carp, shark, spotted golden goatfish, tuna, niger triggerfish, tilapia and northern pike [6,32–41]. In these studies, the main concern is the establishment of valuable extraction processes, and collagen physical and chemical properties, while the biomedical potential is rarely addressed, with reports on biological assessment or biological validation of the extracted collagen typically lacking. This study is intended not only to explore codfish skin as a sustainable and valuable source of collagen, by establishing an extraction methodology to produce high-purity collagen extracts from industrial by-products, but also to address its biomedical potential by characterization according to the guidelines established by the appropriate American Society for Testing and Materials (ASTM) Standard F 2212 [42]—a document designed by ASTM International that defines the methodologies to be used as a reference for the characterization of collagen entitled for use as a component of surgical implants and tissue-engineered medical products, including the use of the extracted codfish collagen as a substrate for human fibroblast cell culture.

2. Results and Discussion

2.1. Electrophoretic Profiling of Acid-Soluble Collagen Extracted from Codfish Skin

To confirm the type of collagen extracted and its purity, SDS-PAGE was performed, which destroyed the structure of the protein and separated the respective components according to their electrophoretic mobility (Figure 1A). Extracted collagen has two clear bands attributed to two different kinds of α-chains, as observed in the case of collagen type I from both rat and bovine species [43].

However, both α-chains of collagen from codfish presented a higher electrophoretic mobility, attributed to lower molecular weights. The molecular weight of α_1 was about 130 kDa for mammal (rat and bovine) collagens and 120 kDa for codfish collagen, while α_2 was about 115 kDa for mammal and 110 kDa for codfish collagen. Skierka et al. studied the extraction processes of collagen from codfish, including skin and backbone [44,45]. Despite the differences in the extraction methodology used and source of collagen between the studies, it was possible to obtain collagen type I with a similar molecular weight as reported. The differences observed in molecular weight compared to mammalian collagen are usually associated with differences in the amino acid composition. The two α_1 chains (α_{1I} and α_{1II}) could not be distinguished because only one band is observed under the electrophoretic conditions employed, as both chains are equal and thus have the same electrophoretic migration pattern [46,47]. Nevertheless, a ratio of α_1/α_2 equal to 2 was observed, indicating that the α_1 band corresponds to double the quantity of chains detected in the α_2 band, which suggests that extracted collagen is type I [43]. In all samples, intramolecular cross-linking β- and γ-components were detected: γ-components were richer in collagen obtained from skin of codfish than in the rat or bovine collagens, while β-components were found in lower amount in the former. Intensive intra- and intermolecular covalent crosslink takes place universally to stabilize the collagen fibrils against proteolytic degradation and promotes the desired tensile strength to the stroma [3]. However, the crosslink of adjacent chains into dimers (β-chains) or trimers (γ-chains), may inhibit the proper folding of collagen protein to a triple-helix conformation, as it may interfere in the degree of freedom of individual monomers. The purification of α-monomers from the collagen may represent an advantage when a triple helix conformation is required.

2.2. Fourier-Transformed Infrared (FTIR) Spectroscopy Analysis

The FTIR spectra of collagen extracted from codfish skin and commercial collagen type I from rat tail and bovine skin are exhibited in Figure 1B. All three spectra share great similarities. They exhibit several features characteristic of collagen molecular organization: amino acids linked together by peptide bounds give rise to infrared active vibration modes amide A and B and amide I, II and III. The characteristic infrared absorption of amide A, associated with N-H stretching vibration, occurred between 3400 and 3440 cm^{-1}; the respective absorption peaks of collagen extracted from codfish, and collagen type I extracted from rat tail and from bovine skin were found at 3329, 3320 and 3324 cm^{-1}, respectively. A shift in the amide A absorption band to lower frequencies appears when a peptide is involved in hydrogen bonds [48–50]. Moreover, band symmetry suggests that a low amount of water is present [51]. Amide B peaks codfish, rat and bovine collagen, representing asymmetrical stretch of CH_2, were found at 2877, 2879 and 2878 cm^{-1}, respectively [52]. The amide I bond is usually observed in the range 1600–1700 cm^{-1} and corresponds to the stretching vibration of C=O along the polypeptide backbone of proteins, which is a sensitive area to protein secondary structure, and therefore used in analysis [48,50,52]. The amide I band in codfish and rat collagen was found at 1649 cm^{-1}, while from bovine collagen was found at 1656 cm^{-1}. The amide II band refers to N-H bending vibration coupled with C-N stretching vibrations and is commonly found at the 1550–1600 cm^{-1} range [51,53]. In codfish and rat collagen, the amide II band was found at 1554 cm^{-1} and from bovine collagen was found at 1546 cm^{-1}. The shift to lower frequencies observed on the band of bovine collagen when compared with collagen obtained from codfish and rat reflects the existence of hydrogen bonds. Amide III bands were found at 1239 cm^{-1} in all samples, reflecting complex intermolecular interactions, including C-N stretching and N-H in-plane bending from amide linkages, as well as absorptions from wagging vibrations from CH_2 groups, from the glycine backbone and proline side-chains [51]. Moreover, the ratio between amide III and the band at 1450 cm^{-1} was close to 1, which is an indication of the presence of the triple helix structure of collagen in powder [49,54]. The similarities between collagen obtained from codfish and commercial collagens suggested that their structures are quite similar in a dry state [55]. The presence of nucleic acids, phospholipids and lipids was identified in all samples at about 1030, 1228 and 1454 cm^{-1}, respectively [52].

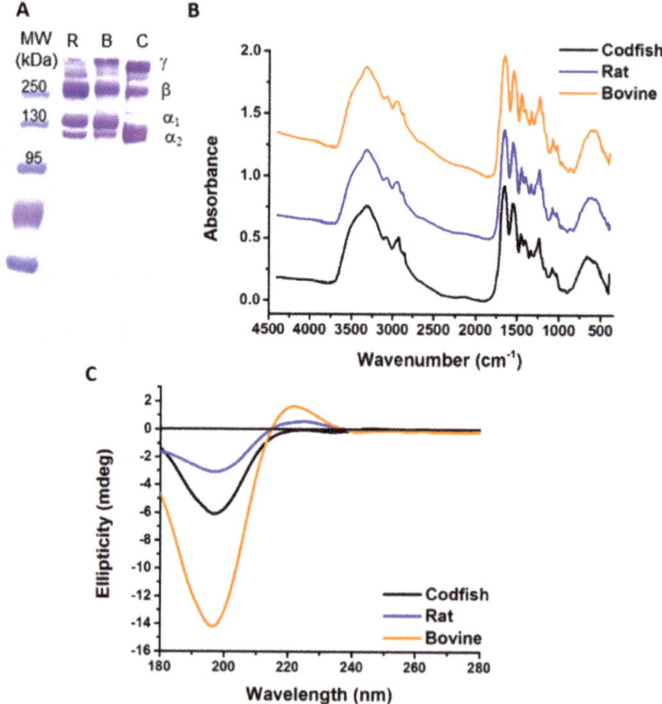

Figure 1. Characterization of collagen extracted from skin of codfish (C) and rat (R) and bovine (B) collagen type I. (**A**) Sodium dodecyl sulfate-polyacrylamide electrophoresis evidences the molecular structure and organization of collagen type I by the presence of monomer $((\alpha_1)_2(\alpha_2))$; dimers (β) and trimers (γ) components are also present. Moreover, similar molecular weights (MW) between the collagens were found. (**B**) Fourier transform infrared spectra of collagens exhibits the main vibrations of collagen molecular organization, amide A, amide B, amide I, amide II and amide III. (**C**) Circular dichroism spectra show a positive band with maximum ellipticity at 225 nm, confirming the presence of triple-helix structure in collagens.

2.3. Extract Purity

Total protein concentration was assessed using a micro BCA kit, while collagen content was quantified using Sircol assay and SDS-PAGE. The protein content of codfish extract was 87.7 ± 3.6%, whereas the collagen content was 75 ± 20%. Syrius red, the dye in the Sircol assay, binds specifically to hydroxyproline in collagen. This amino acid residue is found in marine-origin collagen in lower amounts than in mammalian-origin collagen, which is a standard calibrator for the assay. Therefore, collagen content determined by Sircol assay is not accurate, due to the difference in hydroxyproline contents between both collagen origins, with a higher purity degree expected than that determined [56]. By SDS-PAGE imaging analysis using Image J, collagen purity was determined to be around 90%, in line with the purity found in rat and bovine collagen type I, thus showing it to be a more appropriate methodology to address the purity of marine collagen extracts.

2.4. Circular Dichroism (CD)

It is important to assess whether the extracted collagen type I is in the native–triple helical structure, or in the denatured form–coil structure. Collagen is known to adopt polyproline II-like helical conformation, which is characterized by CD with a positive maximum absorption band at around 222 nm. Herein, the solutions of collagen from codfish skin, rat tail and bovine skin were prepared at 0.1 mg/mL and analysed for conformation using the CD spectroscopy in the far-UV region (Figure 1C). CD spectra of bovine collagen show a preeminent positive band at 222 nm, which is typical of collagen's triple-helical structure, and is also visible, although with less intensity, for rat tail collagen; on the other hand, codfish collagen did not exhibit a clear positive band at 222 nm, which may give an indication of at least partial denaturation of the protein upon extraction. Intermolecular interactions of triple helix are disassociated by dilute acids and by repulsion forces that occur between the same charges of collagen monomers [5]. Moreover, the presence of intramolecular interactions in collagen is highly dependent on the amino acid composition, namely, on the presence of proline and hydroxyproline due to the stereochemical contribution of the pyrrolidine ring and the extra hydrogen bonds that may be formed [57,58].

2.5. Amino Acid Composition

Collagen type I consists of 20 different amino acids organized as three α-chains which wrap around each other forming a characteristic triple-helix conformation. This structure is stabilized by the presence of glycine residue in every three residues, high content of proline (Pro) and hydroxyproline (OHyp), interchain hydrogen bonds, and electrostatic interactions involving lysine and aspartate [59,60]. The amino acid composition of collagen is presented in Table 1. In all samples, glycine was the most abundant amino acid, with 266–333/1000 residues. Glycine content is the principal feature that affects the triple helix formation [61] and accounts for more than 25% of all amino acids in all samples, which is in good agreement with the presence of Gly in each third amino acid residue. This percentage was smaller in the marine-origin collagen. Alanine, proline, hydroxyproline and glutamine were the most common amino acids found in the spiral structure, which agrees with their reported high content [62]. In particular, OHyp and Pro contents in collagen obtained from codfish skin were 39 and 62/1000 residues, respectively. The degree of hydroxylation of proline was calculated to be 46.8%, 46.6% and 38.7% for rat, bovine and codfish collagens, respectively and is correlated with triple helix stability, and thus with the denaturing temperature of the protein [41]. While proline and hydroxyproline were found in lower amounts than in mammalian collagen, serine and methionine were present in higher amounts in fish collagen. Such findings are in agreement with those observed for collagens extracted from other marine-species sources [6,7,35–39,41,49,52,54].

Table 1. Amino acid composition of collagen obtained from the skin of codfish, and of rat and bovine commercial collagen (per 1000 residues).

Amino Acid	Rat	Bovine	Codfish
Alanine	111.16	102.04	91.48
Arginine	42.24	32.86	30.45
Aspartic acid	45.32	36.65	38.82
Cystein	0.99	1.24	1.28
Glutamic acid	73.33	59.43	56.08
Glycine	333.18	296.44	266.12
Histidine	3.61	3.11	5.01
Hydroxylysine	9.33	8.86	6.65
Hydroxyproline	96.06	78.35	39.60
Isoleucine	7.48	6.74	5.61
Leucine	23.29	17.50	16.51
Lysine	27.07	22.20	19.62
Methionine	8.03	7.81	15.04
N-isobutylglycine	12.70	14.33	13.75
Phenylalanine	14.62	11.58	12.70
Proline	109.21	89.89	62.69
Serine	42.74	32.03	53.87
Threonine	18.79	13.20	16.89
Tyrosine	3.76	1.48	2.25
Valinine	17.08	12.86	12.02

2.6. Denaturation Temperature

The structural level of collagen organization in solution depends on the ratio of triple helix:coil structures. Collagen is denatured when 90% of molecules are in coil state and un-denatured when at least 70% of molecules keep their triple helical structure [5].

The denaturation temperature of the collagens solubilized in 20 mM acetic acid was assessed by monitoring the viscosity of those collagen solutions at different temperatures (Table 2). Denaturation temperatures were calculated from the thermal denaturation curve as 18.25 ± 0.40 °C, 35.34 ± 0.03 °C and 40.08 ± 0.01 °C for codfish, bovine and rat origins, respectively. Moreover, 70% of the molecules kept their native structure until 15.77 ± 0.09 °C, 33.19 ± 0.04 °C and 39.65 ± 0.02 °C, respectively. This is in agreement with the tendency suggested by the amino acid content results, where lower amounts of Pro and OHpro are associated with lower denaturing temperature [41].

Table 2. Denaturing temperature (10%) and temperature until which collagen is considered in native structure (30%).

Collagen Origin	Temperature According to % of Native Collagen	
	30	10
Codfish	15.77 ± 0.09	18.24 ± 0.40
Rat	39.65 ± 0.02	40.08 ± 0.01
Bovine	33.19 ± 0.02	35.34 ± 0.03

2.7. Heavy Metals

Metals and other elements can be naturally present in fishes. Mercury (Hg), lead (Pb), cadmium (Cd), and arsenic (As) are of concern due to accumulation in tissues. The content of heavy metal impurities in codfish collagen was determined considering permissible limits. All the elements studied were below the detection limits of the equipment, as follows: $Pb < 2.5$ ppm; $Cd < 0.25$ ppm; $Hg < 0.5$ ppm; and $As < 0.35$ ppm.

The U.S. Pharmacopeia Convention (USP) specifies limits for the amounts of elemental impurities in drug products, including the amount of elemental impurities present in daily dose (permitted daily exposure based on a 50 Kg person) and in large volume parenteral (LVP—daily dose greater than 100 mL). Concentration limits for parenteral drug products with a maximum dose of 10 g/day are Cd 0.25 ppm, Pb 0.5 ppm, As 0.15 ppm, and Hg 0.15 ppm [63]. Moreover, the International Organization for Standardization (ISO) 9917-1 states that As and Pb content in dental water-based cements should be less than 2 and 100 ppm, respectively [64]. The U.S. Food and Drug Administration (FDA) defined the maximum allowed concentration in cosmetics of Hg as 65 ppm and Pb as 10 ppm [65]. However, there is still a need to regulate limits for heavy metal impurities in biomaterials to be used in biomedical applications. In this study, heavy metals were either absent for all cases or, if present, were in undetectable amounts and below the listed limits.

2.8. Cell Viability

The in vitro cytotoxicity of codfish collagen was investigated using lung fibroblast MRC-5 cells by MTS assay, which is based on the mitochondrial activity of vital cells, reflecting their metabolic activity [66]. Cytotoxicity was calculated comparing each condition with a negative control (Figure 2). The different concentrations of collagen tested showed that in the range of 0.01 to 0.05 mg/mL, collagen does not affect cell metabolism (p-value > 0.05). However, at higher concentrations (>0.1 mg/mL), cell metabolism was significantly decreased (p-value < 0.001) in relation to the control. Live/dead assay showed that MRC-5 cells remain viable when seeded in collagen coatings, as no dead cells are observed. However, high collagen density may affect cells' metabolic activity, showed by the decrease of green fluorescence and in accordance with the results from the MTS assay.

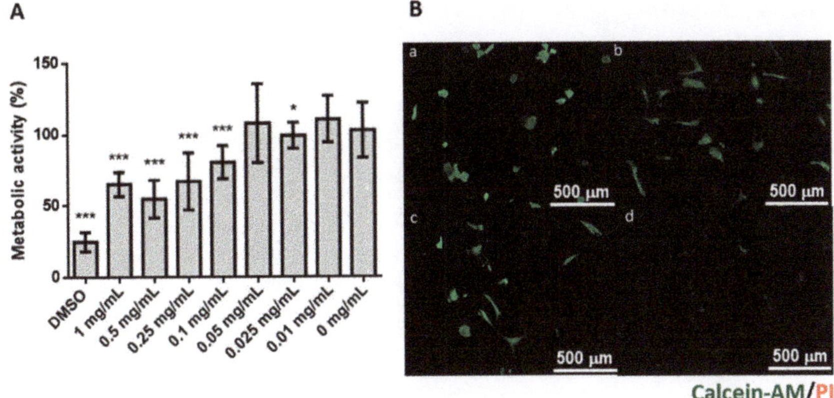

Figure 2. Cytotoxicity evaluation of codfish collagen over MRC-5 human fibroblast cell line. (**A**) Metabolic activity of MRC-5 cell line, measured by MTS assay, when exposed to different concentrations of collagen obtained from codfish skin (0.01–1 mg/mL) in relation to negative control (tissue culture plate). Statistically significant differences were observed: * p-value < 0.05 and *** p-value < 0.001. (**B**) Live/dead staining with calcein-AM (green) and propidium iodide (red): (a) 0 μg/mm^2; (b) 0.5 μg/mm^2; (c) 1.0 μg/mm^2; and (d) 1.5 μg/mm^2.

2.9. Cell Adhesion

Cell adhesion and viability of MRC-5 cell lines cultured onto collagen coatings was assessed by fluorescence microscopy (Figure 3). Cells adhered to the collagen coatings and their morphology was affected by the presence of collagen at different densities. The cell aspect ratio increased with the increase of collagen density in the coating. The higher densities of the adhesive protein may allow a higher number of adhesion points, thus resulting in the observed cell elongation and spindle-like morphology, quantitatively translated as higher values of the aspect ratio shape descriptor. This confirms that codfish collagen kept its cell adhesion properties during the extraction processes, indicating that it could be further used in biomedical applications, namely, those considering the development of collagen-based biomaterials to be used as templates for cell culture.

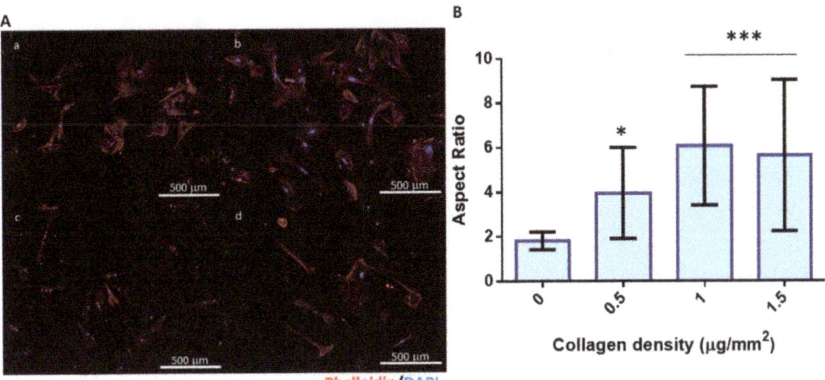

Figure 3. Adhesion of MRC-5 cell line into collagen coatings. Cell morphology was evaluated by (**A**) fluorescence staining of actin–phalloidin (red) and nucleus–DAPI (blue) and (**B**) shape description parameter aspect ratio determined for each condition (a 0 μg/mm^2 or absence of collagen, the control condition; b 0.5 μg/mm^2; c 1.0 μg/mm^2; and d 1.5 μg/mm^2). Statistically significant differences in respect to the control–absence of collagen–were observed: * p-value < 0.05 and *** p-value < 0.001.

3. Materials and Methods

3.1. Preparation of Codfish Skin

Codfish skins were removed from cold preserved fish body parts in an industrial plant and kindly provided by Frigoríficos de Ermida, Lda. (Gafanha, Nazaré, Portugal). Frozen skin was thawed at 4 °C for 24 h, with the scales, remaining flesh and bones then manually removed, and the skin cut into small pieces (1.0 cm × 1.0 cm) and washed with ice-cold deionized water. To remove non-collagenous proteins, the skin was treated with 10 volumes of 0.1 M NaOH solution, changed each 24 h, for 72 h at 4 °C under magnetic stirring. The skin was then rinsed with ice-cold deionized water until reaching pH 7. All procedures were carried out at 4 °C to minimize collagen denaturation.

3.2. Extraction of Acid-Soluble Collagen from Codfish Skin

All the procedures were carried out at 4 °C and collagen was extracted using an acid-base procedure. The skin was soaked in 0.5 M acetic acid with a sample:solution ratio of 1:10 (w/v) for 72 h with continuous stirring. The solution was then strained and centrifuged (Eppendorf 5810R, Hamburg, Germany) at 20,000× g during 1 h. An accurate volume of 50 mM Tris-HCl containing 2.6 M of NaCl at pH 7.4 buffer was added to the supernatant to reach a final concentration of 0.9 M of NaCl to precipitate collagen. The precipitated collagen was removed from solution by centrifugation at 20,000× g for 30 min. A minimal volume of 0.5 M acetic acid was added to the pellet and then dialyzed for 48 h against 10 volumes of 0.1 M acetic acid, 48 h against 0.02 M acetic acid and 48 h against ultrapure water, with the solutions being changed every 24 h. The resulted acid-soluble collagen was freeze-dried for 1 week and stored at −80 °C until further use.

3.3. Codfish Extract Purity

Protein content was evaluated by Micro BCA assay (Fisher Scientific, Rockford, IL, USA). Briefly, 1 mg/mL of collagen solution in 20 mM acetic acid was neutralized with 25 mM of Tris-base containing 3% SDS and denatured at 65 °C during 1 h. After equilibration at room temperature the sample was analysed. Sircol assay (Biocolor, County Antrim, UK) was performed according to the manufacturer's instructions. Collagen type I from rat tail was used as control in Micro BCA and Sircol assays. Extract purity was accessed comparing the protein/collagen content with the initial extract concentration.

3.4. Sodium Dozdecyl Sulfate-Polyacrylamide Gel Electrophoresis (SDS-PAGE)

Codfish skin extract was first dissolved at 1 mg/mL in 20 mM acetic acid at room temperature and then mixed with three-fold-concentrated loading buffer containing 0.1 M 1,4-dithiothreitol (DTT) to a final protein mass of 10 μg. Protein samples were heat-denatured at 65 °C for 30 min and analysed by SDS-PAGE according to Laemmli [67] using 4% stacking and 7.5% resolving gels in a Mini Protean III unit (Bio-Rad Laboratories, Hercules, CA, USA) at 27 mA/gel. Protein bands were stained with Coomassie brilliant Blue R250 and destained with 32% (v/v %) methanol 5.6% (v/v %) acetic acid solution (destaining solution I), and 5% (v/v %) methanol and 7% (v/v %) acetic acid solution (destaining solution II). Rat and bovine collagen were used as controls.

3.5. Fourier Transformed Infrared (FTIR) Spectroscopy

Freeze-dried products were individually mixed with vacuum dried potassium bromide (KBr) and pressed into pellets with a hydraulic press. The infrared spectra were obtained in the wavenumber range between 4400 and 400 cm^{-1} at a resolution of 5 cm^{-1}, using the infrared spectrometer IRPrestige 21 (Shimadzu, Kyoto, Japan). Each spectrum resulted from the average of 50 scans. Extracted collagen was compared with freeze-dried commercially available rat and bovine collagens.

3.6. Protein Conformation

Circular dichroism (CD) spectra of extracts were recorded at 180 to 280 nm on a Jasco J-1500 dichrograph (Jasco Corp., Tokyo, Japan) using a 0.1-cm path length cuvette. Dry collagen was dissolved at 0.1 mg/mL in 5 mM phosphate buffer pH 3. Samples were loaded at 4 °C into precooled CD cuvettes.

3.7. Denaturation Temperature

The denaturation temperature of collagen from codfish skin was evaluated by rheology (Kinexus Pro+, Malvern, UK). The viscosity of 1% (w/v) collagen in 20 mM acetic acid was monitored in the temperature range of 4 °C to 40 °C. Denaturation temperature was determined at 70% and 90% of denatured proteins [68].

3.8. Amino Acid Content

The amino acid content was determined by quantitative amino acid analysis using amino acid analyzer Biochrom 30 (Biochrom Ltd., Cambridge, UK) [8]. Briefly, the samples were hydrolyzed and separated by an ion exchange column. After post-column derivatization by ninhydrin, the samples were analyzed at 2 wavelengths: 440 and 570 nm. An internal standard of norleucine was used to determine the concentrations of amino acids in the sample.

3.9. Heavy Metals Quantification

The presence of lead (Pb), cadmium (Cd), arsenic (As) and mercury (Hg) in the codfish collagen extracts was quantified from codfish collagen according to Official Methods of analysis of AOAC International, 19th Edition (2012). Collagen was digested with 65% nitric acid and 30% hydrogen peroxide in a microwave digester MWS3+ (Berghof, Münster, Germany). Pb and Cd were quantified by atomic absorption spectrophotometry Perkin Elmer Analyst 600 (Perkin Elmer, Waltham, MA, USA) and As and Hg were quantified using atomic absorption spectrophotometry Perkin Elmer Analyst 600 (Perkin Elmer, Waltham, MA, USA) equipped with a Perkin Elmer FIAS 100 analyzer (Perkin Elmer).

3.10. Cell Culture

Human lung fibroblast cell line MRC-5 was grown as monolayer at 37 °C in DMEM low glucose with phenol red (Sigma-Aldrich, St. Louis, MO, USA) supplemented with sodium bicarbonate, 10% fetal bovine serum and 1% antibiotic/antimycotic. Sub-confluent cells in exponential growth phase were detached with TrypLE™ (Gibco, grand Island, NW, USA) and sub-cultured in fresh medium with appropriate cell density.

3.11. MTS Metabolic Activity Assay

Codfish collagen in 20 mM acetic acid with increasing concentration (0–1000 µg/mL) were added to a 48-well plate and evaporated at room temperature in an orbital shaker. Plates were sterilized by 15 min exposure to a 1.2 W UV lamp Omnicure series 2000 EXFO S2000-XLA (Omnicure, Bleichenbach, Germany). MRC-5 cells were seeded at 5×10^4 cells/mL in the collagen containing wells. Wells without collagen or containing 5% DMSO were used as negative and positive controls of cytotoxicity, respectively. Cells were cultured for 24 h at 37 °C in a humidified 5% CO_2 atmosphere. Metabolic activity was measured using a CellTiter 96® kit (Promega Corporation Madison, USA) according to the manufacturer's instructions. Absorbance was recorded at 490 nm in a Microplate Reader-Gen 5 2.0 SYNERGY HT (Biotek, Luzern, Switzerland).

3.12. Cell Adhesion on Collagen Coatings

Cell adhesion was performed on collagen coatings prepared from solutions with different concentrations. Briefly, collagen in 20 mM acetic acid solution was added to 6-well plates and let to dry overnight at room temperature in an orbital shaker, using the solution concentration and volume needed to achieve different collagen densities (0–1.5 $\mu g/mm^2$). After evaporation at room temperature, collagen was crosslinked with 50 mM EDC/NHS (1:1) solution for 24 h at 4 °C. Coatings were washed 10× with 1× phosphate buffer solution (PBS) and sterilized by UV radiation, as described before. MRC-5 cells were seeded at a density of 5×10^4 cells/mL on coverslips with and without collagen coatings. After 16 h incubation at 37 °C and humidified 5% CO_2 atmosphere, cell culture medium was removed, cells were washed with PBS and stained. Cell were permeated with 0.2% Triton-X solution for 5 min at room temperature and then incubated with Phalloidin/DAPI in PBS at 250 ng/mL and 1 $\mu g/mL$, respectively, for 45 min.

Live/dead stain was performed with calcein-AM (2 $\mu g/mL$) and propidium iodide (1 $\mu g/mL$), for 15 min at room temperature. Fluorescent images were acquired in Axio Imager Z1m (Zeiss, Göttingen, Germany), analysed in Zen lite 2.1 (Zeiss, Germany). Shape descriptor parameter aspect ratio was determined using phalloidin signal processing by ImageJ software. Aspect ratio is a shape descriptor corresponding to the ratio between the largest diameter and the smallest diameter perpendicular to it of a round-like particle or body; when the body has nearly a circular shape, the aspect ratio is close to 1, but if the body assumes an elongated shape, for instance caused by stretching, the aspect ratio increases.

3.13. Statistical Analysis

All experiments were performed in triplicate and the results are expressed as mean ± standard deviation. Metabolic activity of MRC-5 cells and shape descriptors showed a non-parametric distribution; therefore Mann-Whitney was used to assess statistically significant differences between results. Difference was considered statistically significant at $p < 0.05$.

4. Conclusions

The present study showed that collagen type I can be obtained from the skin of codfish with a high degree of purity, representing a valuable strategy for the valorization of a marine by-product. Properties such as molecular weight, amino acid composition and molecular structure were close to those of collagen of mammalian origin. The main difference was in regard to the protein denaturation temperature. Collagen with a low denaturation temperature presents poor gelling properties, which limit its application as a gel-forming agent, causing a processing bottlenecking to achieve cohesive gels at physiological temperatures. However, this could be overcome using chemical crosslinking, such as with 1-ethyl-3-(-3-dimethylaminopropyl) carbodiimide/N-hydroxysuccinimide (EDC/NHS), without significant effects on collagen's biological functions. Besides the biomechanical properties, collagen is also involved in a wide range of biological functions. Collagen can specifically interact with particular receptors at the cell surface, such as integrins, discoidin-domain receptors, and glycoprotein VI [42], thus signaling cell adhesion, differentiation and growth, as well as cell survival. Moreover, heavy metals were undetectable and therefore below the regulated limits. Marine-origin collagen may be used as a safe source of adhesion points and molecular modulators that can be used in combination with biopolymers for the development of biomaterials with promising biomedical application.

Author Contributions: Conceptualization and methodology, A.M.C., A.P.M., T.H.S. and R.L.R.; Technical execution, A.M.C.; Discussion of results, A.M.C., A.P.M. and T.H.S.; Writing-Original Draft Preparation, A.M.C.; Writing—Review & Editing, A.P.M. and T.H.S.; Supervision, R.L.R.; Funding Acquisition, R.L.R.

Funding: This research was funded by European Research Council grant agreement ERC-2012-ADG 20120216-321266 for the project ComplexiTE.

Acknowledgments: The authors acknowledge to Leonardo Aires (Frigoríficos da Erminda, Portugal) for the kind contribution of skin from codfish and Ana M. Gomes and Fátima Silva (Catholic University of Portugal–Porto) for the heavy metals analyses.

Conflicts of Interest: The authors declare no conflict of interest.

References

1. Ricard-Blum, S. The collagen family. *Cold Spring Harb. Perspect. Biol.* **2011**, *3*, a004978. [CrossRef] [PubMed]
2. Lee, C.H.; Singla, A.; Lee, Y. Biomedical applications of collagen. *Int. J. Pharm.* **2001**, *221*, 1–22. [CrossRef]
3. Cen, L.; Liu, W.; Cui, L.; Zhang, W.; Cao, Y. Collagen tissue engineering: Development of novel biomaterials and applications. *Pediat. Res.* **2008**, *63*, 492–496. [CrossRef] [PubMed]
4. Maeda, M.; Tani, S.; Sano, A.; Fuijoka, K. Microstructure and release characteristics of the minipellet, a collagen based drug delivery system for controlled release of protein drugs. *J. Control. Release* **1999**, *62*, 313–324. [CrossRef]
5. Albu, M.G.; Titorencu, I.; Ghica, M. Collagen-based drug delivery systems for tissue engineering. In *Biomaterials Applications for Nanomedicine*; Pignatello, R., Ed.; InTechOpen: London, UK, 2011.
6. Pati, F.; Datta, P.; Adhikari, B.; Dhara, S.; Ghosh, K.; Mohapatra, P.K.D. Collagen scaffolds derived from fresh water fish origin and their biocompatibility. *J. Biomed. Mater. Res. Part A* **2012**, *100*, 1068–1079. [CrossRef]
7. Hayashi, Y.; Yanagiguchi, K.; Yamada, S. Fish collagen as a scaffold. *Austin J. Biomed. Eng.* **2015**, *2*, 1029.
8. Barros, A.A.; Aroso, I.M.; Silva, T.H.; Mano, J.F.; Duarte, A.R.C.; Reis, R.L. Water and carbon dioxide: Green solvents for the extraction of collagen/gelatin from marine sponges. *ACS Sustain. Chem. Eng.* **2015**, *3*, 254–260. [CrossRef]
9. Silva, T.H.; Moreira-Silva, J.; Marques, A.L.P.; Domingues, A.; Bayon, Y.; Reis, R.L. Marine origin collagens and its potential applications. *Mar. Drugs* **2014**, *12*, 5881–5901. [CrossRef]
10. Silva, J.C.; Barros, A.A.; Aroso, I.M.; Fassini, D.; Silva, T.H.; Reis, R.L.; Duarte, A.R.C. Extraction of collagen/gelatin from the marine demosponge chondrosia reniformis (nardo, 1847) using water acidified with carbon dioxide—Process optimization. *Ind. Eng. Chem. Res.* **2016**, *55*, 6922–6930. [CrossRef]
11. Pozzolini, M.; Bruzzone, F.; Berilli, V.; Mussino, F.; Cerrano, C.; Benatti, U.; Giovine, M. Molecular characterization of a nonfibrillar collagen from the marine sponge chondrosia reniformis nardo 1847 and positive effects of soluble silicates on its expression. *Mar. Biotechnol.* **2012**, *14*, 281–293. [CrossRef]
12. Pozzolini, M.; Scarfi, S.; Gallus, L.; Castellano, M.; Vicini, S.; Cortese, K.; Gagliani, M.C.; Bertolino, M.; Costa, G.; Giovine, M. Production, characterization and biocompatibility evaluation of collagen membranes derived from marine sponge chondrosia reniformis nardo, 1847. *Mar. Drugs* **2018**, *16*, 111. [CrossRef] [PubMed]
13. Pozzolini, M.; Scarfi, S.; Mussino, F.; Ferrando, S.; Gallus, L.; Giovine, M. Molecular cloning, characterization, and expression analysis of a prolyl 4-hydroxylase from the marine sponge chondrosia reniformis. *Mar. Biotechnol.* **2015**, *17*, 393–407. [CrossRef] [PubMed]
14. Tziveleka, L.A.; Ioannou, E.; Tsiourvas, D.; Berillis, P.; Foufa, E.; Roussis, V. Collagen from the marine sponges axinella cannabina and suberites carnosus: Isolation and morphological, biochemical, and biophysical characterization. *Mar. Drugs* **2017**, *15*, 152. [CrossRef] [PubMed]
15. Ehrlich, H.; Wysokowski, M.; Zoltowska-Aksamitowska, S.; Petrenko, I.; Jesionowski, T. Collagens of poriferan origin. *Mar. Drugs* **2018**, *16*, 79. [CrossRef] [PubMed]
16. Cheng, X.C.; Shao, Z.Y.; Li, C.B.; Yu, L.J.; Raja, M.A.; Liu, C.G. Isolation, characterization and evaluation of collagen from jellyfish rhopilema esculentum kishinouye for use in hemostatic applications. *PLoS ONE* **2017**, *12*, e0169731. [CrossRef]
17. Widdowson, J.P.; Picton, A.J.; Vince, V.; Wright, C.J.; Mearns-Spragg, A. In vivo comparison of jellyfish and bovine collagen sponges as prototype medical devices. *J. Biomed. Mater. Res. Part B Appl. Biomater.* **2018**, *106*, 1524–1533. [CrossRef]
18. Sewing, J.; Klinger, M.; Notbohm, H. Jellyfish collagen matrices conserve the chondrogenic phenotype in two- and three-dimensional collagen matrices. *J. Tissue Eng. Regen. Med.* **2017**, *11*, 916–925. [CrossRef]
19. Exposito, J.Y.; Larroux, C.; Cluzel, C.; Valcourt, U.; Lethias, C.; Degnan, B.M. Demosponge and sea anemone fibrillar collagen diversity reveals the early emergence of a/c clades and the maintenance of the modular structure of type v/xi collagens from sponge to human. *J. Biol. Chem.* **2008**, *283*, 28226–28235. [CrossRef]
20. Nowack, H.; Nordwig, A. Sea-anemone collagen—Isolation and characterization of cyanogen-bromide peptides. *Eur. J. Biochem.* **1974**, *45*, 333–342. [CrossRef]
21. Benayahu, D.; Sharabi, M.; Pomeraniec, L.; Awad, L.; Haj-Ali, R.; Benayahu, Y. Unique collagen fibers for biomedical applications. *Mar. Drugs* **2018**, *16*, 102. [CrossRef]

22. Cozza, N.; Bonani, W.; Motta, A.; Migliaresi, C. Evaluation of alternative sources of collagen fractions from loligo vulgaris squid mantle. *Int. J. Biol. Macromol.* **2016**, *87*, 504–513. [CrossRef]

23. Coelho, R.C.G.; Marques, A.L.P.; Oliveira, S.M.; Diogo, G.S.; Pirraco, R.P.; Moreira-Silva, J.; Xavier, J.C.; Reis, R.L.; Silva, T.H.; Mano, J.F. Extraction and characterization of collagen from antarctic and sub-antarctic squid and its potential application in hybrid scaffolds for tissue engineering. *Mater. Sci. Eng. C* **2017**, *78*, 787–795. [CrossRef] [PubMed]

24. Dai, M.L.; Liu, X.; Wang, N.P.; Sun, J. Squid type ii collagen as a novel biomaterial: Isolation, characterization, immunogenicity and relieving effect on degenerative osteoarthritis via inhibiting stat1 signaling in pro-inflammatory macrophages. *Mater. Sci. Eng. C-Mater. Biol. Appl.* **2018**, *89*, 283–294. [CrossRef]

25. Ferrario, C.; Leggio, L.; Leone, R.; Di Benedetto, C.; Guidetti, L.; Cocce, V.; Ascagni, M.; Bonasoro, F.; La Porta, C.A.M.; Carnevali, M.D.C.; et al. Marine-derived collagen biomaterials from echinoderm connective tissues. *Mar. Environ. Res.* **2017**, *128*, 46–57. [CrossRef] [PubMed]

26. Ovaska, M.; Bertalan, Z.; Miksic, A.; Sugni, M.; Di Benedetto, C.; Ferrario, C.; Leggio, L.; Guidetti, L.; Alava, M.J.; La Porta, C.A.M.; et al. Deformation and fracture of echinoderm collagen networks. *J. Mech. Behav. Biomed. Mater.* **2017**, *65*, 42–52. [CrossRef]

27. Wilkie, I.C.; Emson, R.H.; Young, C.M. Smart collagen in sea lilies. *Nature* **1993**, *366*, 519–520. [CrossRef]

28. Bao, Z.X.; Sun, Y.; Rai, K.; Peng, X.Y.; Wang, S.L.; Nian, R.; Xian, M. The promising indicators of the thermal and mechanical properties of collagen from bass and tilapia: Synergistic effects of hydroxyproline and cysteine. *Biomater. Sci.* **2018**, *6*, 3042–3052. [CrossRef] [PubMed]

29. Fassini, D.; Duarte, A.R.C.; Reis, R.L.; Silva, T.H. Bioinspiring chondrosia reniformis (nardo, 1847) collagen-based hydrogel: A new extraction method to obtain a sticky and self-healing collagenous material. *Mar. Drugs* **2017**, *15*, 380. [CrossRef]

30. Bernhardt, A.; Paul, B.; Gelinsky, M. Biphasic scaffolds from marine collagens for regeneration of osteochondral defects. *Mar. Drugs* **2018**, *16*, 91. [CrossRef] [PubMed]

31. Fernandes-Silva, S.; Moreira-Silva, J.; Silva, T.H.; Perez-Martin, R.I.; Sotelo, C.G.; Mano, J.F.; Duarte, A.R.C.; Reis, R.L. Porous hydrogels from shark skin collagen crosslinked under dense carbon dioxide atmosphere. *Macromol. Biosci.* **2013**, *13*, 1621–1631. [CrossRef]

32. Gómez-Guillén, M.C.; Turnay, J.; Fernández-Díaz, M.D.; Ulmo, N.; Lizarbe, M.A.; Montero, P. Structural and physical properties of gelatin extracted from different marine species: A comparative study. *Food Hydrocoll.* **2002**, *16*, 25–34. [CrossRef]

33. Wang, Y.; Regenstein, J.M. Effect of edta, hcl, and citric acid on ca salt removal from asian (silver) carp scales prior to gelatin extraction. *J. Food Sci.* **2009**, *74*, C426–C431. [CrossRef] [PubMed]

34. Kittiphattanabawon, P.; Benjakul, S.; Visessanguan, W.; Shahidi, F. Isolation and characterization of collagen from the cartilages of brownbanded bamboo shark (chiloscyllium punctatum) and blacktip shark (carcharhinus limbatus). *LWT Food Sci. Technol.* **2010**, *43*, 792–800. [CrossRef]

35. Matmaroh, K.; Benjakul, S.; Prodpran, T.; Encarnacion, A.B.; Kishimura, H. Characteristics of acid soluble collagen and pepsin soluble collagen from scale of spotted golden goatfish (parupeneus heptacanthus). *Food Chem.* **2011**, *129*, 1179–1186. [CrossRef] [PubMed]

36. Fengxiang, Z.; Anning, W.; Zhihua, L.; Shengwen, H.; Lijun, S. Preparation and characterisation of collagen from freshwater fish scales. *Food Nutr. Sci.* **2011**, *2011*.

37. Jeong, H.S.; Venkatesan, J.; Kim, S.K. Isolation and characterization of collagen from marine fish (thunnus obesus). *Biotechnol. Bioprocess Eng.* **2013**, *18*, 1185–1191. [CrossRef]

38. Muralidharan, N.; Jeya Shakila, R.; Sukumar, D.; Jeyasekaran, G. Skin, bone and muscle collagen extraction from the trash fish, leather jacket (odonus niger) and their characterization. *J. Food Sci. Technol.* **2013**, *50*, 1106–1113. [CrossRef]

39. El-Rashidy, A.A.; Gad, A.; Abu-Hussein, A.E.H.G.; Habib, S.I.; Badr, N.A.; Hashem, A.A. Chemical and biological evaluation of egyptian nile tilapia (oreochromis niloticas) fish scale collagen. *Int. J. Biol. Macromol.* **2015**, *79*, 618–626. [CrossRef]

40. Kozlowska, J.; Sionkowska, A.; Skopinska-Wisniewska, J.; Piechowicz, K. Northern pike (esox lucius) collagen: Extraction, characterization and potential application. *Int. J. Biol. Macromol.* **2015**, *81*, 220–227. [CrossRef]

41. Huang, C.Y.; Kuo, J.M.; Wu, S.J.; Tsai, H.T. Isolation and characterization of fish scale collagen from tilapia (oreochromis sp.) by a novel extrusion–hydro-extraction process. *Food Chem.* **2016**, *190*, 997–1006. [CrossRef]

42. International, A. (Ed.) Standard guide for characterization of type i collagen as starting material for surgical implants and substrates for tissue engineered medical products (temps). In *F 2212-08*; ASTM Inetrnational: West Conshohocken, MD, USA, 2008.

43. Gelse, K.; Pöschl, E.; Aigner, T. Collagens—Structure, function, and biosynthesis. *Adv. Drug Deliv. Rev.* **2003**, *55*, 1531–1546. [CrossRef] [PubMed]

44. Skierka, E.; Sadowska, M. The influence of different acids and pepsin on the extractability of collagen from the skin of baltic cod (gadus morhua). *Food Chem.* **2007**, *105*, 1302–1306. [CrossRef]

45. Żelechowska, E.; Sadowska, M.; Turk, M. Isolation and some properties of collagen from the backbone of baltic cod (gadus morhua). *Food Hydrocoll.* **2010**, *24*, 325–329. [CrossRef]

46. Kimura, S.; Ohno, Y. Fish type i collagen: Tissue-specific existence of two molecular forms, (α1)2α2 and α1α2α3, in alaska pollack. *Comp. Biochem. Physiol. Part B Comp. Biochem.* **1987**, *88*, 409–413. [CrossRef]

47. Matsui, R.; Ishida, M.; Kimura, S. Characterization of an α3 chain from the skin type i collagen of chum salmon (oncoorhynchus keta). *Comp. Biochem. Physiol. Part B Comp. Biochem.* **1991**, *99*, 171–174. [CrossRef]

48. Doyle, B.B.; Bendit, E.G.; Blout, E.R. Infrared spectroscopy of collagen and collagen-like polypeptides. *Biopolymers* **1975**, *14*, 937–957. [CrossRef] [PubMed]

49. Singh, P.; Benjakul, S.; Maqsood, S.; Kishimura, H. Isolation and characterisation of collagen extracted from the skin of striped catfish (pangasianodon hypophthalmus). *Food Chem.* **2011**, *124*, 97–105. [CrossRef]

50. Kiew, P.L.; Don, M.M. The influence of acetic acid concentration on the extractability of collagen from the skin of hybrid clarias sp. And its physicochemical properties: A preliminary study. *Focus. Mod. Food Ind.* **2013**, *2*, 123–128.

51. Plepis, A.M.G.; Goissis, G.; Gupta, D.K.D. Dielectric and pyroelectric characterization of anionic and native collagen. *Polym. Eng. Sci.* **1996**, *36*, 2932–2938. [CrossRef]

52. Veeruraj, A.; Arumugam, M.; Ajithkumar, T.; Balasubramanian, T. Isolation and characterization of drug delivering potential of type-i collagen from eel fish evenchelys macrura. *J. Mater. Sci. Mater. Med.* **2012**, *23*, 1729–1738. [CrossRef]

53. Friess, W.; Lee, G. Basic thermoanalytical studies of insoluble collagen matrices. *Biomaterials* **1996**, *17*, 2289–2294. [CrossRef]

54. Pati, F.; Adhikari, B.; Dhara, S. Isolation and characterization of fish scale collagen of higher thermal stability. *Bioresour. Technol.* **2010**, *101*, 3737–3742. [CrossRef] [PubMed]

55. Duan, R.; Zhang, J.; Du, X.; Yao, X.; Konno, K. Properties of collagen from skin, scale and bone of carp (cyprinus carpio). *Food Chem.* **2009**, *112*, 702–706. [CrossRef]

56. Benjakul, S.; Nalinanon, S.; Shahidi, F. Fish collagen. In *Food Biochemistry and Food Processing*; Simpson, B.K., Ed.; John Wiley & Sons, Inc.: Sussex, UK, 2012; pp. 365–387.

57. Piez, K.A.; Gross, J. Amino acid composition of some fish collagens—Relation between composition and structure. *J. Biol. Chem.* **1960**, *235*, 995–998. [PubMed]

58. Ramachan, G.N.; Bansal, M.; Bhatnaga, R.S. Hypothesis on role of hydroxyproline in stabilizing collagen structure. *Biochim. Biophys. Acta* **1973**, *322*, 166–171. [CrossRef]

59. Persikov, A.V.; Ramshaw, J.A.M.; Kirkpatrick, A.; Brodsky, B. Electrostatic interactions involving lysine make major contributions to collagen triple-helix stability. *Biochemistry* **2005**, *44*, 1414–1422. [CrossRef] [PubMed]

60. Fallas, J.A.; Gauba, V.; Hartgerink, J.D. Solution structure of an abc collagen heterotrimer reveals a single-register helix stabilized by electrostatic interactions. *J. Biol. Chem.* **2009**, *284*, 26851–26859. [CrossRef] [PubMed]

61. Piez, K.A. *Extracellular Matrix Biochemistry*; Elsevier: Amsterdam, The Netherlands, 1984.

62. Gauza-Włodarczyk, M.; Kubisz, L.; Włodarczyk, D. Amino acid composition in determination of collagen origin and assessment of physical factors effects. *Int. J. Biol. Macromol.* **2017**, *104*, 987. [CrossRef]

63. Usp chapter <232>, elemental impurities—Limits. In *Second Supplement to USP 35—NF 30*; United States Pharmacopeia: Rockville, MD, USA, 2012.

64. International Organization for Standardization. *Dentistry: Water-Basedcements—Part 1: Powder/Liquid Acid-Base Cements*; 9917-1; ISO: Geneva, Switzerland, 2007; pp. 1–23.

65. U.S. Food and Drug Administration. *Draft Guidance for Industry: Lead in Cosmetic Lip Products and Externally Applied Cosmetics: Recommended Maximum Level*; U.S. Food and Drug Administration: Silver Spring, MD, USA, 2016.

66. Wang, P.; Henning, S.M.; Heber, D. Limitations of mtt and mts-based assays for measurement of antiproliferative activity of green tea polyphenols. *PLoS ONE* **2010**, *5*, e10202. [CrossRef]
67. Laemmli, U.K. Cleavage of structural proteins during the assembly of the head of bacteriophage t4. *Nature* **1970**, *227*, 680–685. [CrossRef]
68. Lepock, J.R. Measurement of protein stability and protein denaturation in cells using differential scanning calorimetry. *Methods* **2005**, *35*, 117–125. [CrossRef] [PubMed]

 *marine drugs*

Article

Physicochemical and Biocompatibility Properties of Type I Collagen from the Skin of Nile Tilapia (*Oreochromis niloticus*) for Biomedical Applications

Wen-Kui Song [1], Dan Liu [2], Lei-Lei Sun [3], Ba-Fang Li [1,*] and Hu Hou [1,*]

[1]   College of Food Science and Engineering, Ocean University of China, No.5, Yu Shan Road, Qingdao 266003, China; sprdice@outlook.com
[2]   College of Chemistry and Chemical Engineering, Ocean University of China, 238 Songling Road, Qingdao 266003, China; liudan090@163.com
[3]   College of Life Science, Yantai University, Yantai 264005, China; leilei.198966@163.com
*   Correspondence: bfli@ouc.edu.cn (B.-F.L.); houhu@ouc.edu.cn (H.H.); Fax: +86-532-82031936 (H.H.)

Received: 29 December 2018; Accepted: 20 February 2019; Published: 26 February 2019

Abstract: The aim of this study is to investigate the physicochemical properties, biosafety, and biocompatibility of the collagen extract from the skin of Nile tilapia, and evaluate its use as a potential material for biomedical applications. Two extraction methods were used to obtain acid-soluble collagen (ASC) and pepsin-soluble collagen (PSC) from tilapia skin. Amino acid composition, FTIR, and SDS-PAGE results showed that ASC and PSC were type I collagen. The molecular form of ASC and PSC is $(\alpha_1)_2\alpha_2$. The FTIR spectra of ASC and PSC were similar, and the characteristic peaks corresponding to amide A, amide B, amide I, amide II, and amide III were 3323 cm^{-1}, 2931 cm^{-1}, 1677 cm^{-1}, 1546 cm^{-1}, and 1242 cm^{-1}, respectively. Denaturation temperatures (Td) were 36.1 °C and 34.4 °C, respectively. SEM images showed the loose and porous structure of collagen, indicting its physical foundation for use in applications of biomedical materials. Negative results were obtained in an endotoxin test. Proliferation rates of osteoblastic (MC3T3E1) cells and fibroblast (L929) cells from mouse and human umbilical vein endothelial cells (HUVEC) were increased in the collagen-treated group compared with the controls. Furthermore, the acute systemic toxicity test showed no acute systemic toxicity of the ASC and PSC collagen sponges. These findings indicated that the collagen from Nile tilapia skin is highly biocompatible in nature and could be used as a suitable biomedical material.

Keywords: Nile tilapia collagen; biomedical application; characterization

---

1. Introduction

Collagen is the main structural protein in the extracellular matrix (ECM), constituting approximately 30% of the whole body protein content in animals [1]. More than 29 different types of collagen have been identified and described. In the human body, most of the collagen is type I [2]. The collagen protein contains triple-helix structures that consist of three almost identical polypeptide chains [3]. Type I collagen is present in bone, skin, dentin, cornea, blood vessels, fibrocartilage, and tendon; it has the unique ability to form fibrils that have high tensile strength and important functions [4,5]. In the past decades, it has been widely used in food manufacturing and the cosmetics industry [6].

Recently, the interest in collagen has become widespread among medicine and tissue engineering because of its predominance in the ECM, excellent biocompatibility, low antigenicity and available methods of isolation from a variety of sources [7,8]. Currently, the major sources of collagen are the tendon or skin of bovine and porcine. However, the prevalence of transferring diseases including foot-and-mouth disease (FMD), bovine spongiform encephalopathy (BSE), and transmissible spongiform encephalopathy (TSE), and the religious barriers of Muslims and Jews [9–11], have limited its application [12]. Therefore, it is essential to find a safety source of collagen for human application. Marine collagen has been isolated and characterized from various marine sources, and can generally be categorized according to source: vertebrates or invertebrates. Vertebrates sources include cat fish [13], silvertip shark [14], salmon [15], yellow tuna [16], and marine mammals such as minke whale [17]. Invertebrates source include jellyfish [18–20], squid [21], and sponges [22–25]. Researchers have demonstrated that similar characteristics exist between marine collagen and mammalian [26–28]. However, some differences exist between collagen extracted from marine sources and collagen extracted from mammals. Compared with the mammalian collagen, marine collagen has lower gelling and melting temperatures, but relatively higher viscosities than equivalent bovine forms [29]. Fish collagens show a similar amino acid distribution to mammalian collagen, with decreased amounts of proline and hydroxyproline, and increased serine, threonine, and in some cases, methionine and hydroxylysine [30]. Compared with mammalian collagen, the difference in the amino acid distribution of fish collagen causes labile cross-links and heat sensitivity [31]. In recent years, marine collagen has been widely used in medicine and tissue engineering fields [32], such as cartilage [33], corneal [34], ligament [35], muscle [36], skin [37], tracheal [38], and vascular [39].

The Nile tilapia (*Oreochromis niloticus*) is a worldwide cultured fish that possesses an important position in China's aquaculture and exports industry [26]. Tilapia skin is a main by-product of its processing, which contains approximately 30% collagen [40]. Our previous study revealed that both acid-soluble collagen (ASC) and pepsin-soluble collagen (PSC) extracted from Nile tilapia skin can be used as raw materials in food and cosmetic preparation [41]. Further, we want to explore whether either is suitable for use in biomedical applications. However, the biocompatibilities of pure collagen extracted from Nile tilapia need be addressed, since the biocompatibilities of fish collagen are profoundly influenced by the molecular composition and arrangement, which is thought to be varied by different extraction methods. In this work, we extract acid-soluble (ASC) and pepsin-soluble (PSC) collagen from Nile tilapia skin, and then describe their physical properties, chemical properties, and biocompatibilities, in order to explore the possibility for applications in biomedical fields.

2. Results and Discussion

2.1. Preparation of ASC and PSC

The yields (dry weight basis, 19.07% and 19.61% respectively) of ASC and PSC were in highly agreement with former reports [42,43]. Comparison of the collagens extract from different fish have been reported, including the skin of paper nautilus (55.2%) [44], bigeye snapper (1.59%) [45], deep-sea redfish (10.3%) [46], Japanese sea bass (40.7%) [47], yellow sea bream (40.1%) [48], and grass carp (45.3%) [43]. In contrast, the yield of collagen from tilapia was different from these reported species. The results suggest that some discrepancies might exist among these fish species. In addition, the yield of ASC was a little bit lower than the yield of PSC, this might be due to the disadvantage of the solubility of cross-links formed through the reaction of aldehyde with lysine and hydroxylysine at telopeptide helical sites [49], which also can be explained by the results of Fourier transform infrared (FTIR) analysis. Besides the effects on solubility, the pepsin that is used in the extraction of PSC might bring other changes, such as the stability and biocompatibility of the resultant collagens.

2.2. Amino Acid Composition

Table 1 shows the amino acid composition of ASC and PSC, which is expressed as residues per 1000 total amino acid residues. According to Table 1, glycine is the most important component in ASC and PSC, with 322 and 343 residues, accounting for about one-third of the total amino acid residues. It is slightly higher than common aquatic organisms carp skin (311 residues), cod skin (308 residues), squid skin (269 residues) [50,51], and very similar to land mammals' bovine skin (320 residues), bovine skin (334 residues), and porcine skin (326 residues) [52,53]. Glycine is the most important amino acid in collagen. All members of the collagen family have a tripeptide (Gly-X-Y) repetitive structure, which plays an important role in the formation of the triple-helix structure. The tripeptides (Gly-X-Y) are repeatedly arranged on each chain of collagen, accounting for about 20–30% of all tripeptide structures. The X position of Gly-X-Y is often occupied by proline, which is consistent with the result in Table 1 (115 and 106 proline residues in ASC and PSC).

Table 1. The amino acid composition of acid-soluble collagen (ASC) and pepsin-soluble collagen (PSC) from Nile tilapia skin.

Amino acid	PSC	ASC
Hydroxyproline	70	86
Aspartic acid	37	40
Threonine	17	15
Serine	28	31
Glutamic acid	93	98
Proline	115	106
Glycine	343	322
Alanine	85	87
Valine	25	22
Methionine	10	9
Isoleucine	16	11
Leucine	20	22
Tyrosine	6	9
Phenylalanine	16	10
Lysine	22	32
Histidine	12	10
Arginine	85	90
Total	1000	1000

The existence of the triple-helix structure is the most direct evidence to distinguish collagen from gelatin. [54]. As is known, electrospinning is a method of stretching a polymer solution to fibers that have a diameter of about several hundred nanometers by electrostatic force. Due to its wide applicability, high efficiency, and simplicity, electrospinning is widely used in the field of tissue engineering scaffold materials.

In addition, it is worth noting that the content of hydroxyproline in tilapia skin is ASC (70 residues) and PSC (86 residues). As a characteristic component of collagen, the content of collagen in raw materials can be measured by the ratio of hydroxyproline. No cystine and tryptophan were detected in both the ASC and PSC of tilapia skin collagen, which was consistent with the characteristics of type I collagen.

2.3. FTIR Analysis

The FTIR spectra of ASC and PSC are exhibited in Figure 1. Each peak in the FTIR spectrums corresponds to the vibration of functional groups in the molecule [55]. The secondary structure of collagen is closely related to different types of hydrogen bonds [56]. By analyzing the FTIR spectrum of ASC and PSC, the different effects of the two extraction methods on the secondary structure of collagen were obtained. At room temperature, ASC and PSC mainly exhibited five absorption peaks at 3323 cm^{-1}, 2931 cm^{-1}, 1677 cm^{-1}, 1546 cm^{-1}, and 1242 cm^{-1}, corresponding to amide A, amide B, amide I, amide II, and amide III, respectively.

The wavenumber of the free N–H stretching vibration was located in the range of 3400–3440 cm^{-1}, and the wavenumber of amide A was measured at 3323 cm^{-1}, indicating that the N–H stretching vibration and the hydrogen bonding are combined [51]. The amide A absorption peak of PSC at 3323.20 cm^{-1} is slightly higher than ASC at the wavenumber of 3327.06 cm^{-1}, indicating that more N–H groups in the ASC are hydrogen-bonded, which suggested that the PSC is slightly weaker than

ASC in structural stability. Both ASC and PSC have a weak absorption peak at the amide B band at 2931.67 cm^{-1}, indicating the asymmetric stretching vibration of –CH$_2$ [57]. Studies have shown that the amide I, amide II, and amide III bands are related to the triple-helix structure of collagen [58]. The amide I bands were attributed to C=O stretching vibration, and the amide I absorption bands of ASC and PSC appeared at 1677.99 cm^{-1} and 1654.78 cm^{-1}, respectively. The red shift of C=O stretching vibration may be caused by the use of the pepsin-degraded part of the telopeptides during the preparation process. The telopeptides plays an important role in the triple-helix structure of collagen, which was attributed to the covalent aldol cross-linking and the collagen fiber formation. Excision of the telopeptides does not completely destroy the natural collagen structure [59], but it leads to an incomplete collagen protein structure and increased solubility [60], which explained why the PSC yields are higher than those of the ASC. From the consideration of biomedical materials applications, much attention should be paid to the effects of the telopeptides on the immunogenicity of collagen. It has been reported that the immunogenicity of collagen's telopeptides was considered the most important factor of the collagen-induced immune response [59]. Previous studies have also suggested that the peptides located at the center of the triple helix of pepsin-treated collagen (from skin of bovine) are the major antigenic sites that cause human immune responses [61].

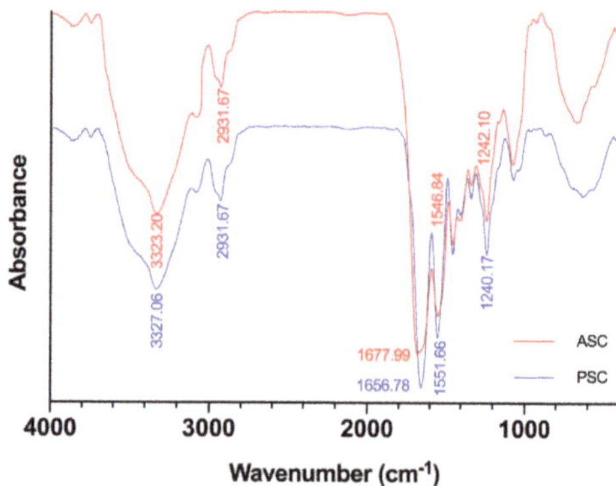

Figure 1. Fourier transform infrared spectroscopy (FTIR) of ASC and PSC from Nile tilapia skin.

Amide II bands produced by N–H bending vibrations and C–N stretching vibrations are usually located in the range of 1550 to 1600 cm^{-1}. Research has shown that the red shift of the amide II peak is related to the hydrogen bond increase of the N–H group [39]. The amide II absorption bands of ASC and PSC were detected at wavenumbers of 1546.84 cm^{-1} and 1551.66 cm^{-1}, respectively. The result indicates that there are more hydrogen bonds between the peptide chains in ASC than PSC.

The amide III bands represent the combination of the C–N stretching vibration and the N–H bending vibration [56]. The amide III absorption bands of ASC and PSC appeared at 1242.10 cm^{-1} and 1240.17 cm^{-1} respectively, which is consistent with previous studies [41].

2.4. Thermal Denaturation Temperatures of ASC and PSC

The tripeptide chains are bound by non-covalent bonds such as hydrogen bonds, which is the basis of the stability of collagen. When the collagen molecules absorbed enough heat from the outside, these non-covalent bonds were destroyed, causing the triple-helix structure to become a random coil structure and destroying the biological properties of collagen. The thermal stability of ASC and PSC were studied by viscosity measurement [62]. According to Figure 2, ASC and PSC have similar curves, and their denaturation temperatures (Td) were 36.1 °C and 34.4 °C, respectively. The Td values can be regarded as the temperature at which the triple-helix structure of collagen is deformed into a random coil structure. The Td values of ASC and PSC from tilapia skin are similar to the collagen extracted from fish living in warm tropical climates such as salmon (29.3 °C) and bigeye snapper

(30.4 °C) [49], and higher than the cold-water fish, such as Baltic herring (15.0 °C) and Argentine salmon (10.0 °C) [63], but lower than terrestrial animals such as bovine (39.7 °C) or porcine (37 °C) [64]. The difference in amino acid composition was the primary cause of the different thermostability of collagen. The loss of the PSC telopeptides has a certain influence on the stability of the triple-helix structure, resulting in lower thermal stability than ASC, which is consistent with the results of amino acid composition analysis and FTIR analysis. Although the thermal stability of tilapia skin collagen is lower than that of terrestrial organisms, it is higher than that of common aquatic organisms, which is an advantage for its application in the field of biomedical materials.

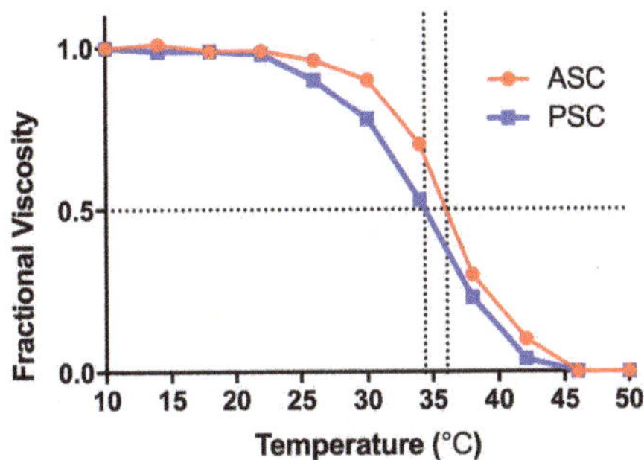

Figure 2. Thermal denaturation curves of ASC and PSC from Nile tilapia skin. The denaturation temperature was determined as the mid-point temperature where viscosity changes reach 0.5.

2.5. SDS-PAGE

SDS-PAGE results present the subunit composition and type of collagen instinctively. As shown in Figure 3, ASC and PSC have very similar protein bands, which have been identified as trimers (γ chains), dimers (β chains), and two alpha chains (α_1 and α_2). Up to now, it has been reported that there are two different collagen trimers in tilapia skin, $(\alpha_1)_2\alpha_2$ and $\alpha_1\alpha_2\alpha_3$, and that the two chains of α_1 and α_2 have the same molecular weight, which cannot be distinguished by electrophoresis [65]. According to Sun [41], the structure of Nile tilapia collagen is $(\alpha_1)_2\alpha_2$ type. The band intensity of α_1 is twice that of α_2, indicating that both ASC and PSC are type I collagen. In addition, the electrophoresis bands are clear, and have no low molecular weight bands, showing that the molecular structure of collagen was well preserved during the extraction process. Although some telopeptides were lost after the treatment of PSC, the banding pattern of PSC was similar to ASC, indicating that the PSC extraction process does not affect the integrity of the triple-helix structure.

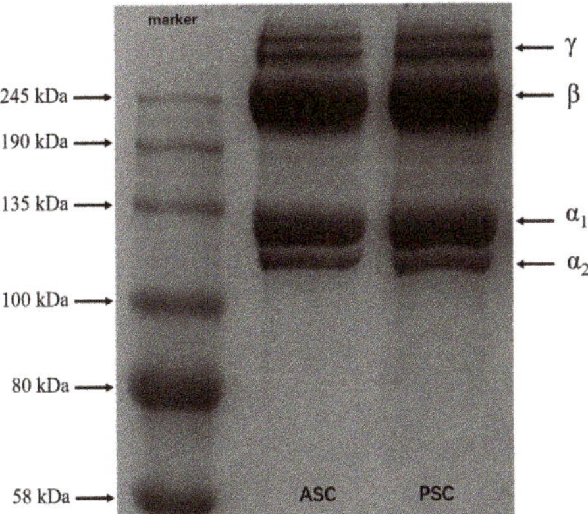

Figure 3. SDS-PAGE patterns of ASC and PSC from Nile tilapia skin.

2.6. Morphology

Figure 4 shows the ultrastructure of the cross-section of the lyophilized collagen sponge. The graphs showed that ASC and PSC exhibit slightly different microscopic morphology. From the 50× magnified image, ASC and PSC display a loose porous network structure, but the pores of ASC show a more uniform and less fiber structure pattern than PSC. As shown in the 400× magnified image, ASC is a dense sheet-like film with uniform alignment, and PSC exhibits irregularly arranged curls.

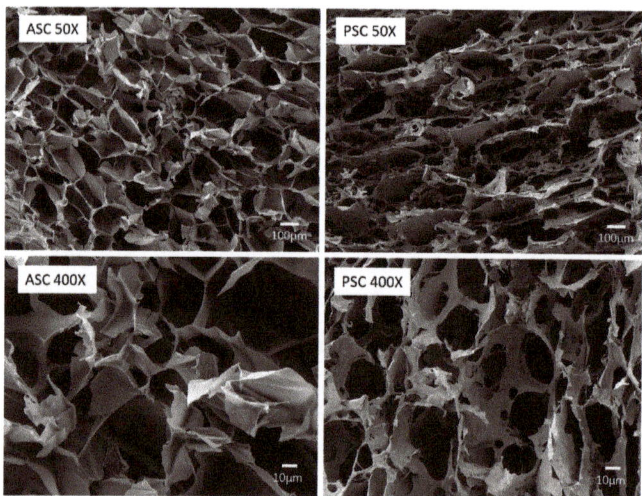

Figure 4. Morphological features of Nile tilapia skin collagen using SEM.

The microstructures determine the physicochemical properties and biofunctionality of the materials; they have a significant value to the application of collagen in biomedical materials. The SEM results of ASC present typical characteristics of aquatic collagen, such as miiuy croaker [56] and Acipenser schrenckii [66], while the microstructures of PSC exhibit a fibrous structure similar to that of terrestrial collagen, such as bovine [67]. This difference may be caused by the structure of PSC being changed under the influence of pepsin, forming more collagen fibers in a non-crosslinked state, which is consistent with the results of FTIR. In the field of biomedical materials, greater porosity facilitates the migration of cells into the interior of the scaffolds, which has certain advantages for promoting wound healing. The lower degree of cross-linking is beneficial to the dissolution and re-processing of collagen, and is suitable for many processes such as electrospinning. Therefore, the microstructures of ASC and PSC indicate that they are appropriate for use in different biomedical material fields.

2.7. Endotoxin Test

Endotoxin, which is also as known as lipopolysaccharide, is found in the outer membrane of Gram-negative bacteria. When endotoxin invades the body, it can cause shock, fever, a fall in blood pressure, and death [68,69]. Endotoxin must be eliminated as much as possible from biomedically used materials. We used an endotoxin denial test to investigate the biosafety of ASC and PSC. The results show that negative results were obtained in endotoxin (below 0.01 EU/mL) in leach liquor of ASC and PSC.

2.8. Cell Proliferation

In order to fully reflect the application potential of ASC and PSC in biomedical materials, MC3T3E1, L929, and human umbilical vein endothelial cells (HUVEC) cells were selected to test its in vitro cytotoxicity. Porcine collagen (PC) was used as a comparison group. According to Figure 5, the addition of collagen significantly promoted cell proliferation compared to the control group. MC3T3E1 and L929 cells had higher proliferation rates in the ASC treatment groups. However, HUVEC had a higher proliferation rate after PSC treatment. Besides, the ASC and PSC treatment groups had higher proliferation rates than the PC group in all three kinds of cells. It has been reported that Nile tilapia collagen contains an antibacterial tilapia piscidin (tp4) that has the ability to stimulate cell proliferation and activate epidermal growth factor (EGF), transforming growth factor (TGF), and vascular endothelial growth factor (VEGF) [70]. In the study of blue shark skin collagen, PSC had a higher proliferation rate for differentiated mouse bone marrow-mesenchymal stem (dMBMS) cells than ASC, while ASC and PSC had no significant difference regarding the proliferation rate of MC3T3E1 cells [71]. The results indicate that both ASC and PSC have potential application in fields requiring wound dressing. The osteogenic properties of ASC make it suitable for applications in the field of bone and cartilage repair, while PSC could be used in areas such as the promotion of angiogenesis or artificial blood vessels.

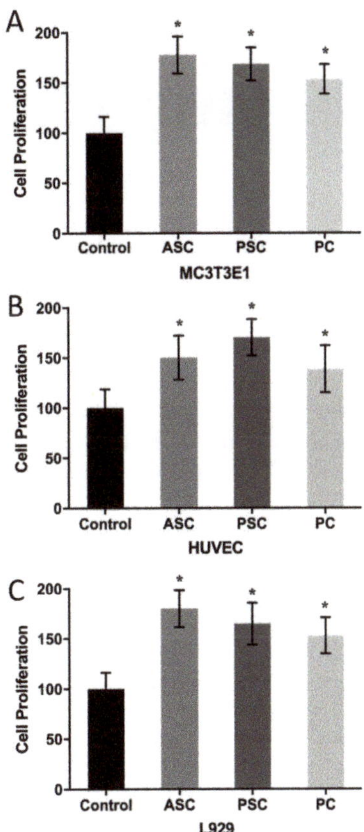

Figure 5. The effect of ASC, PSC, and porcine collagen (PC) on cell proliferation. (**A**): MC3T3E1, (**B**): L929, (**C**): HUVEC. Values with * show significant differences ($p < 0.05$) between groups, as determined by one-way ANOVA.

2.9. Biocompatibility Evaluation

The acute systemic toxicity assay was usually selected to measure the adverse effect of biomedical materials that result either from a single exposure or from multiple exposures in a short period of time. It is an important indicator of biosafety assessment [72]. To evaluate the biosafety of aquatic collagen, porcine collagen was designated as a comparison group. As shown in Figure 6, no significant difference in body weight was observed between the experimental groups and the control group at four hours, 24 h, 48 h, and 72 h after the intraperitoneal injection of the collagen leach liquor. Moreover, the weight of the control group and the experimental groups increased significantly over time, and no dead samples appeared. Furthermore, there were no significant differences between the ASC, PSC, and PC groups. The results showed that ASC and PSC collagen sponges were produced by our process without acute systemic toxicity, and there was not a significant difference from the acute toxicity from commercially available porcine collagen products.

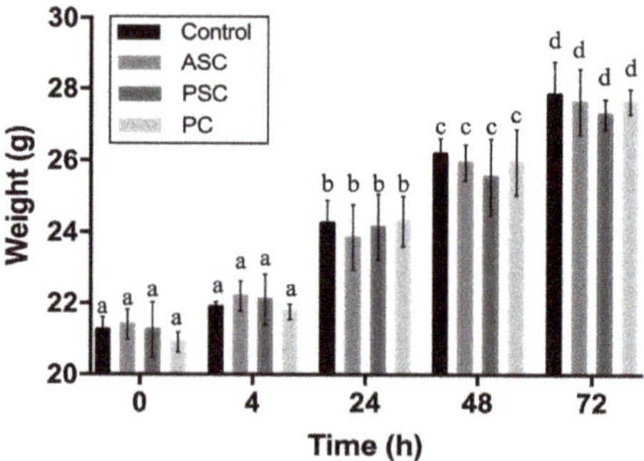

Figure 6. Weight changes of mice after intraperitoneal injection. The different letters in the same group (same type of the bar) represent significant difference ($p < 0.05$).

3. Materials and Methods

3.1. Raw Materials

Nile tilapia skins were procured from Zhenhua Aquatic Product Company (Guangzhou, Guangdong province of China). Frozen skins were thawed with running water and removed from the residuals manually. They were chopped into small pieces and stored at $-20\ ^\circ$C. Porcine collagen (PC) was obtained from Kele Biotech (Chengdu, China). All of the other reagents used were of analytical grade.

3.2. Preparation of ASC

All the procedures were carried out at 4 $^\circ$C to minimize collagen denaturation. An acellular environment was used in the extraction process to reduce the exogenous pyrogen. The pieces of Nile tilapia skins were soaked with 0.1 M of NaOH for 24 h with continuous stirring to remove the non-collagenous proteins. Washing the fish skins repeatedly with cooled deionized water ensured that they were neutralized. Adding 0.5 M of acetic acid in a ratio of 1:50 (w/v) started extraction for two days. Then, they were centrifuged at 10,000 rpm for 30 min at 4 $^\circ$C. NaCl was added to the collected supernatants until a concentration of 0.9 M was reached to salt out the collagen. The precipitated collagen was separated by centrifugation at 10,000 rpm for 30 min at 4 $^\circ$C and dissolved in 0.5 M of acetic acid. Then, the solution was dialyzed for 24 h against 0.1 M of acetic acid in a dialysis membrane with a molecular weight cut-off of 50 kDa, and then for 48 h against ultra-pure water; the water was changed every eight hours. The resulting collagen was freeze-dried for three days and sealed in polythene bags until further use.

3.3. Preparation of PSC

The extraction process of PSC was basically identical to the extraction of ASC except for slightly differences. The tilapia skins were extracted by 0.5 M of acetic acid containing 0.1% (w/v) pepsin for 48 h with stirring. Then, the supernatant was dialyzed against 0.02 M of Na_2HPO_4 for 24 h with solution changed every eight hours before being dialyzed against 0.1 M of acetic acid.

3.4. Extraction Yield of ASC and PSC

The calculation of ASC and PSC yield referred to previous reports [73] and the equation was as follows:

$$\text{yield } (\%) = \text{weight of dried collagen (g)} \times 100 / \text{weight of dried skins (g)}$$

3.5. Amino Acid Composition

The ASC and PSC samples were hydrolyzed by dissolving in 6 M of HCl at 110 °C for 24 h. The solution was analyzed with an amino acid analyzer (835-50, Hitachi, Tokyo, Japan).

3.6. Denaturation Temperature (Td)

The denaturation temperature of collagen from tilapia skin was determined by differential scanning calorimetry (DSC) (Netzsch DSC 200PC, Selb, Bavaria, Germany). A collagen sample with a concentration of 5 mg/mL was dissolved in 0.05 M of acetic acid and sealed in an aluminum pan for scanning. Then, the endothermal curve from 10 °C to 50 °C was obtained at a rate of 5 °C/min in a nitrogen atmosphere.

3.7. Fourier Transform Infrared Spectroscopy (FTIR)

The infrared absorption characteristics of collagen were obtained by an FTIR spectrometer (Tensor 27, BRUKER, Bremen, Germany). First, one-mg collagen samples were mixed with potassium bromide (KBr) at a ratio of 1:100 and pressed into pellets with a manual mechanical squeezing device. The spectra were recorded with a wavenumber range from 4000 to 400 cm^{-1} at the resolution of 2 cm^{-1}.

3.8. Sodium Dodecyl Sulphate-Polyacrylamide Gel Electrophoresis (SDS-PAGE)

The protein molecular mass analysis was studied by SDS-PAGE according to Laemmli [74] with 7.5% resolving gel and 5% stacking gel and 120 voltage using a Bio-Rad electrophoresis. The protein bands were stained with Coomassie Blue R250 and destained with 30% (v/v) methanol and 10% (v/v) acetic acid. The molecular mass of the subunits was analyzed by the location of the bands.

3.9. Scanning Electron Microscopy (SEM)

The morphology of ASC and PSC was observed using a scanning electron microscope (JSM-840, JEOL, Tokyo, Japan) operated at 5 kV. The samples were pasted on a blade and sputter-coated with gold at 30 mA for 15 min. The SEM images were obtained at $50\times$ and $400\times$ magnification.

3.10. Endotoxin Denial Test

An endotoxin denial test was undertaken using a chromogenic end-point tachypleus amebocyte lysate (CE TAL) assay kit (Beijing Solarbio Science & Technology Co., Ltd., Beijing, China). Test samples and standards were prepared according to manufacturer's guidelines. Absorbance was measured at 405 nm using a microplate spectrophotometer (SynergyTM 2, BioTek, Winooski, VT, USA). Data were corrected to exclude background readings. A negative result was defined as a level below 0.01 EU/mL [75].

3.11. Cell Culture

MC3T3E1 and L929 cells were obtained from a cell bank (Procell Life Science & Technology Co., Ltd., Wuhan, China) and HUVEC were received from an institute of zoology (Chinese academy of sciences, Beijing, China). MC3T3E1 was grown as monolayer at 37 °C in a 5% CO_2 incubator and cultured in Dulbecco's modified eagle medium (DMEM) containing 10% (v/v) fetal bovine serum (FBS) and 1% (w/v) penicillin-streptomycin. L929 and HUVEC cells were cultured in MEM and DMEM/F12 (1:1), respectively.

3.12. Cytotoxicity In Vitro

The collagen samples were cut into disks sized for a 48-well plate and sterilized by ^{60}Co for 18 h with an accumulative dose of 20 kGy before using. The cells were seeded a concentration of 1×10^4 cells/well and cultured with a specific medium. After seeding for 48 h, 50 µL of thiazolyl blue tetrazolium bromide (MTT) solution was added to the wells and incubated for four hours at 37 °C/5% CO_2. Wells with collagen were used as negative controls of cytotoxicity. We dissolved the formazan salts using DMSO with shaking for 10 min and replaced the solution with blank wells. The absorbance was measured using a 96-well microplate reader at 490 nm [76].

3.13. Acute Systemic Toxicity Assay

All animal procedures were approved by the ethical committee of animal research in the Ocean University of China, and complied with the requirements of the National Act on the use of experimental animals (China). The acute systemic toxicity of tilapia collagen was investigated according to ISO 10993-11:2009. The collagen samples were cut into 1×1 cm sizes with 0.5-mm thickness and sterilized by ^{60}Co before using. The leach liquor was prepared at a ratio of three cm^2/mL collagen into saline between the total surface area of the materials for 72 h at 37 °C. Then, 20 Kunming (KM) mice (Licensed ID: SCXK 2014-0007) were divided into four groups randomly. Three experimental groups were intraperitoneally injected the leach liquor of ASC, PSC, and PC with a dose of 50 mL/kg respectively, and the blank control group was injected saline with a dose of 50 mL. The weight of mouse was weighed and recorded immediately after injection and at four hours, 24 h, 48 h, and 72 h after injection. The criteria that were used to assess acute systemic toxicity are shown in Table 2.

Table 2. The criteria used to assess acute systemic toxicity.

Acute Systemic Toxicity	Conditions after Treatment with the Test Sample
Negative (−)	None of the five animals showed a significantly greater biological reactivity.
Positive (+)	Two or more of the five animals died.
	Two or more of the five animals showed behavior such as convulsions or prostration.
	Three or more of the five animals showed a body weight loss greater than 10%.

3.14. Statistical Analysis

All of the experiments were replicated in triplicate and the values were expressed as means ± standard deviation (SD). The analyses of multiple groups by one-way ANOVA were using Macintosh GraphPad Prism Version 7, and we considered p-values of less than 0.05 to measure the significant differences.

4. Conclusions

This study showed that ASC and PSC extracted from Nile tilapia skin have typical type I collagen characteristics. No significant differences were found in the amino acid composition and physicochemical properties of ASC and PSC. Both ASC and PSC have a complete triple-helix structure. The thermal stability of PSC was slightly lower than that of ASC, similar to mammals, and higher than cold-water fish. ASC has the effect of promoting osteogenesis and fibroblastation, while PSC is beneficial to the formation of vascular endothelial cells by comparison, and neither ASC nor PSC cause acute systemic toxicity. In summary, collagen extracted from Nile tilapia skin is a biocompatible type I collagen with potential as a biomedical material.

Author Contributions: Conceptualization, W.-K.S.; validation, L.-L.S.; formal analysis, W.-K.S.; resources, D.L. and H.H.; data curation, W.-K.S.; writing—original draft preparation, W.-K.S.; writing—review and editing, H.H.; supervision, B.-F.L.; project administration, H.H.; funding acquisition, B.-F.L.

Funding: This work was supported by National Natural Science Foundation of China (Nos. 31772046 and 31471606), Key Research & Development Plan of Shandong Province (Nos. 2017YYSP015, 2016YYSP005 and 2016YYSP017), and National Key R&D Program of China (2018YFC0311200).

Acknowledgments: The authors acknowledge Yuan-Yuan Wang (Marine Biomedical Research Institute of Qingdao) for the kind contribution of cell cultures.

Conflicts of Interest: The authors declare no conflict of interest.

References

1. Sun, L.; Li, B.; Song, W.; Si, L.; Hou, H. Characterization of Pacific cod (Gadus macrocephalus) skin collagen and fabrication of collagen sponge as a good biocompatible biomedical material. *Process Biochem.* **2017**, *63*, 229–235. [CrossRef]

2. Birk, D.E.; Bruckner, P. Collagen Suprastructures. In *Collagen*; Brinckmann, J., Notbohm, H., Müller, P.K., Eds.; Springer: Berlin/Heidelberg, Germany, 2005; Volume 247, pp. 185–205, ISBN 978-3-540-23272-8.

3. Ogawa, M.; Portier, R.J.; Moody, M.W.; Bell, J.; Schexnayder, M.A.; Losso, J.N. Biochemical properties of bone and scale collagens isolated from the subtropical fish black drum (Pogonia cromis) and sheepshead seabream (Archosargus probatocephalus). *Food Chem.* **2004**, *88*, 495–501. [CrossRef]

4. Sell, S.A.; McClure, M.J.; Garg, K.; Wolfe, P.S.; Bowlin, G.L. Electrospinning of collagen/biopolymers for regenerative medicine and cardiovascular tissue engineering. *Adv. Drug Deliv. Rev.* **2009**, *61*, 1007–1019. [CrossRef] [PubMed]

5. Parmar, P.A.; Chow, L.W.; St-Pierre, J.-P.; Horejs, C.-M.; Peng, Y.Y.; Werkmeister, J.A.; Ramshaw, J.A.M.; Stevens, M.M. Collagen-mimetic peptide-modifiable hydrogels for articular cartilage regeneration. *Biomaterials* **2015**, *54*, 213–225. [CrossRef] [PubMed]

6. Gómez-Guillén, M.C.; Giménez, B.; López-Caballero, M.E.; Montero, M.P. Functional and bioactive properties of collagen and gelatin from alternative sources: A review. *Food Hydrocoll.* **2011**, *25*, 1813–1827. [CrossRef]

7. Tatara, A.M.; Kontoyiannis, D.P.; Mikos, A.G. Drug delivery and tissue engineering to promote wound healing in the immunocompromised host: Current challenges and future directions. *Adv. Drug Deliv. Rev.* **2018**, *129*, 319–329. [CrossRef] [PubMed]

8. Kolacna, L.; Bakesova, J.; Varga, F.; Kostakova, E.; Planka, L.; Necas, A.; Lukas, D.; Amler, E.; Pelouch, V. Biochemical and Biophysical Aspects of Collagen Nanostructure in the Extracellular Matrix. *Physiol. Res.* **2007**, *56*, S51. [PubMed]

9. Chuaychan, S.; Benjakul, S.; Kishimura, H. Characteristics of acid- and pepsin-soluble collagens from scale of seabass (Lates calcarifer). *Lwt Food Sci. Technol.* **2015**, *63*, 71–76. [CrossRef]

10. Ahmad, M.; Benjakul, S. Extraction and characterisation of pepsin-solubilised collagen from the skin of unicorn leatherjacket (Aluterus monocerous). *Food Chem.* **2010**, *120*, 817–824. [CrossRef]

11. Hwang, J.-H.; Mizuta, S.; Yokoyama, Y.; Yoshinaka, R. Purification and characterization of molecular species of collagen in the skin of skate (Raja kenojei). *Food Chem.* **2007**, *100*, 921–925. [CrossRef]

12. Jongjareonrak, A.; Benjakul, S.; Visessanguan, W.; Nagai, T.; Tanaka, M. Isolation and characterisation of acid and pepsin-solubilised collagens from the skin of Brownstripe red snapper (Lutjanus vitta). *Food Chem.* **2005**, *93*, 475–484. [CrossRef]

13. Bama, P.; Vijayalakshimi, M.; Jayasimman, R.; Kalaichelvan, P.T.; Deccaraman, M. Extraction of collagen from cat fish (Tachysurus maculatus) by pepsin digestion and preparation and characterization of collagen chitosan sheet. *Int. J. Pharm. Pharm. Sci.* **2010**, *2*, 5.

14. Jeevithan, E.; Bao, B.; Bu, Y.; Zhou, Y.; Zhao, Q.; Wu, W. Type II Collagen and Gelatin from Silvertip Shark (Carcharhinus albimarginatus) Cartilage: Isolation, Purification, Physicochemical and Antioxidant Properties. *Mar. Drugs* **2014**, *12*, 3852–3873. [CrossRef] [PubMed]

15. Alves, A.; Marques, A.; Martins, E.; Silva, T.; Reis, R. Cosmetic Potential of Marine Fish Skin Collagen. *Cosmetics* **2017**, *4*, 39. [CrossRef]

16. Woo, J.-W.; Yu, S.-J.; Cho, S.-M.; Lee, Y.-B.; Kim, S.-B. Extraction optimization and properties of collagen from yellowfin tuna (Thunnus albacares) dorsal skin. *Food Hydrocoll.* **2008**, *22*, 879–887. [CrossRef]

17. Nagai, T.; Suzuki, N.; Nagashima, T. Collagen from common minke whale (Balaenoptera acutorostrata) unesu. *Food Chem.* **2008**, *111*, 296–301. [CrossRef] [PubMed]

18. Cheng, X.; Shao, Z.; Li, C.; Yu, L.; Raja, M.A.; Liu, C. Isolation, Characterization and Evaluation of Collagen from Jellyfish Rhopilema esculentum Kishinouye for Use in Hemostatic Applications. *PLoS ONE* **2017**, *12*, e0169731. [CrossRef] [PubMed]

19. Bermueller, C.; Schwarz, S.; Elsaesser, A.F.; Sewing, J.; Baur, N.; von Bomhard, A.; Scheithauer, M.; Notbohm, H.; Rotter, N. Marine Collagen Scaffolds for Nasal Cartilage Repair: Prevention of Nasal Septal Perforations in a New Orthotopic Rat Model Using Tissue Engineering Techniques. *Tissue Eng. Part A* **2013**, *19*, 2201–2214. [CrossRef] [PubMed]

20. Pustlauk, W.; Paul, B.; Gelinsky, M.; Bernhardt, A. Jellyfish collagen and alginate: Combined marine materials for superior chondrogenesis of hMSC. *Mater. Sci. Eng. C* **2016**, *64*, 190–198. [CrossRef] [PubMed]

21. Wichuda, J.; Sunthorn, C.; Busarakum, P. Comparison of the properties of collagen extracted from dried jellyfish and dried squid. *Afr. J. Biotechnol.* **2016**, *15*, 642–648. [CrossRef]

22. Pozzolini, M.; Bruzzone, F.; Berilli, V.; Mussino, F.; Cerrano, C.; Benatti, U.; Giovine, M. Molecular Characterization of a Nonfibrillar Collagen from the Marine Sponge Chondrosia reniformis Nardo 1847 and Positive Effects of Soluble Silicates on Its Expression. *Mar. Biotechnol.* **2012**, *14*, 281–293. [CrossRef] [PubMed]

23. Heinemann, S.; Ehrlich, H.; Douglas, T.; Heinemann, C.; Worch, H.; Schatton, W.; Hanke, T. Ultrastructural Studies on the Collagen of the Marine Sponge Chondrosia reniformis Nardo. *Biomacromolecules* **2007**, *8*, 3452–3457. [CrossRef] [PubMed]

24. Nicklas, M.; Schatton, W.; Heinemann, S.; Hanke, T.; Kreuter, J. Preparation and characterization of marine sponge collagen nanoparticles and employment for the transdermal delivery of 17β-estradiol-hemihydrate. *Drug Dev. Ind. Pharm.* **2009**, *35*, 1035–1042. [CrossRef] [PubMed]

25. Swatschek, D.; Schatton, W.; Kellermann, J.; Müller, W.E.; Kreuter, J. Marine sponge collagen: Isolation, characterization and effects on the skin parameters surface-pH, moisture and sebum. *Eur. J. Pharm. Biopharm.* **2002**, *53*, 107–113. [CrossRef]

26. Zeng, S.; Zhang, C.; Lin, H.; Yang, P.; Hong, P.; Jiang, Z. Isolation and characterisation of acid-solubilised collagen from the skin of Nile tilapia (Oreochromis niloticus). *Food Chem.* **2009**, *116*, 879–883. [CrossRef]

27. Exposito, J.-Y.; Valcourt, U.; Cluzel, C.; Lethias, C. The Fibrillar Collagen Family. *Int. J. Mol. Sci.* **2010**, *11*, 407–426. [CrossRef] [PubMed]

28. Hoyer, B.; Bernhardt, A.; Lode, A.; Heinemann, S.; Sewing, J.; Klinger, M.; Notbohm, H.; Gelinsky, M. Jellyfish collagen scaffolds for cartilage tissue engineering. *Acta Biomater.* **2014**, *10*, 883–892. [CrossRef] [PubMed]

29. Eastoe, J.E. The amino acid composition of fish collagen and gelatin. *Biochem. J.* **1957**, *65*, 363–368. [CrossRef] [PubMed]

30. Leuenberger, B.H. Investigation of viscosity and gelation properties of different mammalian and fish gelatins. *Food Hydrocoll.* **1991**, *5*, 353–361. [CrossRef]

31. Berillis, P. Marine collagen: Extraction and applications. In *Research Trends in Biochemistry, Molecular Biology and Microbiology*; Madhukar, S., Ed.; SM Group: Dover, DE, USA, 2015; pp. 1–13.

32. Nagai, N.; Yunoki, S.; Suzuki, T.; Sakata, M.; Tajima, K.; Munekata, M. Application of cross-linked salmon atelocollagen to the scaffold of human periodontal ligament cells. *J. Biosci. Bioeng.* **2004**, *97*, 389–394. [CrossRef]

33. Lin, H.-Y.; Tsai, W.-C.; Chang, S.-H. Collagen-PVA aligned nanofiber on collagen sponge as bi-layered scaffold for surface cartilage repair. *J. Biomater. Sci. Polym. Ed.* **2017**, *28*, 664–678. [CrossRef] [PubMed]

34. Kong, B.; Mi, S. Electrospun Scaffolds for Corneal Tissue Engineering: A Review. *Materials* **2016**, *9*, 614. [CrossRef] [PubMed]

35. Thayer, P.S.; Verbridge, S.S.; Dahlgren, L.A.; Kakar, S.; Guelcher, S.A.; Goldstein, A.S. Fiber/collagen composites for ligament tissue engineering: Influence of elastic moduli of sparse aligned fibers on mesenchymal stem cells: Stiffness of Sparse Aligned Fibers on MSC Differentiation. *J. Biomed. Mater. Res. Part A* **2016**, *104*, 1894–1901. [CrossRef] [PubMed]

36. Takeda, N.; Tamura, K.; Mineguchi, R.; Ishikawa, Y.; Haraguchi, Y.; Shimizu, T.; Hara, Y. In situ cross-linked electrospun fiber scaffold of collagen for fabricating cell-dense muscle tissue. *J. Artif. Organs* **2016**, *19*, 141–148. [CrossRef] [PubMed]

37. Bhowmick, S.; Rother, S.; Zimmermann, H.; Lee, P.S.; Moeller, S.; Schnabelrauch, M.; Koul, V.; Jordan, R.; Hintze, V.; Scharnweber, D. Biomimetic electrospun scaffolds from main extracellular matrix components for skin tissue engineering application—The role of chondroitin sulfate and sulfated hyaluronan. *Mater. Sci. Eng. C* **2017**, *79*, 15–22. [CrossRef] [PubMed]

38. Wu, T.; Zheng, H.; Chen, J.; Wang, Y.; Sun, B.; Morsi, Y.; El-Hamshary, H.; Al-Deyab, S.S.; Chen, C.; Mo, X. Application of a bilayer tubular scaffold based on electrospun poly (L-lactide-co-caprolactone) collagen fibers and yarns for tracheal tissue engineering. *J. Mater. Chem. B* **2017**, *5*, 139–150. [CrossRef]

39. Haghjooy Javanmard, S.; Anari, J.; Zargar Kharazi, A.; Vatankhah, E. In vitro hemocompatibility and cytocompatibility of a three-layered vascular scaffold fabricated by sequential electrospinning of PCL, collagen, and PLLA nanofibers. *J. Biomater. Appl.* **2016**, *31*, 438–449. [CrossRef] [PubMed]

40. Xiaoyan, Y.E.; Zeng, S.; Wenguo, Y.U.; Wenlong, W.U.; Zeng, X.; Huang, L. Study on nutrient components and the extracting condition of the skin gelatin of tilapia. *South China Fish. Sci.* **2008**, *4*, 55–60.

41. Sun, L.; Hou, H.; Li, B.; Zhang, Y. Characterization of acid- and pepsin-soluble collagen extracted from the skin of Nile tilapia (*Oreochromis niloticus*). *Int. J. Biol. Macromol.* **2017**, *99*, 8–14. [CrossRef] [PubMed]

42. Li, J.; Wang, M.; Qiao, Y.; Tian, Y.; Liu, J.; Qin, S.; Wu, W. Extraction and characterization of type I collagen from skin of tilapia (*Oreochromis niloticus*) and its potential application in biomedical scaffold material for tissue engineering. *Process Biochem.* **2018**, *74*, 156–163. [CrossRef]

43. Wang, H.; Liang, Y.; Wang, H.; Zhang, H.; Wang, M.; Liu, L. Physical-Chemical Properties of Collagens from Skin, Scale, and Bone of Grass Carp (*Ctenopharyngodon idella*). *J. Aquat. Food Prod. Technol.* **2014**, *23*, 264–277. [CrossRef]

44. Nagai, T.; Suzuki, N. Preparation and partial characterization of collagen from paper nautilus (*Argonauta argo*, Linnaeus) outer skin. *Food Chem.* **2002**, *76*, 149–153. [CrossRef]

45. Kittiphattanabawon, P.; Benjakul, S.; Visessanguan, W.; Nagai, T.; Tanaka, M. Characterisation of acid-soluble collagen from skin and bone of bigeye snapper (*Priacanthus tayenus*). *Food Chem.* **2005**, *89*, 363–372. [CrossRef]

46. Wang, L.; An, X.; Yang, F.; Xin, Z.; Zhao, L.; Hu, Q. Isolation and characterisation of collagens from the skin, scale and bone of deep-sea redfish (*Sebastes mentella*). *Food Chem.* **2008**, *108*, 616–623. [CrossRef] [PubMed]

47. Nagai, T. Isolation of collagen from fish waste material—Skin, bone and fins. *Food Chem.* **2000**, *68*, 277–281. [CrossRef]

48. Nagai, T.; Suzuki, N. Preparation and characterization of several fish bone collagens. *J. Food Biochem.* **2000**, *24*, 427–436. [CrossRef]

49. Knott, L.; Bailey, A.J. Collagen cross-links in mineralizing tissues: A review of their chemistry, function, and clinical relevance. *Bone* **1998**, *22*, 181–187. [CrossRef]

50. Ikoma, T.; Kobayashi, H.; Tanaka, J.; Walsh, D.; Mann, S. Physical properties of type I collagen extracted from fish scales of Pagrus major and Oreochromis niloticas. *Int. J. Biol. Macromol.* **2003**, *32*, 199–204. [CrossRef]

51. Li, Z.-R.; Wang, B.; Chi, C.; Zhang, Q.-H.; Gong, Y.; Tang, J.-J.; Luo, H.; Ding, G. Isolation and characterization of acid soluble collagens and pepsin soluble collagens from the skin and bone of Spanish mackerel (*Scomberomorous niphonius*). *Food Hydrocoll.* **2013**, *31*, 103–113. [CrossRef]

52. Duan, R.; Zhang, J.; Du, X.; Yao, X.; Konno, K. Properties of collagen from skin, scale and bone of carp (*Cyprinus carpio*). *Food Chem.* **2009**, *112*, 702–706. [CrossRef]

53. Cui, F.; Xue, C.; Li, Z.; Zhang, Y.; Dong, P.; Fu, X.; Gao, X. Characterization and subunit composition of collagen from the body wall of sea cucumber Stichopus japonicus. *Food Chem.* **2007**, *100*, 1120–1125. [CrossRef]

54. Telemeco, T.A.; Ayres, C.; Bowlin, G.L.; Wnek, G.E.; Boland, E.D.; Cohen, N.; Baumgarten, C.M.; Mathews, J.; Simpson, D.G. Regulation of cellular infiltration into tissue engineering scaffolds composed of submicron diameter fibrils produced by electrospinning. *Acta Biomater.* **2005**, *1*, 377–385. [CrossRef] [PubMed]

55. Noreen, R.; Moenner, M.; Hwu, Y.; Petibois, C. FTIR spectro-imaging of collagens for characterization and grading of gliomas. *Biotechnol. Adv.* **2012**, *30*, 1432–1446. [CrossRef] [PubMed]

56. Li, L.-Y.; Zhao, Y.-Q.; He, Y.; Chi, C.-F.; Wang, B. Physicochemical and Antioxidant Properties of Acid- and Pepsin-Soluble Collagens from the Scales of Miiuy Croaker (*Miichthys Miiuy*). *Mar. Drugs* **2018**, *16*, 394. [CrossRef] [PubMed]

57. Muyonga, J.; Cole, C.G.; Duodu, K. Fourier transform infrared (FTIR) spectroscopic study of acid soluble collagen and gelatin from skins and bones of young and adult Nile perch (*Lates niloticus*). *Food Chem.* **2004**, *86*, 325–332. [CrossRef]

58. Heu, M.S.; Lee, J.H.; Kim, H.J.; Jee, S.J.; Lee, J.S.; Jeon, Y.-J.; Shahidi, F.; Kim, J.-S. Characterization of acid- and pepsin-soluble collagens from flatfish skin. *Food Sci. Biotechnol.* **2010**, *19*, 27–33. [CrossRef]

59. Holmes, R.; Kirk, S.; Tronci, G.; Yang, X.; Wood, D. Influence of telopeptides on the structural and physical properties of polymeric and monomeric acid-soluble type I collagen. *Mater. Sci. Eng. C* **2017**, *77*, 823–827. [CrossRef] [PubMed]

60. Beil, W.; Timpl, R.; Furthmayr, H. Conformation dependence of antigenic determinants on the collagen molecule. *Immunology* **1973**, *24*, 13–24. [PubMed]

61. Ellingsworth, L.R.; Delustro, F.; Brennan, J.E.; Sawamura, S.; Mcpherson, J. The human immune response to reconstituted bovine collagen. *J. Immunol.* **1986**, *136*, 877–882. [PubMed]

62. Wu, Q.-Q.; Li, T.; Wang, B.; Ding, G.-F. Preparation and characterization of acid and pepsin-soluble collagens from scales of croceine and redlip croakers. *Food Sci. Biotechnol.* **2015**, *24*, 2003–2010. [CrossRef]

63. Zhang, Y.; Liu, W.; Li, G.; Shi, B.; Miao, Y.; Wu, X. Isolation and partial characterization of pepsin-soluble collagen from the skin of grass carp (*Ctenopharyngodon idella*). *Food Chem.* **2007**, *103*, 906–912. [CrossRef]

64. Veeruraj, A.; Arumugam, M.; Balasubramanian, T. Isolation and characterization of thermostable collagen from the marine eel-fish (*Evenchelys macrura*). *Process Biochem.* **2013**, *48*, 1592–1602. [CrossRef]

65. Kimura, S.; Ohno, Y.; Miyauchi, Y.; Uchida, N. Fish skin type I collagen: Wide distribution of an α3 subunit in teleosts. *Comp. Biochem. Physiol. Part B Comp. Biochem.* **1987**, *88*, 27–34. [CrossRef]

66. Wang, L.; Liang, Q.; Chen, T.; Wang, Z.; Xu, J.; Ma, H. Characterization of collagen from the skin of Amur sturgeon (*Acipenser schrenckii*). *Food Hydrocoll.* **2014**, *38*, 104–109. [CrossRef]

67. León-Mancilla, B.H.; Araiza-Téllez, M.A.; Flores-Flores, J.O.; Piña-Barba, M.C. Physico-chemical characterization of collagen scaffolds for tissue engineering. *J. Appl. Res. Technol.* **2016**, *14*, 77–85. [CrossRef]

68. Brock-Utne, J.G.; Gaffin, S.L. Endotoxins and Anti-endotoxins (Their Relevance to the Anaesthetist and the Intensive Care Specialist). *Anaesth. Intensive Care* **1989**, *17*, 49–55. [CrossRef] [PubMed]

69. Ziegler, E.J.; McCutchan, J.A.; Fierer, J.; Glauser, M.P.; Sadoff, J.C.; Douglas, H.; Braude, A.I. Treatment of Gram-Negative Bacteremia and Shock with Human Antiserum to a Mutant Escherichia coli. *New Engl. J. Med.* **1982**, *307*, 1225–1230. [CrossRef] [PubMed]

70. Huang, H.-N.; Chan, Y.-L.; Wu, C.-J.; Chen, J.-Y. Tilapia Piscidin 4 (TP4) Stimulates Cell Proliferation and Wound Closure in MRSA-Infected Wounds in Mice. *Mar. Drugs* **2015**, *13*, 2813–2833. [CrossRef] [PubMed]

71. Elango, J.; Lee, J.; Wang, S.; Henrotin, Y.; de Val, J.; Regenstein, J.M.; Lim, S.; Bao, B.; Wu, W. Evaluation of Differentiated Bone Cells Proliferation by Blue Shark Skin Collagen via Biochemical for Bone Tissue Engineering. *Mar. Drugs* **2018**, *16*, 350. [CrossRef] [PubMed]

72. Liu, Y.; Liu, Y.; Liao, N.; Cui, F.; Park, M.; Kim, H.-Y. Fabrication and durable antibacterial properties of electrospun chitosan nanofibers with silver nanoparticles. *Int. J. Biol. Macromol.* **2015**, *79*, 638–643. [CrossRef] [PubMed]

73. Wang, J.; Pei, X.; Liu, H.; Zhou, D. Extraction and characterization of acid-soluble and pepsin-soluble collagen from skin of loach (*Misgurnus anguillicaudatus*). *Int. J. Biol. Macromol.* **2018**, *106*, 544–550. [CrossRef] [PubMed]

74. Laemmli, U.K. Cleavage of structural proteins during the assembly of the head of bacteriophage t4. *Nature* **1970**, *227*, 680–685. [CrossRef] [PubMed]

75. Yamamoto, K.; Igawa, K.; Sugimoto, K.; Yoshizawa, Y.; Yanagiguchi, K.; Ikeda, T.; Yamada, S.; Hayashi, Y. Biological Safety of Fish (Tilapia) Collagen. *Biomed Res. Int.* **2014**, 1–9. [CrossRef] [PubMed]

76. Carvalho, A.; Marques, A.; Silva, T.; Reis, R. Evaluation of the Potential of Collagen from Codfish Skin as a Biomaterial for Biomedical Applications. *Mar. Drugs* **2018**, *16*, 495. [CrossRef] [PubMed]

 *marine drugs*

Article

Fish Collagen Surgical Compress Repairing Characteristics on Wound Healing Process In Vivo

Jingjing Chen [1,†], Kaili Gao [1,†], Shu Liu [2], Shujun Wang [3], Jeevithan Elango [1], Bin Bao [1], Jun Dong [5], Ning Liu [1,4,*] and Wenhui Wu [1,4,*]

[1] Department of Marine Bio-Pharmacology, College of Food Science and Technology, Shanghai Ocean University, Shanghai 201306, China; jingjingchen86@163.com (J.C.); m13052325756@163.com (K.G.); srijeevithan@gmail.com (J.E.); bbao@shou.edu.cn (B.B.)

[2] Jiangsu Marine Resources Development Research Institute, Lianyungang 222000, China; jdliushu@163.com

[3] Co-Innovation Center of Jiangsu Marine Bio-Industry Technology, Huaihai Institute of Technology, Lianyungang, 222005, China; shujunwang86@163.com

[4] Shanghai Engineering Research Center of Aquatic-Product Processing & Preservation, Shanghai 201306, China

[5] German Rheumatism Research Center Berlin, 10117 Berlin, Germany; dong@drfz.de

* Correspondence: nliu@shou.edu.cn (N.L.); whwu@shou.edu.cn (W.W.); Tel./Fax: +86-21-61900395 (N.L.); +86-21-61900388 or +86-21-61900364 (W.W.)

† These authors contributed equally to this work.

Received: 12 October 2018; Accepted: 20 November 2018; Published: 8 January 2019

Abstract: The development of biomaterials with the potential to accelerate wound healing is a great challenge in biomedicine. In this study, four types of samples including pepsin soluble collagen sponge (PCS), acid soluble collagen sponge (ACS), bovine collagen electrospun I (BCE I) and bovine collagen electrospun II (BCE II) were used as wound dressing materials. We showed that the PCS, ACS, BCE I and BCE II treated rats increased the percentage of wound contraction, reduced the inflammatory infiltration, and accelerated the epithelization and healing. PCS, ACS, BCE I, and BCE II significantly enhanced the total protein and hydroxyproline level in rats. ACS could induce more fibroblasts proliferation and differentiation than PCS, however, both PCS and ACS had a lower effect than BCE I and BCE II. PCS, ACS, BCE I, and BCE II could regulate deposition of collagen, which led to excellent alignment in the wound healing process. There were similar effects on inducing the level of cytokines including EGF, FGF, and vascular endothelial marker CD31 among these four groups. Accordingly, this study disclosed that collagens (PCS and ACS) from tilapia skin and bovine collagen electrospun (BCE I and BCE II) have significant bioactivity and could accelerate wound healing rapidly and effectively in rat model.

Keywords: collagen; hydroxyproline; fibroblasts proliferation and differentiation; wound healing

---

1. Introduction

Skin trauma especially severe wound is a common clinical problem, and is more challenging to cure. The foremost aim for the treatment of skin defects is to rapidly restore the construction and function of the wound to the levels of normal tissue, involving acute and chronic inflammations, cell division, migration, and differentiation, regeneration and vascularization [1]. In recent years, the mechanism of wound healing properties of the biomaterials is becoming a research hotspot. Wound healing is a multifactorial process that is characterized by angiogenesis, collagen deposition, granulation tissue formation and re-epithelialization. All these phases involve complex biomolecular interactions among cells, soluble cytokines, adherence factors and chemokines. The clinical treatment of skin wound by traditional medicine has a long history from ancient times, however, the major drawback of traditional medicine dressing is less effective and prolonged treatment time. Many researchers focus on finding the new medical tissue engineering materials for wound healing. The medical tissue engineering materials can replace the damaged skin to provide temporary barrier function and avoid the wound being infected [2]. It provides a platform for cellular proliferation, adhesion and differentiation leading to the development of new functional tissues [3,4]. It can promote tissue repair, regeneration, and recovery to accelerate and complete wound healing [5]. Therefore, grafting tissue engineering material for the healing of a full-thickness wound is preferably a suitable model.

As a new tissue engineering material, collagen has good physical and mechanical properties [6], low immunogenicity [7–9], good biocompatibility, and biodegradability [10–12]. Due to minimal inflammation response, cytotoxicity effect, ability to promote cellular growth and good biocompatibility, collagen is the most promising skin substitute or wound dressing biomaterial [13]. Collagen can promote cellular adhesion and proliferation [14,15], collagen synthesis [16] and increase various growth factors [17], in order to accelerate wound-healing process. In earlier reports, collagen sponge from marine fish up-regulated the fibroblasts and keratinocytes growth, proliferation and wound healing potential in rat model [18]. Liane et al. [19] stated that neurotensin-loaded collagen dressings significantly reduced inflammatory cytokine expression, increased fibroblast migration, enhanced collagen I/III expression and deposition. Tian et al. [20] also reported that electrospun tilapia collagen nanofibers could significantly promote the proliferation of human keratinocytes (HaCaTs), stimulate epidermal differentiation and facilitate rat skin regeneration. All these findings claim that collagen is an excellent biomaterial to be used in wound healing purpose.

Biomimetic environment is also essential for tissue regeneration. Electrospun nanofibrous matrix has been proved to be very effective in skin regeneration because of its superiority features including adjustable diameters, porosity, mimic the structure and function of native extracellular matrix (ECM) and high surface-to-volume ratio, which are beneficial for cell adhesion and proliferation [21]. If collagen could be prepared as nanofibers by electrospinning, it might be helpful for its future application.

Our earlier study showed that collagens from Tilapia skin (PCS, ACS) have significant biocompatibility and can be absorbed and degraded by tissues [22]. Bovine collagen electrospun has been confirmed significant biocompatibility and no cytotoxicity. In continuation to our earlier research [23–26], in vivo wound healing properties of tilapia collagen sponges and bovine collagen electrospun were studied in rat models in order to evaluate its mechanism on accelerating wound healing properties.

2. Results

2.1. Macroscopic Observation of the Wounds

Representative images of wound healing process at different time intervals across all the experimental groups are shown in Figure 1. Collagen-treated groups showed faster wound healing process (complete healing after 14 days from wound incision) compared to control group, in detail, the wound regions were covered with epidermis and the wound areas were closed. With the time extending, wound area of each group decreased gradually. On day 3, each group showed different degree of collagen absorption and wound areas of BCS, PCS, ACS, BCE I and BCE II treated groups were obviously smaller than control and woundplast groups; on day 7, the wounds of the groups treated with BCS, PCS, ACS, BCE I, and BCE II contracted further and scabbed, the wounds of control and woundplast groups had only a small amount of granulation tissue; on day 14, dried blood fall-off from the wound and the wounds healed completely. The wounds which were treated with PCS, ACS, BCE I, and BCE II were smooth and no pigmentation. Interestingly, ACS and PCS treated groups exhibited faster healing than control, woundplast, BCS and BCE I groups, however, BCE II treated group had better healing ability than other groups.

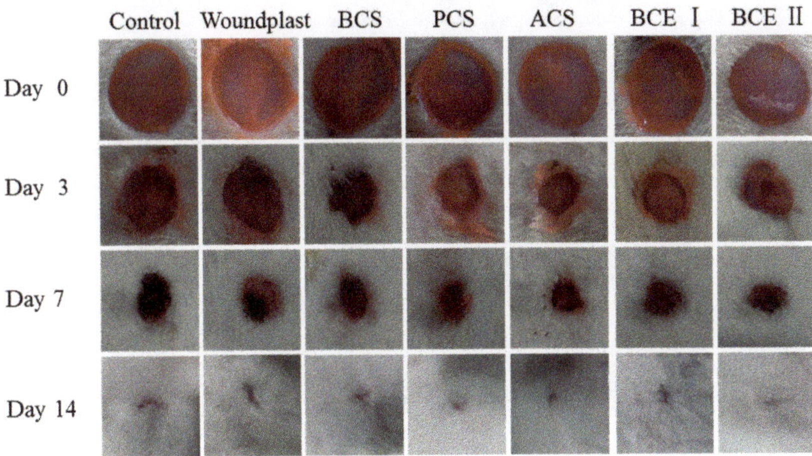

Figure 1. The different stages of wound healing in rats treated with collagens. BCS—bovine collagen sponge, PCS—pepsin soluble collagen sponge, ACS—acid soluble collagen sponge, BCE I—bovine collagen electrospun I, BCE II—bovine collagen electrospun II.

2.2. The Quantification of Aggregate Protein at the Wound Site

An adequate supply with proteins is necessary for consistent wound healing. Therefore, protein content can be used to evaluate the conditions of wound healing [27]. Total protein expression at the wound site of all groups in the excision wound model is shown in Table 1. On day 3 and 7, protein content of the experimental groups was higher than the control and the woundplast groups ($p < 0.05$), and the protein content was high in BCE I and BCE II treated groups compared to other groups on day 7 ($p < 0.05$). On day 14, protein content of the experimental groups was significantly higher than control group ($p < 0.05$). Protein content of ACS treated group was significantly higher than BCS, BCE I, and BCE II treated groups ($p < 0.05$).

Table 1. The effect of each group on total protein content in wound area tissues ($\bar{x} \pm s$, $n = 6$).

Groups	Total Protein (mg/mL)		
	3 Days	**7 Days**	**14 Days**
Control	7.93 ± 0.6 [a]	8.03 ± 1.1 [a]	9.15 ± 0.7 [a]
Woundplast	8.12 ± 0.6 [b]	9.15 ± 0.4 [b]	10.3 ± 0.7 [cd]
BCS	8.61 ± 0.4 [bc]	9.52 ± 1.4 [bc]	9.94 ± 1.1 [b]
PCS	8.48 ± 0.8 [bc]	9.58 ± 0.8 [bc]	11.7 ± 1.3 [bc]
ACS	8.65 ± 0.4 [c]	9.66 ± 1.3 [b]	13.3 ± 0.8 [c]
BCE I	8.79 ± 0.9 [c]	10.2 ± 0.8 [c]	11.4 ± 0.4 [bd]
BCE II	8.85 ± 0.5 [c]	10.5 ± 0.7 [c]	11.3 ± 0.5 [bd]

Note: Different superscript alphabets in each column represent statistical significance ($p < 0.05$).

2.3. Hydroxyproline (Hyp) Content at the Wound Site

Hyp is a specific component of the protein collagen. Therefore, Hyp content might be used as an indicator to determine collagen deposition to measure the speed of wound healing [28]. The significant increase of Hyp in collagen treated group implied faster rate of wound healing process than control group ($p < 0.05$). Indeed, collagen is a major protein of the extracellular matrix and it ultimately contributes to wound healing [29]. Hyp content at the wound site of all groups in the excision wound model is shown in Table 2. Hyp content was higher in ACS and BCE II treated groups (5.87 ± 0.42 mg/g, 5.66 ± 0.12 mg/g tissue) on day 3 and in PCS, ACS, BCE II treated groups on day 7 ($p < 0.05$).

On day 14, PCS and ACS treated groups had high Hyp content (7.32 ± 0.43 mg/g, 7.41± 0.42 mg/g tissue) than control and woundplast treated groups.

Table 2. The effect of each group on the Hydroxyproline content in wound area tissues ($\bar{x} \pm s$, $n = 6$).

Groups	Hydroxyproline Content (mg/g Wet Skin)		
	3 Days	**7 Days**	**14 Days**
Control	5.26 ± 0.33 [a]	6.12 ± 0.29 [a]	6.79 ± 0.39 [a]
Woundplast	5.46 ± 0.41 [ab]	6.38 ± 0.49 [a]	7.29 ± 0.18 [ab]
BCS	5.60 ± 0.24 [ab]	6.29 ± 0.36 [a]	7.02 ± 0.39 [ab]
PCS	5.47 ± 0.36 [ab]	6.55 ± 0.44 [b]	7.32 ± 0.43 [b]
ACS	5.87 ± 0.42 [b]	6.65 ± 0.34 [b]	7.41 ± 0.42 [b]
BCE I	5.42 ± 0.25 [ab]	6.42 ± 0.35 [a]	6.77 ± 0.33 [a]
BCE II	5.66 ± 0.12 [b]	6.68 ± 0.54 [b]	7.22 ± 0.27 [ab]

Note: Different superscript alphabets in each column represent statistical significance ($p < 0.05$).

2.4. Histopathological Examination

In control group, a large number of inflammatory cells appeared in the wound than collagen-treated groups on day 3 (Figure 2). Granulation tissue and fibroblasts activity were more pronounced in collagen-treated groups compared to control group. A large number of inflammatory cells, blood vessels, small amount of collagen fibers and fibroblasts were seen in woundplast and BCS groups. Fibroblast cells were important in the wound site and predominant collagen expression could be seen in PCS, ACS, BCE I, and BCE II treated groups. On day 7, healed regions of the wounds were covered by epithelial tissue with significant fibroblast proliferation, collagen deposition and granulation tissue formation. However, the granulation tissue organization and vascularization in unhealed regions of collagen-treated groups were notably different from the control group. Collagen-treated groups developed collagen deposition and vascularization than the control group. The H&E staining revealed that in collagen-treated wound on day 7, cutaneous appendages like hair, hair follicles and sebaceous glands began to appear, which signify the formation of epidermal layer. A large number of inflammatory cells and collagen fibers and fibroblasts were observed in control group. A large number of fibroblasts and collagen were seen in woundplast and BCS treated groups and skin appendages and skin tissues were gradually formed in collagen treated groups. More fibroblasts and collagen and less inflammatory cells were seen in PCS, ACS, BCE I, and BCE II treated groups than the control group. On day 14, a sufficient number of fully formed skin adnexal and epithelial tissues were present in collagen-treated groups. PCS, ACS, BCE I, and BCE II treated groups showed well-formed stratified epithelial layer, granulation tissue formation and collagen deposition in healed regions than the control, woundplast and BCS groups. As shown in Table 3, a scoring system was used to quantify the pathological results of the H&E stained samples. In this scoring system five criteria were scored. An increasing score denotes the better wound healing. The histological scoring result showed that the PCS, ACS, BCE I, and BCE II treated groups had better wound healing than the control group.

Table 3. Histological evaluation of wound tissue in all groups by H&E staining.

	0	1–3	4–6	7–9
Inflammatory cells	Abundant	Moderate	Scant	Rarely
Fibroblast content	None	Scant	Moderate	Abundant
Re-epithelialization	None	Partial	Thin	Complete
Collagen deposition	None	Scant	Moderate	Abundant
Revascularizations	None	Scant	Moderate	Abundant

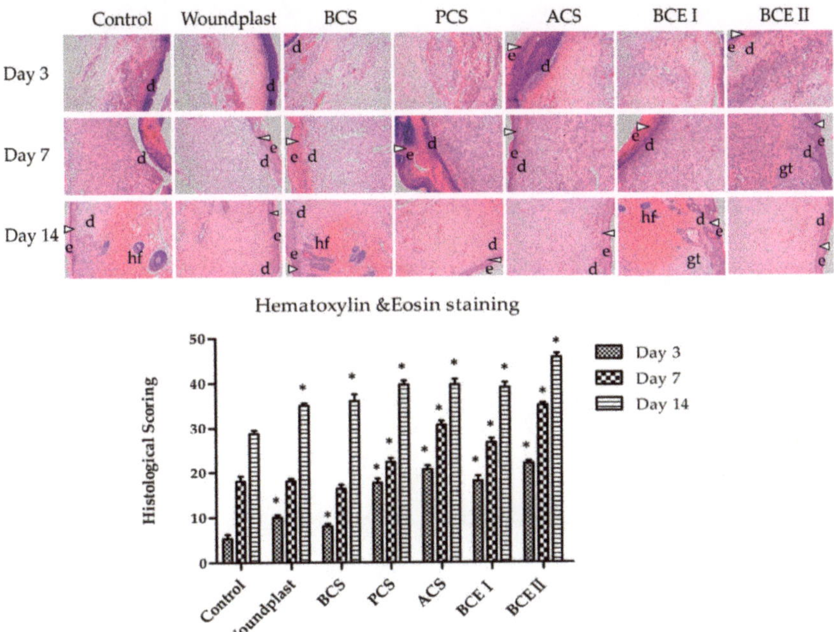

Figure 2. Histological analysis of H&E stained wounded tissues (Magnification, ×100) with histological scoring. The alphabetic letters e, d, gt and hf represent epidermis, dermis, granulation tissue, hair follicle, respectively. (Each bar represents the mean ± SD. * $p < 0.05$: significantly different from the control group.).

2.5. Collagen Promotes the Expression of EGF, FGF, and CD31 in the Wounds

Immuno-histochemical analysis of the reconstituted tissue was shown in Figures 3–5. EGF stands for epidermal growth factor, which can promote proliferation and differentiation of keratinocyte. EGF can induce fibroblasts proliferation and collagen synthesis, resulting in epithelization. Fibroblast growth factor (FGF) can accelerate migration and proliferation of fibroblast, and vascularization. EGF and FGF play an important role in the wound healing process. Prolonging the treated time, the expression of EGF and FGF in control group showed an increasing trend, the expression of EGF and FGF in woundplast and BCS groups increased in the beginning and then decreased, and the expression of EGF and FGF in PCS, ACS, BCE I, and BCE II treated groups showed a decreasing trend. On day 3 and 7, the expression of EGF and FGF in PCS, ACS, BCE I, and BCE II treated groups was higher than control, woundplast and BCS groups. CD31 is a platelet endothelial cells adhesion molecule-1, stands for vessel proliferation. The relative quantity and distribution of CD31 in the construct-treated wound bed is important in wound healing. Brown granules were positive signal in Figure 3. The expression of CD31 showed similar trends same as EGF and FGF of collagen treated groups.

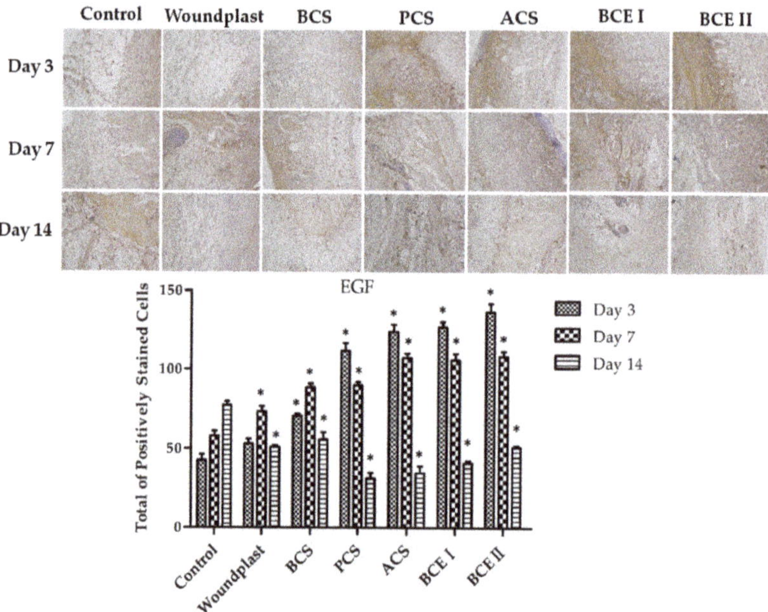

Figure 3. Immunohistochemical analysis of EGF expression in wounded tissues (Magnification, ×100); the histogram shows the total of positively stained cells of EGF in the dermal tissue per group. (Each bar represents the mean ± SD. * $p < 0.05$: significantly different from the control group.).

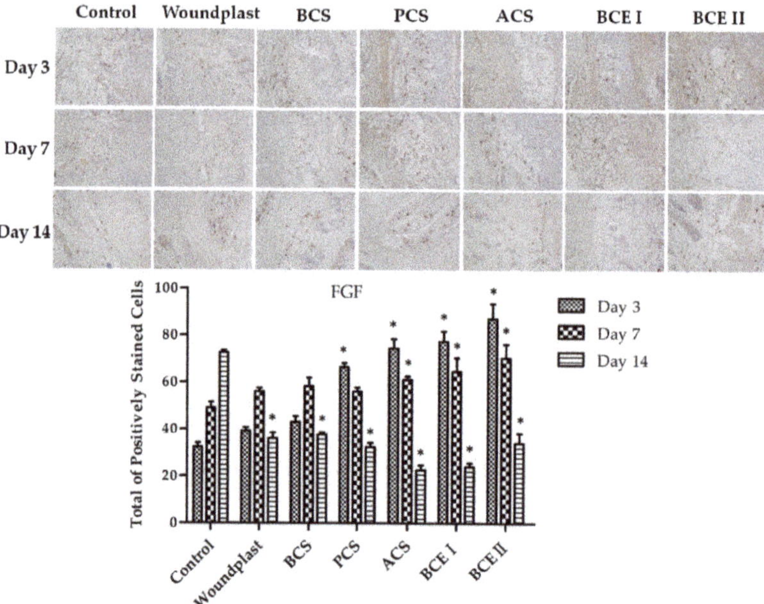

Figure 4. Immunohistochemical analysis of fibroblast growth factor (FGF) expression in wounded tissues (Magnification, ×100); the histogram shows the total of positively stained cells of FGF in the dermal tissue per group. (Each bar represents the mean ± SD * $p < 0.05$: significantly different from the control group.

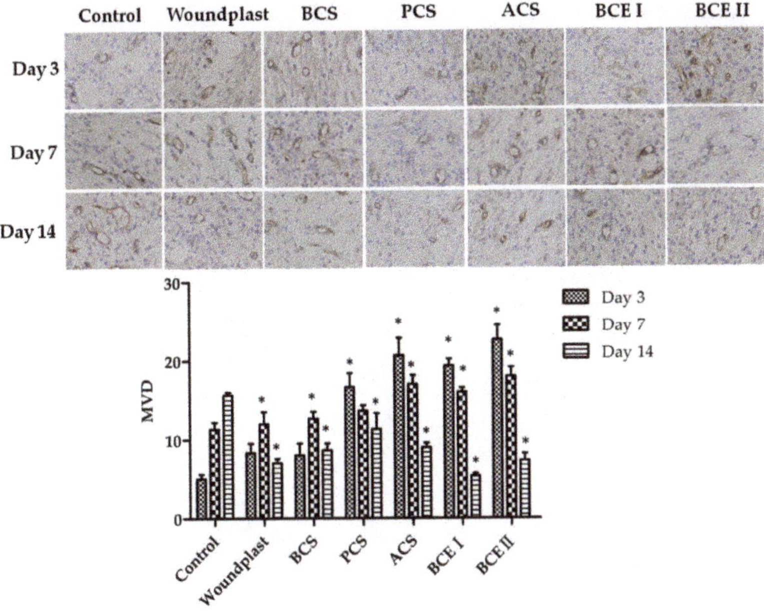

Figure 5. Immunohistochemical analysis of CD31 expression in wounded tissues (Magnification, ×200). The histogram summarizes the microvessel density (MVD), which was determined by immunohistochemical staining for CD31. (Each bar represents the mean ± SD. * $p < 0.05$: significantly different from the control group.).

3. Discussion

As the structural and functional component of dermal extracellular matrix, collagen plays a vital role in wound healing process [18,30–32]. In the present study, PCS, ACS, and electrospun bovine skin collagen nanofibers were successfully fabricated. Due to its unique perforated structure, controllable fiber diameter (50 nm–5 μm), large surface area, high porosity and unique biological property [33], PCS, ACS and BCE I, BCE II were examined by using the full-thickness wound model in SD rats and the total protein, hydroxyproline content, H&E, and immunohistochemical examinations were assessed with control group. In full-thickness wounds in SD rats, PCS, ACS, BCE I and BCE II treated groups revealed significantly higher wound healing ability, total protein and hydroxyproline content, fibroblasts proliferation and collagen synthesis when compared to control, BCS and woundplast groups.

Based on the earlier findings, the possible reasons for accelerating wound healing of PCS, ACS, BCE I, and BCE II were due to its unique structures and the high porosity of the matrix, which could induce fibroblasts proliferation and collagen synthesis, and play crucial roles in re-epithelialization and vascularization in the wound healing process [34,35]. In addition, PCS, ACS, BCE I, and BCE II exhibited good biocompatibility to support the adhesion and proliferation of fibroblasts that lead the deposition and maturation of collagen [9]. Collagen could be completely degraded and absorbed by the wound that helps to formation of fibroblasts and collagen fibers over time in epidermal tissue and therefore, the wounds were replaced with regenerated dermis.

Histopathological and immunohistochemical examinations indicate that PCS, ACS, BCE I, BCE II had a positive effect on neovascularization, inducing fibroblasts proliferation, collagen synthesis, re-epithelialization and regeneration of skin appendages, and consequently led to an increased wound healing ability compared with control, BCS and woundplast groups. Besides, PCS, ACS, BCE I and BCE II were similar to normal skin because of smooth surface along with loose collagen fiber. PCS,

ACS, BCE I, and BCE II could provide a proper microenvironment for fibroblasts attachment in skin due to its three-dimensional features and high porosity.

Vascular endothelial marker CD31, growth factors EGF and FGF are involved in wound healing process [36]. It has been confirmed that PCS, ACS, BCE I, and BCE II significantly induce EGF and FGF expression, which can promote proliferation and differentiation of fibroblasts and keratinocytes. EGF can induce keratinocytes proliferation to impel re-epithelialization of the wound. As chemotactic agent and mitogenic agent of fibroblasts, EGF can promote the proliferation and differentiation of fibroblasts to synthesize collagen. The increased expression of CD31 and FGF revealed the vascularization and wound healing properties of collagens.

In this work, PCS and ACS groups revealed significantly better wound healing ability than woundplast and control groups, slightly higher than BCS group, which might be due to the three-dimensional features and high porosity of PCS and ACS than BCS. However, ACS had a more positive effect than PCS, which was similar to previous report [22]. There was no much difference between BCE I and BCE II treated groups. However, there was less inflammatory cells in BCE II treated group than BCE I, which may be related to the bactericidal effect of chitosan. These effects were probably owed to the biomimetic structure, unique biological property and high porosity of the collagen nanofibers.

4. Materials and Methods

4.1. Materials

Bovine collagen electrospun (BCE) was prepared by Department of Textile Materials College, Donghua University, Shanghai, China, and its biocompatibility and cytotoxicity were confirmed. Bovine collagen sponge (BCS) prepared with polyethylene oxide (PEO) (BCE I) and PEO with chitosan (BCE II) were compared with tilapia collagens. The weight ratio of bovine collagen, PEO and chitosan was 150:17:2. Tilapia skin collagens were prepared by Shanghai Ocean University, Shanghai, China. Briefly, Pepsin soluble collagen sponge (PCS) and acid soluble collagen sponge (ACS) from Tilapia skin were prepared as per our previous method [22]. Both collagens were composed of two α-chains and a β-chain and characterized as type-I collagen [22]. BCS was kindly provided by HaoHai Biological Technology, Shanghai, China, which was characterized as type-I collagen. Woundplast was purchased from Johnson & Johnson (Shanghai, China) and medical gauze was procured from Shanghai Health Materials Factory Co. Ltd., Shanghai, China.

SD rats were purchased from Shanghai Slac Laboratory Animal Co. Ltd., Shanghai, China. Animal study protocols and procedures were approved by the Shanghai Ocean University institutional animal care and use committee (Permit Number: SHOU-DW-2018-054). All methods were employed in accordance with the relevant guidelines and regulations of Scientific and Ethical Care and Use of Laboratory Animals of Shanghai Ocean University.

4.2. Skin Wound Healing in SD Rats

The wound healing experiment was performed as follows: female rats (body weight 200–250 g) were maintained in a pathogen free environment and fed a standard diet. 63 rats were injected intraperitoneal with sodium pentobarbital. Two full-thickness wounds of size 1 cm in diameter were created on the dorsum of SD rats which were 1.5 cm apart from the spine. These wounds were covered with medical gauze, woundplast, BCS, PCS, ACS, BCE I, and BCE II ($n = 6$) to avoid infections. Control group was served without any treatment (only medical gauze). Medical adhesive tape was used to attach the dressings in the wounds. Dressing change was done every two days and rats were kept in individual cages. On days 3, 7, and 14 after surgery, the morphology of the wounds was examined. The skin wounds of animals were photographed and subsequently, the rats were euthanized. Wound tissues were removed by sacrificing three rats each from all groups periodically on the 3rd, 7th, and 14th days of post wound creation and the granulation tissues formed were collected.

4.3. Determination of Total Protein and Hydroxyproline Content

Skin tissue (~90 ± 10 mg) was taken from the wound for determination of total protein and hydroxyproline content. The harvested skin samples collected on days 3, 7, and 14 were washed with ice-cold saline and dried by filter paper. They were then used to determine hydroxyproline and total protein level in specimen [37]. The HYP content in tissue samples was determined using a Tissue hydroxyproline kit, as per the manufacturer's instruction (JianChen Gene Company, Nanjing, China). The healed skin tissues (n = 6) were harvested and cut into pieces and then incubated with tissue lysis buffer for 20 min at 95 °C and homogenized. The tubes were centrifuged (13,000 rpm) at 4 °C and supernatant was collected. Total protein concentration was evaluated using a total protein kit (JianChen Gene Company, Nanjing, China) according to the manufacturer's instructions. The average value was taken from the triplicate readings.

4.4. Histopathological Examination

Harvested wounds together with the surrounding skins were used for the histological evaluation. The harvested samples collected on days 3, 7, and 14 were fixed in 10% buffered formalin solution for 24 h. The tissues were embedded in paraffin and sectioned into 5 μm thick slices for histopathological examination by hematoxylin and eosin (H&E) staining method. Then they were studied by a routine light microscope. The criterion that was studied in histopathological sections consisted of re-epithelialization, collagen deposition, fibroblast content, revascularizations, and inflammatory cells. Analysis of stained skin sections was performed by an experienced pathologist.

4.5. Immuno-Histochemical Examinations

The skin tissues including the wound site were excised and fixed in 10% buffered formalin for more than 24 h, then embedded in paraffin and cut into 5 μm thick slices. After deparaffinization and rehydration, antigen unmasking was performed as follows: the endogenous peroxidase of randomly selected section was inactivated by incubation with 3% hydrogen peroxide/methanol solution at 37 °C for 30 min. The slices were washed three times with PBS for 5 min each wash. In order to recover antigen, these sections were put into 10 mM citrate buffer solution (pH 6.0) and heated at 95 °C for 15 min, and then cool down at room temperature, followed by washing three times with PBS for 5 min each wash. The non-specific binding sites were blocked with 5% goat serum (Gibco, 16210072) for 10 min at 37 °C. After the redundant liquid was discarded, the sections were incubated with the following primary antibodies, respectively: anti-EGF antibody, anti-FGF antibody, and anti-CD31 antibody at 4 °C overnight and washed three times with PBS for 5 min, followed by incubation with biotinylated goat anti-rabbit secondary antibody kit (Santa Cruz, Shanghai, China) at 37 °C for 30 min, and then incubated with streptavidin-HRP for 30 min. The slides were dyeing with a DAB (3,3′-diaminobenzidine) solution, and then counterstained with hematoxylin and following by dehydration with sequential ethanol for sealing and microscope observation. FGF and EGF positive cells were analyzed from three identical areas in the dermal tissue per rat wound tissue section and analyzed for the statistical significance. Individual micro-vessels were counted at 200× magnification (0.152 mm^2/field). For each section, three areas were selected from the vascularity of the wound tissues [38].

4.6. Statistical Analysis

The values were expressed as the mean ± standard deviation (SD). Statistically significant differences ($p < 0.05$) among the different groups were evaluated using Student's *t*-test and one-way analysis of variance (ANOVA) with Tukey's post hoc multiple comparison test. All of the statistical analyses were performed using SPSS 17.0 software.

5. Conclusions

In this study, PCS, ACS, BCE I, and BCE II were successfully fabricated and evaluated for its utility as dermal substitute. PCS, ACS, BCE I, and BCE II treatment increased the wound healing ability, fibroblasts proliferation, collagen synthesis, re-epithelialization and dermal reconstitution in vivo that owed to the biomimetic structure and high porosity of the collagen nanofibers. This study indicated that PCS, ACS, BCE I, and BCE II could accelerate wound healing rapidly and effectively. The overall results of this study suggest that collagen from tilapia and electrospun bovine collagen nanofibers can be used as a promising dermal substitute to treat severe wounds.

Author Contributions: Conceptualization, N.L. and W.W.; Data curation, J.C. and K.G.; Formal analysis, N.L.; Funding acquisition, J.E., B.B. and W.W.; Investigation, J.C., K.G. and B.B.; Methodology, J.E. and J.D.; Project administration, W.W.; Resources, S.L.; Software, J.C.; Supervision, W.W.; Validation, S.W. and W.W.; Visualization, W.W.; Writing—original draft, J.C. and K.G.; Writing—review & editing, J.E., N.L. and W.W.

Funding: This research was funded by the National High Technology Research and Development Program of China [grant number 2011AA09070109], the National Natural Science Foundation of China [grant number 81341082 and 81750110548], the Plan of Innovation Action in Shanghai [grant number 15410722500], and the Jiangsu Provincial Natural Science Foundation [grant number BK20151282]. And the APC was funded by the National Natural Science Foundation of China [grant number 81750110548].

Acknowledgments: The authors acknowledge the financial support from the National High Technology Research and Development Program of China, the National Natural Science Foundation of China, the Plan of Innovation Action in Shanghai and the Jiangsu Provincial Natural Science Foundation.

Conflicts of Interest: The authors declare no conflict of interest.

References

1. Lin, J.; Li, C.; Zhao, Y.; Hu, J.; Zhang, L.M. Co-electrospun nanofibrous membranes of collagen and zein for wound healing. *ACS Appl. Mater. Interfaces* **2012**, *4*, 1050–1057. [CrossRef]

2. De Almeida, E.B.; Cardoso, J.C.; de Lima, A.K.; de Oliveira, N.L.; de Pontes, N.T.; Lima, S.O.; Souza, I.C.L.; de Albuquerque, R.L.C. The incorporation of Brazilian propolis into collagen-based dressing films improves dermal burn healing. *J. Ethnopharmacol.* **2013**, *147*, 419–425. [CrossRef]

3. Banerjee, P.; Mehta, A.; Shanthi, C. Investigation into the cyto-protective and wound healing properties of cryptic peptides from bovine achilles tendon collagen. *Chem.-Biol. Interact.* **2014**, *211*, 1–10. [CrossRef]

4. Castillo-Briceno, P.; Bihan, D.; Nilges, M.; Hamaia, S.; Meseguer, J.; Garcia-Ayala, A.; Farndale, R.W.; Mulero, V. A role for specific collagen motifs during wound healing and inflammatory response of fibroblasts in the teleost fish gilthead seabream. *Mol. Immunol.* **2011**, *48*, 826–834. [CrossRef]

5. Holmer, C.; Praechter, C.; Mecklenburg, L.; Heimesaat, M.; Rieger, H.; Pohlen, U. Anastomotic stability and wound healing of colorectal anastomoses sealed and sutured with a collagen fleece in a rat peritonitis model. *Asian J. Surg.* **2014**, *37*, 35–45. [CrossRef]

6. Jeevithan, E.; Shakila, R.J.; Varatharajakumar, A.; Jeyasekaran, G.; Sukumar, D. Physico-functional and mechanical properties of chitosan and calcium salts incorporated fish gelatin scaffolds. *Int. J. Biol. Macromol.* **2013**, *60*, 262–267. [CrossRef]

7. Lee, C.H.; Chang, S.H.; Chen, W.J.; Hung, K.C.; Lin, Y.H.; Liu, S.J.; Hsieh, M.J.; Pang, J.H.S.; Juang, J.H. Augmentation of diabetic wound healing and enhancement of collagen content using nanofibrous glucophage-loaded collagen/PLGA scaffold membranes. *J. Colloid Interface Sci.* **2015**, *439*, 88–97. [CrossRef]

8. Jeevithan, E.; Jingyi, Z.; Bao, B.; Shujun, W.; JeyaShakila, R.; Wu, W.H. Biocompatibility assessment of type-II collagen and its polypeptide for tissue engineering: Effect of collagen's molecular weight and glycoprotein content on tumor necrosis factor (fas/apo-1) receptor activation in human acute t-lymphocyte leukemia cell line. *RSC Adv.* **2016**, *6*, 14236–14246. [CrossRef]

9. Chattopadhyay, S.; Raines, R.T. Review collagen-based biomaterials for wound healing. *Biopolymers* **2014**, *101*, 821–833. [CrossRef]

10. Jeevithan, E.; Zhang, J.Y.; Wang, N.P.; He, L.; Bao, B.; Wu, W.H. Physico-chemical, antioxidant and intestinal absorption properties of whale shark type-ii collagen based on its solubility with acid and pepsin. *Process Biochem.* **2015**, *50*, 463–472. [CrossRef]

11. Muthukumar, T.; Anbarasu, K.; Prakash, D.; Sastry, T.P. Effect of growth factors and pro-inflammatory cytokines by the collagen biocomposite dressing material containing macrotyloma uniflorum plant extract-in vivo wound healing. *Colloids Surf. B Biointerfaces* **2014**, *121*, 178–188. [CrossRef]

12. Elango, J.; Lee, J.W.; Wang, S.; Henrotin, Y.; de Val, J.; Regenstein, J.M.; Lim, S.Y.; Bao, B.; Wu, W. Evaluation of differentiated bone cells proliferation by blue shark skin collagen via biochemical for bone tissue engineering. *Mar. Drugs* **2018**, *16*, 350. [CrossRef]

13. Ramasamy, P.; Shanmugam, A. Characterization and wound healing property of collagen-chitosan film from sepia kobiensis (hoyle, 1885). *Int. J. Biol. Macromol.* **2015**, *74*, 93–102. [CrossRef]

14. Mahmoud, A.A.; Salama, A.H. Norfloxacin-loaded collagen/chitosan scaffolds for skin reconstruction: Preparation, evaluation and in-vivo wound healing assessment. *Eur. J. Pharm. Sci.* **2016**, *83*, 155–165. [CrossRef]

15. Kim, K.O.; Lee, Y.; Hwang, J.W.; Kim, H.; Kim, S.M.; Chang, S.W.; Lee, H.S.; Choi, Y.S. Wound healing properties of a 3-D scaffold comprising soluble silkworm gland hydrolysate and human collagen. *Colloids Surf. B* **2014**, *116*, 318–326. [CrossRef]

16. Dang, Q.F.; Liu, H.; Yan, J.Q.; Liu, C.S.; Liu, Y.; Li, J.; Li, J.J. Characterization of collagen from haddock skin and wound healing properties of its hydrolysates. *Biomed. Mater.* **2015**, *10*, 015022. [CrossRef]

17. Lee, C.-H.; Chao, Y.-K.; Chang, S.-H.; Chen, W.-J.; Hung, K.-C.; Liu, S.-J.; Juang, J.-H.; Chen, Y.-T.; Wang, F.-S. Nanofibrous rhPDGF-eluting PLGA–collagen hybrid scaffolds enhance healing of diabetic wounds. *RSC Adv.* **2016**, *6*, 6276–6284. [CrossRef]

18. Pal, P.; Srivas, P.K.; Dadhich, P.; Das, B.; Maity, P.P.; Moulik, D.; Dhara, S. Accelerating full thickness wound healing using collagen sponge of mrigal fish (cirrhinus cirrhosus) scale origin. *Int. J. Biol. Macromol.* **2016**, *93*, 1507–1518. [CrossRef]

19. Moura, L.I.F.; Dias, A.M.A.; Suesca, E.; Casadiegos, S.; Leal, E.C.; Fontanilla, M.R.; Carvalho, L.; de Sousa, H.C.; Carvalho, E. Neurotensin-loaded collagen dressings reduce inflammation and improve wound healing in diabetic mice. *Biochim. Biophys. Acta* **2014**, *1842*, 32–43. [CrossRef]

20. Zhou, T.; Wang, N.P.; Xue, Y.; Ding, T.T.; Liu, X.; Mo, X.M.; Sun, J. Electrospun tilapia collagen nanofibers accelerating wound healing via inducing keratinocytes proliferation and differentiation. *Colloids Surf. B* **2016**, *143*, 415–422. [CrossRef]

21. Ye, B.H.; Luo, X.S.; Li, Z.W.; Zhuang, C.P.; Li, L.H.; Lu, L.; Ding, S.; Tian, J.H.; Zhou, C.R. Rapid biomimetic mineralization of collagen fibrils and combining with human umbilical cord mesenchymal stem cells for bone defects healing. *Mater. Sci. Eng. C* **2016**, *68*, 43–51. [CrossRef]

22. Zhang, J.; Jeevithan, E.; Bao, B.; Wang, S.; Gao, K.; Zhang, C.; Wu, W. Structural characterization, in-vivo acute systemic toxicity assessment and in-vitro intestinal absorption properties of tilapia (*Oreochromis niloticus*) skin acid and pepsin solublilized type I collagen. *Process Biochem.* **2016**, *51*, 2017–2025. [CrossRef]

23. Jeevithan, E.; Bao, B.; Zhang, J.Y.; Hong, S.T.; Wu, W.H. Purification, characterization and antioxidant properties of low molecular weight collagenous polypeptide (37 kDa) prepared from whale shark cartilage (*Rhincodon typus*). *J. Food Sci. Technol.* **2015**, *52*, 6312–6322. [CrossRef]

24. Elango, J.; Zhang, J.Y.; Bao, B.; Palaniyandi, K.; Wang, S.J.; Wu, W.H.; Robinson, J.S. Rheological, biocompatibility and osteogenesis assessment of fish collagen scaffold for bone tissue engineering. *Int. J. Biol. Macromol.* **2016**, *91*, 51–59. [CrossRef]

25. Jeevithan, E.; Wu, W.H.; Wang, N.P.; Lan, H.; Bao, B. Isolation, purification and characterization of pepsin soluble collagen isolated from silvertip shark (*Carcharhinus albimarginatus*) skeletal and head bone. *Process Biochem.* **2014**, *49*, 1767–1777. [CrossRef]

26. Jeevithan, E.; Bao, B.; Bu, Y.S.; Zhou, Y.; Zhao, Q.B.; Wu, W.H. Type II collagen and gelatin from silvertip shark (*Carcharhinus albimarginatus*) cartilage: Isolation, purification, physicochemical and antioxidant properties. *Mar. Drugs* **2014**, *12*, 3852–3873. [CrossRef]

27. Nagai, T.; Suzuki, N. Isolation of collagen from fish waste material—Skin, bone and fins. *Food Chem.* **2000**, *68*, 277–281. [CrossRef]

28. Huang, R.; Li, W.Z.; Lv, X.X.; Lei, Z.J.; Bian, Y.Q.; Deng, H.B.; Wang, H.J.; Li, J.Q.; Li, X.Y. Biomimetic LBL structured nanofibrous matrices assembled by chitosan/collagen for promoting wound healing. *Biomaterials* **2015**, *53*, 58–75. [CrossRef]

29. Rousselle, P.; Montmasson, M.; Garnier, C. Extracellular matrix contribution to skin wound re-epithelialization. *Matrix Biol.* **2018**. [CrossRef]

30. Velnar, T.; Bailey, T.; Smrkoli, V. The wound healing process: An overview of the cellular and molecular mechanisms. *J. Int. Med. Res.* **2009**, *37*, 1528–1542. [CrossRef]

31. Moreno-Arotzena, O.; Meier, J.G.; del Amo, C.; Garcia-Aznar, J.M. Characterization of fibrin and collagen gels for engineering wound healing models. *Materials* **2015**, *8*, 1636–1651. [CrossRef]

32. Ramanathan, G.; Thyagarajan, S.; Sivagnanam, U.T. Accelerated wound healing and its promoting effects of biomimetic collagen matrices with siderophore loaded gelatin microspheres in tissue engineering. *Mater. Sci. Eng. C* **2018**, *93*, 455–464. [CrossRef]

33. Ren, K.; Wang, Y.; Sun, T.; Yue, W.; Zhang, H. Electrospun PCL/gelatin composite nanofiber structures for effective guided bone regeneration membranes. *Mater. Sci. Eng. C Mater. Biol. Appl.* **2017**, *78*, 324–332. [CrossRef]

34. Ahearne, M.; Wilson, S.L.; Liu, K.K.; Rauz, S.; El Haj, A.J.; Yang, Y. Influence of cell and collagen concentration on the cell-matrix mechanical relationship in a corneal stroma wound healing model. *Exp. Eye Res.* **2010**, *91*, 584–591. [CrossRef]

35. Shekhter, A.B.; Rudenko, T.G.; Istranov, L.P.; Guller, A.E.; Borodulin, R.R.; Vanin, A.F. Dinitrosyl iron complexes with glutathione incorporated into a collagen matrix as a base for the design of drugs accelerating skin wound healing. *Eur. J. Pharm. Sci.* **2015**, *78*, 8–18. [CrossRef]

36. Jridi, M.; Bardaa, S.; Moalla, D.; Rebaii, T.; Souissi, N.; Sahnoun, Z.; Nasri, M. Microstructure, rheological and wound healing properties of collagen-based gel from cuttlefish skin. *Int. J. Biol. Macromol.* **2015**, *77*, 369–374. [CrossRef]

37. Kwan, K.H.L.; Liu, X.L.; To, M.K.T.; Yeung, K.W.K.; Ho, C.M.; Wong, K.K.Y. Modulation of collagen alignment by silver nanoparticles results in better mechanical properties in wound healing. *Nanomedicine* **2011**, *7*, 497–504. [CrossRef]

38. Liu, N.; Ding, D.; Hao, W.; Yang, F.; Wu, X.; Wang, M.; Xu, X.; Ju, Z.; Liu, J.P.; Song, Z.; et al. Htert promotes tumor angiogenesis by activating VEGF via interactions with the Sp1 transcription factor. *Nucleic Acids Res.* **2016**, *44*, 8693–8703. [CrossRef]

 *marine drugs*

Article

Electrodialysis Extraction of Pufferfish Skin (*Takifugu flavidus*): A Promising Source of Collagen

Junde Chen [1,*], Min Li [2,*], Ruizao Yi [1], Kaikai Bai [1], Guangyu Wang [1], Ran Tan [1], Shanshan Sun [1] and Nuohua Xu [1]

1 Marine Biological Resource Comprehensive Utilization Engineering Research Center of the State Oceanic Administration, The Third Institute of Oceanography of the State Oceanic Administration, Xiamen 361005, China; yiruizao@163.com (R.Y.); kkbai@tio.org.cn (K.B.); 17859733637@163.com (G.W.); tanran@tio.org.cn (R.T.); shshsun123@163.com (S.S.); xunuohua@163.com (N.X.)
2 Plants for Human Health Institutes, North Carolina State University, Kannapolis, NC 28081, USA
* Correspondence: jdchen@tio.org.cn (J.C.); mli33@ncsu.edu (M.L.); Tel.: +86-592-215527 (J.C.); +01-704-250-5479 (M.L.)

Received: 28 November 2018; Accepted: 27 December 2018; Published: 4 January 2019

Abstract: Collagen is widely used in drugs, biomaterials, foods, and cosmetics. By-products of the fishing industry are rich sources of collagen, which can be used as an alternative to collagen traditionally harvested from land mammals. However, commercial applications of fish-based collagen are limited by the low efficiency, low productivity, and low sustainability of the extraction process. This study applied a new technique (electrodialysis) for the extraction of *Takifugu flavidus* skin collagen. We found electrodialysis to have better economic and environmental outcomes than traditional dialysis as it significantly reduced the purification time and wastewater (~95%) while maintaining high extraction yield (67.3 ± 1.3 g/100 g dry weight, $p < 0.05$). SDS-PAGE, amino acid composition analysis, and spectrophotometric characterization indicated that electrodialysis treatment retained the physicochemical properties of *T. flavidus* collagen. Heavy metals and tetrodotoxin analyses indicated the safety of *T. flavidus* collagen. Notably, the collagen had similar thermal stability to calf skin collagen, with the maximum transition temperature and denaturation temperature of 41.8 ± 0.35 and 28.4 ± 2.5 °C, respectively. All evidence suggests that electrodialysis is a promising technique for extracting collagen in the fishing industry and that *T. flavidus* skin collagen could serve as an alternative source of collagen to meet the increasing demand from consumers.

Keywords: skin collagen; electrodialysis; thermal stability; *Takifugu flavidus*

1. Introduction

Collagen is the major structural component of various extracellular matrices in mammalian connective tissues, such as the skin, corneas, cartilage, bone, and blood vessel [1]. Collagen is also widely used in biomaterials, drugs, foods, and cosmetics [2]. Its high biocompatibility as well as the ability to support cell growth and differentiation has made it an important matrix for cell biology, cosmetics, and regenerative medicine [3,4]. Collagen is also widely used as a gelatin precursor in the food industry for formulating emulsions, foams, colloids, and biodegradable films [5]. Annual sales of collagen and its derivatives in the global market have reached billions of dollars [6]. Despite high feasibility and biocompatibility, these mammalian origins potentially limit practical application of collagen due to sociocultural beliefs (e.g., in Muslim countries and India). In addition, the mammalian origins can further restrain collagen application by increasing additional sanitary costs for industrial production because of extensive consumer concerns regarding transmissible diseases from porcine and bovine [7,8].

Fish, a popular dietary ingredient, is a great alternative source of commercial collagen. In addition to its worldwide acceptance by different sociocultural belief systems, fish is less likely to carry mammalian-borne diseases, thus requiring lower sanitary costs for industrial production of collagen [9]. More interestingly, the rapid development of the fishing industry has resulted in a huge amount of collagen-rich by-products, including skin, scales, and bones (50.2–117 million metric tons) [10]. Pufferfish aquaculture is a thriving industry in China, producing more than 14,000 metric tons in 2013 [11]. Studies on fish suggest that pufferfish skin is rich in type I collagen and may serve as an appropriate source of collagen to replace their mammalian counterparts.

While promising, industrial production of fish collagen is limited by the low efficiency, low productivity, and low sustainability of the purification process. Crude fish collagen extract is mainly purified by dialysis through a passive diffusion process in the industry. This method usually takes 3–4 days to isolate pure collagen and often processes less than 1 L of samples [12,13]. Moreover, dialysis of crude fish collagen extract can produce a relatively large amount of acetified wastewater (20 L dialysate/L sample) [14–16], leading to severe environmental stress. An advanced purification technique with better economic and environmental outcome is therefore needed to take full advantage of collagen-rich fish by-products.

Electrodialysis, as an active diffusion technique, appears to be an excellent solution to improve the efficiency, productivity, and sustainability of fish collagen purification. This method can purify charged proteins/peptides by ion-exchange membranes through a stimulated diffusion process under the influence of electric potential difference [17]. A recent study on marine protein hydrolysate showed that electrodialysis could process 3 L of sample within a relative short time (6 h) while requiring just a small amount of dialysis buffer (1 L/L sample) [18].

The aim of this study was to assess the practicality of electrodialysis for the isolation and purification of collagen derived from yellowbelly pufferfish (*Takifugu flavidus*) skin. Considering that this technique has not been applied in the fishing industry, we hypothesized that the application of electrodialysis could not only improve economic and environmental outcomes of fish collagen production but also retain the physicochemical properties of *T. flavidus* collagen. Extraction efficiency, productivity, and wastewater production were also determined in this study, and the quality of *T. flavidus* collagen was systemically evaluated by electrophoresis, spectrophotometric characterization, thermal properties, and solubility.

2. Results and Discussion

2.1. Quality Measurements

Heavy metals can lead to chronic heart disease, cancer, and death [19]. Heavy metals have been banned by law from foods, cosmetics, and drugs [20]. The metal concentrations in pufferfish skin were found to be Pb (0.562 ± 0.052 mg kg^{-1}), Cd (0.042 ± 0.004 mg kg^{-1}), As (0.075 ± 0.002 mg kg^{-1}), and Cr (1.152 ± 0.067 mg kg^{-1}), which are below the permissible limit recommended by the Chinese national standards (Table 1) [20]. Some previous works have suggested that the accumulation of heavy metals in fish depends on the dietary ingestion rate and the concentration of heavy metals in the ingested food [21]. Thus, the low level of heavy metal concentration in pufferfish skin might be due to the artificial feed of lower heavy metals in aquaculture pufferfish. Moreover, the metal concentration in the collagen sample in this study was Pb (0.421 ± 0.015 mg kg^{-1}), Cd (0.023 ± 0.002 mg kg^{-1}), As (0.052 ± 0.001 mg kg^{-1}), and Cr (0.863 ± 0.072 mg kg^{-1}), which are also below the permissible limit recommended by the Chinese national standards (Table 1) [20].

The pufferfish (*T. flavidus*) skin contains 1.7 ± 0.5 MU/g of tetrodotoxin. According to Tani [22], the oral lethal dose of tetrodotoxin for a human being is 10,000 MU, and pufferfish with a tetrodotoxin content of less than 10 MU/g is considered to be nontoxic. Thus, pufferfish (*T. flavidus*) skin is safe for use in food, cosmetics, and pharmaceutical raw materials. Additionally, the content of tetrodotoxin in the collagen sample in this study was below the detection limit, indicating the safety of pufferfish collagen.

Table 1. Heavy metal concentrations (mg kg^{-1} dry weight).

Heavy Metal	Pufferfish Skin		Chinese National Standards [a]	
		Collagen Sample	Food Additive Gelatin (GB 6783)	Maximum Concentrations of Contaminants in Foods (GB 2762)
Pb	0.562 ± 0.052	0.421 ± 0.015	≤1.5	≤0.5
Cd	0.042 ± 0.004	0.023 ± 0.002	-	≤0.1
As	0.075 ± 0.002	0.052 ± 0.001	≤1.0	≤0.1
Cr	1.152 ± 0.067	0.863 ± 0.072	≤2.0	≤2.0

[a] Chinese national standards were obtained from the literature described by Chen et al. (2016).

2.2. Sodium Dodecyl Sulfate Polyacrylamide Gel Electrophoresis (SDS-PAGE) Analyses of T. flavidus Collagen

The electrophoretic pattern of *T. flavidus* collagen was similar to the authentic standard of rat type I collagen (Figure 1), indicating an intact collagen profile after electrodialysis. It consisted of four major protein components with molecular weights of 122, 130, 250, and 310 kDa. The first two components had mass values close to α1 and α2 subunits of rat type I collagen. The ratio of 122/130 kDa components (~1:2) was on a level similar to rat collagen standard and consistent with a previous report on type I collagen extract of *T. rubripes* skin [23]. Therefore, 122 and 130 kDa proteins were identified as α2 and α1 subunits of type I collagen, respectively. On the other hand, the last two protein components (250 and 310 kDa) were tentatively identified as β subunit (dimer) and γ subunit (trimer) of type I collagen, respectively. These two proteins had counterparts with similar molecular weights in rat type I collagen standard (Figure 1) as well as other fish collagen [24]. They were possibly formed via intermolecular and/or intramolecular cross-linking of collagen subunits [25]. As starvation is believed to stimulate collagen cross-linking of fish, the difference in β and γ subunit contents between *T. flavidus* and *T. rubripes* collagen were probably due to variations in both species and feeding conditions [23]. In addition, quantification of stained protein bands showed that α1, α2, β, and γ subunits contributed to 96.1 ± 1.3% of total *T. flavidus* collagen. This indicated that extraction by salting-out electrodialysis was able to produce pure type I collagen for further application.

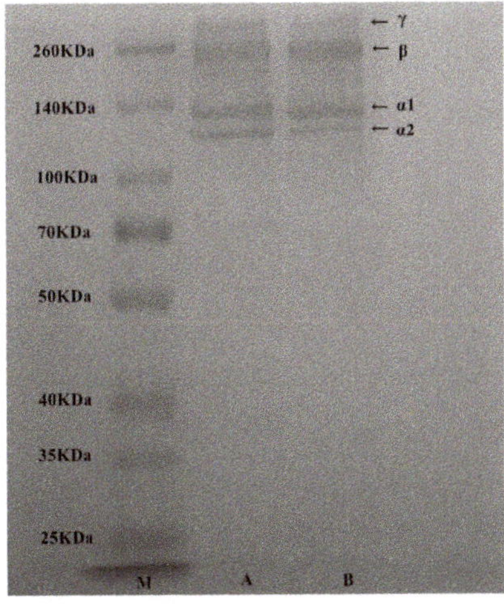

Figure 1. SDS-PAGE of molecular weight standard (lane M), authentic type I collagen standard from rat tail (lane A), and *T. flavidus* collagen extract (lane B).

2.3. Amino Acid Composition of T. flavidus Collagen

T. flavidus collagen demonstrated the characteristic amino acid composition of type I collagen. Gly was the most abundant residue of the collagen, accounting for a quarter of the total amino acid components (Table 2, 268 ± 0.08 residues/1000 amino acid residues). This observation is consistent with the common understanding that Gly content is the highest among all amino acid residues as type I collagen is featured with repeating Gly–Pro–X and/or Gly–X–Hyp sequence (X can be any amino acid residues other than Gly, Pro, and Hyp) [26].

Imino acids (Pro and Hyp) had the second highest amino acid content among all residues of *T. flavidus* collagen (246 ± 0.04 residues/1000 amino acid residues), consistent with a previous study on *T. rubripes* skin collagen (170 residues/1000 amino acid residues) [2,23]. They have been reported to decrease the entropic cost of collagen folding by preorganizing individual poly-Pro II chain [27]. They can also stabilize collagen triple helices via interchain hydrogen bond through hydroxyl groups [28]. The content of imino acids is therefore an important factor modulating collagen thermal stability [27,29]. The imino acid content of *T. flavidus* collagen was higher than many fishes, including big eye snapper, grass carp, and tiger pufferfish (167–195 residues/1000 amino acid residues), but similar to those from their porcine counterparts (220 residues/1000 amino acid residues) [16,24,30]. This suggests that *T. flavidus* collagen might have nutritional values similar to those of common mammalian collagen and could therefore potentially be used as an alternative source of gelatin.

Table 2. Amino acid compositions of *T. flavidus* collagen (residues/1000 amino acid residues).

Amino acids	*T. flavidus*
Glycine	268 ± 0.08
Proline	164 ± 0.05
Hydroxyproline	82.1 ± 0.04
Arginine	32.7 ± 0.12
Hydroxylysine	9.2 ± 0.09
Lysine	37.7 ± 0.08
Alanine	119 ± 0.11
Threonine	39.8 ± 0.03
Valine	31.0 ± 0.09
Serine	14.7 ± 0.13
Isoleucine	13.9 ± 0.07
Leucine	33.8 ± 0.05
Methionine	18.5 ± 0.07
Histidine	13.7 ± 0.06
Phenylalanine	18.1 ± 0.03
Glutamine acid	37.1 ± 0.06
Aspartic acid	60.4 ± 0.05
Cysteine	0.57 ± 0.09
Tyrosine	5.43 ± 0.05

2.4. Spectrophotometric Characterization

The spectrophotometric characterization was in agreement with SDS-PAGE and amino acid composition analyses, further confirming that electrodialysis maintained the physicochemical integrity of type I collagen from *T. flavidus* skin. In Figure 2A, it can be seen that *T. flavidus* collagen and authentic type I collagen standard both have bell shape UV spectra, with the maximum absorption wavelength around 234 nm. This strong absorption can be attributed to peptide bond absorptions by $n \rightarrow \pi^*$ transitions within C=O, COOH, and $CONH_2$ groups of collagen peptide chains [25]. Similar to type I collagen of rat tail, *T. flavidus* collagen also absorbed weakly around 250 and 280 nm. The inability to absorb at higher UV regions is related to the deficiency of Tyr and Phe in *T. flavidus* collagen (<30 residues/1000 residues, Table 2) because Tyr and Phe are the major chromophores responsible for absorption at 251 and 276 nm [31]. This phenomenon suggests a possible deprivation of protease-labile telopeptides from type I collagen [31] during isolation and purification with salting-out electrodialysis extraction. In addition, circular dichroism (CD) analysis was in line with UV results, showing identical spectra between *T. flavidus* collagen and rat type I collagen standard (Figure 2B). Both samples had spectra with a positive amplitude at 221 nm and a negative amplitude at 197 nm. It is also in agreement with recent research on scale collagen of pacific saury [32].

Consistent with UV analysis, *T. flavidus* collagen demonstrated a typical Fourier transform infrared (FTIR) spectrum of type I collagen. Five characteristic peaks were identified in both *T. flavidus* collagen and rat type I collagen: amide A, B, I, II, and III (Figure 2C). The wavenumber of amide A (3311 cm^{-1}) was lower than that of free N–H stretching vibration (3400–3440 cm^{-1}) [33]. This red-shift is in agreement with the extensive distributions of N–H(Gly$\cdots$O=C(X) hydrogen bonds among *T. flavidus* collagen helices during the formation of triple-helical structures [34]. Amide B reflected asymmetrical CH_2 stretching within collagen peptides [35], and it had a wavenumber (2926 cm^{-1}) similar to theoretical values. As an indicator of C=O stretching vibration [33], amide I of *T. flavidus* collagen (1645 cm^{-1}) was modestly red-shifted from the calculated value (~ 1660 cm^{-1}) toward lower wavenumber [33]. This was again a by-product of intercollagen cross-linking by N–H(Gly)$\cdots$O=C(X) hydrogen bonds [34]. N–H bending coupled with C–N stretching vibration actively contributed to amide II formation, inducing FTIR absorption at 1550–1600 cm^{-1} [35]. The amide II wavenumber of *T. flavidus* collagen (1551 cm^{-1}) was at the lower region of this range, further confirming the influence of interhelical hydrogen bonds. In addition, C–H stretching and N–H bending vibrations were detected in *T. flavidus* collagen [36] with evidence on amide III (1242 cm^{-1}) absorption.

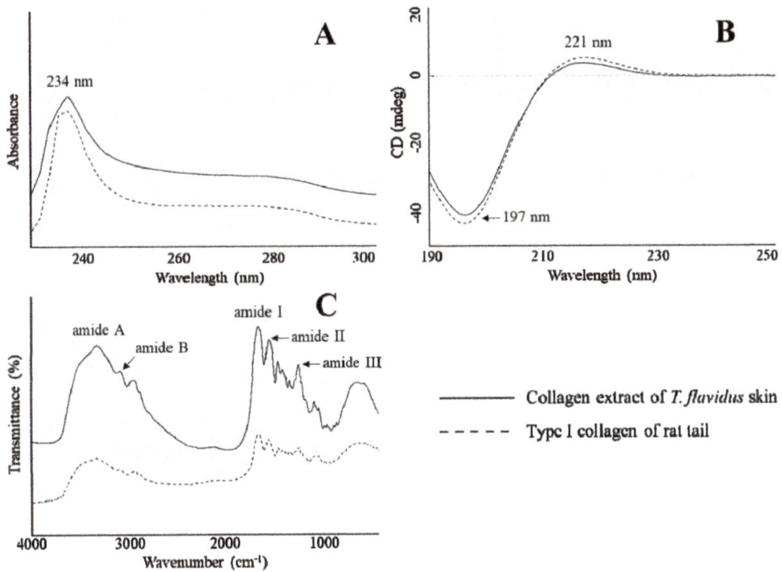

Figure 2. Spectrophotometric characterization of *T. flavidus* collagen extract and type I collagen of rat tail. (**A**) UV spectra; (**B**) circular dichroism (CD) spectra, and (**C**), FTIR spectra.

2.5. Thermal Properties

The thermal stability of *T. flavidus* collagen was characterized by denaturation temperature (T_d) and maximum transition temperature (T_{max}). T_d refers to the temperature at which the triple-helical collagen structure in solution is disintegrated into random coil [16]. The viscosity of *T. flavidus* collagen ($T_d = 28.4 \pm 2.5$ °C, Figure 3A) was found to decrease modestly slower than that of Alaska pollack ($T_d = 16.8$ °C) [37], indicating a milder denaturing process for *T. flavidus* collagen. Furthermore,

T_{max} refers to the temperature at which collagen fiber shrinks to one third of its length. The phase transition involving the conversion of a crystalline triple-helical collagen structure to an amorphous random coil form occurs during the shrinkage process [38]. Differential scanning calorimetry (DSC) analysis of *T. flavidus* collagen confirmed this observation with a T_{max} value of $41.8 \pm 0.35\,°C$ (Figure 3B). The difference between the T_{max} and T_d values of *T. flavidus* collagen was about 13 °C. The T_{max} value of *T. flavidus* collagen was higher than those of bighead carp, bigeye snapper, and grass carp (24.6–33.3 °C) [14,16,24]. It is therefore reasonable to speculate that *T. flavidus* collagen tri-helices are more stable at higher temperature than other fishes. More interestingly, the T_{max} value of *T. flavidus* collagen was similar to calf skin collagen (40.8 °C) [39]. Considering the importance of thermal stability for collagen applications in foods [5] and the aquacultural potential of *T. flavidus* [40], we postulate that collagen from *T. flavidus* could potentially be used as an alternative source of collagen.

The superior thermal stability of *T. flavidus* collagen was likely due to its imino acid content. Imino acids, especially Hyp, have been known for their ability to stabilize collagen tri-helices via intermolecular hydrogen bonds [41]. Our study showed that *T. flavidus* collagen (Table 2, 246 ± 0.04 residues/1000 amino acid residues) had significantly higher imino acid content than the skin collagen of bighead carp, bigeye snapper, grass carp, and *T. rubripes* ($p < 0.05$, 165–193 residues/1000 amino acid residues) [14,16,23,24]. The additional imino acids might form extra hydrogen bonds within *T. flavidus* collagen tri-helices, therefore increasing the molecular stability through rising entropy [42].

In addition, a secondary exothermal peak ($32.9 \pm 0.31\,°C$) was identified in the DSC thermogram (Figure 3B). It was consistent with our observation on the viscosity changes in *T. flavidus* collagen (Figure 3A), indicating a partial denaturation of collagen supramolecular structure due to defibration of thermally unstable hydroxyproline-free sequence in collagen triple helices [28].

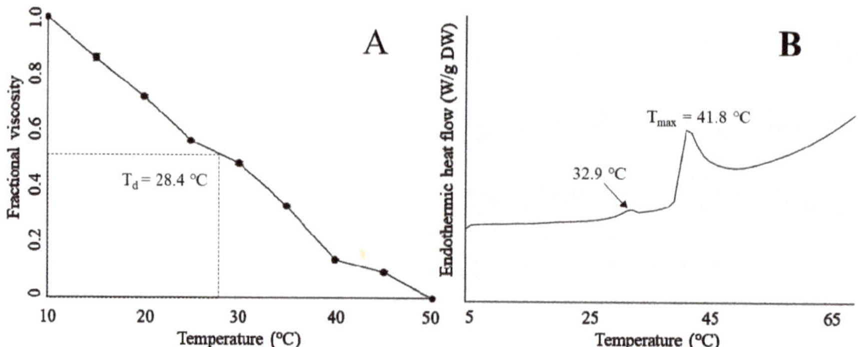

Figure 3. Fractional viscosity (**A**) and Differential scanning calorimetry (DSC) (**B**) of *T. flavidus* collagen extract.

2.6. Relative Solubility

In the absence of NaCl, 0.5 mol/L acetic acid was able to fully dissolve *T. flavidus* collagen at a concentration of 0.16 mg/mL. Increments in NaCl level (≤ 3 g/100 mL) appeared to have a modest impact on the collagen solution, leading to a minor reduction in the relative solubility (99.5%–89.8%, Figure 4A). Higher levels of NaCl (≥ 4 g/100 mL), however, significantly decreased the relative solubility (11.0%–11.7%, $p < 0.05$). This phenomenon is in agreement with an earlier study on catfish skin collagen, which significantly precipitates at high NaCl levels (≥ 4 g/100 mL) [43]. Considering that Na+ and Cl− are the major forces depriving water molecules from hydrophilic amino acid residues of collagen during NaCl-mediated salting-out process [44], it is reasonable to postulate that collagen solubility is affected by the ionic strength of solutions. To avoid collagen aggregation due to potential salting-out events, the ionic strength of 1.01 (equivalent to the solution containing 3 g/100 mL NaCl and 0.5 mol/L acetic acid) could serve as a threshold value for preparing *T. flavidus* collagen in other salt solutions.

The protein content of *T. flavidus* collagen at pH 3.0 (0.04 mg BSAE/mL) was used to determine the relative solubility of the collagen among all pH conditions (Figure 4B). While adjusting pH values within acidic environments could modestly reduce relative solubility to 82.1 ± 0.1% (1 ≤ pH ≤ 3.0, $p < 0.05$), sharper reductions were observed between pH 3.0 and 5.0, with the lowest relative solubility being 37.9 ± 0.01%. This event was also reported by Singh and colleagues in a research involving catfish skin collagen [43]. As ionic strengths of *T. flavidus* collagen solutions (0.50–0.65) were much lower than the threshold value (1.01) during pH adjustments, it might have had a minor impact on the relative solubility of the collagen.

The reduction in relative solubility of *T. flavidus* collagen between pH 3.0 and 5.0 was most likely due to deprotonation of the charged amino acid residues. In acidic environments (pH ≤ 3.0), collagen behaved as a positively charged particle, with most charged amino acid protonated. The net positive charges of collagen assisted in forming hydrogen bonds with water molecules and preventing aggregation with surrounding proteins [45]. However, when the solution pH was increased from 3.0 to 5.0, the collagen lost a large fraction of net positive charges due to deprotonation of the side chains within Asp (pK 3.86) and Glu (pKa 4.25), which accounted for 49.2% of the charged amino acid residues of the collagen (Table 2). Considering the increasing hydrophobic interactions among neutralized collagens [44], it is reasonable to postulate a sharp reduction in relative solubility.

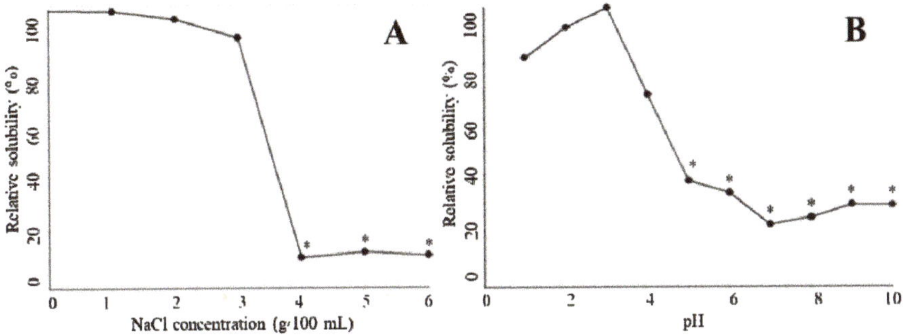

Figure 4. Solubility of *T. flavidus* collagen extract in the presence of varied NaCl concentration (**A**) or pH condition (**B**).

In neutral or basic environments (pH ≥ 6), the relative solubility of *T. flavidus* collagen continuously decreased, with the lowest value found at pH 7.0 (22.7 ± 0.02%, Figure 4B). This observation is in accordance with previous studies on type I collagen [12]. Changes in relative solubility were partially due to the elevated ability of imino acids (Hyp in particular) to facilitate intermolecular cross-linking within collagen tri-helical structures [41,42] as imino acids contributed to 24.6% of the total amino acid compositions of *T. flavidus* collagen (Table 2). It could also be attributed to alterations in total net charges of collagen molecules as electrostatic expulsion is too weak to shield proteins from aggregation around the isoelectric point (pI ≈ 7 for type I collagen) [45].

2.7. Cell Proliferation

The greater the optical density (OD) value indicted, the greater is the number of cells. As shown in Figure 5, the OD value of the negative control and experimental group increased with culture time. However, at different points in time, the OD values of the experimental groups were slightly higher than those of the negative control group, although the difference was not significant. This may be because the cells were cultured at 37 °C, which was above the denaturation temperature of 28.4 ± 2.5 °C for the collagen samples. When the collagen sample was kept at 37 °C for a long time, the collagen degraded, leading to loss of collagen activity. In the next stage of research, one important aim of our research will be to improve the thermal stability of collagen.

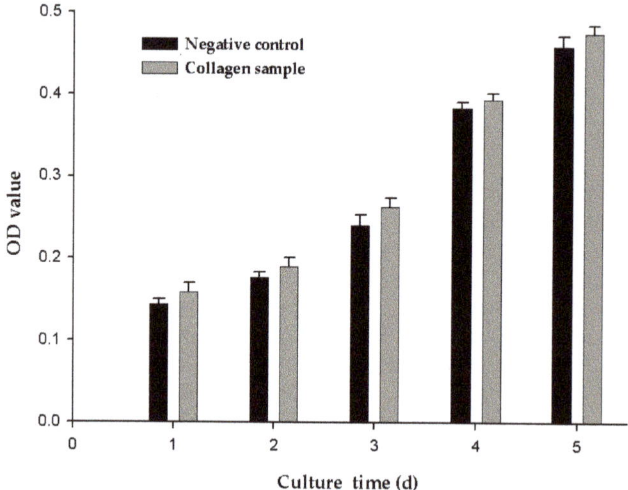

Figure 5. MTT assay detecting the effects of collagen sample on cell proliferation of NIH3T3 cell.

2.8. Extraction of T. flavidus Collagen

To our knowledge, this is the first time electrodialysis was introduced as part of fish collagen purification. The application of electrodialysis significantly increased the extraction yield of fish collagen. Our experiments showed that 67.3 ± 1.3 g of *T. flavidus* collagen could be produced from 100 g of dry skin, which is much higher than the yields of *T. rubripes* collagen (10.7–44.7 g/100 g dried weight, DW) extracted by traditional dialysis [23,30]. This discernible difference could be partially attributed to the variations in collagen contents between the two *Takifugu* species. It is also likely due to electrodialysis-mediated improvements of extraction efficiency as the extraction yield obtained from this study was significantly higher than those reported by previous dialysis research on tilapia (27.2 g/100 g DW) [46] and grass carp (46.6 g/100 g DW) [16].

This study showed that electrodialysis produced more than 10 kg of collagen samples in one purification effort by purifying 100 L of crude collagen solution. It also concentrated the purification step from a traditional 96-h dialysis [13–15] into a 2-h period. Considering that the traditional method could not purify 100 L of crude collagen solution in one effort [12–14], the application of electrodialysis is estimated to significantly reduce instrument and labor time for large-scale production. This demonstrates that electrodialysis has a huge advantage over traditional dialysis for translation into industrial use.

In addition to improving purification efficiency and productivity, electrodialysis is a cost-effective technique with better environmental sustainability relative to the traditional method. In our study, to produce 10 kg of *T. flavidus* collagen, electrodialysis consumed 45.0 L of acetic acid and produced 600 L of waste water (Table 3). In comparison, traditional dialysis is estimated to use 400 times more acetic acid and 600 m of additional dialysis bag [12,14,15] while producing 600 times more waste water to purify the same amount of collagen extract. In total, it would take US$15 to produce 10 kg of collagen using electrodialysis extraction. This indicates a strong potential of the electrodialysis technique to be used in industrial applications.

Table 3. Expenses and instrumental requirements for producing 10 kg of *T. flavidus* collagen by electrodialysis compared to traditional dialysis.

	Traditional Dialysis	**Electrodialysis**
Essential Instruments	Dialysis bag, beakers and stirring hot plates	DSA-II electrodialysis apparatus
Instrument time (h/dialysis)	>96	2.0
Dialysis bag (m)	600	none
Acetic acid (mole)	18,000	45.0
Waste water (L)	360,000	600

Expenses and essential instruments for traditional dialysis was estimated based on the studies of Liu et al. (2012), Matmaroh et al. (2011), Nagai and Suzuki (2000), and Ogawa et al. (2004).

3. Materials and Methods

3.1. Materials and Chemicals

T. flavidus skin was purchased from the local fishery factory (Zhangzhou, China). Upon arrival, the skin was washed with chilled distilled water and then stored at $-20\ ^{\circ}$C until use. Coomassie blue R-250, N,N,N',N'-tetramethylethylenediamime and sodium dodecyl sulfate were purchased from Bio-Rad Laboratories (Hercules, CA, USA). Type I collagen (C3867, rat) was purchased from Sigma-Aldrich Inc. (St. Louis, MO, USA). Electrophoresis loading buffer was obtained from Sangon Biotech Co., Ltd. (Shanghai, China). Other reagents (analytical grade) were purchased from Xiamen Green Reagent Glass Instrument Co. Ltd. (Xiamen, China).

3.2. Isolation and Purification of T. flavidus Collagen

T. flavidus skin (15 kg in dry weight, DW) was incubated in 150 L of 0.1 mol/L sodium bicarbonate for 3 h with continuous stirring and then rinsed with cold water until a neutral pH was reached. The resulting material was hydrolyzed in 120 L of 0.5 mol/L acetic acid for 24 h and then centrifuged at $10,000\times g$ for 15 min to remove non-collagenous proteins and pigments. Skin collagen was salted out from the supernatant in the presence of 0.5 mol/L NaCl, followed by 20 min of $10,000\times g$ centrifugation. The resulting precipitate was dissolved in 0.5 mol/L acetic acid to produce 100 L of crude collagen solution, which was then purified by electrodialysis. The samples were loaded to a DSA-II electrodialyzer (Jiangsu Ritai Environmental Protection Engineering Co., Ltd., Yangzhou, China) equipped with both polyethylene cation-exchange membrane (361BW, Jiangsu Ritai Environmental Protection Engineering Co., Ltd., China) and polyethylene anion-exchange membrane (362BW, Jiangsu Ritai Environmental Protection Engineering Co., Ltd., China). The process conditions of electrodialysis were as follows: feed compartment, 150 L of 3% NaCl (w/v); concentrate compartment, 150 L of distilled water; electric potential, 80 V/cm; flow rate, $\leq 1\ \mathrm{m}^3$/h; electrodialysis duration, 2 h. The purification products were lyophilized and stored at $-20\ ^{\circ}$C until analysis. Isolation and purification were all performed at 4 $^{\circ}$C to prevent microbial growth (Figure 6). The extraction yield was calculated based on the following equation.

$$\text{Yield (g/100 g DW)} = \text{Mass of lyophilized collagen (g)/Mass of dry fish skin (100 g)} \tag{1}$$

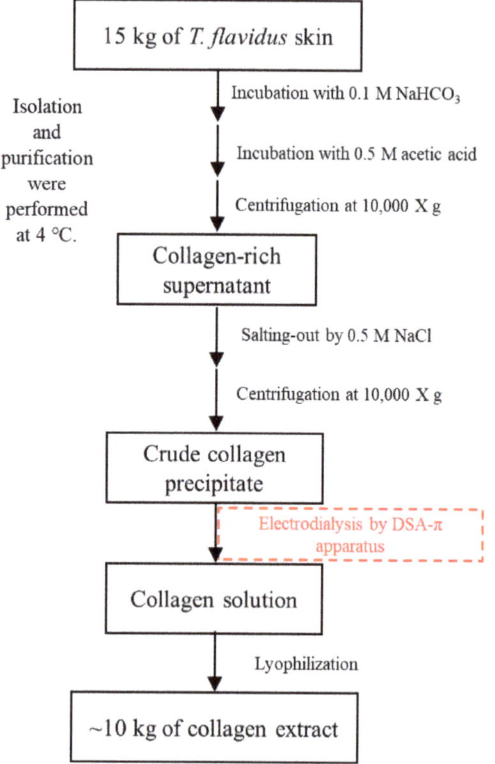

Figure 6. Isolation and purification of collagen from *T. flavidus* skin by electrodialysis.

3.3. Quality Measurements

3.3.1. Heavy Metal

Toxic metals in the pufferfish skin and collagen sample were characterized based on a method described by Kosker et al. [47] with slight modification. Smashed pufferfish skin and collagen sample were dissolved in nitric acid (25% v/v) and then digested using a Milestone ETHOS A microwave digestion instrument (Milestone Srl., Sorisole, Italy). Inductively coupled plasma mass spectrometry (ICP-MS, Thermo Fisher X Series II, Thermo Fisher Scientific Inc., Waltham, MA, USA) was used to identify these metals. Standard solutions for the calibration curve were prepared according to the dilutions of the toxic metals. Prepared solutions of toxic metals had levels of Pb, Cd, As, and Cr within the range of 1–50 mg/L.

3.3.2. Tetrodotoxin

Tetrodotoxin in the pufferfish skin and collagen sample was characterized based on a method described by Wang et al. [48] with slight modification. Smashed pufferfish skin and collagen sample were incubated in 0.5 mol/L acetic acid at 100 °C for 40 min and then centrifuged at $10,000 \times g$ for 15 min. The resulting supernatants were collected and analyzed directly with an enzyme-linked immunosorbent assay (ELISA) test kit (Beijing Zhongnuo Taian Technology Co. Ltd., Beijing, China) for tetrodotoxin. As described in the manual, samples and enzyme-labeled MAb-TTX were added to wells of a microplate in which standard tetrodotoxin had been immobilized as a competitive agent. The absorbance (OD value) was measured at 490 nm. The ratio of absorbance (Ai/Ao, where Ai is the absorbance of standard solution, and Ao is the absorbance of blank with no tetrodotoxin added) was used as the ordinate. The logarithm of the corresponding standard solution concentration was plotted as the abscissa, a standard curve was prepared, and the tetrodotoxin content of the sample was calculated.

3.4. SDS-PAGE

SDS-PAGE was conducted as described in our earlier study [2]. Briefly, 300 µL of 2 µg/µL *T. flavidus* extract was mixed with 100 µL of loading buffer, incubated in boiling water for 3 min, cooled at room temperature, and then centrifuged at $8500 \times g$ for 5 min. Supernatant (5 µL/lane) was characterized by a polyacrylamide gel (4% stacking and 8% running) on a mini vertical electrophoresis system (Bio-Rad Laboratories, US). After electrophoresis, the gel was incubated in fixing solution (methanol:acetic acid:water, 50:10:40) for 30 min, stained by 0.1% Coomassie brilliant blue for 30 min, and then rinsed by 30% methanol containing 10% acetic acid for 30 min. Contents of the collagen subunits were estimated by Quantity One 4.6.0 (Bio-Rad Laboratories, US).

3.5. Amino Acid Composition

Amino acid composition of sample was conducted as described in our earlier study [2]. The *T. flavidus* collagen extract (5 mg) was hydrolyzed with 0.50 mL of 6.0 M hydrochloric acid at 110 °C for 8 h. The hydrolysate was vaporized by a vacuum evaporator, dissolved in 25 mL 0.1 M HCl, and then analyzed by applying the hydrolysate to a HITACHI 835-50 amino acid analyzer (Hitachi, Tokyo, Japan). Contents of the amino acid residues were expressed as residues/1000 residues.

3.6. Spectrophotometric Characterization

Spectra of UV, FTIR, and CD of the purified *T. flavidus* collagen were determined using a method described by Wang et al. [49] with slight modification. UV spectrum (226–300 nm) of the *T. flavidus* collagen extract (1 mg/mL) was determined by a UV-1780 spectrophotometer against 0.5 M acetic acid (blank). FTIR (500–4000 cm^{-1}) of the same extract was measured by a horizontal ATR trough plate crystal cell (PIKE technology Inc., Fitchburg, WI, USA) coupled with a Bruker Model VERTEX 70 FTIR spectrometer (Bruker Co., Karlsruhe, Germany). FTIR spectra in the rage of 500–4000 cm^{-1} with automatic signal gain were collected in 16 scans at a resolution of 4 cm^{-1} prior to analysis by OPUS 6.5 data collection software program (Bruker Co., Karlsruhe, Germany). In addition, CD spectrum (190–260 nm) of the collagen extract (0.25 mg/mL) was produced by accumulating 100 scans at a speed of 100 nm/min with 1 nm interval on a Chirascan spectrometer (Applied Photophysics Limited Inc., Leatherhead, UK).

3.7. Relative Solubility

The effects of pH and NaCl on sample solubility were characterized as described in our earlier study [46].

The relative solubility of the *T. flavidus* collagen extract was determined in the presence of 0–6 g/100 mL NaCl. Briefly, 8 mL of 3 mg/mL collagen extract was mixed with 5 mL of NaCl solution containing 0.5 M acetic acid. The mixture was centrifuged at $20,000 \times g$ and 4 °C for 30 min. The protein concentration of the supernatant was measured according to the calibration curve for bovine serum albumin by Folin assay [50]. The absolute protein concentration of the collagen extract was expressed as mg bovine serum albumin equivalent/mL (mg BSAE/mL). The relative solubility was calculated with Equation (2) below by normalizing the protein concentrations collected at all NaCl levels with values obtained from collagen extracts that received no NaCl treatment.

$$\text{Relative solubility (\%)} = \frac{\text{(Protein concentration of supernatant with NaCl treatment)}}{\text{(Protein concentration of supernatant without NaCl treatment)}} \times 100 \quad (2)$$

The relative solubility of the *T. flavidus* collagen extract was also determined at varied pH values (1–10). The collagen extract (8 mL, 3 mg/mL) was adjusted to designed pH condition by 6.0 M HCl (or NaOH) with a final volume of 10 mL. The resulting solution was centrifuged at 20,000× g and 4 °C for 30 min. The protein concentration of the supernatant was determined by Folin assay as described above. The relative solubility was calculated with Equation (3) below by normalizing the protein concentrations collected at pH conditions with values obtained at pH 3.0.

$$\text{Relative solubility (\%)} = (\text{Protein concentration of supernatant at varied pH})/ \\ (\text{Protein concentration of supernatant at pH 3.0}) \times 100 \tag{3}$$

3.8. DSC

The thermal properties of the sample were characterized by an established DSC method [14] with slight modification. Lyophilized collagen extract was hydrated with 0.05 M acetic acid (1:40 w/v), incubated at 4 °C for 2 days and then characterized using a DSC 2 calorimeter (Mettler-Toledo, Zurich, Switzerland). Heat flow of the hydrated samples was measured between 5 and 75 °C at 1 °C/min using an empty aluminum pan as reference. T_{max} was defined as the peak of the transition curve.

3.9. Viscosity

The viscosity of the sample was characterized based on a method described by Kittiphattanabawon et al. [24] with slight modification. Briefly, protein dispersion (0.1% w/v) was prepared by hydrating the collagen sample in 0.05 M acetic acid at 4 °C for 2 days. An Ostwald's viscometer (Kusano Inc., Tokyo, Japan) loaded with collagen dispersion was incubated in a 15 °C water bath for an extended period. Small temperature increments (2 °C) were then applied to the viscometer at 30-min intervals in a stepwise manner. Viscosity was recorded between 15 and 50 °C prior to each temperature change. Fractional viscosity was calculated with Equation (4). T_d was defined as the temperature for fractional viscosity to reach 0.50.

$$\text{Fractional viscosity} = (\text{Measured viscosity} - \text{minimum viscosity})/ \\ (\text{Maximum viscosity} - \text{minimum viscosity}) \tag{4}$$

3.10. Cell Proliferation

Cell proliferation experiments were performed using a method described by Golser et al. [51] with slight modification. The collagen sample was dissolved in 0.5 mol/L acetic acid and cast into cell culture wells at a final concentration of 0.5 mg/mL. After the collagen was fully dried, the plates were sterilized via UV treatment and seeded with the NIH3T3 fibroblasts. NIH3T3 fibroblasts seeded in the plates without collagen served as a negative control. After adding Dulbecco's Modified Eagle Medium (DMEM) cell culture medium to the plates, the cells were cultured in an incubator at 37 °C. At cell seeded times of 1, 2, 3, 4, and 5 days, 3-(4,5-dimethylthiazol-2-y)-2,5-diphenyltetrazolium bromide (MTT) solution was added to the cell culture medium and incubated for 4 h at 37 °C. The plates were removed, and the culture medium was aspirated. After adding dimethyl sulfoxide (DMSO) for 10 min, the absorbance (OD value) was measured at 490 nm.

3.11. Statistical Analyses

All data in this study are expressed as average ± standard deviation of at least three measurements. One-way ANOVAs coupled with Dunnett's test was performed using SPSS 13.0 to determine the statistical difference at 95% confidence level.

4. Conclusions

Our study, for the first time, introduced electrodialysis for the extraction of fish skin collagen (*T. flavidus*). This cost-effective technique was found to be superior to traditional dialysis with its advanced efficiency (2 h/trial), large capacity (100 L/trial), high extraction yield (67.3 $\pm$ 1.3 g/100 g DW), and better environmental sustainability (6 L waste water/L sample). SDS-PAGE, amino acid composition analysis, and spectrophotometric characterization demonstrated that *T. flavidus* collagen extracted by electrodialysis primarily consisted of type I collagen with a purity of 96.1 $\pm$ 1.3%. ICP-MS analysis demonstrated that the heavy metals of *T. flavidus* collagen were less than the Chinese national standards. ELISA analysis indicated the safety of *T. flavidus* collagen. Notably, the collagen appeared to have better thermal stability than other fish species, with T_{max} and T_d of 41.8 $\pm$ 0.35 and 28.4 $\pm$ 2.5 °C, respectively. This observation can be attributed to its relative higher imino acid content (246 $\pm$ 0.04 residues/1000 amino acid residues). Cell proliferation experiment demonstrated that the cell proliferation rates of the experimental groups were slightly higher than those of the negative control group, but there was no significant difference. In addition, both NaCl level and pH condition were found to affect the relative solubility of *T. flavidus* collagen. All evidence suggests that electrodialysis is a promising technique for fish collagen and that *T. flavidus* skin collagen could serve as an alternative source of collagen to meet the increasing demand from academia and industry.

Author Contributions: J.C. completed the study design, most of the experiments and manuscript preparation; M.L. and R.Y. took part in data analysis; K.B., G.W., R.T., S.S., and N.X. took part in experiments and data analysis.

Funding: This work was gratefully supported by National Natural Science Foundation of China (41676129 & 41106149), and the Scientific Research Foundation of the Third Institute of Oceanography, SOA (2016007).

Conflicts of Interest: The authors declare no conflict of interest.

References

1. Chen, J.D.; Li, L.; Yi, R.Z.; Gao, R.; He, J.L. Release kinetics of Tilapia scale collagen I peptides during tryptic hydrolysis. *Food Hydrocoll.* **2018**, *77*, 931–936. [CrossRef]
2. Chen, J.D.; Liu, Y.; Yi, R.Z.; Li, L.; Gao, R.; Xu, N.H.; Zheng, M.H. Characterization of collagen enzymatic hydrolysates derived from lizardfish (*Synodus fuscus*) scales. *J. Aquat. Food Prod. Technol.* **2017**, *26*, 86–94. [CrossRef]
3. An, B.; Lin, Y.S.; Brodsky, B. Collagen interactions: Drug design and delivery. *Adv. Drug Deliv. Rev.* **2016**, *97*, 69–84. [CrossRef]
4. Ferreira, A.M.; Gentile, P.; Chiono, V.; Ciardelli, G. Collagen for bone tissue regeneration. *Acta Biomater.* **2012**, *8*, 3191–3200. [CrossRef] [PubMed]
5. Gomez-Guillen, M.C.; Gimenez, B.; Lopez-Caballero, M.E.; Montero, M.P. Functional and bioactive properties of collagen and gelatin from alternative sources: A review. *Food Hydrocoll.* **2011**, *25*, 1813–1827. [CrossRef]
6. Wang, G.Y.; Xiao, M.T.; Zhao, P.; Chen, J.D. Progress on collagen aggregates and their aggregation behavior. *Curr. Biotechnol.* **2017**, *7*, 587–593.
7. Jridi, M.; Bardaa, S.; Moalla, D.; Rebaii, T.; Souissi, N.; Sahnoun, Z.; Nasri, M. Microstructure, rheological and wound healing properties of collagen-based gel from cuttlefish skin. *Int. J. Biol. Macromol.* **2015**, *77*, 369–374. [CrossRef]
8. Nicholas, R.A.J.; Ayling, R.D.; McAuliffe, L. Vaccines for Mycoplasma Diseases in Animals and Man. *J. Comp. Pathol.* **2009**, *140*, 85–96. [CrossRef]
9. Venkatesan, J.; Anil, S.; Kim, S.-K.; Shim, M.S. Marine fish proteins and peptides for cosmeceuticals: A review. *Mar. Drugs* **2017**, *15*, 143. [CrossRef]
10. FAO. *The State of World Fisheries and Aquaculture*; Food and Agriculture Organization of the United Nations: Rome, Italy, 2016.
11. Wang, Q.L.; Zhang, H.T.; Ren, Y.Q.; Zhou, Q. Comparison of growth parameters of tiger puffer *Takifugu rubripes* from two culture systems in china. *Aquaculture* **2016**, *453*, 49–53. [CrossRef]
12. Matmaroh, K.; Benjakul, S.; Prodpran, T.; Encarnacion, A.B.; Kishimura, H. Characteristics of acid soluble collagen and pepsin soluble collagen from scale of spotted golden goatfish (*Parupeneus heptacanthus*). *Food Chem.* **2011**, *129*, 1179–1186. [CrossRef]
13. Ogawa, M.; Portier, R.J.; Moody, M.W.; Bell, J.; Schexnayder, M.A.; Losso, J.N. Biochemical properties of bone and scale collagens isolated from the subtropical fish black drum (*Pogonia cromis*) and sheepshead seabream (*Archosargus probatocephalus*). *Food Chem.* **2004**, *88*, 495–501. [CrossRef]

14. Liu, D.S.; Liang, L.; Regenstein, J.M.; Zhou, P. Extraction and characterisation of pepsin-solubilised collagen from fins, scales, skins, bones and swim bladders of bighead carp (*Hypophthalmichthys nobilis*). *Food Chem.* **2012**, *133*, 1441–1448. [CrossRef]

15. Nagai, T.; Suzuki, N. Isolation of collagen from fish waste material—Skin, bone and fins. *Food Chem.* **2000**, *68*, 277–281. [CrossRef]

16. Zhang, Y.; Liu, W.T.; Li, G.Y.; Shi, B.; Miao, Y.Q.; Wu, X.H. Isolation and partial characterization of pepsin-soluble collagen from the skin of grass carp (*Ctenopharyngodon idella*). *Food Chem.* **2007**, *103*, 906–912. [CrossRef]

17. Strathmann, H. Electrodialysis, a mature technology with a multitude of new applications. *Desalination* **2010**, *264*, 268–288. [CrossRef]

18. Suwal, S.; Roblet, C.; Amiot, J.; Doyen, A.; Beaulieu, L.; Legault, J.; Bazinet, L. Recovery of valuable peptides from marine protein hydrolysate by electrodialysis with ultrafiltration membrane: Impact of ionic strength. *Food Res. Int.* **2014**, *65*, 407–415. [CrossRef]

19. Sobihah, N.N.; Zaharin, A.A.; Nizam, M.K.; Juen, L.L.; Kyoung-Woong, K. Bioaccumulation of heavy metals in maricultured fish, *Lates calcarifer* (Barramudi), *Lutjanus campechanus* (red snapper) and *Lutjanus griseus* (grey snapper). *Chemosphere* **2018**, *197*, 318–324. [CrossRef]

20. Chen, S.J.; Chen, H.; Xie, Q.L.; Hong, B.H.; Chen, J.D.; Fang, H.; Bai, K.K.; He, J.L.; Yi, R.Z.; Wu, H. Rapid isolation of high purity pepsin-soluble type I collagen from scales of red drum fish (*Sciaenops ocellatus*). *Food Hydrocoll.* **2016**, *52*, 468–477. [CrossRef]

21. Onsanit, S.; Ke, C.; Wang, X.; Wang, K.J. Trace elements in two marine fish cultured in fish cages in Fujian province China. *Environ. Pollut.* **2010**, *158*, 1334–1342. [CrossRef]

22. Tani, I. *Toxicological Studies of Puffers in Japan*; Teikokutosho: Tokyo, Japan, 1945; p. 12.

23. Nagai, T.; Araki, Y.; Suzuki, N. Collagen of the skin of ocellate puffer fish (*Takifugu rubripes*). *Food Chem.* **2002**, *78*, 173–177. [CrossRef]

24. Kittiphattanabawon, P.; Benjakul, S.; Visessanguan, W.; Nagai, T.; Tanaka, M. Characterisation of acid-soluble collagen from skin and bone of bigeye snapper (*Priacanthus tayenus*). *Food Chem.* **2005**, *89*, 363–372. [CrossRef]

25. Yousefi, M.; Ariffin, F.; Huda, N. An alternative source of type I collagen based on by-product with higher thermal stability. *Food Hydrocoll.* **2017**, *63*, 372–382. [CrossRef]

26. Yu, F.M.; Zong, C.H.; Jin, S.J.; Zheng, J.W.; Chen, N.; Huang, J.; Chen, Y.; Huang, F.F.; Yang, Z.S.; Tang, Y.P.; et al. Optimization of extraction conditions and characterization of pepsin-solubilized collagen from skin of giant croaker (*Nibea japonica*). *Mar. Drugs* **2018**, *16*, 29. [CrossRef] [PubMed]

27. Fontaine-Vive, F.; Merzel, F.; Johnson, M.R.; Kearley, G.J. Collagen and component polypeptides: Low frequency and amide vibrations. *Chem. Phys.* **2009**, *355*, 141–148. [CrossRef]

28. Staicu, T.; Circu, V.; Ionita, G.; Ghica, C.; Popa, V.T.; Micutz, M. Analysis of bimodal thermally-induced denaturation of type I collagen extracted from calfskin. *RSC Adv.* **2015**, *5*, 38391–38406. [CrossRef]

29. Bae, I.; Osatomi, K.; Yoshida, A.; Osako, K.; Yamaguchi, A.; Hara, K. Biochemical properties of acid-soluble collagens extracted from the skins of underutilised fishes. *Food Chem.* **2008**, *108*, 49–54. [CrossRef]

30. Tsukamoto, H.; Yokoyama, Y.; Suzuki, T.; Mizuta, S.; Yoshinaka, R.; Akahane, Y. Isolation of collagen from tiger pufferfish parts and its solubility in dilute acetic acid. *Fish. Sci.* **2013**, *79*, 857–864. [CrossRef]

31. Doyle, R.J.; Bello, J. Ultraviolet absorbance changes accompanying the denaturation of soluble collagen and atelocollagen. *Biochem. Biophys. Res. Commun.* **1968**, *31*, 869–876. [CrossRef]

32. Mori, H.; Tone, Y.; Shimizu, K.; Zikihara, K.; Tokutomi, S.; Ida, T.; Ihara, H.; Hara, M. Studies on fish scale collagen of Pacific saury (*Cololabis saira*). *Mat. Sci. Eng. C Mater.* **2013**, *33*, 174–181. [CrossRef]

33. Abe, Y.; Krimm, S.; Randall, H.M. Normal vibrations of crystalline polyglycine-I. *Biopolymers* **1972**, *11*, 1817–1839. [CrossRef] [PubMed]

34. Veeruraj, A.; Arumugam, M.; Ajithkumar, T.; Balasubramanian, T. Isolation and characterization of collagen from the outer skin of squid (*Doryteuthis singhalensis*). *Food Hydrocoll.* **2015**, *43*, 708–716. [CrossRef]

35. Barth, A.; Zscherp, C. What vibrations tell us about proteins. *Q. Rev. Biophys.* **2002**, *35*, 369–430. [CrossRef] [PubMed]

36. Krimm, S.; Bandekar, J. Vibrational spectroscopy and conformation of peptides, polypeptides, and proteins. *Adv. Protein Chem.* **1986**, *38*, 181–364.

37. Kimura, S.; Ohno, Y. Fish type I collagen: Tissue-specific existence of two molecular forms $(\alpha 1)2$ and $\alpha 1\alpha 2\alpha 3$, in *Alaska pollack*. *Comp. Biochem. Physiol. Part B Comp. Biochem.* **1987**, *88*, 409–413. [CrossRef]

38. Yan, M.Y.; Li, B.F.; Zhao, X.; Ren, G.Y.; Zhuang, Y.L.; Hou, H.; Zhang, X.K.; Chen, L.; Fan, Y. Characterization of acid-soluble collagen from the skin of walleye pollock (*Theragra chalcogramma*). *Food Chem.* **2008**, *107*, 1581–1586. [CrossRef]

39. Komsa-Penkova, R.; Koynova, R.; Kostov, G.; Tenchov, B. Discrete reduction of type I collagen thermal stability upon oxidation. *Biophys. Chem.* **2000**, *83*, 185–195. [CrossRef]

40. Zhang, G.Y.; Shi, Y.H.; Zhu, Y.Z.; Liu, J.Z.; Zang, W.L. Effects of salinity on embryos and larvae of tawny puffer *Takifugu flavidus*. *Aquaculture* **2010**, *302*, 71–75. [CrossRef]

41. Shoulders, M.D.; Raines, R.T. Collagen Structure and Stability. *Annu. Rev. Biochem.* **2009**, *78*, 929–958. [CrossRef]

42. Zanaboni, G.; Rossi, A.; Onana, A.M.T.; Tenni, R. Stability and networks of hydrogen bonds of the collagen triple helical structure: Influence of pH and chaotropic nature of three anions. *Matrix Biol.* **2000**, *19*, 511–520. [CrossRef]

43. Singh, P.; Benjakul, S.; Maqsood, S.; Kishimura, H. Isolation and characterisation of collagen extracted from the skin of striped catfish (*Pangasianodon hypophthalmus*). *Food Chem.* **2011**, *124*, 97–105. [CrossRef]

44. Jiang, F.Z.; Horber, H.; Howard, J.; Muller, D.J. Assembly of collagen into microribbons: Effects of pH and electrolytes. *J. Struct. Biol.* **2004**, *148*, 268–278. [CrossRef] [PubMed]

45. Vojdani, F. Solubility. In *Methods of Testing Protein Functionality*, 1st ed.; Hall, G.M., Ed.; Springer: New York, NY, USA, 1996; pp. 11–60. ISBN 9780751400533.

46. Chen, J.D.; Li, L.; Yi, R.Z.; Xu, N.H.; Gao, R.; Hong, B.H. Extraction and characterization of acid-soluble collagen from scales and skin of tilapia (*Oreochromis niloticus*). *LWT-Food Sci. Technol.* **2016**, *66*, 453–459. [CrossRef]

47. Kosker, A.R.; Ozogul, F.; Durmus, M.; Ucar, Y.; Ozogul, Y.; Boga, E.; Ayas, D. Seasonal changes in proximate composition and mineral-heavy metal content of pufferfish (*Lagocephalus sceleratus*) from northeastern mediterranean sea. *Turk. J. Fish. Aquat. Sci.* **2018**, *18*, 1269–1278. [CrossRef]

48. Wang, X.J.; Yu, R.C.; Luo, X.; Zhou, M.J.; Lin, X.T. Toxin-screening and identification of bacteria isolated from highly toxic marine gastropod *Nassarius semiplicatus*. *Toxicon* **2008**, *52*, 55–61. [CrossRef] [PubMed]

49. Wang, J.; Pei, X.L.; Liu, H.Y.; Zhou, D. Extraction and characterization of acid-soluble and pepsin-soluble collagen from skin of loach (*Misgurnus anguillicaudatus*). *Int. J. Biol. Macromol.* **2018**, *106*, 544–550. [CrossRef] [PubMed]

50. Lowry, O.H.; Rosebrough, N.J.; Farr, A.L.; Randall, R.J. Protein measurement with the Folin phenol reagent. *J. Biol. Chem.* **1951**, *193*, 265–275. [PubMed]

51. Golser, A.V.; Röber, H.G.; Scheibel, S. Engineered collagen: A redox switchable framework for tunable assembly and fabrication of biocompatible surfaces. *ACS Biomater. Sci. Eng.* **2018**, *4*, 2106–2114. [CrossRef]

 *marine drugs*

Article

Collagen Extraction Optimization from the Skin of the Small-Spotted Catshark (*S. canicula*) by Response Surface Methodology

María Blanco [1,2,*], José Antonio Vázquez [1,3], Ricardo I. Pérez-Martín [1,2] and Carmen G. Sotelo [1,2]

1   Grupo de Biotecnología y Bioprocesos Marinos, Instituto de Investigaciones Marinas (CSIC), 36208 Vigo, Spain; jvazquez@iim.csic.es (J.A.V.); ricardo@iim.csic.es (R.I.P.-M.); carmen@iim.csic.es (C.G.S.)
2   Laboratorio de Bioquímica de Alimentos, Instituto de Investigaciones Marinas (CSIC), 36208 Vigo, Spain
3   Laboratorio de Reciclado y Valorización de Materiales Residuales (REVAL), Instituto de Investigaciones Marinas (CSIC), 36208 Vigo, Spain
*   Correspondence: mblanco@iim.csic.es

Received: 29 November 2018; Accepted: 27 December 2018; Published: 9 January 2019

Abstract: The small-spotted catshark is one of the most abundant elasmobranchs in the Northeastern Atlantic Ocean. Although its landings are devoted for human consumption, in general this species has low commercial value with high discard rates, reaching 100% in some European fisheries. The eduction of post-harvest losses (discards and by-products) by promotion of a full use of fishing captures is one of the main goals of EU fishing policies. As marine collagens are increasingly used as alternatives to mammalian collagens for cosmetics, tissue engineering, etc., fish skins represent an excellent and abundant source for obtaining this biomolecule. The aim of this study was to analyze the influence of chemical treatment concentration, temperature and time on the extractability of skin collagen from this species. Two experimental designs, one for each of the main stages of the process, were performed by means of Response Surface Methodology (RSM). The combined effect of NaOH concentration, time and temperature on the amount of collagen recovered in the first stage of the collagen extraction procedure was studied. Then, skins treated under optimal NaOH conditions were subjected to a second experimental design, to study the combined effect of AcOH concentration, time and temperature on the collagen recovery by means of yield, amino acid content and SDS-PAGE characterization. Values of independent variables maximizing collagen recovery were 4 °C, 2 h and 0.1 M NaOH (pre-treatment) and 25 °C, 34 h and 1 M AcOH (collagen extraction).

Keywords: fish discards; fish by-products; collagen; cosmetic applications; experimental designs; response surface methodology

---

1. Introduction

The small-spotted catshark (*Scyliorhinus canicula*) is one of the most abundant elasmobranchs in the Northeastern Atlantic Ocean [1]. Although its landings are sometimes devoted for human consumption (rendering 10% and 16% of fish weight in the form of skin and viscera by-products respectively), it has low commercial value and very often is captured as by-catch resulting in a very high discard rate reaching 100% in some European fisheries. The reduction of post-harvest fish losses (discards and by-products) by the promotion of a full use of fishing captures is one of the main purposes of EU fishing policies [2]. The full use of fishing captures includes the transformation of raw materials for the isolation/production of molecules that could be used in a wide variety of applications, which is indeed one of the approaches included in the "blue growth" strategy of the European Commission. One interesting bioactive compound which could be obtained from fish discards is collagen. Collagen is the main protein present in animal connective tissue and although there are several types of collagens, type I is the most abundant in the skin and bone of teleost fish [3]. Type I collagen, which is a fibrillar collagen, is a heterotrimeric molecule composed of two α1-chains and one α-2 chain with a similar molecular weight of about 100 kDa [4].

Collagens obtained from marine sources include several *Osteichthyes* and *Chondrichthyes* species, jellyfish, mollusks, sponges and sea cucumbers, among others [5–10]. As collagens are being used increasingly as alternatives to mammalian collagens for cosmetics, tissue engineering and other biomedical and pharmaceutical uses, due to safety reasons and ethical or religious constraints, fish skins from discards or by-products represent an excellent and abundant source for obtaining this biomolecule [5,8]. The main difference between marine and mammalian collagen includes a lower content of imino acids (proline and hydroxyproline) in marine collagen, which also influences the lower thermal stability shown by marine collagens [4,6]. In the literature, there is abundant information regarding the extraction of collagen from the skin of different marine species [4,6], however there are only few publications regarding the optimization of the key parameters influencing the process of extraction (temperature, concentration of NaOH and acetic acid and also time of incubation) [11–13]. Thus, having in mind the importance of extraction conditions to achieve a higher collagen yield, and although acid-soluble collagen (ASC) has been obtained previously from the skin of the small-spotted catshark [6], the yield was 52%. It is necessary to study the effects of extraction conditions on trying to obtain a higher recovery of collagen from this species.

The collagen extraction process comprises two main steps: the first step consists of the removal of non-collagen proteins and other impurities such as lipids, calcium etc. from the skin, with the aim of increasing the purity of the collagen extracted. To achieve this objective, 0.1 M NaOH is generally used, with different stirring incubation times [14–16]. The de-proteinized skin is then washed with cold water until it reaches a neutral pH and filtered. The second step in the extraction process consists of an acidic extraction of the previously NaOH treated skin, commonly using 0.5 M AcOH with different incubation times (48 h, 72 h, etc.) [4,5,17]. All procedures are usually done at 4 °C. After centrifugation, supernatant containing the acid-soluble collagen (ASC) is dialyzed and freeze-dried.

As there are several factors influencing the two-step collagen extraction process (time, temperature, NaOH concentration and AcOH concentration) and there is a need to study the optimal conditions of each variable and also the interactions between them, response surface methodology (RSM) has been employed to predict the optimal experimental conditions. RSM is a tool that has been previously used for the optimization of collagen extraction conditions from the skin of different fish species [12,13,18,19], however none of those studies included all the key optimization parameters influencing the two main steps of the extraction process. Thus, this is the first study optimizing the complete process for the extraction of acid-soluble collagen by means of three variables in each optimization stage: temperature, time and chemical treatment (NaOH or AcOH) concentration, from the skin of the small-spotted catshark.

2. Results and Discussion

2.1. Alkaline Pre-Treatment of Skin

The average (±standard deviation (SD)) chemical composition of non-treated skin from the small spotted catshark expressed as dry weight is shown in Table 1.

Table 1. Approximate composition (media ± standard deviation (SD)) expressed as percentage of dry weight of non-treated skin from the small-spotted catshark.

	Composition (%)			
	Moisture	**Protein**	**Lipid**	**Ash**
Non-treated skin	62.22 ± 0.48	69.24 ± 0.67	2.72 ± 0.18	35.13 ± 0.26

Hydroxyproline (HPro) content was used as an estimation of initial collagen content in the non-treated skins, considering that the ratio of HPro in collagen is 12.5 g of HPro/100 g of collagen [20]. Thus, the determined collagen content was 34.22% (g collagen/100 g dried skin). Collagen recovered (g collagen/100 g of collagen in non-treated skins) was estimated in the solid skin residues and in the filtrated liquid for the 20 experiments carried out during the experimental design, from the Kjeldahl determined nitrogen using a factor of 5.4 [21] (Table S1, Supplementary Material).

Experimental data from Table S1 were modelled using second-order equations (Table 2). These polynomial models describe the correlation between variables and the corresponding response followed the general form defined by Equation (1).

Table 2. Second-order equations describing the effect of temperature (T), time (t) and concentration of NaOH (M) on the efficiency of collagen recovery (%) from the skin of the small-spotted catshark. The coefficient of adjusted determination (R^2_{adj}) is also shown. Optimum (opt) values of each independent variable to obtain maximum responses are also shown.

	Polynomial Equations	R^2_{adj}	T_{opt} (°C)	t_{opt} (h)	$NaOH_{opt}$ (M)
Liquid	Collagen (%) = 87.9 + 26.2 × T + 14.4 × NaOH + 12.7 × t − 5.9 × T × NaOH × t − 7.9 × T^2 − 9.6 × $NaOH^2$ − 9.1 × t^2	0.846	25	48	2
Solid	Collagen (%) = 14.4 − 26.5 × T − 17.8 × NaOH − 14.9 × t + 3.7 × T × NaOH + 7.04 × T × t − 4.5 × NaOH × t + 6.6 × T × NaOH × t + 6.9 × T^2 + 11.7 × $NaOH^2$ + 9.3 × t^2	0.811	4	2	0.1

The R^2_{adj} values revealed good agreement among experimental and predicted data described by the second-order equations proposed (a high proportion of variability, more than 81% for both solid residue skins and filtrated liquid, was achieved). The consistency of the polynomial equations was validated since the F1 and F2 ratios from F-Fisher test were satisfied in all cases (data not shown). The results of the multivariate analysis showed significant quadratic terms for temperature, NaOH concentration and time (Student's *t*-test, $p < 0.05$) in the estimated collagen present in both fractions. In the solid fraction, this outcome is graphically translated as a concave surface where the collagen recovery increases with lower temperature, lower concentration of NaOH and low reaction times (Figure 1). The inverse response obtained for temperature, NaOH concentration and time in the filtrated liquid (convex surface) is in agreement with the fact that collagen recovered in the solid fraction is not present in the filtrated liquid fraction. Among the three independent variables, NaOH seems to have a slightly higher effect on collagen recovery in both fractions.

The variables maximizing the recovery of collagen in the solid fraction were 4 °C, 2 h and 0.1 M NaOH. However due to industrial constraints, mainly due to the high cost of low temperature processes, the temperature of 8.3 °C was selected for the next optimization step. Thus, the consensus values for the subsequent acid-soluble collagen extraction step were a temperature of 8.3 °C, a treatment time of 2 h and a NaOH concentration of 0.1 M.

Although previous studies have also shown that a low impact NaOH pre-treatment has a positive effect on the collagen yield, this is the first time that an optimization study has been carried out regarding the skin NaOH pre-treatment. Our results show that as little as two hours of treatment is enough to condition the skin, making it suitable for the posterior acid treatment. Thus, Woo et al. [12] have found that treatment times between 12 h and 36 h and NaOH concentration values between 0.5–1.3 M positively affects the achievement of maximum values of collagen content extracted from yellowfin tuna skin. Zhou and Regenstein [22] found that significant amounts of collagens are lost when pre-treatment conditions include concentration values higher than 0.5 M NaOH, reaction time of 4 days and temperature of 4 °C. Liu et al. [15] have also studied the effect of different alkaline pre-treatment conditions on the acid-soluble collagen obtained from grass carp, concluding that temperature ranges of 4–20 °C for pre-treatment conditions and NaOH concentration between 0.05 and 0.1 M were adequate. Wang et al. [13] also employed 0.1 M NaOH to remove non-collagenous proteins from the skin of grass carp with low temperature (4 °C) but higher reaction time (6 h). These results suggest that the efficiency of alkaline pre-treatment may vary between fish species and also between temperature, time and NaOH concentration conditions, highlighting the importance of specific two-step optimization studies for different species including these three variables.

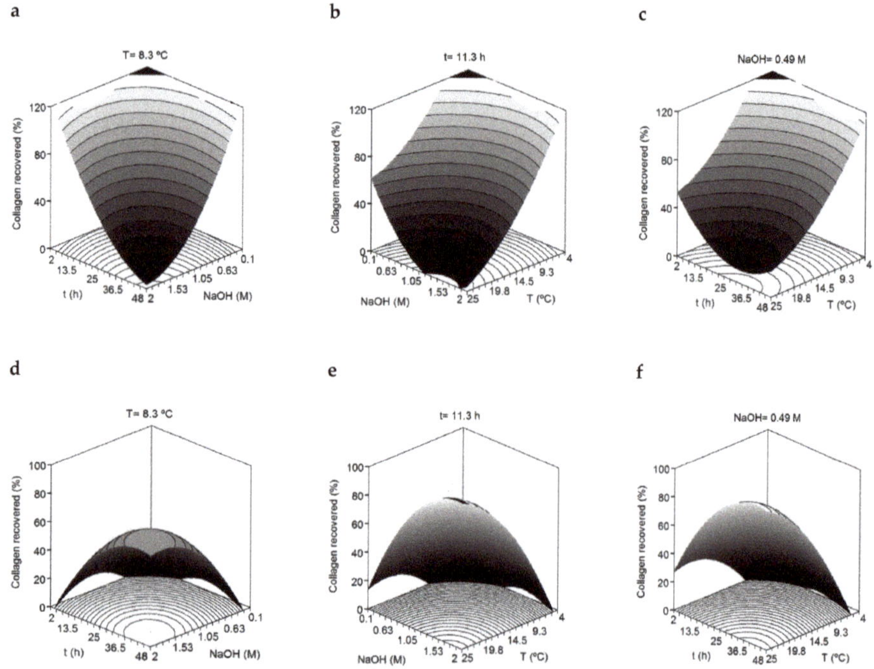

Figure 1. Combined effect of alkali concentration (M), time (t) and temperature (T) on the removal of collagen from the skin of the small-spotted catshark. Collagen recovered in the solid fraction (**a–c**). Collagen recovered in the liquid fraction (**d–f**).

2.2. Acid-Soluble Collagen (ASC) Extraction Stage

The next experiment was designed for the optimization of collagen extraction in acidic media using NaOH pre-treated skin. In this case, the combined effect of acetic acid concentration, temperature and time of processing on collagen production was studied. The average (±SD) chemical composition of NaOH treated skins (under the optimal consensus values obtained in the first experimental optimization stage, expressed as percentage of dry weight) used for this second experimental design was 76.55 ± 1.22% of moisture content; 56.71 ± 0.61% of protein content; 0.59 ± 0.10% of lipid content and 46.49 ± 0.38% of ash content. Compared to the approximate composition of non-treated skins, the protein and lipid content decreased significantly (Kruskal–Wallis test for protein: chi-square = 5.398, d.f. = 1, $p = 0.020$; ANOVA for lipid: $F_{1,4} = 299.483$, $p < 0.01$), confirming the removal of unwanted materials [23]. The significantly higher ash content observed in NaOH treated skin (ANOVA for ash: $F_{1,4} = 1758.801$, $p < 0.01$ is due to the NaOH added. A representation of the lyophilized collagen obtained in some of the 20 experiments is shown in Figure 2. The corresponding amino acid composition from all collagens is summarized in Table 3. In addition, the yields of lyophilized collagen recovered varied between 18.33% and 49.65% and are defined in Table S2.

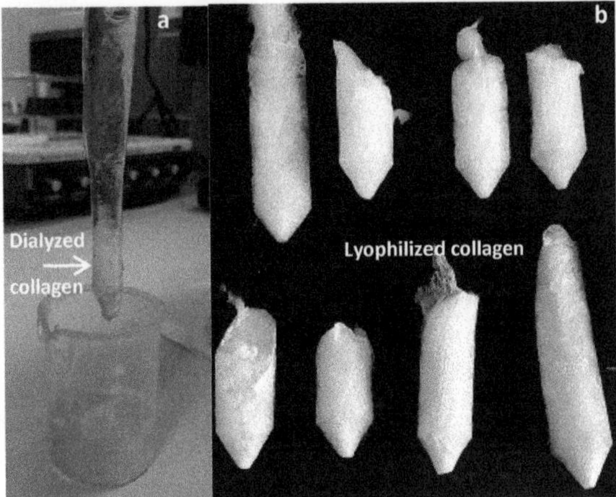

Figure 2. Dialyzed (**a**) and lyophilized (**b**) collagens obtained in each of the 20 experiments developed for the acid-soluble collagen extraction stage experimental design.

Table 3. Hydroxyproline (HPro), Proline (Pro) and Glycine (Gly) content in lyophilized extracted collagen obtained in each of the 20 experiments developed for the acid-soluble collagen extraction stage of the experimental design. Real values of independent variables are indicated, as well as the codified values (in brackets).

N° Experiment	T (°C)	Acetic Acid (M)	t (h)	Micromole in Lyophilized Extracted Collagen		
				OHPro	Pro	Gly
1	8.26 ((−1))	0.36 (−1)	11.33 (−1)	65.12	102.24	367.69
2	20.74 ((1))	0.36 (−1)	11.33 (−1)	151.34	237.70	854.61
3	8.26 ((−1))	0.84 (1)	11.33 (−1)	87.58	137.44	494.40
4	20.74((1))	0.84 (1)	11.33 (−1)	144.44	226.88	815.48
5	8.26 ((−1))	0.36 (−1)	38.67 (1)	81.01	127.18	456.98
6	20.74 ((1))	0.36 (−1)	38.67 (1)	168.79	265.12	952.75
7	8.26 ((−1))	0.84 (1)	38.67 (1)	121.52	191.05	686.78
8	20.74 ((1))	0.84 (1)	38.67 (1)	174.42	274.06	985.54
9	4.00 (−1.682)	0.60 (0)	25.00 (0)	68.59	107.72	387.32
10	25.00 (1.682)	0.60 (0)	25.00 (0)	168.85	265.32	953.85
11	14.50 (0)	0.20 (−1.682)	25.00 (0)	108.93	171.04	614.92
12	14.50 (0)	1.00 (1.682)	25.00 (0)	155.82	244.70	879.74
13	14.50 (0)	0.60 (0)	2.00 (−1.682)	71.32	112.12	403.03
14	14.50 (0)	0.60 (0)	48.00 (1.682)	131.02	205.82	740.10
15	14.50 (0)	0.60 (0)	25.00 (0)	116.16	182.34	655.82
16	14.50 (0)	0.60 (0)	25.00 (0)	131.89	207.09	744.66
17	14.50 (0)	0.60 (0)	25.00 (0)	139.18	218.56	785.62
18	14.50 (0)	0.60 (0)	25.00 (0)	158.14	248.32	892.94
19	14.50 (0)	0.60 (0)	25.00 (0)	141.82	221.77	797.30
20	14.50 (0)	0.60 (0)	25.00 (0)	134.02	210.51	756.67

The dependent variables (responses) evaluated were HPro, Gly, Pro and the sum of Pro + HPro (imino acids) as well as the yield of collagen recovered. Table 4 summarizes the equations obtained from the mathematical modelling and multivariable statistical analysis of the experimental responses mentioned. The accuracy between experimental and theoretical data were remarkable with values of $R^2_{adj} > 0.85$. The robutness of the different response selected and the reproducibility of collagen production was confirmed by the fact that equations and theoretical three-dimensional (3D) surfaces were similar in all cases studied (Figure 3). As in the previous factorial design, the consistency of the equations was also found: all ratios F1–F4 were validated (data not shown). Finally, the values of the independent variables which maximize the recovery of collagen were a temperature of 25 °C, a time of 34 h and a concentration of 1 M acetic acid. Using these optimal extraction conditions, the yield of collagen obtained was 61.24% (g of collagen/100 g of initial collagen in skin), which is higher than that obtained previously [6].

Table 4. Second-order equations describing the effect of temperature (T), time (t) and concentration of AcOH (M) on the collagen recovery by means of HPro, Gly, Pro, HPro + Pro and yield determination, from the skin of the small-spotted catshark. The coefficient of adjusted determination (R^2_{adj}) is also shown. Optimum values of each independent variable to obtain maximum responses are also shown.

Polynomial Equations	R^2_{adj}	T_{opt} (°C)	t_{opt} (h)	$AcOH_{opt}$ (M)
Pro (µmoles) = 214.4 + 52.1 × T + 16.2 × AcOH + 22.8 × t − 17.1 × t²	0.860	25	34.2	1
HPro (µmoles) = 136.5 + 33.1 × T + 10.3 × AcOH + 14.5 × t − 10.9 × t²	0.860	25	34.2	1
Gly (µmoles) = 770.7 + 187.1 × T + 58.3 × AcOH + 81.8 × t − 61.4 × t²	0.860	25	34.2	1
HPro + Pro (µmoles) = 350.9 + 85.2 × T + 26.5 × AcOH + 37.2 × t − 28.0 × t²	0.860	25	34.2	1
Yield (%) = 39.2 + 9.3 × T + 3.1 × AcOH + 4.1 × t − 3.4 × t²	0.853	25	34.2	1

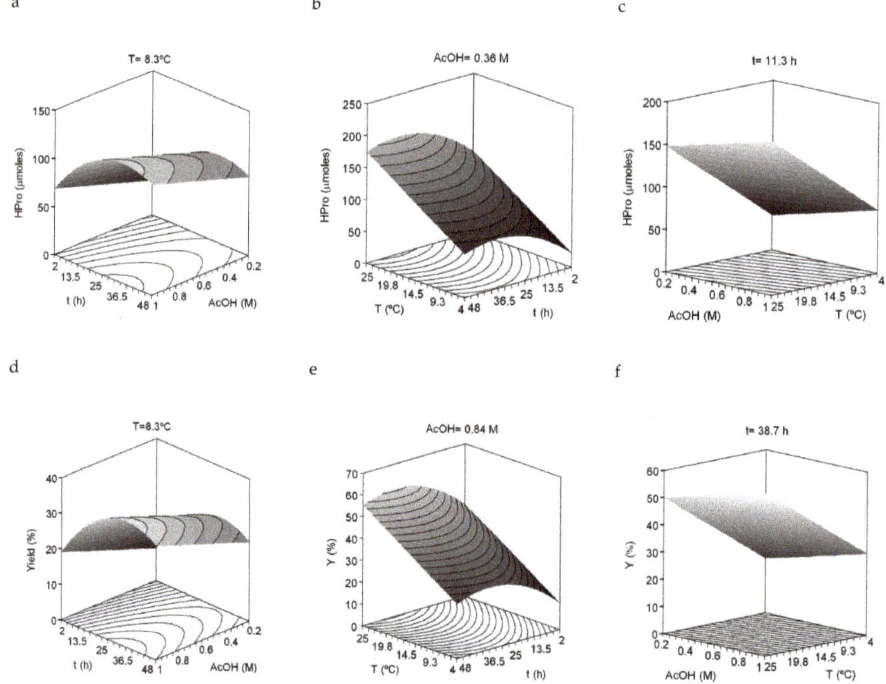

Figure 3. Combined effect of acetic acid (AcOH), time (t) and temperature (°C) on HPro released (**a–c**) and collagen yield (**d–f**) produced from *S. canicula* skins.

Previously published results on acid-soluble collagen extraction from the skin of the small-spotted catshark [6] showed lower yield values as the extraction conditions were different than the optimum values presented here. Several studies have also focused on the extraction and characterization of acid-soluble collagen from different marine fish species, traditionally using 0.5 M acetic acid at around 4 °C [22] without a previous optimization study [24,25]. In recent decades, several manuscripts have addressed the study of the optimization conditions of collagen extraction from different sources;

however, not many of those optimize the complete extraction procedure, including both the alkaline and the acidic stages. Thus, Wang et al. [13] found higher yields of acid-soluble collagen from the skin of grass carp with increased acetic acid concentration (up to 0.5 M) and increased reaction times (up to 32 h), while the optimum temperature differs with different levels of acetic acid or reaction time. The collagen yield reported by these authors was lower than the one obtained in this study. As in the previous alkaline pre-treatment optimization stage, the efficiency of the acidic extraction stage varies between fish species and also with temperature, time and concentration of acetic acid, suggesting the great importance of specific optimization studies for different species including the three variables involved in the process.

As shown in Figure 4, ASC extracted under the different experimental conditions used in this work resulted in similar electrophoretic patterns, which consisted of the typical heterotrimer collagen structure containing two identical α_1 chains (approximately 120 kDa) and one α_2 chain (approximately 110 kDa) in the molecular form of $[\alpha_1(I)]_2\ \alpha_2(I)$, and one β dimer of about 200 kDa [6]. The band intensity of the α_1 chain was not two-fold higher than that of α_2 chain; in fact, the α_2 chain is hardly visible. This fact, together with the high intensity of the β dimer, might suggest the existence of higher crosslink degree between α_2 chain in elasmobranchs This has also been found for other elasmobranchs where the α_2 chains are scarcely visible while the β dimer bands are stronger than in other teleosts [17,26,27]. A γ-component can be also seen in all ASC obtained, similarly to previously reported results by Sotelo et al. [6]. The collagen obtained in some of the experimental conditions (corresponding to Experiments 6, 8, 10 and 12) present a few bands below 100 kDa. These low molecular weight components might be the result of the particular extraction conditions on which those collagens were obtained: the highest temperatures, times and AcOH concentrations or the combination of them (further research on the characterization of those components using ionic exchange chromatography might be interesting, however it exceeds the objectives of this study).

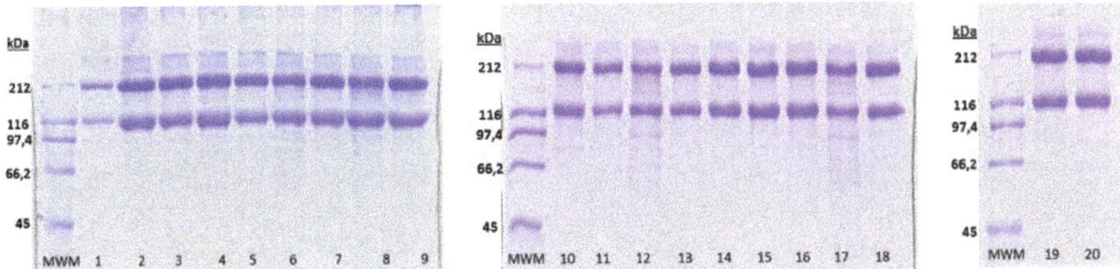

Figure 4. SDS-PAGE (7%) showing acid-soluble collagen (ASC) obtained in each of the 20 experiments developed for the acid-soluble collagen extraction experimental design. MWM: molecular weight marker.

3. Material and Methods

3.1. Biological Samples and Compositional Analysis

Small-spotted catshark (*Scyliorhinus canicula*) individuals obtained approximately 12 h after capture from a local market in Vigo (Northwestern Spain) were manually skinned (ES360570202001 by Galicia Government). Skins were stretched and aligned on top of each other (Figure 5a) in order to select only the central part of the skins (Figure 5b) with the aim of obtaining a homogeneous material. The selected central parts were mechanically cut into small pieces (5 × 5 cm^2) (Figure 5c) and then each of those pieces were manually cut into smaller pieces (0.5 × 0.5 cm^2) (Figure 5d), mixed thoroughly, separated in sealed plastic bags each containing 5 g of skin and stored at −20 °C until used for the experimental designs.

Figure 5. Small spotted catshark skin sampling: Skins stretched and aligned on top of each other (**a**); selected central portions of skins (**b**); homogeneized cutsobtained from the selected central parts of skins using scissors and divided in three for sampling purposes (**c**); small pieces finally obtained using a scalper (**d**).

The chemical composition of the skin was evaluated in triplicate by analyzing crude protein, ash, moisture and fat content. Total nitrogen was determined with the Kjeldahl method [28] in a DigiPREP HT digestor (SCP Science, Baie-d'Urfe, QC, Canada), DigiPREP 500 fully automatic steam distillation (SCP Science, Baie-d'Urfe, QC, Canada) and a TitroLine easy titration unit (Schoot, Mainz, Germany), and crude protein content was calculated as total nitrogen multiplied by 6.25. Fat content was determined by the method of Bligh and Dyer [29]. Moisture was determined after heating the sample at 105 °C for 24 h, and ash content was determined after heating the sample for 24 h at 550 °C [28].

The hydroxyproline content in the skin was determined according to the procedure described in Blanco et al. [30] and used for the estimation of initial collagen content in the untreated skin, considering that the ratio of HPro in collagen is 12.5 g of HPro/100 g of collagen [20].

3.2. Experimental Design and Statistical Analysis

In this work, two experimental designs were performed to analyze the influence of chemical treatment concentration, temperature and time on the extractability of collagen from the skin of the small-spotted catshark. First, the effect of temperature (T), concentration of NaOH (M) and time (t) on the efficiency of removing non-collagenous proteins was studied (alkaline pre-treatment). Then, the effect of temperature (T), concentration of acetic acid (M) and time (t) on the efficiency of extracting acid-soluble collagen (ASC) was optimized (acid-soluble collagen extraction stage). In both cases, the factorial experiments were rotatable second-order designs with six replicates in the center of the experimental domains [31].

3.2.1. Alkaline Pre-Treatment Experimental Design

The conditions of the independent variables studied in the pre-treatment experimental design were: temperature in the range of 4–25 °C, concentration of NaOH in the range of 0.1–2 M and intervals of time between 2–48 h (Table 5). The values of independent variables were selected from previously reported studies to cover a wide range of conditions in order to obtain the values that maximize the isolation of collagen and to reduce the times and concentrations needed for the bioproduction of *S. canicula* collagen. The most common times and NaOH concentration values used in the literature ranged from 1–36 h and 0.05–1.3 M NaOH. The temperature preferentially chosen in the literature for removing other non-collagen proteins ranged from 4–20 °C [11–13].

Table 5. Experimental domain and codification of independent variables in the second-order rotatable designs developed for collagen extraction from *S. canicula* skin.

	Alkaline Pre-Treatment			Acid Extraction		
Coded Values	**T (°C)**	**NaOH (M)**	**t (h)**	**T (°C)**	**AcOH (M)**	**t (h)**
−1.68	4.0	0.10	2.0	4.0	0.20	2.0
−1	8.3	0.49	11.3	8.3	0.36	11.3
0	14.5	1.05	25.0	14.5	0.60	25.0
+1	20.7	1.61	38.7	20.7	0.84	38.7
+1.68	25.0	2.00	48.0	25.0	1.00	48.0

Codification: $V_c = (V_n - V_0)/\Delta V_n$ — Decodification: $V_n = V_0 + (\Delta V_n \times V_c)$
V_n = natural value of the variable to codify — ΔV_n = increment of V_n for unit of V_c
V_0 = natural value in the centre of the domain — V_c = codified value of the variable

The conditions which were maintained as constants were the solid (skin):alkaline solution ratio of 1:10 and high agitation (200 rpm). The reactions were developed in a stirred and thermostated reactor (100 mL). After each of the 20 alkali treatments, the solutions were filtered using a 35 μm membrane. The filtrate was measured, centrifuged and the supernatant collected to be analyzed in terms of collagen content which was determined by means of total nitrogen content according to the Kjeldahl method [28]. The skin residue of the filter was weighed and also analyzed in terms of total nitrogen content. The dependent variable studied was collagen/initial collagen in skin rate, for both the collagen recovered in the skin residues and the collagen measured in the filtered solution.

3.2.2. Acid-Soluble Collagen Extraction Stage Experimental Design

Skins (500 g) obtained as explained in Section 2.1 were introduced in a stirred and thermostated 5 L reactor connected to a pH electrode and a temperature probe (Afora S.A., Barcelona, Spain). Based on the consensus values obtained in the alkaline pre-treatment experimental design, skins were treated with NaOH and then filtered using a 200 μm membrane. The liquid was removed, and the skins were washed with distilled water until neutral pH was achieved (Figure 6), weighed and divided into 5 g sealed plastic bags which were frozen at −20 °C until used for the experimental design. NaOH pre-treated skins (10 g) were used for approximate compositional analysis. Differences in approximate composition between non-NaOH treated skins and NaOH treated skins were statistically analyzed. Prior to analysis, data were checked for normality and homoscedasticity using the Kolmogorov–Smirnov and Levene tests, respectively. The Kolmogorov–Smirnov test showed that protein data and their transformations did not fit with the assumptions of normality. As a consequence, non-parametric statistics were used for these data. Differences in lipids and ashes were compared by one-way ANOVA, with NaOH treated or non-treated skins as the between-subject effect. The Kruskal–Wallis test, the non-parametric equivalent of a one-way ANOVA, was used to examine variations between treated and non-treated skins in protein content. Significance levels were set at $p < 0.05$. Statistical tests were performed with IBM SPSS Statistics 23. 0 (IBM Corp., Armonk, NY, USA).

The conditions of the independent variables studied in the collagen extraction stage of the experimental design were: temperature in the range of 4–25 °C, concentration of acetic acid (AcOH) in the range of 0.2–1 M and intervals of time between 2–48 h (Table 4). The conditions that were maintained as constants were solid (skin residue): alkaline solution ratio of 1:10 and high agitation (200 rpm). The values of independent variables were again selected from previously reported studies to cover a wide range of conditions in order to obtain the values that maximize the isolation of collagen and to reduce the time and concentration needed for the bioproduction of *S. canicula* collagen. The most common times and acetic acid concentration values used in the literature ranged from 1–16 h and

0.1–1 M acetic acid. The temperature preferentially chosen in the literature for the acid extraction of collagen ranged from 4–30 °C [12,13,24].

Figure 6. Filtered and washed NaOH treated skins used for the acid-soluble collagen extraction stage of the experimental design.

The reactions were developed in the same reactor as the pre-treatment experimental design. After each one of the 20 acid experiments, the solutions were filtered using a 200 μm membrane. The filtrated solution was collected, measured, dialyzed and freeze-dried. The freeze-dried collagen was weighed and characterized in terms of collagen content (determined by means of hydroxyproline (HPro), proline (Pro) and glycine (Gly) content), yield of collagen recovered and SDS-PAGE characterization (the content of other amino acid was also analyzed). The dependent variables studied were yield of collagen (as a percentage of collagen recovered/initial collagen in skin) and contents of the collagen characteristic amino acids: HPro, Pro and Gly.

3.2.3. Mathematical Modelling and Statistical Analysis

The experimental results of the factorial designs were modelled by second-order polynomial equations as [31]:

$$Y = b_0 + \sum_{i=1}^{n} b_i X_i + \sum_{\substack{i=1 \\ j>i}}^{n-1} \sum_{j=2}^{n} b_{ij} X_i X_j + \sum_{i=1}^{n} b_{ii} X_i^2 \tag{1}$$

where Y represents the response to be modelled; b_0 is a constant coefficient, b_i is the coefficient of linear effect, b_{ij} is the coefficient of interaction effect, b_{ii} is the coefficient of squared effect, n is the number of variables and X_i and X_j define the independent variables. The statistical significance of the coefficients was verified by means of Student's t-test ($\alpha = 0.05$), goodness-of-fit was established as the adjusted determination coefficient (R^2_{adj}) and the model consistency was determined by the Fisher F test ($\alpha = 0.05$) using the following mean squares ratios (Table 6):

Table 6. Fisher F tests used to check the consistency of polynomial equations.

	The Model is Acceptable When:
F1 = Model/Total error	$F1 \geq F^{num}_{den}$
F2 = (Model + Lack of fitting)/Model	$F2 \leq F^{num}_{den}$
F3 = Total error/Experimental error	$F3 \leq F^{num}_{den}$
F4 = Lack of fitting/Experimental error	$F4 \leq F^{num}_{den}$

F^{num}_{den} are the theoretical values to $\alpha = 0.05$ with the corresponding degrees of freedom for numerator (num) and denominator (den). All fitting procedures, coefficient estimates and statistical calculations were performed on a Microsoft Excel spreadsheet.

3.3. Amino Acid Characterization of Acid-Soluble Collagen

Amino acids were determined in the freeze-dried collagen obtained in each one of the 20 experimental points. For this purpose, acid hydrolysis was conducted with 6 N hydrochloric acid containing 0.1% phenol under an inert atmosphere by heating to 110 °C for 24 h. Then, HCl was removed from the hydrolysate by vacuum. The hydrolysate was resuspended in 20–50 µL of 0.2 M sodium citrate buffer (pH 2.2), to which a known amount of norleucine was added as an internal standard and applied to an automated amino acid analyzer (Biochrom30 Amino Acid Analyzer, Biochrom, UK).

3.4. SDS-PAGE Characterization of Acid-Soluble Collagen

Samples (1 mg/mL) were prepared in sample buffer containing 10.52% glycerol, 21% Sodium Dodecyl Sulfate (SDS) (10%), 0.63% Dithiothreitol (DTT)and 0.5 M Tris HCl (pH 6.8) and heated for 5 min at 100 °C. An aliquot (8 µL) of this mixture was applied to each well in 7% polyacrylamide separating gels. Gels (100 mm × 750 mm × 0.75 mm) were prepared according to the procedure of Laemmli [32] and were subjected to electrophoresis at 20 mA using a Mini-Protean II Cell (Bio-Rad, Hercules, CA, USA). Following electrophoresis, the gels were stained with 0.04% Coomassie Blue in 25% *v*/*v* ethanol and 8% *v*/*v* acetic acid for 2 h. Excess stain was removed with several washes of destaining solvent (25% *v*/*v* ethanol, 8% *v*/*v* acetic acid). Molecular weights of subunits of ASC (acid-soluble collagen) were estimated using molecular weight standards from BIO-RAD (Hercules, CA, USA) SDS-PAGE standards high range: myosin (200 kDa); β-galactosidase (116 kDa); phosphorylase B (97 kDa); bovine serum albumin (66 kDa); ovalbumin (45 kDa).

4. Conclusions

This is the first study optimizing the complete process for the extraction of acid-soluble collagen by means of three variables (temperature, time and chemical treatment concentration) from the skin of the small-spotted catshark using response surface methodology. Two-stage optimizations (alkali pre-treatment and acid extraction) of the collagen extraction process should be accomplished in a species-specific approach due to the variability of collagen extracted from different species (regarding its structure and chemical differences (for example, variations in the amino acid composition)). The variables maximizing the recovery of collagen in the first stage of extraction (alkaline pre-treatment) were 4 °C, 2 h and 0.1 M NaOH. The variables maximizing the recovery of collagen in the second stage of extraction (acid-soluble collagen extraction stage) were 25 °C, 34 h and 1 M AcOH with a yield of 61.24%. The results obtained in this study might be helpful for a potential collagen extraction upscaling study.

Supplementary Materials: The following are available online at http://www.mdpi.com/1660-3397/17/1/40/s1, Table S1: Experimental domains and codification of independent variables in the factorial rotable design executed to study the optimal conditions for removing proteins different of collagen from the skin of small-spotted catshark, Table S2: Experimental domains and codification of independent variables in the factorial rotable design executed to study the optimal conditions for extraction of acid soluble collagen from the skin of small-spotted catshark.

Author Contributions: C.G.S., J.A.V., R.I.P.-M. and M.B. conceived and designed the experiments; M.B. performed the experiments; C.G.S., J.A.V., R.I.P.-M. and M.B. analyzed the data; M.B. wrote the paper. J.A.V. participated in the redaction of the manuscript. C.G.S., J.A.V. and R.I.P.-M. critically revised the manuscript.

Funding: This research was funded by EU INTERREG_POCTEP 2015, 0302_CVMAR_I_1_P; EU-INTERREG Atlantic Area Programme, EAPA_151/2016; Xunta de Galicia, IN607B 2018/19).

Acknowledgments: Authors thanks the financial support received from the projects: CVMar+i (0302_CVMAR_I_1_P, EU INTERREG_POCTEP 2015); BlueHuman (EAPA_151/2016, EU-INTERREG Atlantic Area Programme) and Grupos de Potencial Crecimiento (GAIN, Xunta de Galicia, IN607B 2018/19). The authors are also grateful to Helena Pazó Malvido and Marta Pérez Testa for her technical assistance.

Conflicts of Interest: The authors declare no conflict of interest.

References

1.   Blanco, M. Valorización de Descartes y Subproductos de Pintarroja (*Scyliorhinus canicula*). Ph.D. Thesis, Universidad de Vigo, Vigo, Spain, 2015.

2. Regulation (EU) No 1380/2013 of the European Parliament and of the Council of the European Union. Available online: https://eur-lex.europa.eu/LexUriServ/LexUriServ.do?uri=OJ:L:2013:354:0022:0061:EN:PDF (accessed on 29 November 2018).

3. Shoulders, M.D.; Raines, R.T. Collagen Structure and Stability. *Annu. Rev. Biochem.* **2009**, *78*, 929–958. [CrossRef] [PubMed]

4. Benjakul, S.; Nalinanon, S.; Shahidi, F. Fish collagen. In *Food Biochemistry and Food Processing*, 2nd ed.; Wiley-Blackwell: Oxford, UK, 2012; pp. 365–387.

5. Alves, A.L.; Marques, A.L.; Martins, E.; Silva, T.H.; Reis, R.L. Cosmetic potential of marine fish skin collagen. *Cosmetics* **2017**, *4*, 39. [CrossRef]

6. Sotelo, C.G.; Blanco, M.; Ramos-Ariza, P.; Pérez-Martín, R.I. Characterization of collagen from different discarded fish species of the west coast of the Iberian Peninsula. *J. Aquat. Food Prod. Technol.* **2016**, *25*, 388–399. [CrossRef]

7. Venkatesan, J.; Anil, S.; Kim, S.-K.; Shim, M. Marine Fish proteins and Peptides from Cosmoceuticals: A Review. *Mar. Drugs* **2017**, *15*, 143. [CrossRef] [PubMed]

8. Zhu, B.-W.; Dong, X.-P.; Zhou, D.-Y.; Gao, Y.; Yang, J.-F.; Li, D.-M.; Zhao, X.-K.; Ren, T.-T.; Ye, W.-X.; Tan, H.; et al. Physicochemical properties and radical scavenging capacities of pepsin-solubilized collagen from sea cucumber *Stichopus japonicus*. *Food Hydrocoll.* **2012**, *28*, 182–188. [CrossRef]

9. Swatschek, D.; Schatton, W.; Kellerman, J.; Müller, W.E.; Kreuter, J. Marine sponge collagen: Isolation, characterization and effects on the skin parameters surface-pH, moisture and sebum. *Eur. J. Pharm. Biopharm.* **2002**, *53*, 107–113. [CrossRef]

10. Shen, X.R.; Kurihara, H.; Takahashi, K. Characterization of molecular species of collagen in scallop mantle. *Food Chem.* **2007**, *102*, 1187–1191.

11. Liu, D.; Wei, G.; Li, T.; Hu, J.; Lu, N.; Regenstein, J.M.; Zhou, P. Effects of alkaline pretreatments and acid extraction conditions on the acid-soluble collagen from grass carp (*Ctenopharyngodon idella*) skin. *Food Chem.* **2015**, *172*, 836–843. [CrossRef]

12. Woo, J.W.; Yu, S.J.; Cho, S.M.; Lee, Y.B.; Kim, S.B. Extraction optimization and properties of collagen from yellowfin tuna (*Thunnus albacares*) dorsal skin. *Food Hydrocoll.* **2008**, *22*, 879–887. [CrossRef]

13. Wang, L.; Yang, B.; Du, X.; Yang, Y.; Liu, J. Optimization of conditions for extraction of acid-soluble collagen from grass carp (*Ctenopharyngodon idella*) by response surface methodology. *Innov. Food Sci. Emerg. Technol.* **2008**, *9*, 604–607. [CrossRef]

14. Zhang, M.; Liu, W.; Li, G. Isolation and characterization of collagens from the skin of largefin longbarbel catfish (*Mystus macropterus*). *Food Chem.* **2009**, *115*, 826–831. [CrossRef]

15. Benjakul, S.; Thiansilakul, Y.; Visessanguan, W.; Roytrakul, S.; Kishimura, H.; Prodpran, T.; Meesane, J. Extraction and characterisation of pepsin-solubilised collagen from the skin of bigeye snapper (*Priacanthus tayenus* and *Priacanthusmacracanthus*). *J. Sci. Food Agric.* **2010**, *90*, 132–138. [CrossRef] [PubMed]

16. Liu, D.; Liang, L.; Regenstein, J.M.; Zhou, P. Extraction and characterization of pepsin-solubilized collagen from fins, scales, skins, bones and swim bladders of bighead carp (*Hypophthalamichthys nobilis*). *Food Chem.* **2012**, *133*, 1441–1448. [CrossRef]

17. Kittiphattanabawon, P.; Benjakul, S.; Visessanguan, W.; Kishimura, H.; Shahidi, F. Isolation and characterization of collagen from the skin of brownbanded bamboo shark. *Food Chem.* **2010**, *119*, 1519–1526. [CrossRef]

18. Blanco, M.; Fraguas, J.; Sotelo, C.G.; Pérez-Martín, R.I.; Vázquez, J.A. Production of chondroitin sulphate from head, skeleton and fins of *Scyliorhinus canicula* by-products by combination of enzymatic, chemical precipitation and ultrafiltration methodologies. *Mar. Drugs* **2015**, *13*, 3289–3308. [CrossRef] [PubMed]

19. Murado, M.A.; Montemayor, M.I.; Cabo, M.L.; Vázquez, J.A.; González, M.P. Optimization of extraction and purification process of hyaluronic acid from fish eyeball. *Food Bioprod. Process.* **2012**, *90*, 491–498. [CrossRef]

20. Edwards, C.A.; O´Brien, W.D., Jr. Modified assay for determination of hydroxyproline in a tissue hydrolysate. *Clin. Chim. Acta* **1980**, *104*, 161–167. [CrossRef]

21. Eastoe, J.; Eastoe, B. A method for the determination of total nitrogen in proteins. *Br. Gel. Glue Res. Assoc. Res. Rep.* **1952**, *5*, 1–17.

22. Zhou, P.; Regenstein, J.M. Effects of alkaline and acid pretreatments on Alaska Pollock skin gelatin extraction. *J. Food Sci.* **2005**, *70*, C392–C396. [CrossRef]

23. Regenstein, J.M.; Zhou, P. Collagen and gelation from marine by-product. In *Maximizing the Value of Marine by-Products*, 1st ed.; Shahidi, F., Ed.; CRC Press: Boca Raton, FL, USA, 2007; pp. 279–303.

24. Muyonga, J.H.; Cole, C.G.B.; Duodu, K.G. Characterisation of acid soluble collagen from skins of young and adult Nile perch (*Lates niloticus*). *Food Chem.* **2004**, *85*, 81–89. [CrossRef]

25. Senaratne, S.S.; Park, P.J.; Kim, S.K. Isolation and characterization of collagen from brown backed toadfish (*Lagocephalus gloveri*) skin. *Bioresour. Technol.* **2006**, *97*, 191–197. [CrossRef] [PubMed]

26. Kittiphattanabawon, P.; Benjakul, S.; Visessanguan, W.; Shahidi, F. Isolation and properties of acid and pepsin-soluble collagen from the skin of blacktip shark (*Carcharhinus limbatus*). *Eur. Food. Res. Technol.* **2010**, *230*, 475–483. [CrossRef]

27. Bae, I.; Osatomi, K.; Yoshida, A.; Osako, K.; Yamaguchi, A.; Hara, K. Biochemical properties of acid-soluble collagens extracted from the skin of underutilised fishes. *Food Chem.* **2008**, *108*, 49–54. [CrossRef]

28. Association of Official Analytical Chemistry. *Methods of Analysis*, 15th ed.; Helrich, K., Ed.; Association of Official Analytical Chemistry: Washington, DC, USA, 1990.

29. Bligh, E.G.; Dyer, W.J. A rapid method of total lipid extraction and purification. *Can. J. Biochem. Phys.* **1959**, *37*, 911–917. [CrossRef]

30. Blanco, M.; Vázquez, J.A.; Pérez-Martín, R.I.; Sotelo, C.G. Hydrolysates of fish skin collagen: an opportunity for valorizing fish industry byproducts. *Mar. Drugs* **2017**, *15*, 131. [CrossRef] [PubMed]

31. Box, G.E.; Hunter, J.S.; Hunter, W.G. *Statistics for Experimenters: Design, Innovation, and Discovery*, 2nd ed.; John Wiley & Sons, Inc.: New York, NY, USA, 2005.

32. Laemmli, U.K. Cleavage of structural proteins during the assembly of the head of bacteriophage T4. *Nature* **1970**, *227*, 680–685. [CrossRef]

 *marine drugs* 

Article

Collagen from Cartilaginous Fish By-Products for a Potential Application in Bioactive Film Composite

Emna Ben Slimane [1,2] **and Saloua Sadok** [1,*]

[1]   Laboratory of Blue Biotechnology & Aquatic Bioproducts, Institut National des Sciences et Technologies de la Mer (INSTM), INSTM, 28, rue 2 mars 1934, Salammbô, Tunis 2025, Tunisia; benslimaneemna@gmail.com
[2]   Institut National Agronomique de Tunisie (INAT), Université de Carthage, Tunis 1082, Tunisia
*   Correspondence: salwa.sadok@instm.rnrt.tn; Tel.: +216-99-938-833; Fax: +216-71-735-845

Received: 2 May 2018; Accepted: 21 May 2018; Published: 15 June 2018

Abstract: The acid solubilised collagen (ASC) and pepsin solubilised collagen (PSC) were extracted from the by-products (skin) of a cartilaginous fish (*Mustelus mustelus*). The ASC and PSC yields were 23.07% and 35.27% dry weight, respectively and were identified as collagen Type I with the presence of α, β and γ chains. As revealed by the Fourier Transform Infrared (FTIR) spectra analysis, pepsin did not alter the PSC triple helix structure. Based on the various type of collagen yield, only PSC was used in combination with chitosan to produce a composite film. Such film had lower tensile strength but higher elongation at break when compared to chitosan film; and lower water solubility and lightness when compared to collagen film. Equally, FTIR spectra analysis of film composite showed the occurrence of collagen-chitosan interaction resulting in a modification of the secondary structure of collagen. Collagen-chitosan-based biofilm showed a potential UV barrier properties and antioxidant activity, which might be used as green bioactive films to preserve nutraceutical products.

Keywords: cartilaginous fish by-products; collagen; chitosan; composite films; properties

1. Introduction

During the last decades, the increased consumer awareness of the nutritional value of fish and seafood and the shift towards more processed fishery products in convenient form; has generated larger quantities of by-products accounting for up to 70% of the volume of fish and shellfish [1]. In most cases, such biomasses, which include skin, head, bones and viscera, cause serious economic and ecological issues. However, such biomass is currently of high interest to researchers and industry as it represents a valuable source of compounds with high added value such as proteins, lipids, enzymes, and polysaccharides.

Animal body contains high amount of collagen constituting around 30% of the total amount of protein in vertebrates [2]. Actually, 27 types of collagens have been identified and collagen type I is the most frequent one and is known as fibrillar collagen and plays a structural role by contributing to the molecular architecture, shape and mechanical properties of skin tissues [2,3]. Due to its excellent properties (non-toxicity, low antigenicity and allergenicity, biocompatibility, the ability of film-forming and biodegradability), collagens are utilised in various fields such as medical, pharmaceutical and cosmetics industries, and also as materials for food packaging [4–6].

At the start of its use, collagen was mainly extracted from porcine and bovine sources. Later, people started to show reticence toward this practice due to religious background, beside the proliferation of bovine spongiform encephalopathy (BSE) [7] which represented a source of hazard contamination for the extracted protein [8,9]. As a consequence, several researchers have been interested in marine collagen as an alternative because of the absence of disease transmission and dietary restriction [10].

Collagen has been studied and characterised from various marine sources, mainly from marine invertebrates such as cuttlefish [11], octopus [12], squid [13], jellyfish [14], starfish [15], sea urchin [16], sea cucumber [17] and also of sponges which represent the key of their complex structure and integrity [18]. In marine vertebrate organisms, such interest was rather oriented to fish by-products including scales [19], skin [20,21], swim bladder [22], bone [23] and cartilage [24].

For their extractions, collagens are commonly solubilized in organic acid, generally acetic acid which causes the protonation of collagen polypeptides and consequently the repulsion between the tropo-collagen leading to enhanced collagen solubility [25]. However, such procedure referred to Acid Solubilised Collagen (ASC) gives generally low collagen yield. Therefore, research was oriented to enzymatic extraction to increase collagen solubilisation with pepsin being among the most efficient enzyme. Thus, pepsin provokes not only the cleavage of the collagens teleopeptide region maximising their solubility, but also the hydrolysis of non-collagenous proteins increasing collagen purity. In this case, pepsin maximise the extraction yield of collagens while reducing their antigenicity [26,27].

The common coastal smooth-hound *Mustelus mustelus,* is an abundant species of the genus *Mustelus* in the Mediterranean Sea where it is regularly caught all over the year either as by-catch or as targeted species [28,29]. The consumption of such species generates significant amounts of waste that may be used as source to extract substance of interest such as collagen.

To our knowledge, the extraction of collagen from the skin of *Mustelus mustelus* and its valorisation has never been reported. Therefore, the aim of this work was to extract and characterize collagen using two methods. In a first step, the isolation of collagen was elaborated using acetic acid, which allows a better solubilisation of the molecule followed in a second step, by an enzymatic extraction using pepsin.

Our second objective was to elaborate a biodegradable film using collagen, however following the extraction process (alkali then acid process), the collagen molecule loses its strong mechanical strength compared to the native form [30]. To overcome such issue, we blended collagen with another natural polymer such as chitosan derived from chitin known as the second most abundant polysaccharide after cellulose. Thus, chitosan has attracted much attention for its biodegradability, biocompatibility, bacteriostatic and fungistatic activities as well as for its texturizing properties and its ability to film forming [31]. Therefore, we used chitosan as an adjuvant to elaborate composite film.

2. Results and Discussion

2.1. Collagen Characterization

2.1.1. Collagens Electrophoretic Patterns

The electrophoretic patterns of collagens from smooth-hound skin (ShS) performed under denaturing condition are presented in Figure 1 and showed that there are no differences between the ASC and pepsin solubilised collagen (PSC).

As native collagen molecule is constituted of three polypeptide chains (α-chains) organised in a triple-helix, the denaturising sodium dodecyl sulfate (SDS) break the H-bonds yielding peptides [32]. In both types of ShS-collagen, the α ($\alpha1$, $\alpha2$) and their cross-linked dimer β-chains are the major components with low content of γ-chain. Thus, the electrophoresis mobility and subunit composition may suggest that ASC and PSC isolated from ShS should most likely be classified as type I collagen. The SDS-PAGE, revealed two bands of chains $\alpha1$ and $\alpha2$ with a molecular weight of about 101 kDa and 83 kDa respectively; however with different intensities ($\alpha1$ intensity higher than $\alpha2$ by approximately ratio 2:1). Such results suggest that $\alpha1$ is formed by 2 subunits as collagen type I characterised by the existence of 2 identical subunits of $\alpha1$ and one of $\alpha2$ [33].

Additionally, high molecular weight (MW) components, β-chains were clearly detected in both ShS-ASC and PSC with a mean molecular weight of 226 kDa. Such results are in conformity with several findings reported for other elasmobranches skins of brownbanded bamboo shark [34], skate [21] and shark [35].

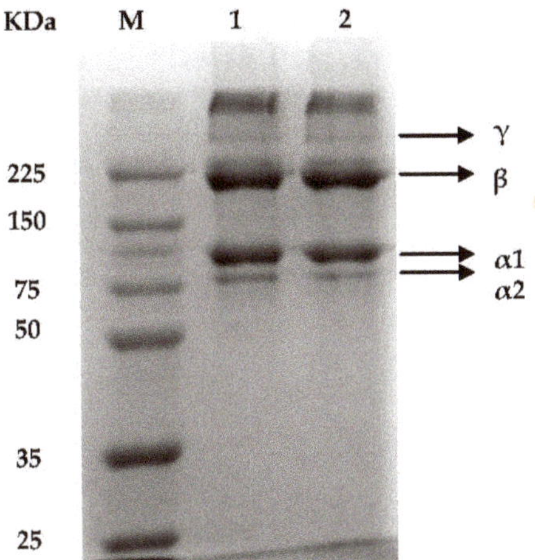

Figure 1. Sodium dodecyl sulfate polyacrylamide gel electrophoresis (SDS-PAGE) of (1): acid soluble collagen (ASC) and (2): pepsin soluble collagen (PSC) from hound-smooth skin M: high molecular weight marker (KDa).

2.1.2. Peptide Mapping

The ShS-ASC and PSC were markedly digested by Lysyl endopeptidase which cleaves peptide bonds at the carboxyl side of lysyl residues [36]. Generally, band intensity of major components α, β and γ of ShS-ASC and PSC decreased after digestion and degraded into smaller peptides with molecular weight ranging from 100 to 13 kDa (Figure 2).

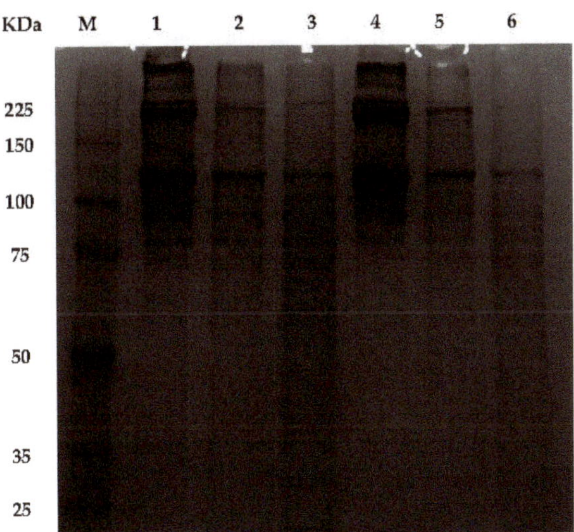

Figure 2. Peptide mapping of acid soluble collagen ASC and pepsin soluble collagen PSC digested by *Lysyl endopeptidase* with different hydrolysis time 1: ASC, 2: ASC-5 min, 3: ASC-25 min, 4: PSC, 5: PSC-5 min, 6: PSC-25 min from hound smooth skin, M: high molecular weight marker (KDa).

When comparing the effect of hydrolysis duration, an enhanced enzymatic hydrolysis was found with an incubation time of 25 min for ShS-ASC and PSC. This was evidenced by the appearance of higher number of peptides bands with low molecular weight (Figure 2, lines 3 and 6); and a decreased band's intensity for PSC collagen. This might be caused by the pepsin action on the telopeptide region inducing its cleavage and thus facilitating the changes in configuration, which may favour the hydrolysis by lysyl endopeptidase.

2.1.3. Viscosity Measurement

The temperature of denaturation (Td) of ShS-ASC and PSC, referred as the temperature at which the variation in viscosity is half completed, was calculated from a plot of temperature-induced variation in viscosity (Figure 3).

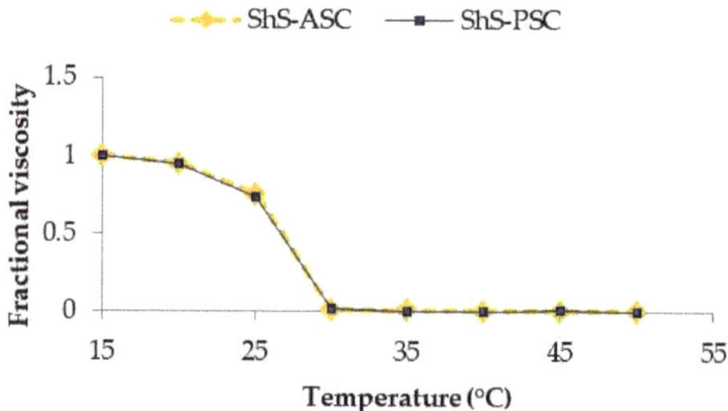

Figure 3. Change in fractional viscosity with temperature of acid soluble collagen (ShS-ASC) and pepsin soluble collagen (ShS-PSC) from the skin of hound smooth.

The viscosities of both collagens were higher at temperature ranging from 15–20 °C then decreased with increased heating up to 25 °C.

The Td's of ASC and PSC were 26.68 °C and 26.66 °C respectively. The similarity in the denaturation temperature may be related to the resemblance of the major peaks wavelength of ASC and PSC [37]. Such values were comparable to Td reported for collagen from other marine species such as the Japanese sea bass (26.5 °C) [23], edible jellyfish exumbrella (26.0 °C) [14], chub mackerel (25.6 °C), bullhead shark (25.0 °C) [23] but lower than porcine skin collagen (37.0 °C) [11]. However, the Td's of both ASC and ASC smooth hound skin were higher than those of Spanish marckel skin (15.12 and 14.66 °C, respectively) [38].

2.1.4. Ultraviolet Spectrophotometric Analysis

The ShS-ASC and PSC collagens exhibited maxima absorbencies at 235 nm and 240 nm respectively (Figure 4), which is in agreement with the maximum absorption of the collagen molecule (230 nm) [39]; principally due to the $n \rightarrow \pi^*$ transitions of the peptide band C=O [40].

Unlike other protein types no peak was detected at 280 nm, suggesting that ASC and PSC collagens have low amount of aromatic residues such as tyrosine and phenylalanine [41]. These results are in line with those previously found in fish skin collagens [34,42–44] and confirm the effectiveness of the alkaline treatment for the removal of non-collagenous proteins.

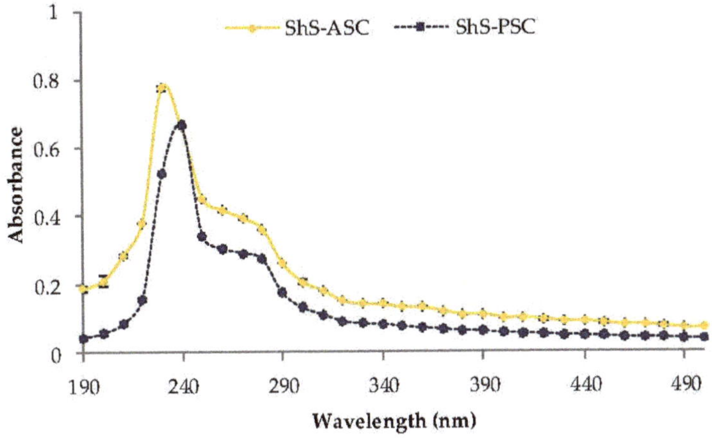

Figure 4. Ultraviloet-Visible spectra of acid soluble (ShS-ASC) and pepsin soluble collagen (ShS-PSC) from the skin of hound smooth.

2.1.5. Fourier Transform Infrared Spectra of Collagens

In order to determine the isolated collagens type; Fourier transform infrared technique was used to detect the vibrational modes of bands and individual chemical groups in the extracted collagens [45].

The ShS-collagen's Fourier Transform Infrared (FTIR) spectra represented in Figure 5, showed that the amide A band of ASC and PSC originated from the stretching vibrations of N–H group were found at 3293.57 and 3296.16 cm^{-1} respectively, although it commonly appears in the range of 3400–3440 cm^{-1} [19,46]. These shifts to lower frequencies means that the collagen NH groups of the samples were involved in hydrogen bonding, which help to stabilize the collagen triple helix structure. The absorption peak of amide B, related to asymmetrical stretch of CH$_2$ [47], appeared at 2942 cm^{-1} for ASC and at 3092.05 cm^{-1} for PSC. Such results are concordant with that reported for the collagen extracted from the skin of splendid squid [13].

The amide I band mainly associated with stretching vibrations of carbonyl groups (C=O bond) along the polypeptide backbone [48], was depicted at 1629.6 cm^{-1} and 1629.5 cm^{-1} for ASC and PSC respectively, this amide is actually a sensitive marker for peptide secondary structure [49].

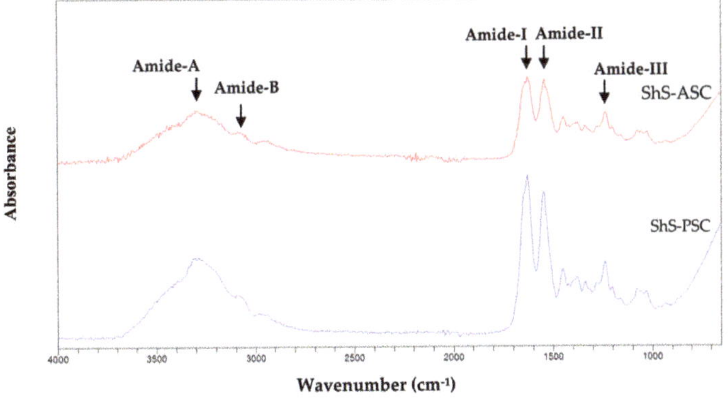

Figure 5. Fourier Transform Infrared spectra of ASC (acid soluble collagen) and PSC (pepsin soluble collagen) from the skin of hound smooth fish.

The PSC and ASC amide II bands were situated at wavenumbers of 1548.95 and 1545.17 cm^{-1} respectively; while the ASC and PSC-amide III bands were located at wavenumbers of 1237.94 and 1239.84 cm^{-1}, respectively. The amide II band represent N=H bending vibrations coupled with C=N stretching vibration [50], and amide III peak reflects intermolecular interactions in collagen, including peaks from C–N stretching and N–H deformation from amide linkages. It is also related to absorptions resulting from wagging vibrations from CH$_2$ groups from the glycine backbone and proline side-chains [51].

The Infra-Red absorption (IR) ratios between amide III and 1454 cm^{-1} band for the ASC and PSC fractions were found around 1 (0.95 and 1.08 respectively); indicating the persistence of the triple helix structure within the extracted collagen [52].

Such detailed description allowed to conclude that the slight differences observed between the ASC and PSC structure may be caused by pepsin treatment which has the effect to remove the telopeptide region, whereas the similarity of IR ratios may suggest that pepsin had no influence on the structure of the collagen triple helix.

2.2. Biofilms Mechanical and Functional Properties

2.2.1. Mechanical Properties

One of the laboratory objectives was to elaborate green edible biofilm using collagen from seafood by-products without any chemical addition and at low collagen percentage taking in consideration the cost of its production. However, when using skin *M. mustelus* collagen solution at 0.1%; the film was too fragile to allow any mechanical properties analysis (Figure 6A). Therefore, blending collagen with another polymer such as chitosan known for its high film-forming ability and lower cost was necessary to enhance the biofilm compactness (Figure 6B). Beside the resulting film showed akin aspect to the pure chitosan film (Figure 6C). The results suggest that the aggregation occurring between the collagen molecules of the film matrix was filled by the dispersed chitosan enhancing the cohesion between the various complexes within the adsorbed layer as shown in other study [53].

(A) **(B)** **(C)**

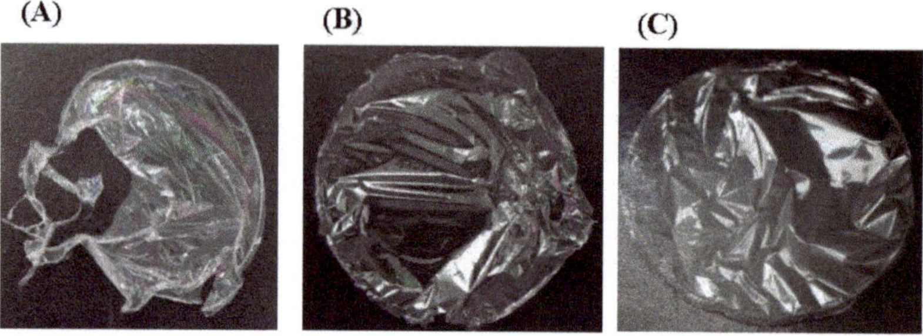

Figure 6. Pictures of (**A**) pure collagen film CO, (**B**) bi-composite Collagen-chitosan film and (**C**) pure chitosan film CH.

Such assumption is reflected by the thickness of the elaborated biofilms which showed different values as summarized in Table 1. The highest thickness value was noted in the pure chitosan film (17.15 μm) and decreased with an increased collagen ratio into the chitosan solution. Similar result were reported for the composite films using chitosan and collagen from the unicorn leatherjacket skin [54]. In the present study, higher chitosan ratio (75%) had no significant effect ($p > 0.05$) on film's thickness (Table 1) which showed similar smoothness and compactness. Such results suggest that the chitosan charge density was sufficiently high at the 50% ratio to assure the complexation between protein and polysaccharide knowing that the degree of compactness of the gel network is regulated by the polysaccharide charge density [55].

Table 1. Thickness and mechanical properties (TS: Tensile strength and EAB: elongation at break) of CH and composite films C_{50} and C_{75}.

	TS (MPa)	EAB (%)	Thickness (µm)
CH	70.52 ± 3.39 [a]	4.25 ± 0.63 [a]	17.15 ± 1.41 [a]
C_{50}	55.42 ± 8.6 [b]	5.67 ± 0.51 [b]	15.66 ± 1.63 [b]
C_{75}	66.28 ± 2.7 [a,b]	4.49 ± 0.23 [b]	16.07 ± 1.10 [b]

All values are mean $\pm$ standard deviation; [a,b] different superscripts in the same column indicate significant differences ($p < 0.05$). CH: pure chitosan film; C_{50}: Collagen-chitosan film 50%:50%, C_{75}: Collagen-chitosan film 25%:75%.

The tensile strength (TS) was also affected by degree of protein/polysaccharide ratio. Thus, the pure chitosane film value was 70.52 MPa and decreased with the addition of collagen. The increased chitosan proportion in composite films increased significantly ($p < 0.05$) the tensile strength from 55.42 MPa for C_{75} to 66.28 MPa for C_{50}. This is not only due to the interactions between chitosan and collagen molecules by electrostatic force but also by hydrogen bonding [56,57]. Similarly, the percentage of elongation at break (EAB) increased significantly ($p < 0.05$) with the incorporation of collagen in composite films from 4.25% in pure chitosane films to 4.49–5.67% in C_{75} and C_{50} respectively. This phenomenon is attributed to the hydrophilic properties of collagen which provides a certain increase in the hydration degree of the film giving an upper elongation at break value [58].

The comparison of the mechanical properties of the obtained composite films with those found in literature gives contradictory interpretations (Table 2) due to several factors including species origin (habitat, diet), the collagen amino acid composition [59], the protocol of extraction and the polymers ratio (collagen/chitosan).

Similarly, the mechanical properties of chitosan-based films are affected by various parameters such as the chitosan deacetylation degree, their molecular weight, as well as the conditions of film preparation (pH of the film-forming solution, the water content, and the drying conditions) [60–63].

In this study, composite films (C_{50}, C_{75}) exhibited tensile strength of similar or higher values than commercial films (LDPE 13%, HDPE 26%, Hydroxypropyl cellulose 15%) and collagen-chitosan-based films reported in other studies (Table 2). However, the elongations at break values of the composite films (C_{50}, C_{75}) were much lower than those of commercial films, since there is an inverse relationship between TS and EAB [64].

Table 2. Table summarising tensile strength (TS) and elongation at break (EAB) values of biofilms reported in other works and commercial films.

Film	TS(MPa)	EAB (%)	Reference
	70.52	**4.25**	**Present study**
Pure Chitosane	51.04	2.25	[65]
	5.8	17.3	[54]
Pure Collagen	25.3	14.7	[54]
	2.3	2.2	[66]
Collagen: Chitosan C50	**55.42**	**5.67**	**Present study**
Collagen: Chitosan C75	**66.28**	**4.49**	**Present study**
Collagen from shark catfish skin: Chitosane	8.16	14.3	[67]
collagen from jumbo squid: chitosan (15:85)	35.5	12.3	[58]
Polyester	178	85	
Polyvinyl chloride (PVC)	93	30	
Low-density polyethylene (LDPE)	13	500	[68]
High-density polyethylene (HDPE)	26	300	
Hydroxypropyl cellulose	15	33	

2.2.2. Water Solubility

The highest water solubility was observed in pure collagen films (32.14%) and decreased significantly ($p < 0.05$) with the incorporation of increased chitosan percentage (24.55% and 17.64% for the C_{50} and C_{75} films respectively; Table 3).

Table 3. Color properties and solubility of collagen films where L*: luminance/brightness, a*: red/green, b*: yellow/blue and ΔE*: total difference in colour.

	L*	a*	b*	ΔE*	Film Solubility (%)
CO	97.93 ± 0.001 [a]	0.07 ± 0.2 [a]	2.13 ± 0.01 [a]	0.35 ± 0.02 [a]	32.14 ± 2.3 [a]
C_{50}	97.76 ± 0.002 [b]	−0.09 ± 0.06 [b]	2.63 ± 0.002 [b]	0.85 ± 0.009 [b]	24.55 ± 1.88 [b]
C_{75}	97.30 ± 0.0002 [c]	−0.19 ± 0.09 [c]	2.62 ± 0.003 [b]	0.99 ± 0.008 [c]	17.64 ± 2 [c]
CH	97.26 ± 0.001 [c]	−0.13 ± 0.03 [d]	3.25 ± 0.03 [c]	1.56 ± 0.08 [d]	13.29 ± 1.02 [d]

All values are mean ± standard deviation; [a–d] different superscripts in the same column indicate significant differences ($p < 0.05$). CO: pure collagen film; CH: pure chitosan film; C_{50}: Collagen-chitosan film 50%:50%, C_{75}: Collagen-chitosan film 25%:75%.

Actually, the film's resistance to water is owed to the hydrophobic nature of chitosan molecule and to the covalent bond "amide bond" which has the effect of reducing the polarity of the films [54].

Indeed, an edible film must have both good resistance to water in order to preserve the integrity of the product [69] and a good ability to dissolve when ingested by the consumer and degrade naturally if it released into the environment [70].

However, the increase of water resistance of composite films could not be perceived as an advantage since high solubility cannot shield the product from humidity and water loss [71].

2.2.3. Optical Properties—Colour, Opacity and Light Transmittance of the Films

The film colour is a key element in the consumer's appreciation of the product since this parameter has a direct influence on the product appearance, especially when the film is to be used for packaging. For the various elaborated films, the highest ($p < 0.05$) L*-value (Lightness) and a*-value (redness/greenness) and the lowest b*-value (yellowness) and ΔE* (colour difference) were recorded in pure collagen films (Table 2). Thus, incorporation of chitosan induced a significant decrease in the lightness ($p < 0.05$), particularly in the films with the highest concentration of chitosan C_{75}, making them more yellowish. This may be due to the reaction of Maillard which took place between the carbonyls groups of chitosan and collagen amino groups [72].

The highest b*-value was observed in pure chitosan film (CH) and as described by Kurek et al. [73], this parameter (b*-value) defines the natural colour of chitosan, Yellow, which is related to the presence of β-1-4 linked 2-amino-2-deoxy-D-glucopyranose repeating units [74].

In addition to the colour of the film, transparency is also a very influential parameter in relation to the acceptability of the product. Generally, a clear film is more attractive clearly displaying the contents of the product. For all films, the light transmission in UV-Visible range was negligible at 200 nm, regardless of types and concentrations of chitosan. Collagen film exhibit the highest transmission at 280 nm but after the addition of chitosan, transmission decreased from 72.6 % for collagen pure film to 63.3% and 50.6% for composite films C_{50} and C_{75} respectively ($p < 0.05$).

The results show that when increasing the concentration of chitosan, composite films have better UV barrier properties. For instance, it makes these collagen-chitosan films usable as preventive materials against loss of nutrients and discoloration caused by the lipid oxidation [75]. The transmission of visible light at 400–800 nm, was superior to 80% in pure collagen film (CO), and was significantly higher ($p < 0.05$) than that of the collagen-chitosan composite films (Figure 7).

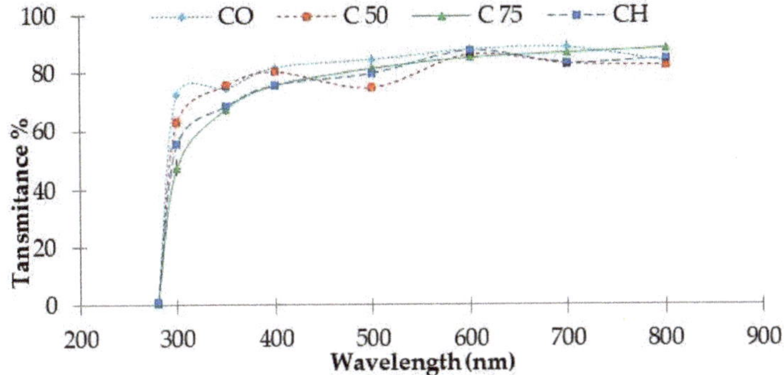

Figure 7. Optical transmission spectra of collagen films. CO: pure collagen film; CH: pure chitosan film; C_{50}: Collagen-chitosan film 50%:50%, C_{75}: Collagen-chitosan film 25%:75%.

2.2.4. Fourier Transform Infrared Spectra of Composite Films

For the composite biofilms, FTIR was used to detect the new interactions between collagen and chitosan and to identify the nature of the new linkages between both molecules.

The FTIR spectra revealed the characteristics of the specific bands corresponding to functional groups in all films (Figure 8).

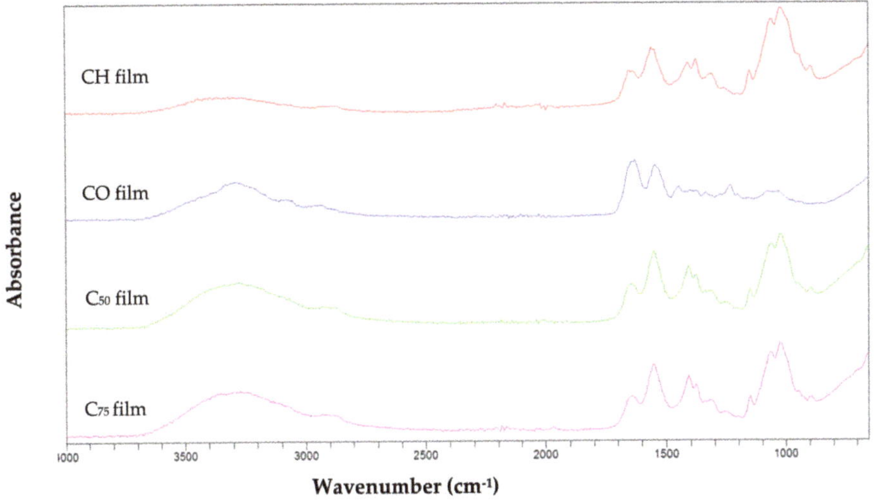

Figure 8. Fourier transform infrared spectra of different collagen films, CO: pure collagen film; CH: pure chitosan film; C_{50}: Collagen-chitosan film 50%:50%, C_{75}: Collagen-chitosan film 25%:75%.

The CO, C_{50}, and C_{75} composite films; displayed amide I bands at the wavenumbers of 1641.2, 1645.5 and 1642 cm^{-1} respectively. The shift to higher wavenumber was a result of a structural rearrangements occurring in the film structure with a strong affinity between the collagen and chitosan.

The amide I peak in CH film at the wavenumber of 1642.6 cm^{-1} was assigned to C=O stretching of N-acetyl group [76]. The amide II band was found at approximately the same wavenumber in CO film (1543.9 cm^{-1}) and C_{50} film (1543.6 cm^{-1}) but at lower peak compared to C_{75} (1550.61 cm^{-1}) and CH films (1550.8 cm^{-1}).

At amide III region, the addition of the chitosan had as result a significant decrease in the wavenumber from 1239.9 for CO film to 1016.9 and 1022.7 cm^{-1} for C_{50} and C_{75} films respectively. This shift induced by chitosan addition suggested some interaction between the CH_2 side chains of collagen molecule with that of chitosan molecules [67].

As shown in Figure 8, amide A peak of collagen film decreased after incorporation of chitosan from 3294.5 to 3250.4 cm^{-1} for C_{50} film and to 3288.5 cm^{-1} for C_{75} film, this suggests the loss of hydrogen bonding between water and collagen by chitosan interaction [77].

Similarly, a shift to lower frequency was noticed for amide B from 2941.9 for collagen film to 2932.2 and 2925.42 cm^{-1} for collagen-chitosan films C_{50} and C_{75}. As previously reported, when the CH group of a peptide was involved in a hydrogen bond with other polymer, the position of amide B moved to lower frequency.

The FTIR spectra clearly indicated that interactions between the two polymers have taken place and the secondary structure of collagen had been changed by chitosan incorporation.

2.2.5. Radical Scavenging Activity of Films

The radical scavenging activities, DPPH (1,1-diphenyl-2-picrylhydrazyl) of the different type of films are showed in Table 4.

Table 4. Radical scavenging activity DPPH (1,1-diphenyl-2-picrylhydrazyl)of the different biofilms.

Films	DPPH (%)	Reference
Pure Collagen CO	30.88 ±0.03 [a]	This study
Cuttlefish Gelatin	30.99	[78]
Chitosan CH	24.13± 0.75 [b]	This study
	0.17	[79]
Collagen: Chitosan C50	23.91± 1.15 [b]	This study
Collagen: Chitosan C75	19.77± 0.25 [b]	
Gelatin film + henna extract	61.86	[78]
Chitosan + Eucalyptus globulus essential oil	23.03–43.62	[80]
Chitosan + xanthan gum + fish protein hydrolysate	1.7–2.46	[79]

All values are mean ± standard deviation; [a,b] different superscripts in the same column indicate significant differences ($p < 0.05$). C50: Collagen-chitosan film 50%:50%, C75: Collagen-chitosan film 25%:75%.

Pure collagen film exhibited the highest radical scavenger with 30.8%. However, the addition of chitosan into the collagen solution, induced a significant decrease ($p < 0.05$) of the DPPH radical-scavenging ability of the composite films to values of 23.91% and 19.77% for C50, C75, respectively, when compared to pure collagen film. This result might be explained by the reaction that took place between residual free amino groups of chitosan and free radicals which may form stable macromolecular radicals and ammonium groups [81].

The scarcity of data on the DPPH- scavenging activity of collagen-based films, did not allow a comparative study. However, when comparing with other work such as cuttlefish gelatin-based films, CO film exhibited similar DPPH- scavenging values (Table 4). Regarding the composite films (C50, C75), scavenging activity was less or similar to other composite films (Table 4).

3. Experimental Section

3.1. Raw Materials

Smooth-hound (*Mustelus mustelus*) were purchased from a local market in La Goulette, Tunisia and brought immediately to the laboratory. Fish were thoroughly washed with cold tap water and manually de-skinned as occurs in the marked. The cleaned fish skins were cut into pieces (approximately 1 cm × 1 cm) and subjected to a pre-treatment for collagen extraction.

3.2. Pretreatment of Fish Skin

To remove pigments and non-collagenous proteins; the fish skins pieces (FSPs) were immersed into a solution of NaOH (0.1 M, ratio skin: solution 1:10) during 48 h with a daily solution changing. To reach neutral pH, the FSPs were washed with cold distilled water, then soaked for 24 h in butanol solution (10%, ratio 1:10) to eliminate fat and then thoroughly washed with cold water.

3.3. Collagen Extraction

Acid solubilised collagen and pepsin solubilised collagen were isolated from hound-smooth skin following the method proposed by Nagai and Suzuki [23] to which we introduced some modification as detailed in the following paragraph. All preparations were conducted in a cold room at 4 °C.

3.3.1. Acid Extraction

To extract collagen, the pre-treated FSPs were suspended in a solution of acetic acid (0.5 M, ratio 1:10) for 3 days with a continuous stirring, filtered and the residue was subjected to a second extraction under the same conditions. Following this step NaCl was added to both supernatants to a final concentration of 2.0 M to precipitate collagen. The pellets were recovered by centrifugation ($7000\times g$, during 1 h) then re-dissolved in acetic acid (0.5 M).

For purification, the resulting acetic acid solution was dialysed (bag cut-off of 14 kDa) against an acetic acid (0.1 M) solution, then against distilled water during 48 h in each case. The purified extract was thereafter lyophilized (Christ, Alpha 2–4 LD plus, Osterode am Harz, Germany), and the resulting collagen was referred as acetic ASC.

3.3.2. Enzymatic Extraction

Un-dissolved materials (residue 2) resulting from the previous steps (Figure 9) were washed with cold distilled water and re-suspended in acetic acid (0.5 M) containing 1% pepsin (w/w) at a ratio of 1:10 (w/v) then incubated for 72 h at 4 °C. The filtrate was recovered in two steps for precipitation, dialysis and freeze drying as explained in ASC purification. The resulting collagen was called PSC. The collagens were stored at -20 °C until analyses.

3.4. Extracted Collagen Characterisation

3.4.1. Sodium Dodecyl Sulfate Gel Electrophoresis

The determination of the collagen's electrophoretic profiles was performed according to Laemmli [82] method. The extracted collagens ASC and PSC (1 mg/mL) were dissolved in 0.02 M sodium phosphate buffer (pH 7.2) containing urea (3.5 M) and sodium dodecyl sulfate (SDS = 1% w/v). The solubilised samples were then mixed with a Tris HCl buffer (0.5 M, pH 6.8) containing 10% (w/v) SDS, 50% (v/v) glycerol and 5% (v/v) b-mercaptoethanol (b-ME), at 1:1 (v/v) ratio. Electrophoresis was carried on a polyacrylamide gel made of 7.5% running gel and 4% stacking gel. Following 150 min of electrophoretic migration, the protein bands were stained with Coomassie brilliant blue R-250 (0.1%) in methanol and acetic acid (45%, 10% v/v respectively). After that the gel was finally distained with methanol and acetic acid (10%, 10% v/v respectively). High molecular weight markers (Biorad, CA, USA) were used to estimate collagen molecular weights.

3.4.2. Peptide Mapping

The ASC and PSC peptide mappings were determined according to the method of Kittiphattanabawon et al. [83] with a slight modification. To solubilise collagen, samples (6 mg) were suspended in 0.1 M sodium phosphate buffer (pH 7.2) containing 0.5% SDS, then heated at 100 °C for 5 min. The digestion of collagen in solutions (300 µL) was realized by adding 200 µL of Lysyl endopeptidase (from *Achromobacter lyticus*, 10 µg/mL buffer) and incubating at 37 °C for 5 min and 25 min. The proteolysis was stopped by incubating samples in boiling water for 3 min, then SDS-PAGE was realized using 12% running gel and 4% stacking gel.

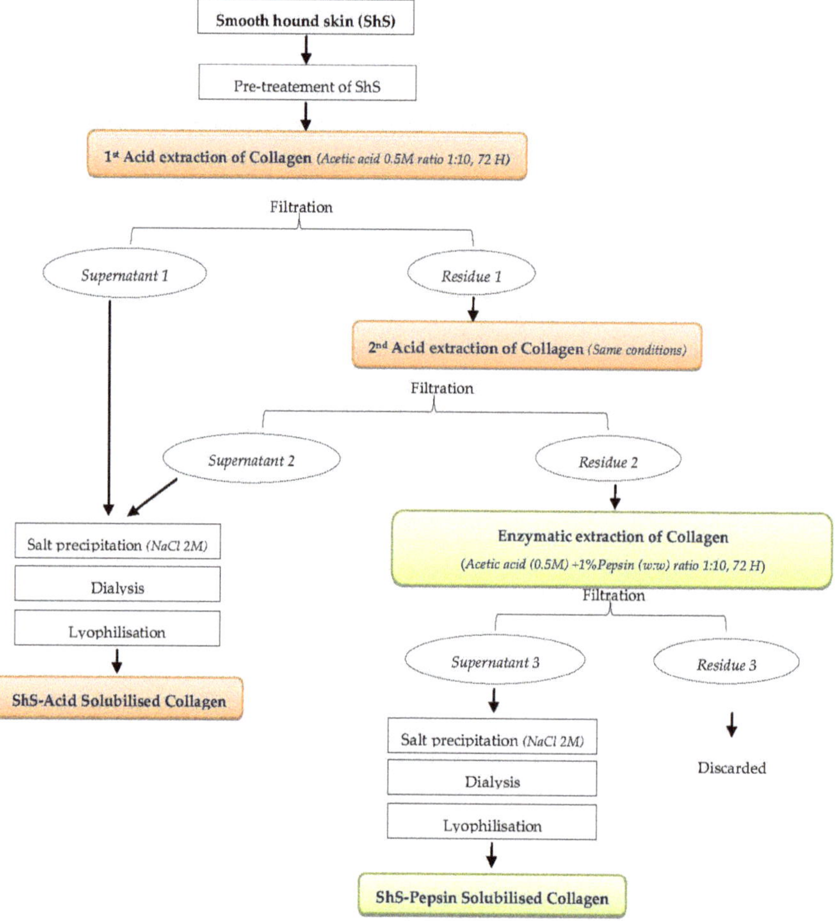

Figure 9. Collagen extraction process from smooth hound skin (ShS).

3.4.3. Viscosity Measurement

The ASC and PSC viscosities at different temperatures were measured according to Kimura and Ohno [84] procedure with some modification. The Ubbelohde viscometer (AVS 470, SI Analytics, Weilheim, Germany) was first filled with one of the collagen solution (1 mg/mL 0.1 M acetic acid), immersed in the water bath at 15 °C for 20 min; then temperature was increased stepwise up to 50 °C. The collagen solution viscosities were measured in each step of a 5 °C-temperature interval maintained for 20 min. The fractional viscosity of ASC and PSC at a designated temperature was calculated according to the formula below:

$$\text{Fractional viscosity} = \frac{(\text{maximum viscosity} - \text{measured viscosity})}{(\text{maximum viscosity} - \text{minimum viscosity})}$$

3.4.4. Ultraviolet Spectrophotometric Analysis

The UV absorption spectrum of collagen samples was measured with an UV-spectrophotometer (LLG Labware, model unispec, Meckenheim, Germany). The spectrum of ASC and PSC samples (5 mg/mL) in acetic acid (0.5 M) were identified by scanning the wavelength between 190 and 500 nm. The baseline was set with 0.5 M acetic acid.

3.5. Preparation of Polymer Composite Films and Characterisation

To prepare the film forming solution, lyophilized PSC collagen was suspended at 0.1 (*w*/*v*) in 4% aqueous commercial vinegar solution with continuous stirring overnight at 4 °C.

The film-forming solution of chitosan extracted from shrimp waste (mean degree of deacetylation = 80%), was solubilised in a 4% aqueous commercial vinegar solution (1%, *w:v*) with an overnight continuous homogenisation at room temperature.

Once both polymer solutions were prepared, different films were prepared: pure CO, CH and mixture of collagen-chitosan solution with different ratios C_{50} (50% collagen, 50% chitosan), and C_{75} (25% collagen, 75% chitosan). For a homogeneous suspension, the collagen-chitosan solution was stirred continuously for at least 30 min at room temperature; the solution was then poured into Petri dishes and dried at room temperature. Finally, the dried films were peeled off manually and stored at maintained at 55% relative humidity in a desiccator before analysis.

3.5.1. Thickness

The thickness of films was measured using a Thickness Tester (Thwing-Albert Instrument, ProGage, NJ, USA). Different locations on each film sample were used for determination of thickness.

3.5.2. Mechanical Properties

The film's stress-strain properties (tensile strength TS and elongation at break point EAB) were measured in accordance to ISO 527-1 [85]. The test was performed using the universal testing machine (Lloyd Instruments Ltd, LRX plus Series, West Sussex, UK).

Films were cut to width 15 mm and conditioned at 23 ± 2 °C and ~50 ± 10% RH 25 ± 0.5 °C for 24 h before measurement. Initial grip separation distance was set to 30 mm and mechanical crosshead speed to 100 mm/min.

3.5.3. Water Solubility

The film water solubility was tested using the procedure of Gómez Estaca, et al. [86] with slight modification. Film sections of 1×1 cm^2 (*n* = 6, in each case) were placed in an oven at 105 °C until constant weights were reached (W_i). The films were then immersed in water for 24 h at room temperature with gentle shaken. After filtration, the residual films were dried again at 105 °C for 24 h (W_f). The solubility of the films was calculated as:

$$\text{Solubility } (\%) = \frac{\text{Wi} - \text{Wf}}{\text{Wi}} \times 100$$

where, W_i = Initial weight of the film, W_f = Weight of the un-dissolved dried films residue.

3.5.4. Light Barrier Properties

Using an UV-Visible Spectrophotometer (LLG Labware, model unispec, Meckenheim, Germany), the light barrier properties of the different polymer films were measured by exposing them to light absorption at wavelengths ranging from 200 nm to 800 nm.

3.5.5. Colour Properties

The colour of the films was determined using a CIE colorimeter (Konica minolta Sensing, CR 410, Japan), and was expressed as: L*: luminance/brightness, a*: red/green and b*: yellow/blue. DE* (total difference in colour) was calculated using the equation below [87]:

$$\Delta E* = \sqrt{(\Delta L*)^2 + (\Delta a*)^2 + (\Delta b*)^2}$$

where $\Delta L*$, $\Delta a*$ and $\Delta b*$ are the differences between the corresponding colour parameter of the sample and the white standard.

3.5.6. Fourier Transform Infrared Spectroscopy

The FTIR analyses of collagens ASC and PSC extracted from smoothhound skin and of different films were performed using a Cary 630 FTIR spectrophotometer (Agilent Technologies, Santa Clara, CA, USA) within the wavenumber ranging between 4000 and 400 cm^{-1}. In each case, the sample was placed directly on the FTIR spectrometer fitted with an Agilent diamond ATR sample.

3.5.7. 1,1-Diphenyl-2-picrylhydrazyl Radical Scavenging Ability

The free radical-scavenging ability of films was measured as reported by Shimada, et al. [88] with slight modification. Films were cut in small pieces and dissolved in acetic acid solution (0.5 M) at 5 mg/mL, then 500 μL 0.1 mM DPPH solution was added to 500 μL of each film sample and kept in the dark for 30 min. Absorbance was measured at 571 nm and DPPH radical scavenging ability was calculated with the following equation:

$$DPPH(\%) = \frac{(A\ control - A\ sample)}{A\ control} \times 100$$

where A_{sample} and $A_{control}$ were the absorbencies of sample and control group, respectively.

3.6. Statistical Analysis

The experiments (carried out at least in triplicate) were presented as mean ± standard deviation. Statistical interpretation of the results was performed by One way ANOVA and LSD (least significant differences) tests (using $p < 0.05$ level of significance to compare mean values) using the software SPSS 22.0 (SPSS 22.0 for Windows, SPSS Inc., Chicago, IL, USA).

4. Conclusions

In a first step the type of collagen extracted from the hound-smooth skin was identified. Thus, electrophoretic patterns revealed that both extracted ASC and PSC are mostly composed of α (α1, α2) and β-chains with low content of γ-chain suggesting that hound-smooth skin collagen should most likely be classified as type I collagen.

Fourier transform infrared investigations showed that the secondary structure and the triple helical structure of ASC and PSC were both maintained intact even after the enzymatic hydrolysis with pepsin. Similar denaturation temperatures were found for ASC and PSC (26.68 °C and 26.66 °C respectively).

In a second step, collagen was used to elaborate film but with the incorporation of chitosan. The addition of chitosan increased the films mechanical strength and reduced its water solubility. The FTIR spectra clearly indicated that interactions between both polymers occurred and the secondary structure of collagen triple helix have been changed by the addition of chitosan.

The addition of 50% chitosan into collagen films was sufficient to obtain an edible film with good mechanical properties, suitable solubility and antioxidant activity. Owing to its anti-UV properties, such collagen-chitosan film could be used as a protective material to preserve nutraceutical products.

Author Contributions: E.B.S., and S.S. conceived and designed the experiments; E.B.S. conducted the experimental analysis and wrote the manuscript; S.S. supervised, validate the data analyses and co-wrote the manuscript. All authors reviewed the manuscript.

Acknowledgments: This work was conducted within the project "Biotechnologie Marine Vecteur d'Innovation et de Qualité-BIOVecQ PS1.3_08" co-financed by the cross-border IEVP Italy-Tunisia program and the ministry of Higher Education and Scientific Research-Tunisia.

Conflicts of Interest: The authors declare no conflict of interest.

References

1. FAO. The State of World Fisheries and Aquaculture 2016. In *Contributing to Food Security and Nutrition for All*; FAO: Rome, Italy, 2016; p. 224, ISBN 978-92-5-209185-1.
2. Birk, D.E.; Bruckner, P. Collagen suprastructures. *Collagen* **2005**, *247*, 185–205. [CrossRef]
3. Kadler, K. Extracellular matrix 1: Fibril-forming collagens. *Protein Profile* **1995**, *2*, 491–619. [PubMed]
4. Cavallaro, J.F.; Kemp, P.D.; Kraus, K.H. Collagen fabrics as biomaterials. *Biotechnol. Bioeng.* **1994**, *43*, 781–791. [CrossRef] [PubMed]
5. Hood, L.L. Collagen in sausage casings. In *Advances in Meat Research Volume 4: Collagen as Food*; Pearson, A.M., Dutson, T.R., Bailey, A.J., Eds.; Van Nostrand Reinhold Company: New York, NY, USA, 1987; Volume 4, pp. 109–129.
6. Stainsby, G. Gelatin gels. In *Collagen as Food: Volume 4. Advances in Meat Research*; Pearson, A.M., Dutson, T.R., Bailey, A.J., Eds.; Van Nostrand Reinhold Company: New York, NY, USA, 1987; Volume 4, pp. 209–222.
7. Wells, G.A.; Scott, A.C.; Johnson, C.T.; Gunning, R.F.; Hancock, R.D.; Jeffrey, M.; Dawson, M.; Bradley, R. A novel progressive spongiform encephalopathy in cattle. *Vet. Rec.* **1987**, *31*, 419–420. [CrossRef]
8. Jongjareonrak, A.; Benjakul, S.; Visessanguan, W.; Nagai, T.; Tanaka, M. Isolation and characterisation of acid and pepsin-solubilised collagens from the skin of Brownstripe red snapper (*Lutjanus vitta*). *Food Chem.* **2005**, *93*, 475–484. [CrossRef]
9. Moon, S.H.; Lee, Y.J.; Rhie, J.W.; Suh, D.S.; Oh, D.Y.; Lee, J.H.; Kim, Y.J.; Kim, S.M.; Jun, Y.J. Comparative study of the effectiveness and safety of porcine and bovine atelocollagen in Asian nasolabial fold correction. *J. Plast. Surg. Hand Surg.* **2015**, *49*, 147–152. [CrossRef] [PubMed]
10. Ahmad, M.; Nirmal, N.P.; Chuprom, J. Molecular characteristics of collagen extracted from the starry triggerfish skin and its potential in the development of biodegradable packaging film. *RSC Adv.* **2016**, *6*, 33868–33879. [CrossRef]
11. Nagai, T.; Yamashita, E.; Taniguchi, K.; Kanamori, N.; Suzuki, N. Isolation and characterisation of collagen from the outer skin waste material of cuttlefish (*Sepia lycidas*). *Food Chem.* **2001**, *72*, 425–429. [CrossRef]
12. Nagai, T.; Nagamori, K.; Yamashita, E.; Suzuki, N. Collagen of octopus *Callistoctopus arakawai* arm. *Int. J. Food Sci. Technol.* **2002**, *37*, 285–289. [CrossRef]
13. Kittiphattanabawon, P.; Nalinanon, S.; Benjakul, S.; Kishimura, H. Characteristics of Pepsin-Solubilised Collagen from the Skin of Splendid Squid (*Loligo formosana*). *J. Chem.* **2015**, *8*. [CrossRef]
14. Nagai, T.; Ogawa, T.; Nakamura, T.; Ito, T.; Nakagawa, H.; Fujiki, K.; Nakao, M.; Yano, T. Collagen of edible jellyfish exumbrella. *J. Sci. Food Agric.* **1999**, *79*, 855–858. [CrossRef]
15. Lee, K.J.; Hee, Y.P.; Kim, Y.K.; Park, J.I.; Ho, D.Y. Biochemical Characterization of Collagen from the Starfish *Asteria samurensis*. *J. Korean Soc. Appl. Biol. Chem.* **2009**, *52*. [CrossRef]
16. Nagai, T.; Suzuki, N. Partial characterization of collagen from purple sea urchin (*Anthocidaris crassispina*) test. *Int. J. Food Sci. Technol.* **2000**, *35*, 497–501. [CrossRef]
17. Zhong, M.; Chen, T.; Hu, C.; Ren, C. Isolation and Characterization of Collagen from the Body Wall of Sea Cucumber *Stichopus monotuberculatus*. *J. Food Sci.* **2015**, *80*, 671–679. [CrossRef] [PubMed]
18. Ehrlich, H.; Wysokowski, M.; Żółtowska-Aksamitowska, S.; Petrenko, I.; Jesionowski, T. Collagens of Poriferan Origin. *Mar. Drugs* **2018**, *16*, 79. [CrossRef] [PubMed]
19. Chuaychan, S.; Benjakul, S.; Kishimura, H. Characteristics of acid- and pepsin-soluble collagens from scale of seabass (*Lates calcarifer*). *Lwt-Food Sci. Technol.* **2015**, *63*, 71–76. [CrossRef]
20. Ahmad, M.; Benjakul, S. Extraction and characterisation of pepsin-solubilised collagen from the skin of unicorn leatherjacket (*Aluterus monocerous*). *Food Chem.* **2010**, *120*, 817–824. [CrossRef]

21. Hwang, J.H.; Mizuta, S.; Yokoyama, Y.; Yoshinaka, R. Purification and characterization of molecular species of collagen in the skin of skate (*Raja kenojei*). *Food Chem.* **2007**, *100*, 921–925. [CrossRef]

22. Liu, D.; Zhang, X.; Li, T.; Yang, H.; Zhang, H.; Regenstein, J.M.; Zhou, P. Extraction and characterization of acid- and pepsin-soluble collagens from the scales, skins and swim-bladders of grass carp (*Ctenopharyngodon idella*). *Food Biosci.* **2015**, *9*, 68–74. [CrossRef]

23. Nagai, T.; Suzuki, N. Isolation of collagen from fish waste material-skin, bone and fins. *Food Chem.* **2000**, *68*, 277–281. [CrossRef]

24. Jeevithan, E.; Bao, B.; Bu, Y.; Zhou, Y.; Zhao, Q.; Wu, W. Type II Collagen and Gelatin from Silvertip Shark (*Carcharhinus albimarginatus*) Cartilage: Isolation, Purification, Physicochemical and Antioxidant Properties. *Mar. Drugs* **2014**, *12*, 3852. [CrossRef] [PubMed]

25. Benjakul, S.; Nalinanon, S.; Shahidi, F. Fish collagen. In *Food Biochemistry and Food Processing*, 2nd ed.; Simpson, B.K., Nollet, L.M.L., Toldra, F., Benjakul, S., Paliyath, G., Hui, Y.H., Eds.; John Wiley and Sons: Hoboken, NJ, USA, 2012; pp. 365–387.

26. Cao, H.; Xu, S.Y. Purification and characterization of type II collagen from chick sternal cartilage. *Food Chem.* **2008**, *108*, 439–445. [CrossRef] [PubMed]

27. Lin, Y.K.; Liu, D.C. Effects of pepsin digestion at different temperatures and times on properties of telopeptide-poor collagen from bird feet. *Food Chem.* **2006**, *94*, 621–625. [CrossRef]

28. Bradai, M.N.; Saidi, B.; Enajjar, S.; Bouain, A. The Gulf of Gabe's: A spot for the Mediterranean elasmobranches. In *The Proceedings of the Workshop on Mediterranean Cartilaginous Fish with Emphasis on Southern and Eastern Mediterranean*; Basusta, N., Keskin, C., Serena, F., Seret, B., Eds.; Turkish Marine Research Foundation: Istanbul, Turkey, 2006; pp. 107–117.

29. Costantini, M.; Bernardini, M.; Cordone, P.; Guilianini, P.G.; Orel, G. Observations on fishery, feeding habits and reproductive biology of *Mustelus mustelus* (Chondrichtyes, Triakidae). *North. Adriat. Sea Biol. Mar. Mediterr.* **2000**, *7*, 427–432.

30. Kim, H.W.; Song, D.H.; Choi, Y.S.; Kim, H.Y.; Hwang, K.E.; Park, J.H.; Kim, Y.J.; Choi, J.H.; Kim, C.J. Effects of Soaking pH and Extracting Temperature on the Physicochemical Properties of Chicken Skin Gelatin. *Korean J. Food Sci. Anim.* **2012**, *32*, 316–322. [CrossRef]

31. Rinaudo, M. Chitin and chitosan: Properties and applications. *Prog. Polym. Sci.* **2006**, *31*, 603–632. [CrossRef]

32. Damodaran, S. *Food Proteins and Their Applications*, 1st ed.; Marcel Dekker Inc.: New York, NY, USA, 1997; pp. 356–359.

33. Matmaroh, K.; Benjakul, S.; Prodpran, T.; Encarnacion, A.B.; Kishimura, H. Characteristics of acid soluble collagen and pepsin soluble collagen from scale of spotted golden goatfish (*Parupeneus heptacanthus*). *Food Chem.* **2011**, *129*, 1179–1186. [CrossRef] [PubMed]

34. Kittiphattanabawon, P.; Benjakul, S.; Visessanguan, W.; Kishimura, H.; Shahidi, F. Isolation and Characterisation of collagen from the skin of brownbanded bamboo shark (*Chiloscyllium punctatum*). *Food Chem.* **2010**, *119*, 1519–1526. [CrossRef]

35. Nomura, Y.; Yamano, M.; Shirai, K. Renaturation of α1 chains from shark skin collagen type I. *J. Food Sci.* **1995**, *60*, 1233–1236. [CrossRef]

36. Jekel, P.A.; Weijer, W.J.; Beintema, J.J. Use of endoproteinase Lys-C from Lysobacter enzymogenes in protein sequence analysis. *Anal. Biochem.* **1983**, *134*, 347–354. [CrossRef]

37. Heu, M.S.; Lee, J.H.; Kim, H.J.; Jee, S.J.; Lee, J.S.; Jeon, Y.J.; Shahidi, F.; Kim, J. Characterization of acid- and pepsin-soluble collagens from flatfish skin. *Food Sci. Biotechnol.* **2010**, *19*, 27–33. [CrossRef]

38. Li, Z.; Wang, B.; Chi, C.; Zhang, Q.; Gong, Y.; Tang, J.; Luo, H.; Ding, G. Isolation and characterization of acid soluble collagens and pepsin soluble collagens from the skin and bone of Spanish mackerel (*Scomberomorous niphonius*). *Food Hydrocoll.* **2013**, *31*, 103–113. [CrossRef]

39. Sampath Kumar, N.S.; Nazeer, R.A.; Jaiganesh, R. Wound Healing Properties of Collagen from the Bone of Two Marine Fishes. *Int. J. Pept. Res. Ther.* **2012**, *18*, 185–192. [CrossRef]

40. Heredia, A.; Colin-Garcia, M.; Pena-Rico, M.A.; Aguirre Beltran, L.F.L.; Gracio, J.; Contreras-Torres, F.F.; Rodriguez-Galvan, A.; Bucio, L.; Basiuk, V.A. Thermal, infrared spectroscopy and molecular modeling characterization of bone: An insight in the apatite-collagen type I interaction. *Adv. Biol. Chem.* **2013**, *3*, 215–223. [CrossRef]

41. Rao, K.P.; Alamelu, S. Effect of crosslinking agent on the release of an aqueous marker from liposomes sequestered in collagen and chitosan gels. *J. Memb. Sci.* **1992**, *71*, 161–167. [CrossRef]

42. Kozlowska, J.; Sionkowska, A.; Skopinska-Wisniewska, J.; Piechowicz, K. Northern pike (*Esox lucius*) collagen: Extraction, characterization and potential application. *Int. J. Biol. Macromol.* **2015**, *81*, 220–227. [CrossRef] [PubMed]

43. Nalinanon, S.; Benjakul, S.; Kishimura, H. Collagens from the skin of arabesque greenling (*Pleurogrammus azonus*) solubilized with the aid of acetic acid and pepsin from albacore tuna (*Thunnus alalunga*) stomach. *J. Sci. Food Agric.* **2010**, *90*, 1492–1500. [CrossRef] [PubMed]

44. Jia, Y.; Wang, H.; Wang, H.; Li, Y.; Wang, M.; Zhou, J. Biochemical Properties of Skin Collagens Isolated from Black Carp (*Mylopharyngodon piceus*). *Food Sci. Biotechnol.* **2012**, *21*, 1585–1592. [CrossRef]

45. Belbachir, K.; Noreen, R.; Gouspillou, G.; Petibois, C. Collagen types analysis and differentiation by FTIR spectroscopy. *Anal. Bioanal. Chem.* **2009**, *395*, 829–837. [CrossRef] [PubMed]

46. Doyle, B.B.; Bendit, E.G.; Blout, E.R. Infrared spectroscopy of collagen and collagen-like polypeptides. *Biopolymers* **1975**, *14*, 937–957. [CrossRef] [PubMed]

47. Muyonga, J.H.; Cole, C.G.B.; Duodu, K.G. Characterisation of acid soluble collagen from skins of young and adult Nile perch (*Lates niloticus*). *Food Chem.* **2004**, *85*, 81–89. [CrossRef]

48. Payne, K.J.; Veis, A. Fourier transform IR spectroscopy of collagen and gelatin solutions: Deconvolution of the amide I band for conformational studies. *Biopolymers* **1988**, *27*, 1749–1760. [CrossRef] [PubMed]

49. Surewicz, W.K.; Mantsch, H.H. New insight into protein secondary structure from resolution-enhanced infrared spectra. *Biochim. Biophys. Acta* **1988**, *952*, 115–130. [CrossRef]

50. Krimm, S.; Bandekar, J. Vibrational Spectroscopy and Conformation of Peptides, Polypeptides, and Proteins. *Adv. Protein Chem.* **1986**, *38*, 181–364. [CrossRef] [PubMed]

51. Jackson, M.; Choo, L.P.; Watson, P.H.; Halliday, W.C.; Mantsch, H.H. Beware of connective tissue proteins: Assignment and implications of collagen absorptions in infrared spectra of human tissues. *Biochim. Biophys. Acta* **1995**, *1270*, 1–6. [CrossRef]

52. Plepis, A.M.D.; Goissis, G.; DasGupta, D.K. Dielectric and pyroelectric characterization of anionic and native collagen. *Polym. Eng. Sci.* **1996**, *36*, 2932–2938. [CrossRef]

53. Ganzevles, R.A. Protein/Polysaccharide Complexes at Air/Water Interfaces. Ph.D. Thesis, Wageningen University, Wageningen, the Netherlands, 2007.

54. Ahmad, M.; Nirmal, N.P.; Danish, M.; Chuprom, J.; Jafarzedeh, S. Characterisation of composite films fabricated from collagen/chitosan and collagen/soy protein isolate for food packaging applications. *RSC Adv.* **2016**, *6*, 82191–82204. [CrossRef]

55. Zhang, S.; Zhang, Z.; Vardhanabhuti, B. Effect of charge density of polysaccharides on self-assembled intragastric gelation of whey protein/polysaccharide under simulated gastric conditions. *Food Funct.* **2014**, *5*, 1829–1838. [CrossRef] [PubMed]

56. Charulatha, V.; Rajaram, A. Influence of different crosslinking treatments on the physical properties of collagen membranes. *Biomaterials* **2003**, *24*, 759–767. [CrossRef]

57. Gomez-Estaca, J.; Gomez-Guillén, M.C.; Fernandez-Martín, F.; Montero, P. Effects of gelatin origin, bovine-hide and tuna-skin, on the properties of compound gelatin chitosan films. *Food Hydrocoll.* **2011**, *25*, 1461–1469. [CrossRef]

58. Uriarte-Montoya, M.H.; Arias-Moscoso, J.L.; Plascencia-Jatomea, M.; Santacruz-Ortega, H.; Rouzaud-Sández, O.; Cardenas-Lopez, J.L.; Marquez-Rios, E.; Ezquerra-Brauer, J.M. Jumbo squid (*Dosidicus gigas*) mantle collagen: Extraction, characterization, and potential application in the preparation of chitosan–collagen biofilms. *Bioresour. Technol.* **2010**, *101*, 4212–4219. [CrossRef] [PubMed]

59. O'Sullivan, A.; Shaw, N.B.; Murphy, S.C.; Van de Vis, J.W.; Van Pelt-Heerschap, H.; Kerry, J.P. Extraction of Collagen from Fish Skins and Its Use in the Manufacture of Biopolymer Films. *J. Aquat. Food Prod. Technol.* **2006**, *15*, 21–32. [CrossRef]

60. Kołodziejska, I.; Piotrowska, B. The water vapour permeability, mechanical properties and solubility of fish gelatin–chitosan films modified with transglutaminase or 1-ethyl-3-(3-dimethylaminopropyl) carbodiimide (EDC) and plasticized with glycerol. *Food Chem.* **2007**, *103*, 295–300. [CrossRef]

61. Caner, C.; Vergano, P.J.; Wiles, J.L. Chitosan film mechanical and permeation properties as affected by acid, plasticizer, and storage. *J. Food Sci.* **1998**, *63*, 1049–1053. [CrossRef]

62. Rivero, S.; García, M.A.; Pinotti, A. Composite and bi-layer films based on gelatin and chitosan. *J. Food Eng.* **2009**, *90*, 531–539. [CrossRef]

63. Butler, B.L.; Vergano, P.J.; Testin, R.F.; Bunn, J.M.; Wiles, J.L. Mechanical and barrier properties of edible chitosan films as affected by composition and storage. *J. Food Sci.* **1996**, *61*, 953–955. [CrossRef]

64. Rhim, J.W.; Gennadios, A.; Handa, A.; Weller, C.L.; Hanna, A. Solubility, tensile, and color properties of modified soy protein isolate films. *J. Agric. Food Chem.* **2000**, *48*, 4937–4941. [CrossRef] [PubMed]

65. Liu, J.; Liu, S.; Chen, Y.; Zhang, L.; Kan, J.; Jin, C. Physical, mechanical and antioxidant properties of chitosan films grafted with different hydroxybenzoic acids. *Food Hydrocoll.* **2017**, *71*, 176–186. [CrossRef]

66. Elango, J.; Bu, Y.; Bin, B.; Geevaretnam, J.; Robinson, J.S.; Wu, W. Effect of chemical and biological cross-linkers on mechanical and functional properties of shark catfish skin collagen films. *Food Biosci.* **2016**, *17*, 42–51. [CrossRef]

67. Elango, J.; Robinson, J.S.; Geevaretnam, J.; Rupia, E.J.; Arumugam, V.; Durairaj, S.; Wenhui, W. Physicochemical and rheological properties of composite shark catfish (*Pangasius pangasius*) skin collagen films integrated with chitosan and calcium salts. *J. Food Biochem.* **2016**, *40*, 304–315. [CrossRef]

68. Lacroix, M.; Cooksey, K. Edible films and coatings from animal origin proteins. In *Innovations in Food Packaging*, 1st ed.; Jung, H., Ed.; Elsevier: New York, NY, USA, 2005; pp. 301–317.

69. Perez-Gago, M.B.; Krochta, J.M. Water vapor permeability, solubility and tensile properties of heat denatured versus native whey protein films. *J. Food Sci.* **1999**, *64*, 1034–1037. [CrossRef]

70. Pitak, N.; Rakshit, S.K. Physical and antimicrobial properties of banana flour/chitosan biodegradable and self sealing films used for preserving fresh-cut vegetables. *LWT-Food Sci. Technol.* **2011**, *44*, 2310–2315. [CrossRef]

71. Gontard, N.; Guilbert, S.; Cuq, J.L. Water and glycerol as plasticizer affect mechanical and water vapor barrier properties of an wheat gluten film. *J. Food Sci.* **1993**, *58*, 206–211. [CrossRef]

72. Prodpran, T.; Benjakul, S.; Artharn, A. Properties and microstructure of protein-based film from round scad (*Decapterus maruadsi*) muscle as affected by palm oil and chitosan incorporation. *Int. J. Biol. Macromol.* **2007**, *41*, 605–614. [CrossRef] [PubMed]

73. Kurek, M.; Descours, E.; Galic, K.; Voilley, A.; Debeaufort, F. How composition and process parameters affect volatile active compounds in biopolymer films. *Carbohydr. Polym.* **2012**, *88*, 646–656. [CrossRef]

74. Pereda, M.; Ponce, A.G.; Marcovich, N.E.; Ruseckaite, R.A.; Martucci, J.F. Chitosan-gelatin composites and bi-layer films with potential antimicrobial activity. *Food Hydrocoll.* **2011**, *25*, 1372–1381. [CrossRef]

75. Martins, J.T.; Cerqueira, M.A.; Vicente, A.A. Influence of a-tocopherol on physicochemical properties of chitosan-based films. *Food Hydrocoll.* **2012**, *27*, 220–227. [CrossRef]

76. Liu, M.; Zhou, Y.; Zhang, Y.; Yu, C.; Cao, S. Preparation and structural analysis of chitosan films with and without sorbitol. *Food Hydrocoll.* **2013**, *33*, 186–191. [CrossRef]

77. Sionkowska, A. Effects of solar radiation on collagen and chitosan films. *J. Photochem. Photobiol. B* **2006**, *82*, 9–15. [CrossRef] [PubMed]

78. Jridi, M.; Sellimi, S.; Lassoued, K.B.; Beltaief, S.; Souissi, N.; Mora, L.; Toldra, F.; Elfeki, A.; Nasri, M.; Nasri, R. Wound healing activity of cuttlefish gelatin gels and films enriched by henna (*Lawsonia inermis*) extract. *Colloids Surf. A Physicochem. Eng. Asp.* **2017**, *512*, 71–79. [CrossRef]

79. De Morais Lima, M.; Bianchini, D.; Guerra Dias, A.; Da Rosa Zavareze, E.; Prentice, C.; Da Silveira Moreira, A. Biodegradable films based on chitosan, xanthan gum,and fish protein hydrolysate. *J. Appl. Polym. Sci.* **2017**, *134*. [CrossRef]

80. Hafsa, J.; Smach, M.; Ben Khedher, M.R.; Charfeddine, B.; Limem, K.; Majdoub, H.; Rouatbi, S. Physical, antioxidant and antimicrobial properties of chitosan films containing Eucalyptus globulus essential oil. *LWT-Food Sci. Technol.* **2016**, *68*, 356–364. [CrossRef]

81. Xie, W.; Xu, P.; Liu, Q. Antioxidant activity of water-soluble chitosan derivatives. *Bioorg. Med. Chem. Lett.* **2001**, *11*, 1699–1701. [CrossRef]

82. Laemmli, U.K. Cleavage of structural proteins during the assembly of the head of bacteriophage T4. *Nature* **1970**, *227*, 680–685. [CrossRef] [PubMed]

83. Kittiphattanabawon, P.; Benjakul, S.; Visessanguan, W.; Nagai, T.; Tanaka, M. Characterisation of acid-soluble collagen from skin and bone of bigeye snapper (*Priacanthus tayenus*). *Food Chem.* **2005**, *89*, 363–372. [CrossRef]

84. Kimura, S.; Ohno, Y. Fish type I collagen: Tissue-species existence of two molecular forms, (a1)2a2 and a1a2a3, in alaska pollack. *Comp. Biochem. Physiol.* **1987**, *88*, 409–413.

85. ISO. Plastiques—Détermination des propriétés en traction—Partie 1: Principes généraux. In *ISO 527-1*; ISO: Geneva, Switzerland, 2012; p. 23.

86. Gómez Estaca, J.; Montero, P.; Fernández, M.F.; Gómez-Guillén, M.C. Physico-chemical and film-forming properties of bovine-hide and tuna-skin gelatin: A comparative study. *J. Food Eng.* **2009**, *90*, 480–486. [CrossRef]

87. Gennadios, A.; Weller, C.L.; Hanna, M.A.; Froning, G.W. Mechanical and Barrier Properties of Egg Albumen Films. *J. Food Sci.* **1996**, *61*, 585–589. [CrossRef]

88. Shimada, K.; Fujikawa, K.; Yahara, K.; Nakamura, T. Antioxidative properties of xanthan on the antioxidation of soybean oil in cyclodextrin emulsion. *J. Agric. Food Chem.* **1992**, *40*, 945–948. [CrossRef]

 *marine drugs* 

Article

Evaluation of Differentiated Bone Cells Proliferation by Blue Shark Skin Collagen via Biochemical for Bone Tissue Engineering

Jeevithan Elango [1,†], Jung Woo Lee [1,2,†], Shujun Wang [3], Yves Henrotin [4], José Eduardo Maté Sánchez de Val [5], Joe M. Regenstein [6], Sun Young Lim [2], Bin Bao [1,*] and Wenhui Wu [1,3,7,*]

1   Department of Marine Bio-Pharmacology, College of Food Science and Technology, Shanghai Ocean University, Shanghai 201306, China; srijeevithan@gmail.com (J.E.); truthwo@naver.com (J.W.L.)
2   Division of Marine Bioscience, Korea Maritime and Ocean University, Busan 606791, Korea; sylim@kmou.ac.kr
3   Co-Innovation Center of Jiangsu Marine Bio-industry Technology, Huaihai Institute of Technology, Lianyungang 222005, China; shujunwang86@163.com
4   Bone and Cartilage Research Unit, Arthropôle Liège, University of Liège, CHU Sart-Tilman, 4000 Liège, Belgium; yhenrotin@ulg.ac.be
5   Department of Biomaterials Engineering, Universidad Católica San Antonio de Murcia, 30107 Murcia, Spain; jemate@ucam.edu
6   Department of Food Science, Cornell University, Ithaca, NY 14853-7201, USA; jmr9@cornell.edu
7   Laboratory of Quality and Safety Risk Assessment for Aquatic Products on Storage and Preservation, Ministry of Agriculture, Shanghai 201306, China
*   Correspondence: bbao@shou.edu.cn (B.B.); whwu@shou.edu.cn (W.H.W.); Tel.: +86-21-61900388 (B.B. & W.H.W); Fax: +86-21-61900364 (B.B. & W.H.W.)
†   These authors equally contributed to this work.

Received: 22 August 2018; Accepted: 17 September 2018; Published: 25 September 2018

Abstract: Collagen from a marine resource is believed to have more potential activity in bone tissue engineering and their bioactivity depends on biochemical and structural properties. Considering the above concept, pepsin soluble collagen (PSC) and acid soluble collagen (ASC) from blue shark (*Prionace glauca*) skin were extracted and its biochemical and osteogenic properties were investigated. The hydroxyproline content was higher in PSC than ASC and the purified collagens contained three distinct bands α_1, α_2, and β dimer. The purity of collagen was confirmed by the RP-HPLC profile and the thermogravimetric data showed a two-step thermal degradation pattern. ASC had a sharp decline in viscosity at 20–30 °C. Scanning electron microscope (SEM) images revealed the fibrillar network structure of collagens. Proliferation rates of the differentiated mouse bone marrow-mesenchymal stem (dMBMS) and differentiated osteoblastic (dMC3T3E1) cells were increased in collagen treated groups rather than the controls and the effect was dose-dependent, which was further supported by higher osteogenic protein and mRNA expression in collagen treated bone cells. Among two collagens, PSC had significantly increased dMBMS cell proliferation and this was materialized through increasing RUNX2 and collagen-I expression in bone cells. Accordingly, the collagens from blue shark skin with excellent biochemical and osteogenic properties could be a suitable biomaterial for therapeutic application.

Keywords: blue shark collagen; osteogenic activity; Runx2; differentiated mesenchymal stem cell; osteoblast; proliferation

1. Introduction

There are three major types of protein present in fish, such as sarcoplasmic (25–30%), myofibrillar (66–77%), and stroma protein (1–5%). Among them, stromal proteins are mostly located in the interstitial space of muscle cells and the extracellular membrane. One of the most available stromal proteins in fish is collagen. It is extracted from different part of fish such as bones, scales, skin, swim bladder, and fins. Currently, collagen from the marine organism has been widely investigated due to its excellent biological activity with low or no side effect as an alternative to mammalian collagen [1,2]. Collagen is graded based on its biochemical, functional, and rheological characteristics, which may depend on fish species and the manufacturing method. Based on the quality and bioactivity, fish collagen can be used as a drug in the biomedical industries or as a food supplement in food industries.

In tissue engineering, fish collagen is used in the form of films, scaffolds, sponge, hydrogel, microspheres, and composite biopolymers. Wang et al. [3] studied the hemostatic properties of type I collagen to treat tissue burns. In another study, type I collagen 3D matrix was used as a potential biomaterial for heart regeneration [4]. It is well known that the growth and characteristic of cells such as proliferation, adhesion, and maturation were enhanced by collagen labeling [5]. It was found that the inflammatory response of tilapia collagen biomaterials in rabbits was similar to that of collagen from porcine or polyethylene [6]. In addition, the collagen from jellyfish could induce an immune response similar to that of bovine collagen [7]. Sugahara et al. [8] reported that jellyfish collagen produced by pepsin digestion up-regulated the production of immunoglobulin IgM in the human hybridoma cells and IgM and IgG in human peripheral blood lymphocytes due to the presence of telopeptides. Our recent research confirmed the biocompatible and non-immunogenic effect of acid soluble- and pepsin soluble-collagen isolated from shark cartilages and tilapia skin and these collagens did not elicit immune response in in-vivo and in-vitro models [9,10]. In that sense, collagen from shark can be used as a potential biomaterial for bone tissue engineering.

Indeed, shark collagen has been widely used in bone tissue engineering application due to its excellent biocompatible, osteoconductive, osteoinductive, and natural bone biomimetic properties. The collagen material used in bone tissue engineering not only regulates morphological properties but also maintains appropriate cues for regulating cellular processes during bone formation. It was reported that the collagen-based biomaterial implanted in rabbit induced neotendon and neoligament formation [11,12]. Our previous study revealed the osteogenic potential of whale shark bone collagen on mesenchymal stem cells and primary osteocytes [13]. In another study, composite scaffold 3D matrices were prepared by mixing acid soluble collagen from blue shark (*Prionace glauca*) skin with calcium phosphates from the teeth of *Prionace glauca* and *Isurus oxyrinchus* and tested the effect of a composite collagen scaffold 3D matrix in the proliferation of osteoblast-like cells, Saos-2 [14]. However, the osteogenic properties of pure collagen isolated from blue shark skin need to be addressed before being used as a biomaterial in bone tissue engineering, since the osteogenic potential of fish collagen is profoundly influenced by the molecular composition and arrangement, which is thought to be varied by different extraction methods. Considering the above hypothesis, for the first time, we explored the osteogenic response of type I collagens (Acid and Pepsin soluble) isolated from blue shark (*Prionace glauca*) skin using differentiated mouse bone marrow-mesenchymal stem (dMBMS) and differentiated osteoblastic (dMC3T3E1) cells and their relationship with the biochemical and functional properties of collagen in the present study.

2. Results and Discussion

2.1. Hydroxyproline and Collagen Content of Raw Materials

The quantification of total collagen content from hydroxyproline (Hyp) was determined as per the earlier method [15]. In general, the PSC had a high content of Hyp than ASC, similarly the collagen content was also high in PSC. The hydroxyproline and collagen contents of PSC and ASC were about 120 and 106.36 mg/g and, 896.738 and 793.49 mg/g, respectively (Figure 1). The present result was in accordance with the hydroxyproline content of collagen from silvertip shark (113 and 105 mg/g) and brownbanded bamboo shark (103.71 mg/g) [1,16]. The hydroxyproline content may depend on the seasonal variations and body temperature of the fish species.

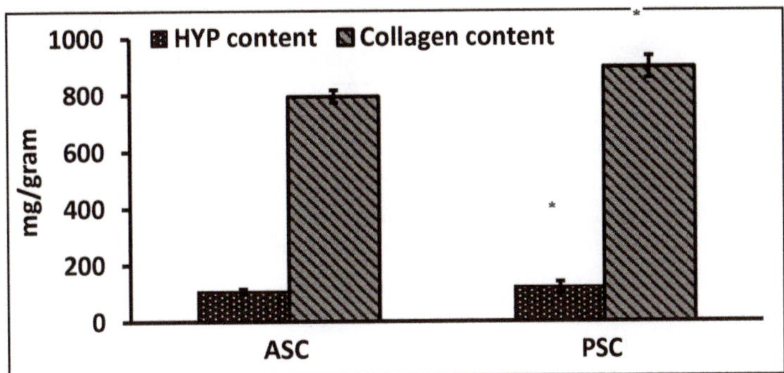

Figure 1. Hydroxyproline and collagen content of blue shark skin collagen. ASC-Acid soluble collagen, PSC-Pepsin soluble collagen. * $p < 0.05$ vs. ASC.

2.2. Amino Acid Content

The amino acid profiles of blue shark collagens (PSC and ASC) were similar in trend and expressed as residues per 1000 total residues (Table 1). Both ASC and PSC had glycine (392 and 387) as a foremost amino acid, followed by alanine (144–130) and proline (132–124). The amino acid contents of glycine, asparagine, proline, and hydroxyl-proline were high in general. The sum of Pro and Hyp (imino acid) content was higher in PSC (222) than ASC (213). The high content of alanine, glycine, hydroxyproline, and proline were the typical amino acids that represent the high content of collagen in blue shark skin. The present finding of collagen amino acid composition was similar to other fish collagen [16,17]. The differing amino acid composition between PSC and ASC was due to the removal of some non-collagen amino acid and the breakdown of certain specific amino acid residues at the telopeptide region during pepsin hydrolysis.

Table 1. The amino acid composition of ASC and PSC from blue shark skin.

Amino Acids	ASC	PSC
Glycine	392.44	387.12
Alanine	144.59	130.88
Proline	132.98	124.46
Hydroxyproline	80.45	98.28
Glutamic acid	69.23	71.50
Aspartic acid	36.73	40.51
Arginine	28.64	26.25
Leucine	24.44	13.83
Serine	21.75	26.25
Lysine	14.84	15.95
Phenylalanine	13.63	14.02
Methionine	10.17	12.03
Threonine	10.04	8.19
Histidine	5.39	6.65
Valine	7.61	8.38
Isoleucine	3.79	9.38

2.3. Molecular Weight Analysis

The electrophoretic pattern of the skin collagen was shown in Figure 2A. The electrophoretic pattern of ASC and PSC was comparable, which was comprised of three distinct bands, α_1, α_2 and β chains. However, there was some smaller molecular weight component observed in the

PSC electrophoretic pattern, which was due to the hydrolysis of pepsin at the telopeptide region. The present finding of the collagen molecular pattern (α_1, α_2 and β) was a characteristic type I collagen electrophoretic profile and was comparable with other teleost fish species [18].

2.4. Viscosity and Solubility

Upon increasing temperature, the viscosity of ASC and PSC had decreased gradually (Figure 2B). The viscosity of PSC was gradually decreased in respect to temperature, which was similar to other fish collagen [19,20], whereas the viscosity of ASC had a sharp decline at 20–30 °C. The collagens had lost their viscosity against the temperature due to the breaking of the collagen molecule hydrogen bonds that lead separate triple helix chains into random coils or distinct chains during heating at high temperature. Therefore, the secondary structure highly influences the viscosity of collagen. In this study, PSC was more viscous than ASC due to the high content of imino acid, which is believed to be responsible for the stability of collagen [19].

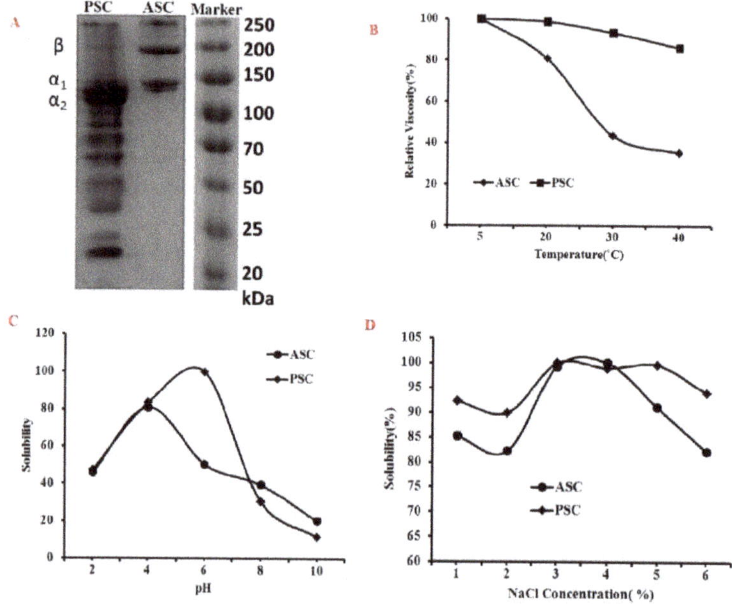

Figure 2. Electrophoretic profile (**A**), viscosity (**B**) and solubility against pH (**C**) and salt (**D**) of blue shark skin collagen. ASC-Acid soluble collagen, PSC-pepsin soluble collagen.

The solubility of collagen against pH was presented in Figure 2C. The results showed that ASC had high solubility in acidic conditions at pH 4.0, whereas the maximum solubility PSC was observed at pH 6.0. The solubility of collagen against salt, NaCl, showed that the maximum solubility of PSC and ASC was observed at 3% and 4% NaCl concentration, respectively (Figure 2D). Similar findings on salt solubility (3–4%) of eel fish skin collagen were earlier reported by Veeruraj et al. [17]. In general, the collagen tends to precipitate at their iso-ionic point (pH 5–8), which leads to decrease the solubility [18]. Also, the solubility of collagen against pH is closely related with the molecular composition and ionic state of the functional groups present in collagen [15].

2.5. UV Absorbance

The UV absorbance pattern of blue shark skin collagen was presented in Figure 3A. Both collagens, PSC and ASC, had a UV maximum absorption at 225 nm. It was reported that the major functional groups (-COOH, C=O and CO-NH$_2$) of collagen tend to absorb UV light [21]. An earlier study revealed that the purity of collagen was indirectly measured by the absence of UV absorption at 280 nm since the non-collagen proteins tend to absorb maxima at this particular UV wavelength [16].

2.6. Fourier Transform Infrared Spectroscopy (FTIR)

The FTIR peak of blue shark skin collagen was presented in Figure 3B. As shown, the primary peaks such as Amide-A and Amide-B were observed at 3302.67 and 3301.20 cm^{-1}, and 2972.57 and 2972.36 cm^{-1} for ASC and PSC, respectively. Amide-A and Amide-B peaks are generally produced by N-H stretch joined with the H$_2$ bond in a carbonyl group of peptide chains and an asymmetric stretching vibration of NH$_3{}^+$ with =C-H, respectively. The major spectral wavelength observed in blue shark skin collagen was similar to earlier studies [17,20,22,23]. In this study, the shifting of Amide-A peak in PSC to lower wavelength was due to the hydrogen bond vibration of the NH group.

The other major peaks like Amide-1 and Amide-2 occurred at 1632.28–1632.06 cm^{-1} and 1547.89–1547.25 cm^{-1} for blue shark skin ASC and PSC, respectively. The secondary structure of collagen is closely represented by Amide-1 peak, which is due to the C=O stretching vibration of peptides. Therefore, any shifting in the amide-1 peak directly reflects the changes in the secondary structure of a protein. In this study, the amide-1 peak of PSC shifted to a lower wavelength compared to ASC, which denotes the secondary structural changes of PSC by pepsin digestion.

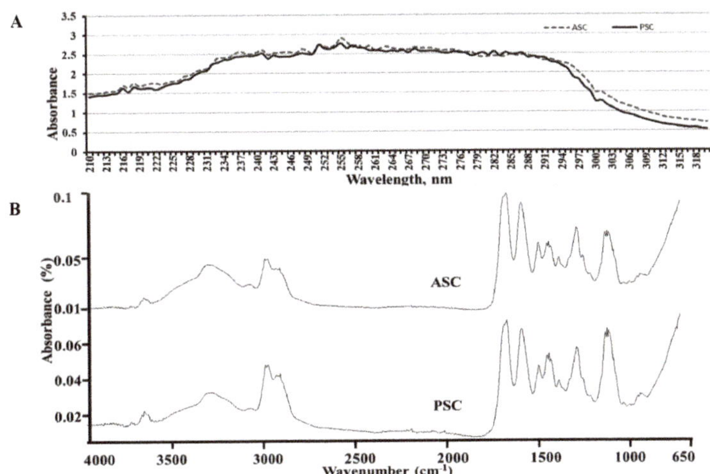

Figure 3. UV absorption (**A**) and Fourier transform infrared spectroscopy (FTIR) pattern (**B**) of blue shark skin collagen. ASC-Acid soluble. collagen, PSC-Pepsin soluble collagen.

2.7. Reversed-Phase High-Performance Liquid Chromatography (RP-HPLC)

In order to confirm the purity, the extracted blue shark skin collagens were eluted through RP-HPLC column and compared with standard bovine collagen. As shown in Figure 4, there were two main chromatogram peaks eluted at 3.427 and 19.0 min for ASC and 3.407 and 22.980 min for PSC, respectively. The first peak observed at 3.4 was the acetic acid peak. The present observation was in accordance with a previous report [24]. Compared to the standard bovine collagen, the blue shark fish collagen showed a dissimilar chromatographic pattern. We speculated that this might be due to the difference in their molecular arrangements between mammalian and fish collagen. Also, the RP-HPLC pattern confirmed that extracted collagen was pure and free from non-collagen protein and other contaminants, further justifying the SDS-PAGE data of collagen (Figure 2A).

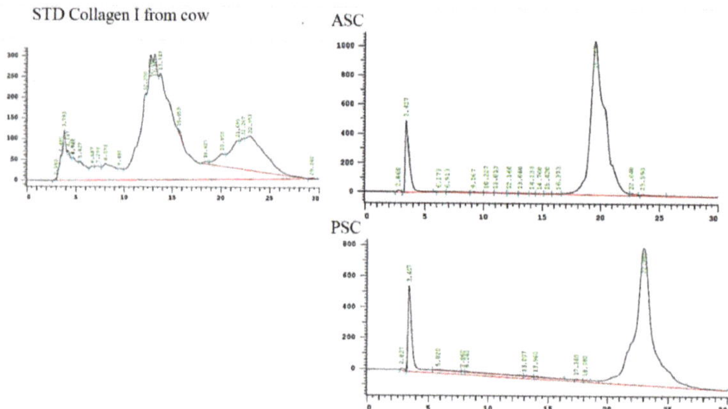

Figure 4. RP-HPLC elution profile of blue shark skin collagen. ASC-Acid soluble collagen, PSC-Pepsin soluble collagen.

2.8. Thermal Properties

Thermal properties such as the material weight loss towards temperature and the thermal disintegration configuration of the blue shark skin collagens were evaluated by the thermogravimetric analyzer (TGA). From the TGA curves, 50% weight loss in ASC and PSC were observed at 339 and 342 °C, respectively (Figure 5A). At high temperature (650 °C), the mass weight of ASC and PSC were about 22.53 and 23.64, respectively.

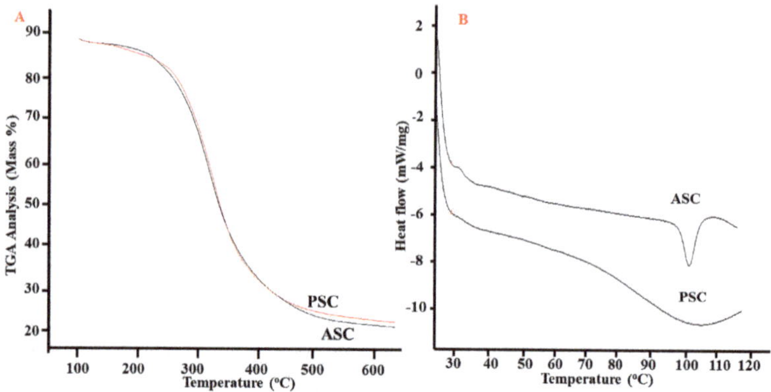

Figure 5. Thermogravimetric analysis (TGA) (**A**) and differential scanning calorimetry (DSC) (**B**) of blue shark skin collagen. ASC-Acid soluble collagen, PSC-pepsin soluble collagen.

ASC and PSC from blue shark skin had two endothermic peaks, which denoted the thermal denaturation and melting temperature. The denaturation temperature of ASC and PSC was about 28.3 and 29.8 °C, respectively (Figure 5B). The degree of hydroxylation and certain specific amino acid residues such as Gly-Pro-Hyp might determine the denaturation temperature of collagen [15]. It was observed that imino acids play a major role in the thermal properties of collagen. For instance, the hydroxyproline mainly contributes to the stability of the helix chain [25]. In this study, PSC had better thermal properties than ASC due to the high content of iminoacids.

2.9. Morphalogical Characteristics

The field emission scanning electron microscope was used to analyse the morphological structures of the blue shark skin collagens (PSC and ASC). From the microphotography, both collagens were rich in a fibrillar network with many flaky orientation layers (Figure 6). The present finding was in accordance with the morphology of other fish collagens reported earlier [10,17]. However, the morphology was regular and an organized fibrillar network in PSC, while ASC had irregular fibrillar interconnectivity and wall morphology. It was opioned that the morphalogical features of collagen such as fibril interconnectivity, shape and wall morphology were important factors in tissue engineering, since they may influence the growth, migration, proliferation, differentiation, and maturation of cells [26]. In recent research, the collagen from the same species (blue shark skin) was used to formulate a scaffold 3D matrix and tested the proliferation and mineralisation effect on Saos-2 cells, and further confirmed that the scaffold prepared from blue shark skin collagen could support Saos-2 cell attachment and osteoblast-like cells formation [14]. In the present study, the morphological features confirmed the aptness of blue shark skin collagens in bone tissue engineering.

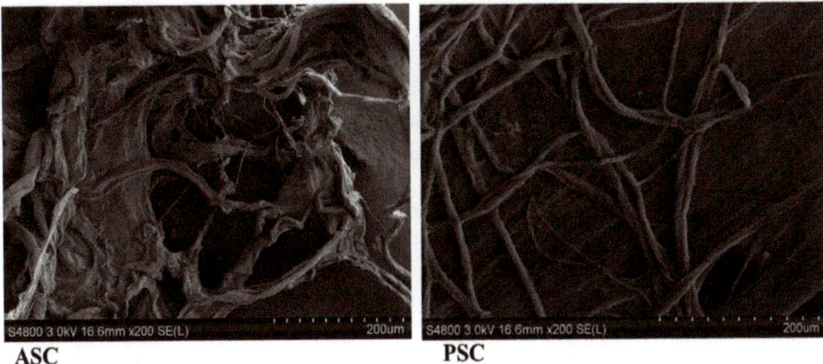

Figure 6. Morphological features of blue shark skin collagens using field emission scanning electron microscope. ASC-Acid soluble collagen, PSC-Pepsin soluble collagen. SEM image with 200 micro meter magnifications.

2.10. Effect of Collagen on Bone Cells

The effect of collagens on the dMBMS and dMC3T3E1 cell growth was determined and presented in Figure 7. As shown in Figure 7, the collagens treated dMBMS and dMC3T3E1 cells showed a higher cell proliferative rate than the control cells and the rate was accelerated with increasing collagen concentration in a dose-dependent manner. Remarkably, the osteogenic effect of the collagen on the bone cell was maxima at a high concentration (50 ng/mL). Among the two cells, dMBMS cells had a higher proliferation rate in PSC treated group. Whereas dMC3T3TE1 cells showed similar cell proliferation in both ASC and PSC treated groups. A recent study by Diogo et al. [14] reported that the composite blue shark skin collagen-calcium phosphate scaffold crosslinked with 12.5% EDC/NHS accelerated Saos-2 cells metabolic activity and supported osteoblast-like cells formation, however, this study was conducted using acid soluble collagen. Conversely, in the present study, the collagen was extracted by two different methods using acetic acid and pepsin, respectively, and the proliferative effect of freeze dried collagens was tested using dMBMS and dMC3T3E1 cells.

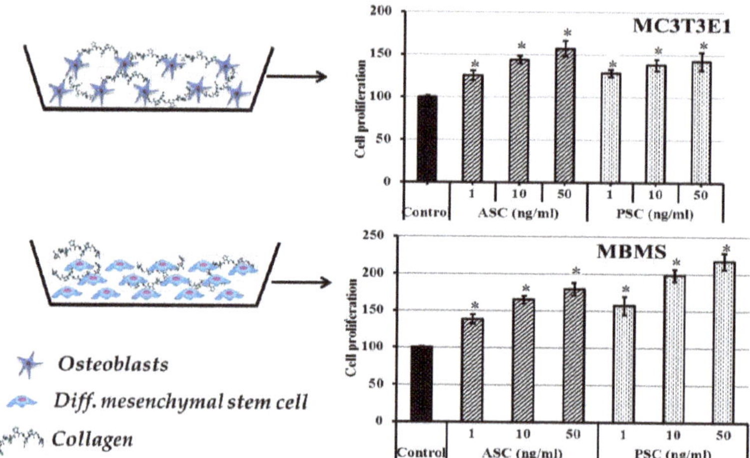

Figure 7. The effect of blue shark skin collagen on differentiated bone cells proliferation. MBMS: differentiated mouse bone marrow mesenchymal stem cells, MC3T3E1- differentiated-osteoblasts, ASC-Acid soluble collagen, PSC-pepsin soluble collagen. For differentiation, MBMS and MC3T3E1 cells were cultured with osteoblast differentiation medium for 21 days. * $p < 0.05$ vs control.

The levels of osteogenic mRNA expression of alkaline phosphatase, collagen 1 aplha1 and Runx2 were significantly increased in the collagen treated cells than the control cells, however, the osteocalcin mRNA expression was not significantly altered between the collagen and control cells (Figure 8A). Confocal laser scanning microscope showed a high number of mature bone cells in collagen treated cells compared to control cells (Figure 8B), which further substantiated the osteogenic stimulatory activities of blue shark skin collagens. Similar to the present study, type I collagen and its peptides from rat tail showed osteogenic stimulatory activities on a mesenchymal stem cell [27,28]. Recently, Chiu et al. [29] reported that the MBMS cell expressed a high the level of integrin $\alpha_2\beta_1$ complex upon collagen treatment. In support of the proliferation result, the levels of osteogenic mRNA of alkaline phosphatase and collagen 1 were increased in collagen treated cells. Runx mRNA expression of collagen treated cells revealed the potential osteogenic activity of collagen. These findings further justify the increased proliferation rate of collagen treated cells. Recent studies claimed that certain amino acids such as glutamine, alanine, asparagine, and glycine of collagen triggered new bone cell formation through the initiation of FAK-JNK signaling pathway via RUNX2 in MBMS cell [29,30].

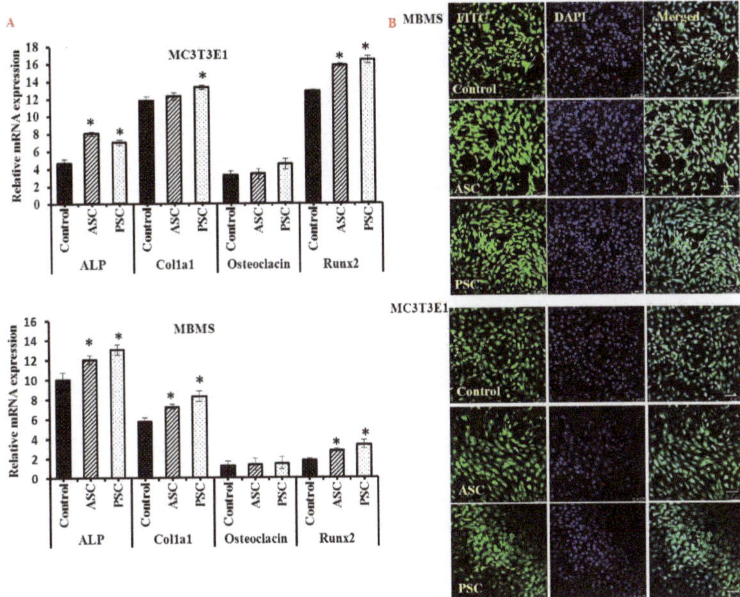

Figure 8. mRNA (**A**) and confocal images (**B**) of blue shark collagen treated bone cells. Scale bars: 75 micrometers. MBMS: differentiated mouse bone marrow mesenchymal stem cells, MC3T3E1-differentiated osteoblasts, ASC-Acid soluble collagen, PSC-pepsin soluble collagen. For differentiation, MBMS and MC3T3E1 cells were cultured with osteoblast differentiation medium for 21 days. * $p < 0.05$ vs. control.

The western blot analysis of the osteogenic protein expression of collagen treated cells was in agreement with the proliferation, mRNA expression, and microscopic results, which indicated the higher osteogenic protein expression of collagen I alpha I and Runx2 in collagen treated cells than in the control (Figure 9A).

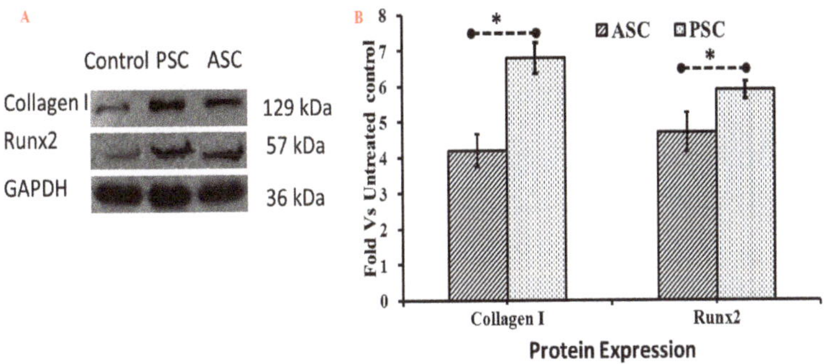

Figure 9. Osteogenic regulatory protein expressions (Collagen I and Runx2) of bone cells (differentiated MBMS) treated with blue shark skin collagen by western blot assay. (**A**) Level of osteogenic protein expression of bone cells (**B**) Fold changes of osteogenic protein expression compared to control (GAPDH). ASC-Acid soluble collagen, PSC-Pepsin soluble collagen. For differentiation, MBMS cells were cultured with osteoblast differentiation medium for 21 days, * $p < 0.05$.

In addition, the cells treated with PSC showed higher osteogenic regulatory protein expressions than ASC treated cells (Figure 9B). Interestingly, the Runx2 protein was highly expressed in collagen treated cells, especially in PSC treated cells. It was reported that the collagen could interact with integrin alpha1 beta 2 of the mesenchymal stem cells and trigger FAK/JNK signals through Runx2 during osteoblast differentiation [29]. In that sense, the blue shark skin collagen accrued osteoblast differentiation through upregulating the Runx2 protein expression in dMBMS cells. These findings further confirmed the admirable osteogenic properties of blue shark skin collagens.

3. Materials and Methods

3.1. Extraction, Purification, and Total Collagen Content of Fish Collagen

The raw material, blue shark skin (*Prionace glauca*), was procured from a private fish processing plant, M/s. Yueqing Biological Health Care Product Co., Ltd. Zhejiang, China and chopped into small pieces. They were then homogenized in PBS prior to extraction. The pretreatment was done in order to remove water-soluble protein, non-collagen protein, and mineral content using distilled water, NaOH and ethylenediaminetetraacetic acid (EDTA), respectively. The pretreated samples were used for the extraction of acid soluble collagen (ASC) and pepsin soluble collagen (PSC) using acetic acid and acetic acid containing 1% pepsin, respectively, as per our previous method [10]. The extracted samples were purified using Sephadex G-100 column chromatography coupled with UV spectroscopy with the absorbance of 230 nm [1]. The purified samples were freeze-dried using a lyophilizer. The hydroxyproline and total collagen content of the purified samples were determined [31].

3.2. Molecular Mass by SDS-PAGE and Amino Acid Profile

The protein molecular mass analysis was done by SDS-PAGE using 10% running gel, 4% stacking gel, 20 microliter sample loading volume (10 micrograms per lane) and 120 voltage as per our earlier method [10]. After the run, the gels were stained with Coomassie blue R-250 in methanol and acetic acid for 30 min followed by detaining with methanol and acetic acid [32]. The amino acid profile was determined by hydrolyzing the samples in 6M hydrochloric acid at 110 °C for 24 h and then injected (aliquot of 0.4 mL) into an amino acid analyzer (Hitachi L-8800, Tokyo, Japan) and expressed as residues per 1000 residues [1].

3.3. Viscosity and Solubility

The viscosity and solubility of the collagen samples were done as per our earlier method [1]. For viscosity, the samples were heated from 5 to 40 °C followed by injection to a Brookfield LVDV-II+P viscometer (Brookfield Engineering Laboratories Ltd., Middleboro, MA, USA) to measure relative viscosity. The percentage of collagen solubility was tested against different pH (2–10) and salt concentrations (1–6%). The relative solubility was measured from the protein content [33].

3.4. Absorbance UV Maxima and FTIR Spectra

The maximum UV absorbance of collagen was done using a UV-spectrophotometer, scanning at a different wavelength from 210 to 318 nm. For the FTIR spectra, the collagen samples were mixed with kBr to produce a disk and scanned at different wavelengths using a Fourier transform infrared spectrometer (Nicolet 6700, Thermofisher Scientific Inc., Waltham, MA, USA) equipped with a DLaTGS detector. The absorption intensity of each peak was calculated as mentioned in our previous method [1].

3.5. RP-HPLC

The purity of extracted collagen was confirmed by RP-HPLC as per our earlier method [34]. In brief, a collagen sample dissolved in acetic acid was loaded on a ZORMAX 300 SM-C18 (Agilent, Shanghai, China) column of LaChrom Elite HPLC system (Hitachi, Tokyo, Japan) equipped with an

L-2130 pump, L-2300 column oven, L-2400 UV detector and L-2200 auto-sampler; and eluted with a two gradient solvent system using acetonitrile and trifluoroacetic acid with the peak absorbance at 230 nm using the HITACHI D-2000 Elite workstation software (Hitachi, Tokyo, Japan).

3.6. Thermal Stability

The thermal stability of collagen was determined from differential scanning calorimetry (DSC) and TGA analysis [35]. In brief, the DSC spectrum was obtained by scanning the collagen samples from 20 to 120 °C using a differential scanning calorimeter (Model-DSC822e, Mettler-Toledo GmbH, Greifensee, Switzerland). For TGA Analysis, the samples were scanned using a TG 209 F1 analyzer (NETZSCH-Geratebau GmbH, Selb, Germany) from 0 to 650 °C at a rate of 10 °C min^{-1} in a nitrogen atmosphere.

3.7. Morphological Analysis

The morphology of the blue shark skin colalgens, PSC and ASC was captured using a field emission scanning electron microscope (FE-SEM S-3400 N, Hitachi, Tokyo, Japan) operated at 5 kV. The samples were sputter-coated with gold to produce 10 nm thin layer and moved to the SEM chamber for image capturing at 20 kV voltage.

3.8. Effect of Collagen on Osteoblastogenesis

3.8.1. Cell Culture

Mouse bone marrow-mesenchymal stem (MBMS) and MC3T3E1 cells were received from the cell bank (Shanghai Zhong Qiao Xin Zhou Biotechnology Co., Ltd., Shanghai, China) and were cultured in stem cell culture medium and DMEM, respectively, comprising 10% FBS and antibiotics (1% penicillin/streptomycin) at 37 °C in a CO_2 incubator. Mesenchymal stem cell growth supplement was added with the culture medium to induce MBMS cell growth.

3.8.2. Osteogenic Differentiation

The MBMS and MC3T3E1 cells with a density of 5×10^4 cells/well were seeded in 6 well microtiter culture plate and cultured with osteoblast differentiation medium composed of growth supplements (Shanghai Zhong Qiao Xin Zhou Biotechnology Co., Ltd., Shanghai, China) for 21 days as reported in our previous protocol [13]. The culture medium was replaced every three days.

3.8.3. Proliferation Assay

After differentiation, the cells were trypsinized using EDTA/Trypsin solution and re-seeded in 96 well plates. Then the cells were treated with 1, 10, and 50 ng/mL concentration of blue shark skin collagens. Cells grown without collagen samples served as the control. The proliferative effect of collagen on bone cells was measured using a cell counter (Invitrogen, Countess II Automated Cell Counter, ThermoFisher Scientific, Shanghai, China) at 3 days after treatment.

3.8.4. mRNA Expression

The bone cells were treated with collagen as mentioned above and the total mRNA was isolated using the TRIzol method as per our earlier protocol [13]. Briefly, a Trizol lysed cell monolayer was mixed with chloroform to obtain RNA and the RNA was precipitated using isopropanol. After thorough washing with 70% ethanol, the RNA was converted to first strand cDNA using an Invitrogen kit as per the manufacturer's instructions. The percentage of osteogenic mRNA expression (ALP, Runx2, osteocalcin, Col1a1, and beta-actin) (Table 2) of blue shark collagen treated bone cells was determined using an ABI 7500 Fast Real-Time PCR System (Applied Biosystems, Shanghai, China) [13]. The cDNA template was mixed with SYBR Green Fast qPCR RT Master Mix (Invitrogen, Shanghai,

China) and target primers as per the manufacture's instructions. The relative mRNA expression of the target gene was calculated by subtracting the mRNA expression of a house-keeping gene, beta-actin.

Table 2. List of primers and its sequence used in this study.

S.No	Primers Name	Primers Sequence
1	Alkaline phosphatase (ALP)	5′-TCC TGA CCA AAA ACC TCA AAG G-3′
		5′-TGC TTC ATG CAG AGC CTG C-3′
2	Osteocalcin	5′-CTC ACA GAT GCC AAG CCC-3′
		5′-CCA AGG TAG CGC CGG AGT CT-3′
3	Collagen 1 alpha 1 (Col1a1)	5′-GCG AAG GCA ACA GTC GCT-3′
		5′-CTT GGT GGT TTT GTA TTC GAT GAC-3′
4	Runx2	5′-CCA CCA CTC ACT ACC ACA CG-3′
		5′-TCA GCG TCA ACA CCA TCA TT-3′
5	Beta-actin	5′-CTG GCA CCA CAC CTT CTA CA-3′
		5′-GGT ACG ACC AGA GGC ATA CA-3′

3.8.5. Immunocytochemistry

For immunocytochemistry, cells were grown in the confocal disc (Cat no. 150682, ThermoFisher Scientific, Shanghai, China) with collagen, washed with ice-cold PBS, fixed with 4% paraformaldehyde for 15 min and permeabilized at room temperature with 0.1% Triton X-100 for 15 min. Then, the cells were incubated with fluorescence stains such as fluorescein isothiocyanate (FITC) and 4′,6-diamidino-2-phenylindole (DAPI). The images were captured using a confocal laser scanning microscope (Leica TCS SP8, Leica Microsystems CMS GmbH, Wetzlar, Germany).

3.9. Western Blot Analysis

The osteogenic effect of the ASC and PSC was further tested in dMBMS cells using the western blot method, as an empirical study. In brief, the cells were seeded at a cell density of 1×10^5 in 6 well plates and cultured with 1 mL culture medium for control. For collagen treatment, the samples (50 ng/mL) were mixed with 1 mL cultured medium and cultured for three days. Then, the treated and untreated cells were harvested using a lysis buffer with 30 s sonication and centrifuged at 12,000 rpm at 4 °C for 15 min for the extraction of total cellular protein. The protein content was quantified using a BCA kit as per the manufacturer's instructions (Thermo Fisher Scientific, Shanghai, China) and then was separated by 12% SDS-PAGE and transferred to PVDF nitrocellulose membrane (Invitrogen, Shanghai, China) using iblot-2 dry blotting system (Invitrogen, Shanghai, China). A protein transferred membrane blocked with 5% BSA-PBST were incubated with primary antibodies such as anti-GAPDH, anti-Col1α2 and anti-Runx2 (ABcam, Shanghai, China) overnight at 4 °C. Then the membrane was incubated with secondary goat anti-rabbit IgG-HRP (Abcam) for 1 h at 37 °C and exposed to an enhanced chemiluminescent reagent. Images were captured with a Universal Hood II Gel Doc System (Bio-Rad, Rochester, NY, USA).

3.10. Statistical Analysis

All the experiments were conducted three times and the data were presented as mean with standard error based on descriptive statistics. One way ANOVA ($p < 0.05$) and post hoc test were adapted using the statistical software SPSS version 16.0 (SPSS, Chicago, IL, USA).

4. Conclusions

In this study, two types of collagens, PSC and ASC were isolated and purified from blue shark skin, which contained three distinct chains, 2 alpha, and beta chains. The amino acid composition and the FTIR secondary structure of PSC varied with ASC due to pepsin hydrolysis. Among two collagens, PSC had high thermal stability, which might be due to high imino acid content. Most importantly, the obtained collagens promoted osteoblast cell growth and upregulated collagen synthesis in bone cells, which is a most desirable property of biomaterials for the treatment of a bone disorder. Proliferation of differentiated osteogenic cells by blue shark skin collagen may be achieved through activation of Runx2 dependent FAK/JNK signaling pathway. Overall, the blue shark skin collagens with good biochemical and osteogenic properties may be considered as a potential drug in biomedical application. Conversely, the in vivo evaluation of blue shark skin collagen and its molecular interaction with bone cells needs to be addressed by further research in order to understand the actual mechanism of the action.

Author Contributions: J.E. and J.W.L. carried out the laboratory works, compiled data, and wrote the manuscript. Y.H., S.W., J.E.M.S.d.V., S.Y.L. and J.M.R. revised the manuscript and provided valuable advice on interpretation of data. W.W. and B.B. designed and directed the study, inferred the data, and revised the manuscript.

Funding: This research was funded by National Natural Science Foundation of China, project grant number 81750110548, the National High Technology Research and Development Program of China, project grant number 2011AA09070109), and the Plan of Innovation Action in Shanghai project grant number 15410722500 and 17490742500.

Acknowledgments: The authors gratefully acknowledge the funding support received from the National Natural Science Foundation of China (Project Grant No. 81750110548), an International Young Scientist Research Fellowship, the National High Technology Research and Development Program of China (Project Grant No. 2011AA09070109), and the Plan of Innovation Action in Shanghai (Project Grant No. 15410722500 and 17490742500). The authors also thank Shanghai Ocean University, Shanghai, China for providing the necessary facilities to undertake this work.

Conflicts of Interest: The authors declare no conflict of interest.

References

1. Jeevithan, E.; Bao, B.; Bu, Y.; Zhou, Y.; Zhao, Q.; Wu, W. Type II collagen and gelatin from silvertip shark (*Carcharhinus albimarginatus*) cartilage: Isolation, purification, physicochemical and antioxidant properties. *Mar. Drugs* **2014**, *12*, 3852–3873. [CrossRef] [PubMed]

2. Jongjareonrak, A.; Benjakul, S.; Visessanguan, W.; Nagai, T.; Tanaka, M. Isolation and characterisation of acid and pepsin-solubilised collagens from the skin of Brownstripe red snapper (*Lutjanus vitta*). *Food Chem.* **2005**, *93*, 475–484. [CrossRef]

3. Wang, X.; You, C.; Hu, X.; Zheng, Y.; Li, Q.; Feng, Z.; Sun, H.; Gao, C.; Han, C. The roles of knitted mesh-reinforced collagen-chitosan hybrid scaffold in the one-step repair of full-thickness skin defects in rats. *Acta Biomater.* **2013**, *9*, 7822–7832. [CrossRef] [PubMed]

4. Kawaguchi, N.; Hatta, K.; Nakanishi, T. 3D-culture system for heart regeneration and cardiac medicine. *BioMed Res. Int.* **2013**, *2013*, 895967. [CrossRef] [PubMed]

5. Spector, M. Novel cell-scaffold interactions encountered in tissue engineering: Contractile behavior of musculoskeletal connective tissue cells. *Tissue Eng.* **2002**, *8*, 351–357. [CrossRef] [PubMed]

6. Sugiura, H.; Yunoki, S.; Kondo, E.; Ikoma, T.; Tanaka, J.; Yasuda, K. In vivo biological responses and bioresorption of tilapia scale collagen as a potential biomaterial. *J. Biomater. Sci. Polym. Ed.* **2009**, *20*, 1353–1368. [CrossRef] [PubMed]

7. Song, E.; Yeon, K.S.; Chun, T.; Byun, H.J.; Lee, Y.M. Collagen scaffolds derived from a marine source and their biocompatibility. *Biomaterials* **2006**, *27*, 2951–2961. [CrossRef] [PubMed]

8. Sugahara, T.; Ueno, M.; Goto, Y.; Shiraishi, R.; Doi, M.; Akiyama, K.; Yamauchi, S. Immunostimulation effect of jellyfish collagen. *Biosci. Biotechnol. Biochem.* **2006**, *70*, 2131–2137. [CrossRef] [PubMed]

9. Jeevithan, E.; Jingyi, Z.; Bao, B.; Shujun, W.; JeyaShakila, R.; Wu, W.H. Biocompatibility assessment of type-II collagen and its polypeptide for tissue engineering: Effect of collagen's molecular weight and glycoprotein content on tumor necrosis factor (Fas/Apo-1) receptor activation in human acute T-lymphocyteleukemia cell line. *RSC Adv.* **2016**, *6*, 14236–14246.

10. Zhang, J.; Jeevithan, E.; Bao, B.; Wang, S.; Gao, K.; Zhang, C.; Wu, W.H. Structural characterization, in-vivo acute systemic toxicity assessment and in-vitro intestinal absorption properties of tilapia (*Oreochromis niloticus*) skin acid and pepsin solublilized type I collagen. *Process. Biochem.* **2016**, *51*, 2017–2025. [CrossRef]

11. Kato, Y.P.; Dunn, M.G.; Zawadsky, J.P.; Tria, A.J.; Silver, F.H. Regeneration of Achilles tendon with a collagen tendon prosthesis: Results of a one year implantation study. *J. Bone Jt. Surg.* **1991**, *73*, 561–574. [CrossRef]

12. Dunn, M.G.; Tria, A.J.; Bechler, J.R.; Ochner, R.S.; Zawadsky, J.P.; Kato, Y.P.; Silver, F.H. Anterior cruciate ligament reconstruction using a composite collagenous prosthesis. A biomechanical and histologic study in rabbits. *Am. J. Sports Med.* **1992**, *20*, 507–515. [CrossRef] [PubMed]

13. Jeevithan, E.; Sanchez, C.; De Val, J.E.M.S.; Henrotin, Y.; Wang, S.; Motaung, K.S.C.M.; Guo, R.; Wang, C.; Robinson, J.; Regenstein, J.M.; et al. Cross-talk between primary osteocytes and bone marrow macrophages for osteoclastogenesis upon collagen treatment. *Sci. Rep.* **2018**, *8*, 5318.

14. Diogo, G.S.; Senra, E.L.; Pirraco, R.P.; Canadas, R.F.; Fernandes, E.M.; Serra, J.; Pérez-Martín, R.I.; Sotelo, C.G.; Marques, A.P.; González, P.; et al. Marine collagen/apatite composite scaffolds envisaging hard tissue applications. *Mar. Drugs* **2018**, *16*, 269. [CrossRef] [PubMed]

15. Kittiphattanabawon, P.; Benjakul, S.; Visessanguan, W.; Nagai, T.; Tanaka, M. Characterisation of acid-soluble collagen from skin and bone of bigeye snapper (*Pricanthus tayenus*). *Food Chem.* **2005**, *89*, 363–372. [CrossRef]

16. Kittiphattanabawon, P.; Benjakul, S.; Visessanguan, W.; Shahidi, F. Isolation and characterization of collagen from the cartilages of brownbanded bamboo shark (*Chiloscyllium punctatum*) and blacktip shark (*Carcharhinus limbatus*). *LWT-Food Sci. Technol.* **2010**, *43*, 792–800. [CrossRef]

17. Veeruraj, A.; Arumugam, M.; Balasubramanian, T. Isolation and characterization of thermostable collagen from the marine eel-fish (*Evenchelys macrura*). *Process. Biochem.* **2013**, *48*, 1592–1602. [CrossRef]

18. Bae, I.; Osatomi, K.; Yoshida, A.; Osako, K.; Yamaguchi, A.; Hara, K. Biochemical properties of acid-soluble collagens extracted from the skins of underutilised fishes. *Food Chem.* **2008**, *108*, 49–54. [CrossRef]

19. Ogawa, M.; Moody, M.W.; Portier, R.J.; Bell, J.; Schexnayder, M.A.; Losso, J.N. Biochemical properties of black drum and sheepshead seabream skin collagen. *J. Agric. Food Chem.* **2003**, *51*, 8088–8092. [CrossRef] [PubMed]

20. Wang, L.; Liang, Q.; Chen, T.; Wang, Z.; Xu, J.; Ma, H. Characterization of collagen from the skin of Amur sturgeon (*Acipenser schrenckii*). *Food Hydrocoll.* **2014**, *38*, 104–109. [CrossRef]

21. Edwards, C.A.; O'Brien, W.D., Jr. Modified assay for determination of hydroxyproline in a tissue hydrolyzate. *Clin. Chim. Acta* **1980**, *104*, 161–167. [CrossRef]

22. Liang, Q.; Wang, L.; Sun, W.; Wang, Z.; Xu, J.; Ma, H. Isolation and characterization of collagen from the cartilage of Amur sturgeon (*Acipenser schrenckii*). *Process. Biochem.* **2014**, *49*, 318–323. [CrossRef]

23. Muyonga, J.H.; Cole, C.G.B.; Duodu, K.G. Characterisation of acid soluble collagen from skins of young and adult Nile perch (*Lates niloticus*). *Food Chem.* **2004**, *85*, 81–89. [CrossRef]

24. Ho, H.O.; Lin, C.W.; Sheu, M.T. Characterization of collagen isolation and application of collagen gel as a drug carrier. *J. Control. Release* **1997**, *44*, 103–112. [CrossRef]

25. Wong, D.W. Proteins. In *Mechanism and Theory in Food Chemistry*, 2nd ed.; Springer: New York, NY, USA, 1989.

26. Zhang, M.; Liu, W.T.; Li, G.Y. Isolation and characterization of collagens from the skin of largefinlongbarbel catfish (*Mystus macropterus*). *Food Chem.* **2009**, *115*, 826–831. [CrossRef]

27. Hennessy, K.M.; Pollot, B.E.; Clem, W.C.; Phipps, M.C.; Sawyer, A.A.; Culpepper, B.K.; Bellis, S.L. The effect of collagen I mimetic peptides on mesenchymal stem cell adhesion and differentiation, and on bone formation at hydroxyapatite surfaces. *Biomaterials* **2009**, *30*, 1898–1909. [CrossRef] [PubMed]

28. Gao, C.; Harvey, E.J.; Chua, M.; Chen, B.P.; Jiang, F.; Liu, Y.; Li, A.; Wang, H.; Henderson, J.E. MSC-seeded dense collagen scaffolds with a bolus dose of VEGF promote healing of large bone defects. *Eur. Cells Mater.* **2013**, *26*, 195–207. [CrossRef]

29. Chiu, L.H.; Lai, W.F.; Chang, S.F.; Wong, C.C.; Fan, C.Y.; Fang, C.L.; Tsai, Y.H. The effect of type II collagen on MSC osteogenic differentiation and bone defect repair. *Biomaterials* **2014**, *35*, 2680–2991. [CrossRef] [PubMed]

30. Tuckwell, D.S.; Ayad, S.; Grant, M.E.; Takigawa, M.; Humphries, M.J. Conformation dependence of integrin-type II collagen binding Inability of collagen peptides to support alpha 2 beta 1 binding, and mediation of adhesion to denatured collagen by a novel alpha 5 beta 1-fibronectin bridge. *J. Cell Sci.* **1994**, *107*, 993–1005. [PubMed]

31. Bergman, I.; Loxley, R. Two improved and simplified methods for the spectrophotometric determination of hydroxyproline. *Anal. Chem.* **1963**, *35*, 1961–1965. [CrossRef]
32. Laemmli, U.K. Cleavage of structural proteins during the assembly of the head of bacteriophage T4. *Nature* **1970**, *227*, 680. [CrossRef] [PubMed]
33. Lowry, O.H.; Rosebrough, N.J.; Farr, A.L.; Randall, R.J. Protein measurement with the Folin phenol reagent. *J. Biol. Chem.* **1951**, *193*, 265–275. [PubMed]
34. Yongshi, B.; Jeevithan, E.; Jingyi, Z.; Bin, B.; Ruihua, G.; Krishnamoorthy, P.; Jeya Shakila, R.; Jeyasekaran, G.; Joe, M.R.; Wenhui, W. Immunological effects of collagen and collagen peptide from blue shark cartilage on 6T-CEM cells. *Process. Biochem.* **2017**, *57*, 219–227.
35. Rochdi, A.; Foucat, L.; Renou, J.P. NMR and DSC studies during thermal denaturation of collagen. *Food Chem.* **2000**, *69*, 295–299. [CrossRef]

Review

Facial Bone Reconstruction Using both Marine or Non-Marine Bone Substitutes: Evaluation of Current Outcomes in a Systematic Literature Review

Marco Cicciù [1,*], Gabriele Cervino [1], Alan Scott Herford [2], Fausto Famà [3], Ennio Bramanti [1], Luca Fiorillo [1], Floriana Lauritano [1], Sergio Sambataro [4], Giuseppe Troiano [5] and Luigi Laino [6]

[1] Department of Biomedical and Dental Sciences and Morphological and Functional Imaging, Messina of University, 98100 Messina, Italy; gcervino@unime.it (G.C.); enniobramanti@gmail.com (E.B.); lucafiorillo@live.it (L.F.); flauritano@unime.it (F.L.)

[2] Department of Maxillofacial Surgery, Loma Linda University, Loma Linda, CA 92354, USA; aherford@llu.edu

[3] Department of Human Pathology, University of Messina, 98100 Messina, Italy; famafausto@yahoo.it

[4] Private Practive COS Center, 95100 Catania, Italy; ssambataro@centrodiortodonzia.it

[5] Department of Clinical and Experimental Medicine, University of Foggia, 71121 Foggia, Italy; giuseppe.troiano@unifg.it

[6] Multidisciplinary Department of Medical-Surgical and Odontostomatological Specialties, University of Campania "Luigi Vanvitelli", 80121 Naples, Italy; luigi.laino@unicampania.it

* Correspondence : acromarco@yahoo.it or mcicciu@unime.it; Tel.: +39-0902216920; Fax: +39-0902216921

Received: 2 November 2017; Accepted: 22 December 2017; Published: 13 January 2018

Abstract: The aim of the present investigation was to systematically analyse the literature on the facial bone reconstruction defect using marine collagen or not and to evaluate a predictable treatment for their clinical management. The revision has been performed by searched MEDLINE and EMBASE databases from 2007 to 2017. Clinical trials and animal in vitro studies that had reported the application of bone substitutes or not for bone reconstruction defect and using marine collagen or other bone substitute material were recorded following Preferred Reporting Items for Systematic Reviews and Meta-Analyses (PRISMA) guidelines. The first selection involved 1201 citations. After screening and evaluation of suitability, 39 articles were added at the revision process. Numerous discrepancies among the papers about bone defects morphology, surgical protocols, and selection of biomaterials were found. All selected manuscripts considered the final clinical success after the facial bone reconstruction applying bone substitutes. However, the scientific evidence regarding the vantage of the appliance of a biomaterial versus autologous bone still remains debated. Marine collagen seems to favor the dimensional stability of the graft and it could be an excellent carrier for growth factors.

Keywords: bone grafting; bone biocompatible materials; bone regeneration; marine collagen

---

1. Introduction

The management of large facial bone defects is a current challenge for clinicians and surgeons. Treatment success is frequently related to the size of the defect, the quality of the soft tissue covering the defect, the decision of reconstructive method and the choice of the grafted material [1–3].

Numerous bone grafts' regenerative procedures are currently available for having complete regenerative processes after bone trauma, or for favoring healing between two bones across a diseased joint, and also for obtaining new clinical function or aesthetic on site affected by disease, infection, or resection [2–5].

Facial bone augmentation procedure using autologous bone is a reliable technique, as confirmed by several studies; however, this treatment choice has integration advantages associated with several disadvantages. Autogenous bone harvested from the patient's extra oral or intra oral sites even considered the gold standard, at the same time is related with surgical intra and post operative complications, biological cost, and patient morbidity, pain and discomfort at the bone grafting area [1–7].

Recently, great interest has been directed towards the application of synthetic three-dimensional biomaterials scaffolds as bone substitutes used for facial large bone defect regeneration in order to have a substantial quantity of material and to avoid a second surgery site. Those synthetic bone substitutes materials should be non-toxic, compatible with the biological systems, and bio absorbable. The biomaterial has to be a macroscopic structure that is easy for surgeons to handle. Its microstructure should be able to promote cell adhesion, proliferation and new bone formation [4–8]. The fundamental key parameters for an excellent biomaterial are related to its capability on replacing the natural bone extracellular matrix. Secondly, it should be able to recall the osteo-genic cells in order to lay down the bone tissue matrix, and then the biomaterial should guarantee a sufficient vascularization to meet the growing tissue nutrient supply and clearance needs. Therefore, after being placed in situ, the microscopic features of the biomaterial should influence the host by releasing osteogenic and/or genic growth factors [5–10].

Nowadays, the tissue engineering approaches to the facial bone regeneration are connected with biomaterial matrices/scaffold that favorably interact with cells. The potential benefits of using recent tissue engineering findings is fundamental today in order to limit donor site morbidity, reducing operative time, and replacing the anatomical microstructure in an attempt to restore physiological craniofacial functions [8–11]. Currently, advances in computer-aided modeling and biomaterials manufacturing help the craniofacial surgery field, which is frequently confronted with the rebuilding of three-dimensional anatomic structures on function and aesthetics. Three-dimensional models of bone defects can be realized from patient computed tomography scans, creating a customized scaffold that interacts with the defect site and re-builds the complex anatomical features [4,5,8,12,13].

Recently, the use of bone substitutes obtained from marine origins is being considered high attractive by the industry as an alternative source. In this specific field, the marine collagen can be obtained from numerous sources. Type I collagen is obtained predominantly from skin, tendon, bone and muscle (epimysium), which is the most abundant type of collagen. Marine fish collagens find applications in numerous biomedical fields. Besides its mechanical elastic properties, marine collagen exhibited good absorption characteristics with interconnectivity between pores, which allowed human Mesenchymal Stem Cells (hMSCs) to adhere and proliferate, being a good base for osteogenic differentiation [3,7,12].

The aim of this review is to screen recent papers about biomaterials applied for facial bone reconstruction in order to give the clinicians' valuable suggestions about the possibility of replacing autologous bone graft for large reconstruction defects.

This systematic review also aimed to evaluate the potential of reconstructive marine biomaterials uses like scaffolds for growth factor in order to provide better results in comparison to others.

2. Materials and Methods

2.1. Application Protocol and Website Recording Data

A protocol including the investigation methods and the inclusion criteria for the current revision was submitted in advance and documented on the Center for Review and Dissemination (CRD) York website PROSPERO, an international prospective register of systematic reviews. The parameters and the analytic structure of the present work can be visualized relating the CRD id and code: Application number: CRD 74603.

The data of this systematic investigation observed the Preferred Reporting Items for Systematic Review accordingly with the (PRISMA) statement [14].

2.2. Target Questions

The next spotlight questions processed the following guidelines, parameters, and possible aims of the Patient Intervetion Control Outcome (PICO) study design:

- What are the overall treatment outcomes of reconstructive procedures using bone substitutes in the place of autologous bone?
- As an alternative focused question, does marine collagene use provide beneficial clinical outcomes applied as scaffolds for growth factors?

2.3. Search Strategy

The investigation method followed the examinations of electronic databases and searches by hand. A search of five electronic databases, including Ovid MEDLINE, PubMed, EMBASE, and Dentistry and Oral Sciences Source, and Biomaterials, for relevant studies published in the English language from February 2010 to September 2017 was carried out.

Digital and searches by hand were then performed in facial bone reconstruction, biomaterials and bone graft scaffold journals from February 2010 to September 2017, including the following: (1) Journal of Craniofacial Surgery; (2) Journal Biomed. Mater. Research A; (3) International Journal of Oral and Maxillofacial Implants; (4) Clinical Oral Implants Research; (5) Implant Dentistry; (6) International Journal of Oral and Maxillofacial Surgery; (7) Journal of Oral and Maxillofacial Surgery; (8) Journal of Dental Research; (9) Biomed. Research International; (10) International Journal of Prosthodontics; (11) Acta Biomaterial.; (12) Journal of Clinical Periodontology; (13) Clinical Implant Dentistry and Related Research; (14) European Journal of Oral Implantology; (15) Facial Plastic Surgery; (16) Materials; (17) Marine Drugs; (18) Biomaterials; (19) Br. J. Oral Maxillofacial Surgery; (20) Aesthetic Plastic Surgery; (21) Ophthal. Plast. Reconstr. Surg.; (22) J. Craniomaxillofacial Surg.; (23) Arch Facial Plast Surg.; (24) Head and Face Medicine; (25) Talanta; (26) Eur. Cell Materials; (27) Mater. Sci. Eng. C Mater. Biol. Appl.; (28) Int. J. Biol. Macromol.; (29) Adv. Food Nutr. Res.; and (30) Macromol. Bioscience. The search was limited to English language articles. In-depth research of the reference lists in the recorded manuscripts was performed in order to add significant studies and to increase the sensitivity of the revision.

2.4. Collection Data

The Medical Subject Headings (MeSH) was applied for finding the keywords used in the present revision. The selected key words: "facial" or "face" and "bone" or "bone graft" and "reconstruction" or "biomaterial" or "biomaterials" and "marine" and "collagen" were recorded for collecting the data.

2.5. Manuscript Selections

Two independent reviewers of two different universities (Messina and Naples) singularly analyzed the obtained papers in order to select inclusion and exclusion criteria as follows. Reviewers correlated their evaluations and analyzed differences through comparing the manuscripts and consulting a third experienced senor independent reviewer (Loma Linda University) when consensus could not be reached. For the stage of the full-text articles revision, a complete independent dual analysis was performed.

2.6. Manuscripts Selection

The manuscripts selected in the analysis involved experimental and clinical research on humans and animals published in the English language. Letters, editorials, case reports, and PhD theses were excluded.

2.7. Research Classifications

The method of classification included all human prospective and retrospective clinical studies, split mouth cohort studies, case-control papers, and case series manuscripts, animal investigations and literature review published between February 2010 and September 2017, on biomaterial used for facial bone reconstruction.

2.8. Statement of the Problem

The sentence case of "facial bone defect reconstructive graft" was searched over each selected papers; moreover, authors investigated if there was a documented bone reconstructive surgery and a biomaterial graft placed in situ [1–10].

2.9. Exclusion and Inclusion Criteria

The full texts of all studies related to the main revision topics were obtained for comparing the inclusion parameters:

- Investigated surgical bone regenerative procedures in patients with vertical and horizontal defect of the jaws.
- Studies involving animals in which the created bone defects were vertical and horizontal.
- Clinical human prospective or retrospective follow-up research and trials, cohort studies, case-control investigations, and case series papers with at least six months follow-up
- Animal or in vitro studies

The following exclusion parameters for manuscripts were decided as follows:

- Research treating patients with general specific diseases, heart disease, bloody pressure disease, virus infected patients, osteoporosis, immunologic disorders, uncontrolled diabetes mellitus, or other surgical risk related systemic conditions;
- Not enough information regarding the selected topic;
- Articles published prior to 1 February 2010;
- No access to the title and abstract number in the English language.

2.10. Strategy for Collecting Data

After the first literature analysis, the entire manuscript titles list was highlighted to exclude irrelevant publications, case reports and the non-English language publications. Then, research was not selected based on data obtained from screening just the abstracts. The final selection was performed reading the full texts of the papers in order to approve each study's eligibility, based on the inclusion and exclusion criteria.

2.11. Record of the Extracted and Collected Data Extraction

The results and conclusions of the selected full text papers were used for assembling the data, according to the aims and themes of the present revision, as listed onwards.

The following parameters were used as a method for assembling the data and then organized following the schemes:

"Author (Year)"—revealed the first author and the year of publication;

"Type of study"—indicated the method of the research;

"Sample size"—described the number of patients, animals or models examined;

"Bone substitute/membrane"—described types of bone grafts and membranes used for regeneration;

"Follow-up"—yes/no described the duration of the observed outcomes;

"Bone graft histology"—yes/no described the presence of the graft at 6–9 months follow-up control.

2.12. Risk of Bias Assessment

The grade of bias risk was independently considered, and in duplicate by the two independent reviewers at the moment of data extraction process. According to Moher and Higgins, this revision followed the Cochrane Collaboration's two-part tool for assessing risk of bias and PRISMA statement [14,15].

Potential causes of bias were investigated:

- Selection bias;
- Performance bias and detection bias;
- Attrition bias;
- Reporting bias;
- Examiner blinding, examiner calibration, standardized follow-up description, standardized residual graft measurement, standardized radiographic assessment.

In this way, the possible random sequence generation, the possible allocation concealment, the possibility of blinding of participants and personnel, the possible presence of having incomplete outcome data and other biases were all considered and evaluated.

This method applied by the two reviewers was valuable for giving to each study a level of bias. Then, the selected papers were classified with low, moderate, high and unclear risk.

3. Results

3.1. Manuscript Collection

Manuscript revision and recording data process followed the PRISMA flow diagram (Figure 1). The initial electronic and hand search retrieved 1197 citations and four more papers from Dentistry, PubMed MEDLINE and Oral Sciences Source with a total of 1201 selected papers. Furthermore, 617 papers were excluded because they were published prior to 1 February 2010. Then, titles and abstracts were evaluated, and 52 articles were selected as having significant data regarding "facial bone reconstruction biomaterials" topics. In addition, 48 articles were determined as full-text papers, 39 of which were incorporated in this work. Some research was excluded due to being classified as single case reports presented ($n = 4$), weak methods or being way off-topic ($n = 5$).

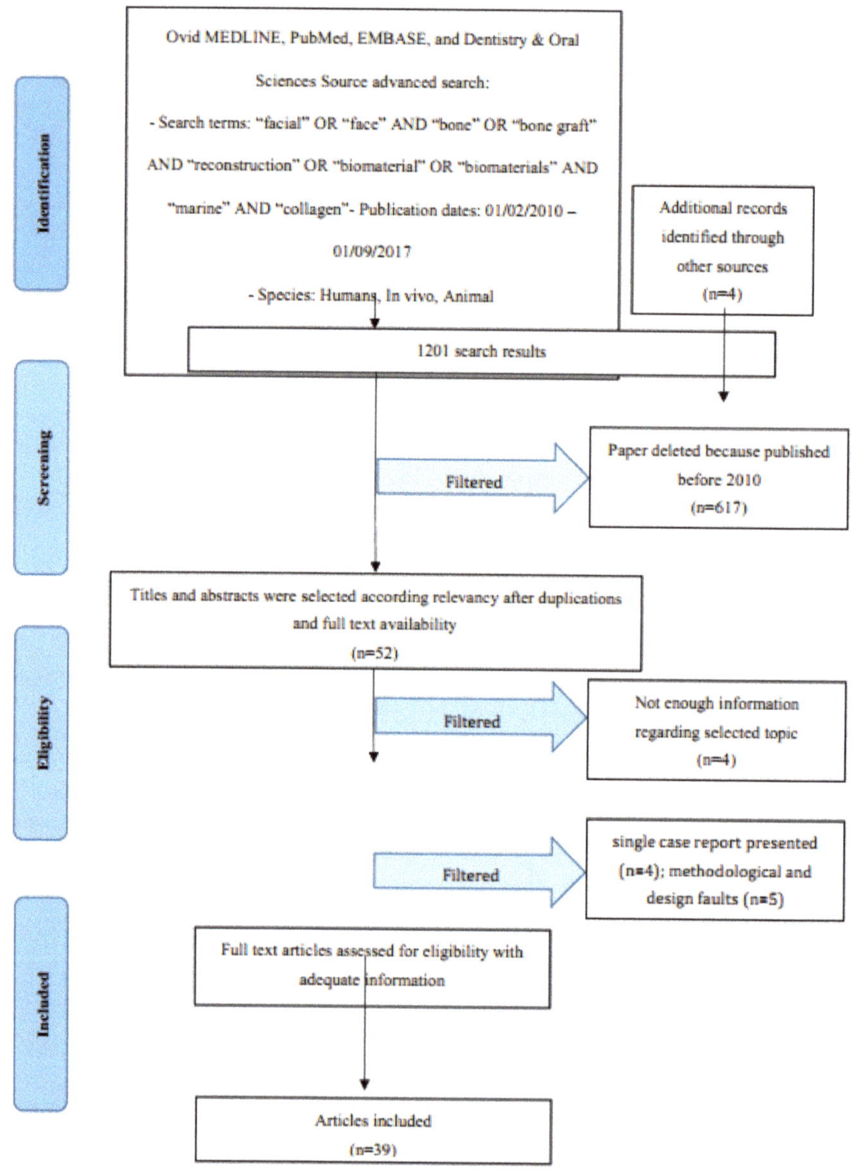

Figure 1. PRISMA flow diagram.

3.2. Study Characteristics

After the study selection, a new division related to the kind of bone graft has been performed:

- Autlogous Bone: Thirteen studies [16–28];
- Homologous Bone: Five studies [29–33];
- Xenograft or Allograft material and bone derived from animal or synthetic origin: Ten studies [34–43];
- Marine collagen material: Eleven studies [44–55].

3.3. Risk of Bias within Studies

Evaluation on the total risk of bias for each selected paper, and the majority of the manuscripts were allocated as unclear risk [20–23,27–29,34–36]. Three research papers were considered as having a low risk of bias [32,37,39], where another one was classified as moderate risk [30].

3.4. Risk of Bias across Studies

Numerous limitations have arisen from the present revision. Current analysis of the data extracted from studies written in English only could introduce a publication bias. The main limitation of the revision is related to the different kinds of biomaterials used for the same final objective, having a bone graft material able to be predictable and safe in the reconstructive and regenerative bone tissue procedures. Regarding the bias, some selected papers have a relatively short follow-up period and, when in clinical study, included relatively small numbers of treated patients. Moreover, the presented data underlined high heterogeneity and several differences in each study method, selections of the cases, and final treatment results.

It is also important to note that choice of biomaterial for doing reconstruction surgery of the facial bones is a convoluted technique and its success is related to numerous parameters, comprehending patients' general health conditions, oral hygiene habits, bone defect size, surgical procedures, operator skill, and various other factors that are not possible to fit within the frames of systematic literature. Table 1 resumes the studies selected and their results.

Table 1. Studies findings and results kind of bone graft and possible clinical application.

Author and Year	In Vivo vs. In Vitro	Kind of Graft	Follow up after 12 Months
Janner et al., 2016	In Vivo	Autologous	Yes
Emodi et al., 2017	In Vivo	Autologous	Yes
Du Toit et al., 2016	In Vivo	Autologous	Yes
Nadon et al., 2015	In Vivo	Autologous	Yes
Bande et al., 2014	In Vivo	Autologous	Yes
O'Connell et al., 2014	In Vivo	Autologous	No
Gultekin et al., 2016	In Vivo	Autologous	Yes
Zhang et al., 2014	In Vivo	Autologous	Yes
Nkenke et al., 2014	In Vivo	Autologous	Yes
Cicciù et al., 2014	In Vivo	Autologous	No
Nary Filho et al., 2014	In Vivo	Autologous	Yes
Koerdt et al., 2013	In Vivo	Autologous	Yes
Pereira R et al., 2013	In Vivo	Autologous	Yes
Krasny et al., 2015	In Vivo	Homologous	Yes
Schlee et al., 2014	In Vivo	Homologous	No
Monje et al., 2014	In Vivo	Homologous	No
Fretwurst et al., 2014	In Vivo	Homologous	No
Sbordone et al., 2014	In Vivo/In Vitro	Homologous	Yes
Scheyer et al., 2016	In Vivo/In Vitro	Xenograft/Synthetic	No
Le et al., 2016	In Vivo/In Vitro	Xenograft/Synthetic	Yes
Fienitz et al., 2017	In Vivo/In Vitro	Xenograft/Synthetic	N/A
You et al., 2016	In Vivo/In Vitro	Xenograft/Synthetic	No
De oliveira et al., 2016	In Vivo/In Vitro	Xenograft/Synthetic	Yes
Ghanaati et al., 2014	In Vivo/In Vitro	Xenograft/Synthetic	Yes
Peng et al., 2016	In Vivo/In Vitro	Xenograft/Synthetic	Yes
Klein et al., 2014	In Vivo/In Vitro	Xenograft/Synthetic	No
Figueiredo et al., 2013	In Vivo/In Vitro	Xenograft/Synthetic	No
Kim et al., 2011	In Vivo/In Vitro	Xenograft/Synthetic	No
Lin et al., 2010	In Vivo/In Vitro	Marine Collagen	N/A
Hayashi et al., 2012	In Vivo/In Vitro	Marine Collagen	N/A
Senni et al., 2013	In Vitro	Marine Collagen	N/A
Fernandes et al., 2013	In Vitro	Marine Collagen	N/A

Table 1. *Cont.*

Author and Year	In Vivo vs. In Vitro	Kind of Graft	Follow up after 12 Months
Yamamoto et al., 2014	In Vitro	Marine Collagen	N/A
Hayashi et al., 2014	In Vivo/In Vitro	Marine Collagen	N/A
Silva et al., 2014	In Vivo/In Vitro	Marine Collagen	N/A
Jridi et al., 2015	In Vivo/In Vitro	Marine Collagen	N/A
Derkus et al., 2016	In Vivo/In Vitro	Marine Collagen	N/A
Raftery et al. 2016	In Vivo/In Vitro	Marine Collagen	N/A
Coelho et al. 2017	In Vivo/In Vitro	Marine Collagen	N/A

3.5. Autogenous Bone

Autogenous bone graft is still considered the "gold" standard graft by the international community. The selected papers underlined the autogenous bone graft osteoinductivity properiety and its good integration in the treated bone defect; however, its limited availability due to the donnor site has been underlined.

Janner et al. demonstrated how the addition of autogenous bone chips and the presence of the collagen membrane increased bone formation. Wound protection from mechanical noxa during early healing may be critical for bone formation within the grafted area, but the presence of the chips guaranteed the cells osteoinduction [16]. The predictable use of autogenous bone graft has been investigated by Emodi et al. underlining the possibility of costhcondral graft application for rebuilding a mandibular condyle by using a three-dimensional printing template [17]. Moreover, the capability of autogenous graft has been proved in a split mouth study performed by Du Toit et al. Authors compared split-mouth human bone biopsy specimens derived from Platelet Rich Fibrin (PRF) and autogenous bone with bone that had healed without intervention, concluding how the quality of newly formed bone is the same in the two groups [18]. Autogenous bone graft for repairing cleft palate has been investigated by Nadon et al. The six-month outcomes of all examined patients were excellent in terms of both bone graft stability and closure of the oronasal fistulae [19]. In a retrospective study, Baden et al. evaluated the stability of bone grafted from maxillary sinus for repairing orbital trauma. The choice of autogenous bone is demonstrated to be a predictable choice for reducing the floor of orbit manually to the proper position, which helps to decrease the orbital floor defect [20]. Extra oral autogenous bone graft, and specifically the iliac bone graft for repairing the orbital floor fractures, has been retrospectively evaluated by O'Connell et al. concluding how isolated orbital blow-out fractures may be safely and predictably reconstructed using autogenous iliac crest bone [21]. Volumetric changes after autogenous ramus block bone grafting (RBG) or guided bone regeneration (GBR) in horizontally deficient maxilla before implant placement have been retrospectively evaluated by Gultekin et al. Authors stated how the two techniques are both predictable and demonstrated that the autogenous bone works have long-term dimensional stability for the next dental implant placement [22]. Zahng et al. retrospectively evaluated the outcomes of using autogenous coronoid process like bone grafts ($n = 32$) and compared with costochondral grafts ($n = 28$) in the condylar reconstruction in case of temporomandibular joint (TMJ) ankylosis. The clinical outcomes in both groups were satisfactory and comparable concluding how the autogenous coronoid process grafting may therefore be a good alternative for condylar reconstruction in patients with ankylosis of the TMJ [23]. Nkenke made a revision investigating numerous published papers about the morbidity related to the autogenous bone graft harvesting and the follow-up graft resorption and dental implant survival in the grafted sites. The author made conclusions about how the all-autologous grafts are predictable because there is no significative difference on bone resorption related to the grafting sites [24]. Cicciù et al. published a significative paper about the combination of autologous bone and growth factors applied to the mandibular continuity defects reconstruction. Authors concluded that the use of rhBMP-2 without concomitant autogenous bone grafting materials in large critical-sized mandibular defects secondary to a large mandibular tumor produced excellent regeneration of the treated area [25]. Nary Filho et al. have analyzed the possibility of autogenous bone graft infection and exposure after the regenerative surgery. The autogenous bone has less chance of causing infection due to its high possibility of integration with the host. However, all of the regenerative procedures should guarantee the covering of the grafted bone once fixed [26]. Koerdt et al. evaluated the revascularization processes in autogenous bone grafts from the iliac crest to the alveolar ridge. Even if the resorption prediction is not predictable, the immunohistochemical investigation performed showed blood vessels between the graft and the alveolar ridge [27]. Pereira Rdos et al. analyzed the management of orbital fracture with autologous bone graft in their study. A computed tomography scan shows excellent bone healing at the anterior and posterior parts of the medial orbital wall reconstruction [28].

3.6. Allogeneic Bone

Allogeneic bone graft used for facial reconstruction has been considered as a good alternative option in the place of autogenous bone. However, a short number of investigations recorded long-term clinical success and totally safe procedures for the treated patients.

Krasny et al. recently demonstrated how the fresh, frozen, radiation sterilized, allogeneic bone blocks constitute good and durable bone-replacement material allowing effective and long-lasting reconstruction of the atrophied alveolar ridge to support durable, implant-based, prosthetic restoration [29]. Schlee et al. underlined the concept about bone grafting with allogeneic allografts yielding equivalent results to autologous grafting, and patients appreciate the omission of bone harvesting from donor sites [30]. Monje et al. investigated the feasibility by means of survival rate, histologic analysis, and causes of failure of allogeneic block grafts for augmenting the atrophic maxilla. In their conclusions, the authors stated how the atrophied maxillary reconstruction with allogeneic bone block grafts represents a reliable option as shown by low block graft failure rate, minimal resorption, and high implant survival rate [31]. Fretwurst et al. performed a histological and biochemical evaluation of four allogeneic bone blocks concluding how this kind of graft can absolutely stimulate the newly bone formation and can be considered a valid alternative to the autologous bone graft [32]. Sbordone et al. in their study analyzed the potential bone volume changes after sinus augmentation using blocks of autogenous iliac bone or freeze-dried allogeneic bone (FDBA) from the hip. Authors concluded how application of FDBA for the short-term sinus grafting procedure showed an outcome close to that reported for autogenous bone [33].

3.7. Xenograft and Synthetic Bone

The bone grafts derived from animal or by synthetic production can be classified as biomaterials with osteoconductivity properties and currently have been used for being scaffolds for growth factor application in the bone defects. Scheyer et al. just published a randomized controlled multicenter clinical trial in which different biomaterials for the preservation of the bone volume after tooth extraction were tested. Authors stated how, at six months follow up, the xenograft collagen and autologous bone graft can give no significant difference on the socket modifications after extraction [34]. Le et al. evaluated 14 patients affected by soft tissue recessions around implant-supported restorations in the maxillary central or lateral incisor area. In their records, authors concluded that the use of the allograft and xenogeneic collagen significantly favored the alveolar volume conditioning hard and soft tissue dimensions in the aesthetic zone of the anterior maxilla [35]. Fienitz et al. histologically and radiologically compared a sintered and a no sintered bovine bone graft used in the sinus lift surgeries. The authors affirmed that both xenogeneic materials showed comparable results regarding the possibility of having new bone formation [36].

You et al. evaluated the effects of the bilayer bone augmentation technique for the treatment of dehiscence-type defects around implants and evaluated the role as a membrane of the xenogeneic bone using a histological method for evaluating the new bone formation. The results of this study showed the osteogenic effect of autogenous bone and the effect of mechanical support for prolonged space maintenance of xenogeneic collagen membrane applied for the treatment of dehiscence-type bone defects around implants [37].

De Oliveira et al. studied the regenerative results of the addition of bone marrow aspirate concentrate, using a single or double centrifugation protocol, to a xenogeneic bone graft in sinus floor elevation. This pilot study indicates that the clinical use of bone marrow aspirate concentrate, obtained by either a single or double centrifugation process, combined with a xenograft result in more adequate bone repair when used in the bone regenerative surgery [38]. Ghanaati et al. studied the structure of two allogeneic bone blocks and three xenogeneic bone grafts, which are used in dental and orthopedic surgery, and histologically analysis have been performed. The final findings affirmed that, even manufactory declared blocks were free of organic/cellular remnants, authors' histological analysis revealed that bone blocks did contain such remnants. Moreover, such specimens might be able to induce an immune response within the recipient [39]. Peng et al. analyzed the influence of platelet rich plasma (PRP) associated with xenograft for managing peri-implant bone defects. The results indicate how the PRP associated with the bovine-derived xenograft in the small bone defect can favor the bone healing [40]. Klein et al. published a systematic histomorphometric analysis of two human bone biopsy specimens analyzed at five-years follow up after a bone regenerative procedure using a xenogeneic bovine bone substitute material. Authors demonstrated a completed bony integration without extensive resorption of the biomaterial particles [41]. Figuerido et al. evaluated the chemical and structural features of a xenogeneic and an alloplastic material highlighting the in vivo inflammatory response.

The in vivo results analyzing the data extracted from the inflammatory infiltrates revealed that the grade of inflammation is not severe, particularly in terms of collagen production and formation of fibrous capsule [42].

Kim et al. investigated the efficacy of the alveolar ridge preservation technique using collagen sponge and xenograft after extraction. The results indicated that, in the ridge preservation using collagen sponge and xenograft, xenograft prevents the horizontal resorption of the alveolar ridge, and the upper collagen sponge blocks the infiltration of soft tissues to the lower area, and thus it has the advantage of the enhancement of bone fill [43].

3.8. Marine Collagen and Derived Bone

Marine collagen and marine derived bone substitutes as an alternative to autologous bone are quickly advancing, especially since the service of tissue engineering is researching biomaterial with low cost and high availability. Marine collagen should be a true alternative source of collagen. Marine species present a distinct advantage as a lower known risk of transmission to humans of infection-causing agents and are thought to be far less associated with cultural and religious concerns regarding the human use of marine derived products. Moreover, a clarification of the marine collagen origin is important in order to underline the microscopical features of the final material used for bone defect repairing. Not only marine invertebrate collagen sources, but also marine vertebrate ones reflect several similarity with human collagens [44]. Fish would be rich sources of collagen in terms of its production and application in various biological process. Marine organisms like coral or sponge are rich in mineralized porous structures and their microstructures seem to replace the human bone features. Evaluating the source of collagen extraction, jellyfish and invertebrate collagen are obtained from mesoglea, following a methodology based on solubilization in acetic acid solution, typically during three days. Nowadays, the collagen is considered the major constituent of the extracellular matrices of all animal and metazoans. For this reason, collagen derived from marine sponges can be evaluated as available substitutes for uses like scaffolds in the bone regenerative procedures [44–52].

Recently, Lin et al. developed a novel scaffold, derived from fish scales, as an alternative functional material with sufficient mechanical strength for medical regenerative applications. Fish scales, which are usually considered marine wastes, were acellularized, decalcified and fabricated into collagen scaffolds. The scanning electron microscope (SEM) was used for imaging the microstructure of the scaffold. The highly centrally-oriented micropatterned structure of the scaffold was beneficial for efficient nutrient and oxygen supply to the cells cultured in the three-dimensional matrices, and therefore it is useful for high-density cell seeding and spreading [45]. Hayashi et al. evaluated the biomedical application of chitosan and collagen from marine products and advantages and disadvantages of regeneration medicine, demonstrating that the properties of biocompatibility and biodegradation of fish atelocollagen are suitable for the scaffolds in regenerative medicine [46]. Senny et al. investigated the HE800 exopolysaccharide (HE800 EPS) secreted by a deep sea hydrothermal bacterium displays an interesting glycosaminoglycan-like feature resembling hyaluronan, and the results of the study proved how the HE800 EPS family can be considered as an innovative biotechnological source of glycosaminoglycan-like compounds useful to design biomaterials and drugs for tissue engineering applications [47]. Fernandes-Silva et al. analyzed the possibility to fabricate marine collagen porous structures cross linked with genipin under high pressure carbon dioxide. By the in vitro data results of their investigation, authors concluded that cell culture tests performed with a chondrocyte-like cell line showed good cell adherence and proliferation, which is a strong indication of the potential of these scaffolds to be used in tissue cartilage tissue engineering [48]. Yamamoto et al. published an in vitro and in vivo biological study of medical materials to investigate the safety and the predictable results on applying fish collagen for regenerative procedure. The extract of fish collagen gel was examined to clarify its sterility and demonstrated that atelocollagen prepared from tilapia is a promising biomaterial for use as a scaffold in regenerative medicine [49]. Hayashi et al. demonstrated the contributions for a proteomic view of chitosan nanoparticle to hepatic cells, the promotion of D-glucosamine to transfection efficiency, and chitin application as skin substitutes. Moreover, the latter showed the contributions for hydroxyapatite-gelatin nanocomposite, genipin modification of dentin collagen, dentin phosphophoryn/collagen composite for dental biomaterial, and biological safety of fish collagen [50]. Silva et al. evaluated all of the available forms of marine collagen and their potential application in regenerative medicine. Authors concluded that marine collagen could be considered a valuable source of collagen [51]. Jiridi et al. studied the structural and rheological properties of collagen-based gel obtained from cuttlefish skin, and to investigate its ability to enhance wound healing, demonstrating that cuttlefish collagen based gel might be useful as a wound healing agent [52]. Derkus et al. described the sonochemical isolation of nano-sized spherical hydroxyapatite (nHA) from egg shell and application towards thrombin aptasensing. The data of the presented paper reflected how, for clinical application of the developed aptasensor, thrombin levels in blood and cerebrospinal fluid (CSF) samples obtained from patients with Multiple Sclerosis, Myastenia Gravis, Epilepsy, Parkinson, polyneuropathy and healthy donors were analyzed using both the aptasensor and commercial ELISA kit. The results showed that the proposed system is a promising candidate for clinical analysis of thrombin [53].

Raffery et al. tried to determine if the incorporation of chitosan into collagen scaffolds could improve the mechanical and biological properties of the scaffold. In addition, the study assessed if collagen, derived from salmon skin (marine), can provide an alternative to collagen derived from bovine tendon (mammal) for tissue engineering applications. The data results underlined how the collagen–chitosan composites showed similar results to the bovine one. Moreover, a clear support to stem cell differentiation towards both bone and cartilage tissue was demonstrated. The collagen obtained from the bovine bone resulted in a versatile scaffold incorporating the marine biomaterial chitosan and showing great potential as appropriate platforms for promoting orthopaedic tissue repair while the use of salmon skin-derived collagen may be more suitable in the repair of soft tissues such as skin [54].

Coelho et al. performed an investigation in which collagen has been isolated from the skins of the squids using acid-based and pepsin-based protocols, with the higher yield being obtained from I. The produced collagen was selected for evaluating its biomedical potential, exploring its incorporation on poly-ε-caprolactone (PCL) 3D printed scaffolds for the development of hybrid scaffolds for tissue engineering, exhibiting hierarchical features [55].

4. Discussion

Nowadays, the autologous bone is considered to be the "gold standard" by the scientific community, as its use is a predictable practice. As widely described above, it increases the formation of new bone near defects. The main advantage, moreover, is to have an osteoinductive power, and to offer increased resistance to infections due to its biological activity and the revascularization of the graft, so this is a biological active tissue. In any case, in the case of autologous bone regenerations, the patient's morbidity should be considered, due to the donor site, such as the inability to access large amounts of bone tissue [1,8,12–28]. However, allogeneic bone grafting is a viable alternative that has the advantage of not requiring a donor site, in the event that no alloplastic and autologous mixture grafts are used. The alloplastic material can be subjected to different chemical-physical treatment, satisfying also the requirements of larger graft surgeries, having bone blocks available [29–34]. This biomaterial also has the advantage of being able to stimulate the formation of new bone. Moreover, tissue grafts from other species or synthetic biomaterial with osteoconductivity can be available. These are mainly used as a scaffold for the regeneration of bone defects. Even if this type of material is "free cell tissue", it could induce an immune response within the recipient [2,7,35–43]. Finally, marine collagen and derived bone can be considered a viable alternative, thanks mainly to its microscopic structure that would favor the metabolism of new bone tissue and hence its formation. It is similar to those found in the human spongiform bone and thus has osteoconductive abilities [45–55]. The porosity characteristics reflect the optimal ones, and the compression resistance is high but has little resistance to tension. As already mentioned by some studies, the ability of some cell lines to adhere to this material and proliferate on it has emerged. It is an easy-to-find material and is considered commercially as a waste material. Its production involves decellularization and then decalcification having biocompatibility properties, reducing the risk of transmitting infectious agents to humans. Marine collagen can be derived from different animal genera, from marine sponges to salmon skin or fish scales. Surely, a production of these materials using existing 3D printing technologies could revolutionize the field of biomaterials and used materials for bone facial defects.

Furthermore, Rahman published a paper just recently that demonstrated the potential for marine calcifiers to generate new drugs. Among the different sources of polysaccharides, algal polysaccharides such as chitin and collagen could play an important role in future development of tissue engineering, bone regeneration, and much more. In light of these emerging findings, in the near future, established techniques might also be potentially useful for isolating skeletal proteins from similar marine calcifiers for drug discovery [56]. Moreover, the isolation, biochemical and biophysical features of the collagen from the marine sponges Axinella cannabina and suberites carnosus were analyzed by Tzivileka et al. Authors demonstrated how marine collagen can be considered a valuable and safe alternative to the common collagen used in the current biomedical application [57]. Moreover, Hermann Ehrlich performed several experiments demonstrating how chitin and collagen are valid alternative template scaffolds in the field of mineralization [58–60]. Therefore, in a monograph published in 2015, the author classified significant information regarding the modern knowledge on biomineralization, biomimetics and materials science with a deep investigation about marine vertebrates. For the first time in scientific literature, the author gives the most coherent analysis of the nature, origin and evolution of biomaterials and biopolymers isolated from marine sources. Moreover, the variety of marine vertebrate organisms (fish, reptilian, birds and mammals) and within their unique hierarchically organized structural formations has been highlighted [61]. This can be a reference for performing future studies and research about the possibility of using chitin and collagen marine in the field of the biomedicine.

Limitations

Even if a comprehensive and complete investigation of the effects of surgical therapies had been performed, there were some limitations to this systematic review. Our findings could not provide the ideal material for the bone reconstruction technique of the face. The choice of the materials depend on the size of the defect, the skill of the surgeons and the donor host in the case of autogenous bone.

Moreover, there could be potential language bias in this systematic review, as we only considered literature written in English.

5. Conclusions

The first aim of this revision is to highlight the evolving role of biomaterial used like bone graft in the place of autologous bone during craniofacial bone reconstruction and regenerative procedures. Discussion highlighted how the ideal biomaterial scaffold is still far from being realized because of it it still being hard to reproduce the design, the micro shapes and the chemical features of the autologous bone. While human clinical applications are limited to date, great promise seems to start from scaffolds originating from the lab or obtained from the sea. Specifically, marine collagen replacing the human bone features can be optimally used for being a safe scaffold on large bone reconstruction defects of the facial area.

Author Contributions: Marco Cicciù wrote and revise the whole paper; Alan Scott Herford revised the final manuscript, the grammar and the English style; Gabriele Cervino, Luca Fiorillo, Ennio Bramanti collected the data; Floriana Lauritano, Fausto Famà checked the data and prepared the tables; Sergio Sambataro checked the form for the submission and prepared the references style and order; Luigi Laino and Giuseppe Troiano abalyzed the data and compared the results for each group.

Conflicts of Interest: The authors report no conflicts of interest related to this study.

References

1. Herford, A.S.; Tandon, R.; Stevens, T.W.; Stoffella, E.; Cicciù, M. Immediate distraction osteogenesis: The sandwich technique in combination with rhBMP-2 for anterior maxillary and mandibular defects. *J. Craniofac. Surg.* **2013**, *24*, 1383–1387. [CrossRef] [PubMed]
2. Boccaccini, A.R.; Maquet, V. Bioresorbable and bioactive polymer/Bioglass composites with tailored pore structure for tissue engineering applications. *Compos. Sci. Technol.* **2003**, *63*, 2417–2429. [CrossRef]
3. Zhang, C.Y.; Lu, H.; Zhuang, Z.; Wang, X.P.; Fang, Q.F. Nano-hydroxyapatite/poly(L-lactic acid) composite synthesized by a modified in situ precipitation: Preparation and properties. *J. Mater. Sci.* **2010**, *21*, 3077–3083. [CrossRef] [PubMed]
4. Herford, A.S.; Lu, M.; Akin, L.; Cicciù, M. Evaluation of a porcine matrix with and without platelet-derived growth factor for bone graft coverage in pigs. *Int. J. Oral Maxillofac. Implants* **2012**, *27*, 1351–1358. [PubMed]
5. Cicciù, M.; Herford, A.S.; Stoffella, E.; Cervino, G.; Cicciù, D. Protein-signaled guided bone regeneration using titanium mesh and Rh-BMP2 in oral surgery: A case report involving left mandibular reconstruction after tumor resection. *Open Dent. J.* **2012**, *6*, 51–55. [CrossRef] [PubMed]
6. Hong, L.; Wang, Y.L.; Jia, S.R.; Huang, Y.; Gao, C.; Wan, Y.Z. Hydroxyapatite/bacterial cellulose composites synthesized via a biomimetic route. *Mater. Lett.* **2006**, *60*, 1710–1713. [CrossRef]
7. Nge, T.T.; Sugiyama, J. Surface functional group dependent apatite formation on bacterial cellulose microfibrils network in a simulated body fluid. *J. Biomed. Mater. Res. A* **2007**, *81*, 124–134. [CrossRef] [PubMed]
8. Herford, A.S.; Cicciù, M. Recombinant human bone morphogenetic protein type 2 jaw reconstruction in patients affected by giant cell tumor. *J. Craniofac. Surg.* **2010**, *21*, 1970–1975. [CrossRef] [PubMed]
9. Fama, F.; Cicciu, M.; Sindoni, A.; Nastro-Siniscalchib, E.; Falzea, R.; Cervino, G.; Polito, F.; De Ponte, F.; Gioffre-Florioa, M. Maxillofacial and concomitant serious injuries: An eight-year single center experience. *Chin. J. Traumatol.* **2017**, *20*, 4–8. [CrossRef] [PubMed]
10. De Ponte, F.S.; Falzea, R.; Runci, M.; Siniscalchi, E.N.; Lauritano, F.; Bramanti, E.; Cervino, G.; Cicciu, M. Histomorhological and clinical evaluation of maxillary alveolar ridge reconstruction after craniofacial trauma by applying combination of allogeneic and autogenous bone graft. *Chin. J. Traumatol.* **2017**, *20*, 14–17. [CrossRef] [PubMed]
11. Hollister, S.J. Porous scaffold design for tissue engineering. *Nat. Mater.* **2005**, *4*, 518–524. [CrossRef] [PubMed]
12. Hollister, S.J.; Lin, C.Y.; Saito, E.; Lin, C.Y.; Schek, R.D.; Taboas, J.M. Engineering craniofacial scaffolds. *Orthod. Craniofac. Res.* **2005**, *8*, 162–173. [CrossRef] [PubMed]

13. Hong, P.; Boyd, D.; Beyea, S.D.; Bezuhly, M. Enhancement of bone consolidation in mandibular distraction osteogenesis: A contemporary review of experimental studies involving adjuvant therapies. *J. Plast. Reconstr. Aesthet. Surg.* **2013**, *66*, 883–895. [CrossRef] [PubMed]

14. Moher, D.; Liberati, A.; Tetzlaff, J.; Altman, D.G.; PRISMA Group. Preferred reporting items for systematic reviews and meta-analyses: The PRISMA statement. *J. Clin. Epidemiol.* **2009**, *62*. [CrossRef] [PubMed]

15. Higgins, J.P.; Altman, D.G.; Gotzsche, P.C.; Jüni, P.; Moher, D.; Oxman, A.D.; Savovic, J.; Schulz, K.F.; Weeks, L.; Sterne, J.A. The Cochrane Collaboration's tool for assessing risk of bias in randomised trials. *BMJ* **2011**, *343*, d5928. [CrossRef] [PubMed]

16. Janner, S.F.; Bosshardt, D.D.; Cochran, D.L.; Chappuis, V.; Huynh-Ba, G.; Jones, A.A.; Buser, D. The influence of collagen membrane and autogenous bone chips on bone augmentation in the anterior maxilla: A preclinical study. *Clin. Oral Implants Res.* **2016**. [CrossRef] [PubMed]

17. Emodi, O.; Shilo, D.; Israel, Y.; Rachmiel, A. Three-dimensional planning and printing of guides and templates for reconstruction of the mandibular ramus and condyle using autogenous costochondral grafts. *Br. J. Oral Maxillofac. Surg.* **2017**, *55*, 102–104. [CrossRef] [PubMed]

18. Du Toit, J.; Siebold, A.; Dreyer, A.; Gluckman, H. Choukroun Platelet-Rich Fibrin as an Autogenous Graft Biomaterial in Preimplant Surgery: Results of a Preliminary Randomized, Human Histomorphometric, Split-Mouth Study. *Int. J. Periodontics Restor. Dent.* **2016**, *36*, 75–86. [CrossRef] [PubMed]

19. Nadon, F.; Chaput, B.; Périssé, J.; de Bérail, A.; Lauwers, F.; Lopez, R. Interest of Mineralized Plasmatic Matrix in Secondary Autogenous Bone Graft for the Treatment of Alveolar Clefts. *J. Craniofac. Surg.* **2015**, *26*, 2148–2151. [CrossRef]

20. Bande, C.R.; Daware, S.; Lambade, P.; Patle, B. Reconstruction of Orbital Floor Fractures with Autogenous Bone Graft Application from Anterior Wall of Maxillary Sinus: A Retrospective Study. *J. Maxillofac. Oral Surg.* **2015**, *14*, 605–610. [CrossRef] [PubMed]

21. O'Connell, J.E.; Hartnett, C.; Hickey-Dwyer, M.; Kearns, G.J. Reconstruction of orbital floor blow-out fractures with autogenous iliac crest bone: A retrospective study including maxillofacial and ophthalmology perspectives. *J. Craniomaxillofac. Surg.* **2015**, *43*, 192–198. [CrossRef] [PubMed]

22. Gultekin, B.A.; Bedeloglu, E.; Kose, T.E.; Mijiritsky, E. Comparison of Bone Resorption Rates after Intraoral Block Bone and Guided Bone Regeneration Augmentation for the Reconstruction of Horizontally Deficient Maxillary Alveolar Ridges. *Biomed. Res. Int.* **2016**. [CrossRef] [PubMed]

23. Zhang, W.; Gu, B.; Hu, J.; Guo, B.; Feng, G.; Zhu, S. Retrospective comparison of autogenous cosotochondral graft and coronoid process graft in the management of unilateral ankylosis of the temporomandibular joint in adults. *Br. J. Oral Maxillofac. Surg.* **2014**, *52*, 928–933. [CrossRef] [PubMed]

24. Nkenke, E.; Neukam, F.W. Autogenous bone harvesting and grafting in advanced jaw resorption: Morbidity, resorption and implant survival. *Eur. J. Oral Implantol.* **2014**, *7* (Suppl. S2), 203–217.

25. Cicciù, M.; Herford, A.S.; Cicciù, D.; Tandon, R.; Maiorana, C. Recombinant human bone morphogenetic protein-2 promote and stabilize hard and soft tissue healing for large mandibular new bone reconstruction defects. *J. Craniofac. Surg.* **2014**, *25*, 860–862. [CrossRef] [PubMed]

26. Nary Filho, H.; Pinto, T.F.; de Freitas, C.P.; Ribeiro-Junior, P.D.; dos Santos, P.L.; Matsumoto, M.A. Autogenous bone grafts contamination after exposure to the oral cavity. *J. Craniofac. Surg.* **2014**, *25*, 412–414. [CrossRef] [PubMed]

27. Koerdt, S.; Siebers, J.; Bloch, W.; Ristow, O.; Kuebler, A.C.; Reuther, T. Immunohistochemial study on the expression of von Willebrand factor (vWF) after onlay autogenous iliac grafts for lateral alveolar ridge augmentation. *Head Face Med.* **2013**, *11*, 40. [CrossRef] [PubMed]

28. Pereira Rdos, S.; Jorge-Boos, F.B.; Hochuli-Vieira, E.; da Rocha, H.V., Jr.; Homsi, N.; de Melo, W.M. Management of pure medial orbital wall fracture with autogenous bone graft. *J. Craniofac. Surg.* **2013**, *24*, e475–e477. [CrossRef] [PubMed]

29. Krasny, M.; Krasny, K.; Fiedor, P.; Zadurska, M.; Kamiński, A. Long-term outcomes of the use of allogeneic, radiation-sterilised bone blocks in reconstruction of the atrophied alveolar ridge in the maxilla and mandible. *Cell Tissue Bank.* **2015**, *16*, 631–638. [CrossRef] [PubMed]

30. Schlee, M.; Dehner, J.F.; Baukloh, K.; Happe, A.; Seitz, O.; Sader, R. Esthetic outcome of implant-based reconstructions in augmented bone: Comparison of autologous and allogeneic bone block grafting with the pink esthetic score (PES). *Head Face Med.* **2014**, *28*, 10–21. [CrossRef] [PubMed]

31. Monje, A.; Pikos, M.A.; Chan, H.L.; Suarez, F.; Gargallo-Albiol, J.; Hernández-Alfaro, F.; Galindo-Moreno, P.; Wang, H.L. On the Feasibility of Utilizing Allogeneic Bone Blocks for Atrophic Maxillary Augmentation. *Biomed. Res. Int.* **2014**, *2014*. [CrossRef] [PubMed]

32. Fretwurst, T.; Spanou, A.; Nelson, K.; Wein, M.; Steinberg, T.; Stricker, A. Comparison of four different allogeneic bone grafts for alveolar ridge reconstruction: A preliminary histologic and biochemical analysis. *Oral Surg. Oral Med. Oral Pathol. Oral Radiol.* **2014**, *118*, 424–431. [CrossRef] [PubMed]

33. Sbordone, C.; Toti, P.; Guidetti, F.; Califano, L.; Pannone, G.; Sbordone, L. Volumetric changes after sinus augmentation using blocks of autogenous iliac bone or freeze-dried allogeneic bone. A non-randomized study. *J. Craniomaxillofac. Surg.* **2014**, *42*, 113–118. [CrossRef] [PubMed]

34. Scheyer, E.T.; Heard, R.; Janakievski, J.; Mandelaris, G.; Nevins, M.L.; Pickering, S.R.; Richardson, C.R.; Pope, B.; Toback, G.; Velásquez, D.; et al. A randomized, controlled, multicentre clinical trial of post-extraction alveolar ridge preservation. *J. Clin. Periodontol.* **2016**, *43*, 1188–1199. [CrossRef] [PubMed]

35. Le, B.; Borzabadi-Farahani, A.; Nielsen, B. Treatment of Labial Soft Tissue Recession Around Dental Implants in the Esthetic Zone Using Guided Bon Regeneration With Mineralized Allograft: A Retrospective Clinical Case Series. *J. Oral Maxillofac. Surg.* **2016**, *74*, 1552–1561. [CrossRef] [PubMed]

36. Fienitz, T.; Moses, O.; Klemm, C.; Happe, A.; Ferrari, D.; Kreppel, M.; Ormianer, Z.; Gal, M.; Rothamel, D. Histological and radiological evaluation of sintered and non-sintered deproteinized bovine bone substitute materials in sinus augmentation procedures. A prospective, randomized-controlled, clinical multicenter study. *Clin. Oral Investig.* **2017**, *21*, 787–794. [CrossRef] [PubMed]

37. You, D.J.; Yoon, H.J. Bone Regeneration with Bilayer Bone Augmentation Technique for the Treatment of Dehiscence-Type Defects Around Implants: A Preliminary Study in Dogs. *Int. J. Oral Maxillofac. Implants* **2016**, *31*, 318–323. [CrossRef] [PubMed]

38. De Oliveira, T.A.; Aloise, A.C.; Orosz, J.E.; de Mello, E.; Oliveira, R.; de Carvalho, P.; Pelegrine, A.A. Double Centrifugation Versus Single Centrifugation of Bon Marrow Aspirate Concentrate in Sinus Floor Elevation: A Pilot Study. *Int. J. Oral Maxillofac. Implants* **2016**, *31*, 216–222. [CrossRef] [PubMed]

39. Ghanaati, S.; Barbeck, M.; Booms, P.; Lorenz, J.; Kirkpatrick, C.J.; Sader, R.A. Potential lack of "standardized" processing techniques for production of allogeneic and xenogeneic bone blocks for application in humans. *Acta Biomater.* **2014**, *10*, 3557–3562. [CrossRef] [PubMed]

40. Peng, W.; Kim, I.; Cho, H.; Seo, J.; Lee, D.; Jang, J.; Park, S. The healing effect of platelet-rich plasma on xenograft in peri-implant bone defects in rabbits. *Maxillofac. Plast. Reconstr. Surg.* **2016**, *38*, 61–65. [CrossRef] [PubMed]

41. Klein, M.O.; Kämmerer, P.W.; Götz, H.; Duschner, H.; Wagner, W. Long-term bony integration and resorption kinetics of a xenogeneic bone substitute after sinus floor augmentation: Histomorphometric analyses of human biopsy specimens. *Int. J. Periodontics Restor. Dent.* **2013**, *33*, e101–e110. [CrossRef] [PubMed]

42. Figueiredo, A.; Coimbra, P.; Cabrita, A.; Guerra, F.; Figueiredo, M. Comparison of a xenogeneic and an alloplastic material used in dental implants in terms of physico-chemical characteristics and in vivo inflammatory response. *Mater. Sci. Eng. C Mater. Biol. Appl.* **2013**, *33*, 3506–3513. [CrossRef] [PubMed]

43. Kim, Y.K.; Yun, P.Y.; Lee, H.J.; Ahn, J.Y.; Kim, S.G. Ridge preservation of the molar extraction socket using collagen sponge and xenogeneic bone grafts. *Implant. Dent.* **2011**, *4*, 267–272. [CrossRef] [PubMed]

44. Zahng, X.; Vecchio, K.S. Conversion of natural marine skeletons as scaffolds for bobe tissue engineering. *Front. Mater. Sci.* **2013**, *7*, 103–117. [CrossRef]

45. Lin, C.C.; Ritch, R.; Lin, S.M.; Ni, M.H.; Chang, Y.C.; Lu, Y.L.; Lai, H.J.; Lin, F.H. A new fish scale-derived scaffold for corneal regeneration. *Eur. Cell Mater.* **2010**, *26*, 50–57. [CrossRef]

46. Hayashi, Y.; Yamada, S.; Yanagi Guchi, K.; Koyama, Z.; Ikeda, T. Chitosan and fish collagen as biomaterials for regenerative medicine. *Adv. Food Nutr. Res.* **2012**, *65*, 107–120. [PubMed]

47. Senni, K.; Gueniche, F.; Changotade, S.; Septier, D.; Sinquin, C.; Ratiskol, J.; Lutomski, D.; Godeau, G.; Guezennec, J.; Colliec-Jouault, S. Unusual glycosaminoglycans from a deep sea hydrothermal bacterium improve fibrillar collagen structuring and fibroblast activities in engineered connective tissues. *Mar. Drugs* **2013**, *11*, 1351–1369. [CrossRef] [PubMed]

48. Fernandes-Silva, S.; Moreira-Silva, J.; Silva, T.H.; Perez-Martin, R.I.; Sotelo, C.G.; Mano, J.F.; Duarte, A.R.; Reis, R.L. Porous hydrogels from shark skin collagen crosslinked under dense carbon dioxide atmosphere. *Macromol. Biosci.* **2013**, *13*, 1621–1631. [CrossRef] [PubMed]

49. Yamamoto, K.; Igawa, K.; Sugimoto, K.; Yoshizawa, Y.; Yanagiguchi, K.; Ikeda, T.; Yamada, S.; Hayashi, Y. Biological safety of fish (tilapia) collagen. *Biomed. Res. Int.* **2014**, *2014*, 630757. [CrossRef] [PubMed]

50. Hayashi, Y.; Yamauchi, M.; Kim, S.K.; Kusaoke, H. Biomaterials: Chitosan and collagen for regenerative medicine. *Biomed. Res. Int.* **2014**, *2014*, 690485. [CrossRef] [PubMed]

51. Silva, T.H.; Moreira-Silva, J.; Marques, A.L.; Domingues, A.; Bayon, Y.; Reis, R.L. Marine origin collagens and its potential applications. *Mar. Drugs* **2014**, *12*, 5881–5901. [CrossRef] [PubMed]

52. Jridi, M.; Bardaa, S.; Moalla, D.; Rebaii, T.; Souissi, N.; Sahnoun, Z.; Nasri, M. Microstructure, rheological and wound healing properties of collagen-based gel from cuttlefish skin. *Int. J. Biol. Macromol.* **2015**, *77*, 369–374. [CrossRef] [PubMed]

53. Derkus, B.; Arslan, Y.E.; Emregul, K.C.; Emregul, E. Enhancement of aptamer immobilization using egg shell-derived nano-sized spherical hydroxyapatite for thrombin detection in neuroclinic. *Talanta* **2016**, *158*, 100–109. [CrossRef] [PubMed]

54. Raftery, R.M.; Woods, B.; Marques, A.L.P.; Moreira-Silva, J.; Silva, T.H.; Cryan, S.A.; Reis, R.L.; O'Brien, F.J. Multifunctional biomaterials from the sea: Assessing the effects of chitosan incorporation into collagen scaffolds on mechanical and biological functionality. *Acta Biomater.* **2016**, *43*, 160–169. [CrossRef] [PubMed]

55. Coelho, R.C.G.; Marques, A.L.P.; Oliveira, S.M.; Diogo, G.S.; Pirraco, R.P.; Moreira-Silva, J.; Xavier, J.C.; Reis, R.L.; Silva, T.H.; Mano, J.F. Extraction and characterization of collagen from Antarctic and Sub-Antarctic squid and its potential application in hybrid scaffolds for tissue engineering. *Mater. Sci. Eng. C Mater. Biol. Appl.* **2017**, *78*, 787–795. [CrossRef] [PubMed]

56. Rahman, M.A. An Overview of the Medical Applications of Marine Skeletal Matrix Proteins. *Mar. Drugs* **2016**, *14*, 167. [CrossRef] [PubMed]

57. Tziveleka, L.-A.; Ioannou, E.; Tsiourvas, D.; Berillis, P.; Foufa, E.; Roussis, V. Collagen from the Marine Sponges *Axinella cannabina* and *Suberites carnosus*: Isolation and Morphological, Biochemical, and Biophysical Characterization. *Mar. Drugs* **2017**, *15*, 152. [CrossRef] [PubMed]

58. Ehrlich, H.; Ilan, M.; Maldonado, M.; Muricy, G.; Bavestrello, G.; Kljajic, Z.; Carballo, J.L.; Shiaparelli, S.; Ereskovsky, A.V.; Schupp, P.; et al. Three-Dimensional chitin-based scaffolds from Verongida sponges (Demospongiae: Porifera). Part I. Isolation and identification of chitin. *Int. J. Biol. Macromol.* **2010**, *47*, 132–140. [CrossRef] [PubMed]

59. Ehrlich, H.; Kaluzhnaya, O.V.; Brunner, E.; Tsurkan, M.V.; Ereskovsky, A.; Ilan, M.; Tabachnick, K.R.; Bazhenov, V.V.; Paasch, S.; Kammer, M.; et al. Identification and first insights into the structure and biosynthesis of chitin from the freshwater sponge *Spongilla lacustris*. *J. Struct. Biol.* **2013**, *183*, 474–483. [CrossRef] [PubMed]

60. Ehrlich, H.; Keith Rigby, J.; Botting, J.P.; Tsurkan, M.V.; Werner, C.; Schwille, P.; Petrášek, Z.; Pisera, A.; Simon, P.; Sivkov, V.N.; et al. Discovery of 505-million-year old chitin in the basal demosponge *Vauxia gracilenta*. *Sci. Rep.* **2013**, *3*, 3497. [CrossRef] [PubMed]

61. Ehrlich, H. *Biological Materials of Marine Origin: Vertebrates*; Springer: Berlin, Germany, 2015; p. 594.

 *marine drugs*

Article

Effects of Composite Supplement Containing Collagen Peptide and Ornithine on Skin Conditions and Plasma IGF-1 Levels—A Randomized, Double-Blind, Placebo-Controlled Trial

Naoki Ito *, Shinobu Seki and Fumitaka Ueda

Pharmaceutical and Healthcare Research Laboratories, Research and Development Management Headquarters, FUJIFILM Corporation, 577, Ushijima, Kaisei-machi, Ashigarakami-gun, Kanagawa 258-8577, Japan; shinobu.seki@fujifilm.com (S.S.); fumitaka.ueda@fujifilm.com (F.U.)

* Correspondence: naoki.a.ito@fujifilm.com

Received: 31 October 2018; Accepted: 29 November 2018; Published: 3 December 2018

Abstract: Aging-associated changes of skin conditions are a major concern for maintaining quality of life. Therefore, the improvement of skin conditions by dietary supplementation is a topic of public interest. In this study, we hypothesized that a composite supplement containing fish derived-collagen peptide and ornithine (CPO) could improve skin conditions by increasing plasma growth hormone and/or insulin-like growth factor-1 (IGF-1) levels. Twenty-two healthy Japanese participants were enrolled in an 8-week double-blind placebo-controlled pilot study. They were assigned to either a CPO group, who were supplemented with a drink containing CPO, or an identical placebo group. We examined skin conditions including elasticity and transepidermal water loss (TEWL), as well as plasma growth hormone and IGF-1 levels. Skin elasticity and TEWL were significantly improved in the CPO group compared with the placebo group. Furthermore, only the CPO group showed increased plasma IGF-1 levels after 8 weeks of supplementation compared with the baseline. Our results might suggest the novel possibility for the use of CPO to improve skin conditions by increasing plasma IGF-1 levels.

Keywords: collagen peptide; ornithine; skin elasticity; transepidermal water loss; growth hormone; insulin-like growth factor-1

---

1. Introduction

The major role of skin is the protection of our body from external stimuli. Skin also plays an important role in maintaining homeostasis, including protection against loss of moisture and adjustment of body temperature [1]. Furthermore, aging-associated changes of skin conditions, such as the increased wrinkles and decreased skin elasticity, are a major concern for maintaining quality of life. Skin is composed of epidermis, dermis, and subcutaneous tissue. Collagen and elastin in the dermis maintain the structure of skin and create its elasticity [2,3]. Notably, the fibrous protein collagen, which plays a major role in maintaining the mechanical strength of skin, constitutes the majority of the dermis. The collagen molecule is formed by the three polypeptides, named alpha chains. The alpha chains are composed of high levels of glycine, hydroxyproline, proline and alanine [4]. Thus, collagen has unique amino acid composition. These collagen molecules assemble to form collagen fibrils by cross linking. Then, collagen fibrils assemble to form large collagen fiber. There are several types of collagen, and type I and type III collagen are the most abundant collagen in the skin [5]. Because collagen synthesis decreases with aging, decreased collagen content is one of the major causes of aging-associated changes of skin conditions [6,7]. As skin aging and nutrition states are linked [8], the improvement of skin conditions by dietary supplementation is a topic of increasing public interest [9].

The effects of growth hormone, a peptide hormone secreted from the pituitary gland (especially during the first few hours of sleep), cover a broad range of biological phenomena including cell growth, proliferation, regeneration, and metabolism [10–12]. For skin, severe growth hormone deficiency results in early aging, such as wrinkling and dryness [13]. In addition to the direct effects of growth hormone to several tissues, it also elicits indirect effects mediated by insulin-like growth factor-1 (IGF-1) [12]. Plasma IGF-1 levels correlate with plasma growth hormone levels, as production of IGF-1 by the liver is stimulated by growth hormone [14,15]. Similar to growth hormone, IGF-1 activates cell growth in several tissues including bone [16], muscle [17], and skin [18,19], whereby it contributes to both epidermal or dermal skin development and maintenance. Secretion of growth hormone is decreased with aging, suggesting that aging-associated changes of skin conditions are mediated, at least in part, by decreased levels of growth hormone, or its associated decrease of IGF-1 [20]. Thus, these previous studies suggest that aging-associated changes of skin conditions might be improved by increasing growth hormone and/or IGF-1 levels.

In recent years, many products containing collagen or a denatured form of collagen have been used for a wide variety of purposes, including cosmetics and food. With regard to food or supplements, the effects of collagen have been controversial, as orally ingested native collagen or its partially hydrolyzed form, gelatin, are not efficiently absorbed [21]. However, several lines of evidence revealed the beneficial role of collagen-derived small peptides, which exhibit high absorbability compared with native collagen [21,22], for a wide variety of tissues including bone [23], joint [24], muscle [25], tendon [26], and skin [27–29] in humans. Collagen has been isolated from many marine, brackish water and freshwater sources such as fishes [30–32] and mollusks [33–35]. Compared with collagen peptide derived from land animals, collagen peptide derived from these aquatic sources has unique molecular and biological properties for amino acid composition, antioxidant activity, neuroprotective activity and anti-skin aging activity, because of low temperature and/or high salt condition in the surrounding environment [36–39]. Furthermore, a previous study showed that collagen derived from sea- and freshwater-rainbow trout had quite similar amino acid composition and molecular weight properties [40]. Collagen derived from two marine demosponges, *Axinella cannabina* and *Suberites carnosus*, collected from the Aegean and the Ionian Seas, respectively, had low imino acid content, and showed lower or similar denaturation temperatures compared with collagen derived from other marine organisms such as tropical fish [41], suggesting the universal properties of collagen derived from aquatic sources. In addition to collagen peptide, ornithine is a non-essential, non-protein amino acid contained in various foods such as freshwater clams. Recent studies have highlighted ornithine as a functional food for improving sleep quality [42] and recovery from fatigue [43,44] in human.

The beneficial effects of collagen peptide for skin conditions have been analyzed by several groups both in rodents [39] and humans [27–29,45]. However, to our knowledge, the effects of collagen peptide, ornithine or the combined effects of collagen peptide and ornithine (CPO) on the increase of growth hormone and/or IGF-1 levels, and the subsequent improvements of skin conditions have not been investigated. A previous study showed the increase of plasma growth hormone levels after ingestion of ornithine [46], though its relationship to the improvements of skin conditions have not been investigated. In this study, we anticipated the combinational effects of fish-derived collagen peptide and ornithine on skin conditions, and plasma growth hormone and/or IGF-1 levels in healthy Japanese people. We hypothesized that orally administered CPO induced an increase of plasma growth hormone and/or IGF-1 levels, which exerted subsequent improvements of skin conditions including elasticity, moisture and transepidermal water loss (TEWL).

2. Results

2.1. Participants

Forty participants were recruited from the Osaka area, of whom 22 participants (aged from 31 to 48 years, 18 females and 4 males) exhibiting low skin moisture and elasticity were enrolled. Participants were recruited from September to October 2017. Included participants were assigned to either the CPO group ($n = 11$) or placebo group ($n = 11$). All participants completed the study. One participant in the CPO group was excluded from analysis because of aberrant blood ureic acid levels both before and after supplementation (Figure 1).

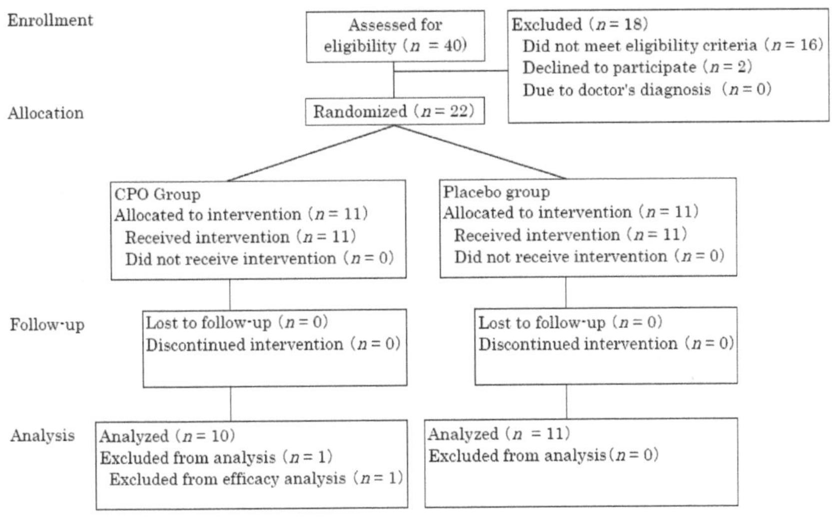

Figure 1. Flow diagram of participants.

Finally, 21 participants (aged from 31 to 48 years, 17 females and 4 males) were analyzed. Thus, per protocol set analysis was performed. Statistically significant differences between originally included participants and finally analyzed participants were not observed for baseline scores including age, skin elasticity, and moisture. This study consisted of an 8-week administration period from October to December 2017. CPO and placebo groups were matched according to age, gender, body mass index (BMI), skin elasticity, and moisture at baseline (Tables 1 and 2).

Table 1. Baseline characteristics of participants who completed 8-week test.

	Placebo ($n = 11$)	Collagen Peptide and Ornithine (CPO) ($n = 10$)	*p* Value
Age (mean ± standard deviation (SD))	40.4 ± 5.2	40.0 ± 6.8	0.89
Female, *n* (%)	9 (81.8)	8 (80.0)	
Body mass index (BMI) (mean ± SD)	20.6 ± 1.8	21.2 ± 2.2	0.55

Average ingestion rate was $100 \pm 0\%$ and $99.8 \pm 0.5\%$ in CPO and placebo groups, respectively. All subjects had a more than 98% ingestion rate. No statistically significant differences were observed for ingestion rate between CPO group and placebo groups.

2.2. Skin Conditions

The aim of this study was to evaluate the effects of CPO on skin conditions including elasticity, moisture, and TEWL. These skin conditions were measured at baseline and after 8 weeks of supplementation (Table 2). TEWL was significantly attenuated in the CPO group compared with the placebo group after 8 weeks of supplementation. For skin elasticity and VISIA analysis, no statistically significant differences were observed between raw values for CPO and placebo groups. However, placebo group showed a decrease of elasticity, from 0.818 ± 0.05 at baseline to 0.779 ± 0.06 after 8 weeks of supplementation. On the other hand, CPO groups showed an increase of elasticity, from

0.766 ± 0.07 at baseline to 0.784 ± 0.07 after 8 weeks of supplementation (Table 2). By analyzing the changes from the baseline, we observed statistically significant differences between the placebo and CPO groups (−0.039 ± 0.047 in placebo group, and 0.018 ± 0.065, Figure 2A). In addition to elasticity, a significantly reduced change in the number of skin pores was also observed after 8 weeks of supplementation in the CPO group compared with the placebo group (Figure 2B).

Table 2. Score of skin conditions.

Skin Conditions	Group	Week 0 Mean ± SD	Week 8 Mean ± SD
Moisture (A.U.)	Placebo	57.3 ± 13.3	60.9 ± 10.7
	CPO	58.5 ± 7.4	60.3 ± 8.2
TEWL (g/m^2 × h)	Placebo	14.4 ± 4.3	16.2 ± 3.4
	CPO	11.3 ± 4.1	11.5 ± 4.2 *
Elasticity	Placebo	0.818 ± 0.05	0.779 ± 0.06
	CPO	0.766 ± 0.07	0.784 ± 0.07
Skin-pH	Placebo	5.96 ± 0.23	6.00 ± 0.24
	CPO	6.03 ± 0.24	6.15 ± 0.23
DermaLab Collagen Score	Placebo	45.9 ± 17.9	40.9 ± 14.6
	CPO	41.1 ± 15.2	38.9 ± 13.4
VISIA Spots	Placebo	99.9 ± 38.1	96.1 ± 37.6
	CPO	83.3 ± 39.0	75.1 ± 34.6
VISIA Wrinkles	Placebo	172.3 ± 100.5	186.9 ± 124.5
	CPO	186.2 ± 108.2	209.5 ± 106.9
VISIA pores	Placebo	670.2 ± 328.8	745.8 ± 354.2
	CPO	848.4 ± 404.6	763.5 ± 452.7
VISIA Texture	Placebo	1972.0 ± 1171.7	1755.8 ± 1172.9
	CPO	1773.4 ± 1276.9	1610.5 ± 1243.7
VISIA Porphyrins	Placebo	719.9 ± 709.6	607.6 ± 585.6
	CPO	694.4 ± 402.4	868.2 ± 648.8
VISIA Spots	Placebo	193.3 ± 63.7	202.0 ± 56.9
	CPO	199.7 ± 56.4	209.9 ± 43.1
Red Areas	Placebo	49.5 ± 14.6	49.3 ± 14.8
	CPO	42.9 ± 17.2	36.4 ± 17.3
Brown Spots	Placebo	123.5 ± 58.5	125.3 ± 61.1
	CPO	104.1 ± 47.4	109.6 ± 50.5

* $p < 0.05$ vs. placebo.

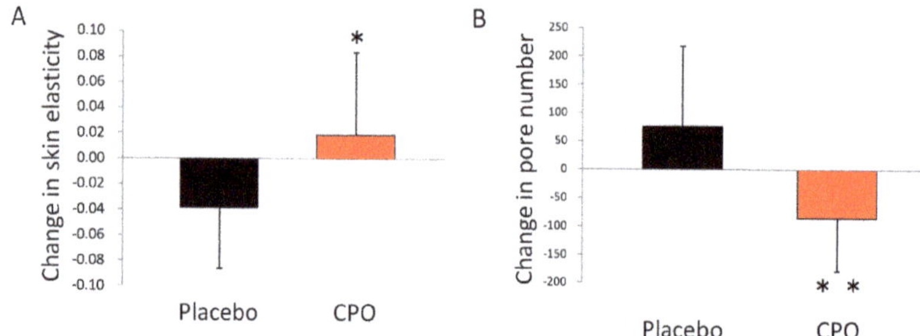

Figure 2. Dietary supplementation with collagen peptide and ornithine increased skin elasticity, and decreased the number of pores. Change in skin elasticity from baseline (**A**), and change in the number of pores from baseline (**B**). * $p < 0.05$ and ** $p < 0.01$ by unpaired *t*-test. Error bars indicate SD.

In addition, only the placebo group showed significantly decreased collagen scores compared with baseline using the DermaLab test (placebo group, from 45.9 ± 17.9 to 40.9 ± 14.6, $p = 0.027$; CPO group, from 41.1 ± 15.2 to 38.9 ± 13.4, $p = 0.097$). No statistically significant differences in moisture or superficial pH were observed between CPO and placebo groups.

2.3. Plasma Growth Hormone and Insulin-Like Growth Factor-1 (IGF-1) Levels

As aging-associated changes of skin condition are partially mediated by decreased levels of growth hormone [13], we analyzed plasma levels of growth hormone before and 30, 60, 120, 150, 180 and 240 min after supplementation. Growth hormone levels before supplementation were more than two-fold higher in the CPO group compared with the placebo group (0.65 ± 0.49 ng/mL in the placebo group vs. 1.41 ± 1.70 ng/mL in the CPO group, $p = 0.778$), suggesting that comparison of growth hormone levels between CPO and placebo groups was unworthy of evaluation. Indeed, we observed no significant differences of plasma growth hormone levels in the CPO group compared with the placebo group. We also evaluated IGF-1 levels at baseline and after 8 weeks of supplementation because IGF-1 levels reflect growth hormone secretion [15]. No statistically significant difference between CPO and placebo groups was observed. However, a statistically significant increase of IGF-1 levels from the baseline was observed only in the CPO group (Figure 3), suggesting that the improvements of skin conditions by CPO were mediated by the increase of IGF-1 levels, at least in part.

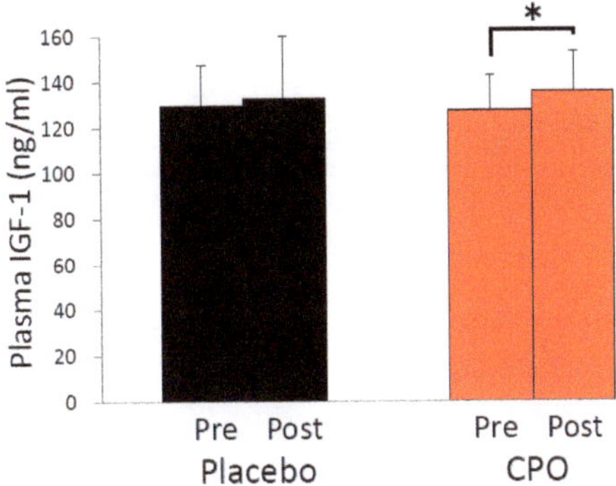

Figure 3. Dietary supplementation with collagen peptide and ornithine increased plasma insulin-like growth factor-1 (IGF-1) levels. Comparison of plasma IGF-1 levels before and after 8-week supplementation with collagen peptide and ornithine (CPO) (right) or with placebo (left). * $p < 0.05$ by paired *t*-test. Error bars indicate SD.

2.4. Clinical Safety

We observed neither adverse events nor severe changes in scores for general biochemical examination of blood or hematologic tests. Adverse events related to the ingestion of CPO were not observed. Thus, safety concerns were not observed.

3. Discussion

To our knowledge, our study is the first report to demonstrate the combined effects of CPO on skin conditions, plasma growth hormone and IGF-1 levels. We found that dietary supplementation of CPO improved skin elasticity and TEWL. TEWL was increased in the placebo group, while the CPO group showed no increase, indicating that the seasonal increase of TEWL was prevented by CPO (Table 2). As TEWL is linked to the barrier function of skin [1], this result suggests a protective effect of CPO for skin barrier function. In addition to TEWL, we observed a statistically significant difference between placebo and CPO groups in the changes of elasticity from baseline. Elasticity in the placebo group was decreased, while the CPO group showed increased elasticity, suggesting that the seasonal decrease in elasticity was prevented by CPO (Figure 2A). Furthermore, we observed increased IGF-1 levels only in the CPO group, suggesting that the improvements of skin conditions were mediated, at least in part, by increased IGF-1 levels (Figure 3).

The effects of collagen peptide on the improvements of skin moisture, elasticity, wrinkles, ultraviolet-induced erythema and ultraviolet-induced pigmented spots were previously revealed by several groups [28,29,45,47–49]. Furthermore, it has been previously reported that oral administration of marine collagen peptide derived from the skin of Nile Tilapia enhanced the process of wound healing [32]. Marine collagen peptide derived from Chum Salmon also promoted cutaneous wound healing [50,51]. Our study reinforced the beneficial effects of fish-derived collagen peptide to maintain or improve skin conditions such as elasticity and TEWL. Furthermore, we performed combined administration of collagen peptide and ornithine. Similar to collagen peptide, a previous study showed that ornithine enhanced wound healing effects by upregulating collagen synthesis in mice [52], suggesting that both collagen peptide and ornithine contribute to the improvements of skin conditions. In light of the independent effects of collagen peptide and ornithine, we hypothesized that skin conditions would be improved by the synergistic effects of CPO to increase growth hormone and/or IGF-1 levels, as described in the Introduction. Indeed, we observed the increased IGF-1 levels only in the CPO group. Between-group differences for TEWL and elasticity reinforced our hypothesis. However, because the effects of ornithine on skin conditions have not been investigated in humans, we could not conclude that the improvements of skin conditions observed in this study were derived from either the sole effects of collagen peptide or ornithine, or the synergistic effects of CPO. Furthermore, because we did not evaluate the sole effects of collagen peptide in this study, we could not conclude that the previously observed improvements of skin conditions including moisture and TEWL by collagen peptide [28,29,45,47–49] was enhanced by co-administration of ornithine. A comparison of the sole effects of ornithine, collagen peptide, and CPO is required to evaluate the synergist effects of CPO in the future.

As described in the Materials and Methods section, we employed a cutometer with a 6-mm diameter probe, suggesting that the improvement of skin elasticity reflected the state of relatively deep skin areas, such as the dermis. Generally, improvement of skin barrier function leads to the attenuation of TEWL, which results in subsequent improvement of the dermis environment [53,54]. In addition to this general understanding, increased IGF-1 levels in the CPO group suggested that the attenuation of TEWL occurred through the improvement of the dermal environment, which can result in the activation of dermal fibroblasts. Increased collagen scores in the CPO group, as measured by the DermaLab test, support the notion of CPO improving the dermal environment. Furthermore, previous studies have shown that the increased IGF-1 levels or treatment with collagen peptides leads to the activation of dermal fibroblasts [55,56]. Activated dermal fibroblasts construct the firm structure of the basement membrane, which is required for stable adherence of epidermal cells to the basement membrane. This stable adherence maintains an adequate balance between proliferation and differentiation of epidermal cells, leading to the enhancement of barrier function and subsequent attenuation of TEWL [57]. However, further study is required to determine how CPO improved skin elasticity and TEWL, as well as the relationship between improvements of elasticity and TEWL. Furthermore, we focused on the improvement of skin elasticity exclusively in the neck because only a thin muscle is present under neck skin [58]. However, as the effects of IGF-1 would not be restricted only to neck skin, we suspect that the positive effects observed in this study would be applicable to other areas of skin.

In this study, we hypothesized that CPO improved skin condition by increasing the secretion of growth hormone and/or IGF-1. Indeed, we observed increased plasma IGF-1 levels, which reflect increased secretion of growth hormone in the CPO group [15]. However, we did not observe an apparent increase of growth hormone levels immediately after CPO supplementation. One possibility for this result is that CPO enhanced secretion of growth hormone levels during the night, as we required participants to take CPO before bed time and growth hormone is secreted during non-rapid eye movement sleep. Thus, analysis for the effect of CPO on growth hormone secretion during sleep merits future investigation to potentially explain increased IGF-1 levels elicited by CPO supplementation.

Our study has several limitations. Even though the sample size was limited, which rated this study as a pilot trial, we observed improvement of skin elasticity, TEWL and increase of IGF-1 levels by CPO, suggesting the strong effects of CPO. Ornithine is found in freshwater clams, a traditional food for Japanese people. Furthermore, fish dishes are favored by Japanese people. We prohibited participants from continuously ingesting a functional food with identical or similar effects as the active ingredient of the test food. However, we did not estimate the exact dietary intake of CPO by participants; thus, the effects of CPO were potentially underestimated or overestimated. We hypothesized that combined supplementation of CPO elicited increased growth hormone and/or IGF-1 levels, which was followed by the improvement of the skin condition. In fact, we observed improvements of both elasticity and TEWL. However, the only intra-group difference was observed for increased plasma IGF-1 levels in the CPO group. Thus, precise mechanisms underlying how CPO improved the condition of skin, the specific contributions of CPO, and potential synergistic effects of CPO were not elucidated. Further analysis or a large-scale study is required to examine how CPO influences skin conditions.

4. Materials and Methods

4.1. Study Design, Randomization and Blinding

This study was a randomized, double-blind, placebo-controlled, parallel-group comparison trial to evaluate the effects of CPO dietary supplementation on skin conditions, plasma growth hormone and IGF-1 levels in healthy Japanese participants. Equal numbers of participants were allocated to active and placebo groups. This study, which was approved by the Kenshokai Ethical Review Board (Approved Number: 20170927-1) and registered in the UMIN Clinical Trials Registry (ID: UMIN000028924), followed the Declaration of Helsinki and Ethical Guidelines for Medical and Health Research Involving Human Subjects. Participants, clinicians, and practitioners were blinded. Practitioners performed intervention, outcome measurements, and analysis. Clinicians performed safety evaluation. According to our previous independent trial to evaluate skin TEWL or minimum erythema dose by dietary supplementation of astaxanthin in 10 healthy people [9,59], we set the required sample size as 10. The evaluation of skin elasticity was the primary outcome, while secondary outcome measures included other skin conditions such as skin moisture and TEWL, as well as plasma levels of growth hormone and IGF-1, and safety evaluation. Participants were enrolled and randomly allocated into the CPO or placebo group using a random number table with consideration of sex, age, skin elasticity, and moisture by practitioners. Allocation was concealed until all participants completed the tests.

4.2. Participants

Participants aged from 30 to less than 50 years in the Osaka area were included in this study. Every participant received an explanation of the objectives and details of this study, and provided written informed consent themselves. This study consisted of an 8-week ingestion period from October to December 2017. Participants meeting the following criteria were included in the study: (1) aged from 30 to 49 years at the time informed consent was provided; (2) exhibited relatively low levels of blood IGF-1; (3) exhibited relatively low skin moisture and skin elasticity; (4) BMI was less than 25; (5) capable of visiting the administrative facility on every inspection day; and (6) provided written informed consent for involvement in this trial themselves. Participants with the following criteria were excluded from the study: (1) continuous ingestion of a functional food or supplement; (2) continuous ingestion of a functional food or quasi-medicine with identical or similar effects as the active ingredient in test food; (3) frequent ingestion of food which rich in the same active ingredient in test food or ingestion of these kinds of food during the 3 days before and after trial initiation, or during the last 3 days of the trial; (4) worked a night shift or day and night shifts; (5) receiving medical treatment or prophylactic treatment, or diagnosed with the need for medical treatment; (6) presence of skin disease or abnormality in skin condition, such as atopic dermatitis; (7) exhibited apparent change of skin condition that was not related to the intake of test food at the end of the trial compared with the initiation; (8) history of severe disease or abnormality in glucose metabolism, lipid metabolism, liver function, kidney function, or cardiovascular system function including heart, respiratory tract, endocrine system and nerve system function, or psychiatric disorder; (9) exhibiting anemia, or felt sick as a result of blood collection; (10) history of alcoholism or drug addiction; (11) risk of food allergy; (12) exhibited apparent abnormality in blood test or were positive for hepatitis B antigen or hepatitis C virus antibody in trial duration, including the screening period; (13) pregnant or lactating when informed consent was provided, or hoped to become pregnant during the trial; (14) involvement in another trial within 4 weeks prior to this trial, or participation in another concurrent trial; and (15) otherwise judged to be inappropriate for this trial by the clinician responsible for this trial.

4.3. Supplement Formulation

One 30-mL CPO drink contained 10 g of fish-derived collagen peptide (FUJIFILM, Tokyo, Japan), 400 mg of ornithine, and other ingredients including vitamin C, acidifier, and sweetener. To prepare fish-derived collagen peptide, gelatin was extracted from the fish scales, almost all from Tilapia, by hot water extraction. Gelatin was then digested by food-processing protease to prepare fish-derived collagen peptide. Total amount of hydroxyproline in fish-derived collagen peptide, which was analyzed by the hydrolysis of collagen peptide and subsequent high performance liquid chromatography, was 1100 mg in one CPO drink. The weight-average molecular weight of the collagen peptide was 2000 to 3000 Da. This drink contained less than 9 mg of potassium and 0.3 mg of magnesium. The placebo drink contained the same amount of vitamin C, acidifier, and sweetener, but did not contain CPO. One bottle was administered every day before bedtime for 8 weeks. On the test initiation day, the test drink was administered in the daytime because of the measurement of growth hormone levels. CPO and placebo drinks, including the bottle, were indistinguishable by shape, taste, or color. The dosage of collagen peptide was determined according to the previous report that showed the dose-dependent increase of blood hydroxyproline content by the intake of 2, 10 and 25 g of collagen peptide in humans [60]. In addition, the dose and the duration of this study were based on our previous study focusing on the effects of collagen peptide (UMIN: 000016587). The dosage of ornithine was based on an earlier study focusing on the effects of ornithine on sleep quality [42].

4.4. Evaluation of Skin Condition

For evaluation of skin conditions, we measured skin elasticity using a Cutometer MPA580 with a 6-mm diameter probe (Courage and Khazaka Electronic GmbH, Cologne, Germany). A 6-mm diameter probe was used, as opposed to a 2- or 4-mm diameter probe, to analyze functional changes in relatively deep areas of skin, such as the dermis. Furthermore, we analyzed skin elasticity at the neck because only thin muscles, such as the sternocleidomastoid muscle, are under the skin, making the neck a suitable area to analyze skin elasticity compared with the cheek, which mount on facial muscle [58]. To evaluate skin moisture, TEWL, and superficial pH at the cheek, a Corneometer (Courage and Khazaka Electronic GmbH, Cologne, Germany), VAPO SCAN AS-VT100RS (ASCH JAPAN Co., Tokyo, Japan), and Skin-pH-Meter PH905 (Courage & Khazaka Electronic GmbH, Cologne, Germany) were used, respectively. Skin conditions were also evaluated by VISIA Evolution (Canfield Scientific Ltd., Parsippany, NJ, USA), which analyzed spots, wrinkles, pores, texture, porphyrins, UV spots, red areas and brown spots from the picture of the participant's face [61], and DermaLab (Cortex Technology, Hadsund, Denmark), which analyzed echo-graphic density of subcutaneous tissue including epidermis, dermis and subcutaneous fat and calculated collagen score from its ultrasound image. All measurements of skin condition were evaluated in a testing room with stable temperature $(21 \pm 1\ ^\circ C)$ and humidity $(50 \pm 5\%)$.

4.5. Blood Sampling and Safety Evaluation

Serum was obtained at baseline and after 8 weeks of supplementation, and blood IGF-1 levels were evaluated. To analyze the increases of plasma growth hormone levels, serum was obtained at 0, 60, 120, 150, 180, and 240 min after ingestion of test food. Plasma IGF-1 and growth hormone levels were measured by LSI Medience Co, Ltd. (Osaka, Japan) General biochemical examination of blood and hematologic tests were performed for safety evaluation.

4.6. Statistical Analysis

All results were presented as mean $\pm$ standard deviation (SD). Normality was analyzed by the Shapiro–Wilk test. If the data showed a normal distribution, differences between CPO and placebo groups were assessed by an unpaired *t*-test, and intra-group changes were evaluated by paired *t*-test. If the data did not show a normal distribution, we performed a Wilcoxon signed-rank test to analyze inter-group differences, and Wilcoxon rank sum test to analyze between-group differences. No additional analysis was performed. Probabilities less than 5% (* $p < 0.05$ and ** $p < 0.01$) were considered to be statistically significant. Statistical analyses were performed with JMP (version 13).

Author Contributions: N.I. and S.S. designed the concept of this study. N.I. interpreted the results and prepared the manuscript. F.U. supervised this study. All authors discussed the results and commented on the manuscript.

Funding: This research received no external funding.

Acknowledgments: We thank DRC Corporation for clinical management and analysis of the results as a practitioner for this study. We also thank Yuri Okano (CIEL Corporation), who supervised this study, and Yuriko Oda, Ayano Imai and Yoshiyuki Shirakura (FUJIFILM Corporation) for valuable discussion.

Conflicts of Interest: N.I., S.S. and F.U. are employees of FUJIFILM Corporation, a sponsor and funder of this study. As described in Author Contributions, these employees designed the concept of this study, interpreted the results and prepared the manuscript.

References

1. Fore, J. A review of skin and the effects of aging on skin structure and function. *Ostomy Wound Manag.* **2006**, *52*, 24–37.
2. Weihermann, A.C.; Lorencini, M.; Brohem, C.A.; de Carvalho, C.M. Elastin structure and its involvement in skin photoageing. *Int. J. Cosmet. Sci.* **2017**, *39*, 241–247. [CrossRef] [PubMed]
3. Ganceviciene, R.; Liakou, A.I.; Theodoridis, A.; Makrantonaki, E.; Zouboulis, C.C. Skin anti-aging strategies. *Dermatoendocrinology* **2012**, *4*, 308–319. [CrossRef] [PubMed]
4. Kittiphattanabawon, P.; Nalinanon, S.; Benjakul, S.; Kishimura, H. Characteristics of pepsin-solubilised collagen from the skin of splendid squid (Loligo formosana). *J. Chem.* **2015**. [CrossRef]
5. Berillis, P. The Role of Collagen in the Aorta's Structure. *Open Circ. Vasc. J.* **2013**. [CrossRef]
6. Mays, P.K.; McAnulty, R.J.; Campa, J.S.; Laurent, G.J. Age-related changes in collagen synthesis and degradation in rat tissues. Importance of degradation of newly synthesized collagen in regulating collagen production. *Biochem. J.* **1991**, *276*, 307–313. [CrossRef]
7. Varani, J.; Dame, M.K.; Rittie, L.; Fligiel, S.E.G.; Kang, S.; Fisher, G.J.; Voorhees, J.J. Decreased collagen production in chronologically aged skin: Roles of age-dependent alteration in fibroblast function and defective mechanical stimulation. *Am. J. Pathol.* **2006**, *168*, 1861–1868. [CrossRef] [PubMed]
8. Schagen, S.K.; Zampeli, V.A.; Makrantonaki, E.; Zouboulis, C.C. Discovering the link between nutrition and skin aging. *Dermatoendocrinology* **2012**, *4*, 298–307. [CrossRef]
9. Ito, N.; Seki, S.; Ueda, F. The protective role of astaxanthin for UV-induced skin deterioration in healthy people—A randomized, double-blind, placebo-controlled trial. *Nutrients* **2018**, *10*, 817. [CrossRef]
10. Van Cauter, E.; Plat, L. Physiology of growth hormone secretion during sleep. *J. Pediatr.* **1996**, *128*, S32–S37. [CrossRef]
11. Bartke, A. Growth hormone and aging: A challenging controversy. *Clin. Interv. Aging* **2008**, *3*, 659–665. [CrossRef] [PubMed]

12. Sonntag, W.E.; Csiszar, A.; De Cabo, R.; Ferrucci, L.; Ungvari, Z. Diverse roles of growth hormone and insulin-like growth factor-1 in mammalian aging: Progress and controversies. *J. Gerontol. A Biol. Sci. Med. Sci.* **2012**, *67*, 587–598. [CrossRef] [PubMed]

13. Tanriverdi, F.; Karaca, Z.; Unluhizarci, K.; Kelestimur, F. Unusual effects of GH deficiency in adults: A review about the effects of GH on skin, sleep, and coagulation. *Endocrine* **2014**, *47*, 679–689. [CrossRef] [PubMed]

14. Laron, Z. Insulin-like growth factor 1 (IGF-1): A growth hormone. *Mol. Pathol.* **2001**, *54*, 311–316. [CrossRef] [PubMed]

15. Ribeiro-Oliveira, A.; Faje, A.; Barkan, A. Postglucose growth hormone nadir and insulin-like growth factor-1 in naïve-active acromegalic patients: Do these parameters always correlate? *Arq. Bras. Endocrinol. Metabol.* **2011**, *55*, 494–497. [CrossRef] [PubMed]

16. Guntur, A.R.; Rosen, C.J. IGF-1 regulation of key signaling pathways in bone. *Bonekey Rep.* **2013**, *2*. [CrossRef] [PubMed]

17. Schiaffino, S.; Mammucari, C. Regulation of skeletal muscle growth by the IGF1-Akt/PKB pathway: Insights from genetic models. *Skelet. Muscle* **2011**, *1*. [CrossRef] [PubMed]

18. Sadagurski, M.; Yakar, S.; Weingarten, G.; Holzenberger, M.; Rhodes, C.J.; Breitkreutz, D.; Leroith, D.; Wertheimer, E. Insulin-like growth factor 1 receptor signaling regulates skin development and inhibits skin keratinocyte differentiation. *Mol. Cell. Biol.* **2006**, *26*, 2675–2687. [CrossRef] [PubMed]

19. Lewis, D.A.; Travers, J.B.; Somani, A.K.; Spandau, D.F. The IGF-1/IGF-1R signaling axis in the skin: A new role for the dermis in aging-associated skin cancer. *Oncogene* **2010**, *29*, 1475–1485. [CrossRef]

20. Rudman, D.; Feller, A.G.; Nagraj, H.S.; Gergans, G.A.; Lalitha, P.Y.; Goldberg, A.F.; Schlenker, R.A.; Cohn, L.; Rudman, I.W.; Mattson, D.E. Effects of human growth hormone in men over 60 years old. *N. Engl. J. Med.* **1990**, *323*, 1–6. [CrossRef]

21. Sontakke, S.B.; Jung, J.H.; Piao, Z.; Chung, H.J. Orally available collagen tripeptide: Enzymatic stability, intestinal permeability, and absorption of gly-pro-hyp and pro-hyp. *J. Agric. Food Chem.* **2016**, *64*, 7127–7133. [CrossRef] [PubMed]

22. Mari, W.K.; Muneshige, S.; Shin, K.; Yasuki, T.; Hideyuki, S.; Fumiki, M.; Hitoshi, S.; Yuji, F.; Michio, K. Absorption and effectiveness of orally administered low molecular weight collagen hydrolysate in rats. *J. Agric. Food Chem.* **2010**, *58*, 835–841. [CrossRef]

23. König, D.; Oesser, S.; Scharla, S.; Zdzieblik, D.; Gollhofer, A. Specific collagen peptides improve bone mineral density and bone markers in postmenopausal women—A randomized controlled study. *Nutrients* **2018**, *10*, 97. [CrossRef] [PubMed]

24. Kumar, S.; Sugihara, F.; Suzuki, K.; Inoue, N.; Venkateswarathirukumara, S. A double-blind, placebo-controlled, randomised, clinical study on the effectiveness of collagen peptide on osteoarthritis. *J. Sci. Food Agric.* **2014**, *95*, 702–707. [CrossRef] [PubMed]

25. Zdzieblik, D.; Oesser, S.; Baumstark, M.W.; Gollhofer, A.; König, D. Collagen peptide supplementation in combination with resistance training improves body composition and increases muscle strength in elderly sarcopenic men: A randomised controlled trial. *Br. J. Nutr.* **2015**, *114*, 1237–1245. [CrossRef]

26. Minaguchi, J.; Koyama, Y.; Meguri, N.; Hosaka, Y.; Ueda, H.; Kusubata, M.; Hirota, A.; Irie, S.; Mafune, N.; Takehana, K. Effects of ingestion of collagen peptide on collagen fibrils and glycosaminoglycans in Achilles tendon. *J. Nutr. Sci. Vitaminol.* **2005**, *51*, 169–174. [CrossRef]

27. Asserin, J.; Lati, E.; Shioya, T.; Prawitt, J. The effect of oral collagen peptide supplementation on skin moisture and the dermal collagen network: Eevidence from an ex vivo model and randomized, placebo-controlled clinical trials. *J. Cosmet. Dermatol.* **2015**, *14*, 291–301. [CrossRef]

28. Inoue, N.; Sugihara, F.; Wang, X. Ingestion of bioactive collagen hydrolysates enhance facial skin moisture and elasticity and reduce facial ageing signs in a randomised double-blind placebo-controlled clinical study. *J. Sci. Food Agric.* **2016**, *96*, 4077–4081. [CrossRef]

29. Sugihara, F.; Inoue, N.; Wang, X. Clinical effects of ingesting collagen hydrolysate on facial skin properties: -A randomized, placebo-controlled, double-blind trial. *Jpa. Pharmacol. Ther.* **2015**, *43*, 67–70.

30. Silva, T.H.; Moreira-Silva, J.; Marques, A.L.P.; Domingues, A.; Bayon, Y.; Reis, R.L. Marine origin collagens and its potential applications. *Mar. Drugs* **2014**, *12*, 5881–5901. [CrossRef]

31. Muthumari, K.; Anand, M.; Maruthupandy, M. Collagen extract from marine finfish scales as a potential mosquito larvicide. *Protein J.* **2016**. [CrossRef] [PubMed]

32. Hu, Z.; Yang, P.; Zhou, C.; Li, S.; Hong, P. Marine collagen peptides from the skin of Nile Tilapia (Oreochromis niloticus): Characterization and wound healing evaluation. *Mar. Drugs* **2017**, *15*, 102. [CrossRef] [PubMed]

33. Xuan Ri, S.; Hideyuki, K.; Koretaro, T. Characterization of molecular species of collagen in scallop mantle. *Food Chem.* **2007**. [CrossRef]

34. Mizuta, S.; Tanaka, T.; Yoshinaka, R. Comparison of collagen types of arm and mantle muscles of the common octopus (Octopus vulgaris). *Food Chem.* **2003**. [CrossRef]

35. Kołodziejska, I.; Sikorski, Z.E.; Niecikowska, C. Parameters affecting the isolation of collagen from squid (Illex argentinus) skins. *Food Chem.* **1999**. [CrossRef]

36. Wang, L.; An, X.; Yang, F.; Xin, Z.; Zhao, L.; Hu, Q. Isolation and characterisation of collagens from the skin, scale and bone of deep-sea redfish (Sebastes mentella). *Food Chem.* **2008**. [CrossRef] [PubMed]

37. Wang, B.; Wang, Y.M.; Chi, C.F.; Luo, H.Y.; Deng, S.G.; Ma, J.Y. Isolation and characterization of collagen and antioxidant collagen peptides from scales of croceine croaker (Pseudosciaena crocea). *Mar. Drugs* **2013**, *11*, 4641–4661. [CrossRef]

38. Xu, L.; Dong, W.; Zhao, J.; Xu, Y. Effect of marine collagen peptides on physiological and neurobehavioral development of male rats with perinatal asphyxia. *Mar. Drugs* **2015**, *13*, 3653–3671. [CrossRef]

39. Tanaka, M.; Koyama, Y.; Nomura, Y. Effects of collagen peptide ingestion on UV-B-induced skin damage. *Biosci. Biotechnol. Biochem.* **2009**, *73*, 930–932. [CrossRef]

40. Lee, J.K.; Kang, S.I.; Kim, Y.J.; Kim, M.J.; Heu, M.S.; Choi, B.D.; Kim, J.S. Comparison of collagen characteristics of sea- and freshwater-rainbow trout skin. *Food Sci. Biotechnol.* **2016**. [CrossRef]

41. Tziveleka, L.A.; Ioannou, E.; Tsiourvas, D.; Berillis, P.; Foufa, E.; Roussis, V. Collagen from the marine sponges Axinella cannabina and Suberites carnosus: Isolation and morphological, biochemical, and biophysical characterization. *Mar. Drugs* **2017**, *15*, 152. [CrossRef] [PubMed]

42. Miyake, M.; Kirisako, T.; Kokubo, T.; Miura, Y.; Morishita, K.; Okamura, H.; Tsuda, A. Randomised controlled trial of the effects of L-ornithine on stress markers and sleep quality in healthy workers. *Nutr. J.* **2014**, *13*, 53. [CrossRef] [PubMed]

43. Demura, S.; Morishita, K.; Yamada, T.; Yamaji, S.; Komatsu, M. EVect of L-ornithine hydrochloride ingestion on intermittent maximal anaerobic cycle ergometer performance and fatigue recovery after exercise. *Eur. J. Appl. Physiol.* **2011**, *111*, 2837–2843. [CrossRef] [PubMed]

44. Sugino, T.; Shirai, T.; Kajimoto, Y.; Kajimoto, O. l-Ornithine supplementation attenuates physical fatigue in healthy volunteers by modulating lipid and amino acid metabolism. *Nutr. Res.* **2008**, *28*, 738–743. [CrossRef] [PubMed]

45. Koyama, Y.I.; Kuwaba, K.; Kondo, S.; Tsukada, Y. Supplemental ingestion of collagen peptide suppresses ultraviolet-induced erythema—A randomized double-blind placebo-controlled study. *Jpn. Pharmacol. Ther.* **2014**, *42*, 781–790.

46. Evain-Brion, D.; Donnadieu, M.; Roger, M.; Job, J.C. Simultaneous study of somatotrophic and corticotrophic pituitary secretions during ornithine infusion test. *Clin. Endocrinol.* **1982**, *17*, 119–122. [CrossRef]

47. Kim, D.U.; Chung, H.C.; Choi, J.; Sakai, Y.; Lee, B.Y. Oral intake of low-molecular-weight collagen peptide improves hydration, elasticity, and wrinkling in human skin: A randomized, double-blind, placebo-controlled study. *Nutrients* **2018**, *10*, 826. [CrossRef]

48. Sugihare, F.; Inoue, N. Clinical effects of collagen hydrolysates ingestion on UV-induced pigmented spots of human skin: A preliminary study. *Heal. Sci.* **2012**, *28*, 153–156.1.

49. Proksch, E.; Schunck, M.; Zague, V.; Segger, D.; Degwert, J.; Oesser, S. Oral intake of specific bioactive collagen peptides reduces skin wrinkles and increases dermal matrix synthesis. *Skin Pharmacol. Physiol.* **2014**, *27*, 113–119. [CrossRef]

50. Zhang, Z.; Wang, J.; Ding, Y.; Dai, X.; Li, Y. Oral administration of marine collagen peptides from Chum Salmon skin enhances cutaneous wound healing and angiogenesis in rats. *J. Sci. Food Agric.* **2011**. [CrossRef]

51. Wang, J.; Xu, M.; Liang, R.; Zhao, M.; Zhang, Z.; Li, Y. Oral administration of marine collagen peptides prepared from chum salmon (Oncorhynchus keta) improves wound healing following cesarean section in rats. *Food Nutr. Res.* **2015**, *59*. [CrossRef] [PubMed]

52. Shi, H.P.; Fishel, R.S.; Efron, D.T.; Williams, J.Z.; Fishel, M.H.; Barbul, A. Effect of supplemental ornithine on wound healing. *J. Surg. Res.* **2002**, *106*, 299–302. [CrossRef]

53. Darlenski, R.; Kazandjieva, J.; Tsankov, N. Skin barrier function: Morphological basis and regulatory mechanisms. *J. Clin. Med.* **2011**, *4*, 36–45.

54. Goad, N.; Gawkrodger, D.J. Ambient humidity and the skin: The impact of air humidity in healthy and diseased states. *J. Eur. Acad. Dermatol. Venereol.* **2016**, *30*, 1285–1294. [CrossRef] [PubMed]

55. Sanchez, A.; Blanco, M.; Correa, B.; Perez-Martin, R.I.; Sotelo, C.G. Effect of fish collagen hydrolysates on type I collagen mRNA levels of human dermal fibroblast culture. *Mar. Drugs* **2018**, *16*, 144. [CrossRef]

56. Ghahary, A.; Tredget, E.E.; Shen, Q.; Kilani, R.T.; Scott, P.G.; Houle, Y. Mannose-6-phosphate/IGF-II receptors mediate the effects of IGF-1-induced latent transforming growth factor beta 1 on expression of type I collagen and collagenase in dermal fibroblasts. *Growth Factors* **2000**, *17*, 167–176. [CrossRef]

57. Hegde, S.; Raghavan, S. A skin-depth analysis of integrins: Role of the integrin network in health and disease. *Cell Commun. Adhes.* **2013**, *20*, 155–169. [CrossRef]

58. Allen, E.; Bhimji, S.S. *Anatomy, Head and Neck, Thyroid*; StatPearls Publishing LLC.: Petersburg, FL, USA, 2018.

59. Ayano, I.; Yuriko, O.; Shinobu, S.; Hiroyuki, S.; Takashi, K.; Koh, M.; Masami, S. Effects of capsule containing astaxanthin on skin condition in healthy subjects—A randomized, double-blind, parallel-group, placebo-controlled study. *Jpn. Pharmacol. Ther.* **2016**, *44*, 1209–1216.

60. Shigemura, Y.; Kubomura, D.; Sato, Y.; Sato, K. Dose-dependent changes in the levels of free and peptide forms of hydroxyproline in human plasma after collagen hydrolysate ingestion. *Food Chem.* **2014**. [CrossRef]

61. Goldsberry, A.; Hanke, C.W.; Hanke, K.E. VISIA system: A possible tool in the cosmetic practice. *J. Drugs Dermatol.* **2014**, *13*, 1312–1314.

MDPI

St. Alban-Anlage 66

4052 Basel

Switzerland

Tel. +41 61 683 77 34

Fax +41 61 302 89 18

www.mdpi.com

Marine Drugs Editorial Office

E-mail: marinedrugs@mdpi.com

www.mdpi.com/journal/marinedrugs